P9-DXK-422

Wavelets and Subband Coding

PRENTICE HALL SIGNAL PROCESSING SERIES

Alan V. Oppenheim, Editor

ANDREWS & HUNT *Digital Image Restoration*
BRACEWELL *Two Dimensional Imaging*
BRIGHAM *The Fast Fourier Transform and Its Applications*
BURDIC *Underwater Acoustic System Analysis, 2/E*
CASTLEMAN *Digital Image Processing*
COHEN *Time-Frequency Analysis*
CROCHIERE & RABINER *Multirate Digital Signal Processing*
DUDGEON & MERSEREAU *Multidimensional Digital Signal Processing*
HAYKIN *Advances in Spectrum Analysis and Array Processing, Vols. I, II, & III*
HAYKIN, ED. *Array Signal Processing*
JOHNSON & DUDGEON *Array Signal Processing*
KAY *Fundamentals of Statistical Signal Processing*
KAY *Modern Spectral Estimation*
KINO *Acoustic Waves: Device, Imaging, and Analog Signal Processing*
LIM *Two-Dimensional Signal and Image Processing*
LIM, ED. *Speech Enhancement*
LIM & OPPENHEIM, EDS. *Advanced Topics in Signal Processing*
MARPLE *Digital Spectral Analysis with Applications*
McCLELLAN & RADER *Number Theory in Digital Signal Processing*
MENDEL *Lessons in Estimation Theory for Signal Processing Communications and Control, 2/E*
NIKIAS & PETROPULU *Higher-Order Spectra Analysis*
OPPENHEIM & NAWAB *Symbolic and Knowledge-Based Signal Processing*
OPPENHEIM, WILLSKY, WITH YOUNG *Signals and Systems*
OPPENHEIM & SCHAFER *Digital Signal Processing*
OPPENHEIM & SCHAFER *Discrete-Time Signal Processing*
PHILLIPS & NAGLE *Digital Control Systems Analysis and Design,3/E*
PICINBONO *Random Signals and Systems*
RABINER & GOLD *Theory and Applications of Digital Signal Processing*
RABINER & SCHAFER *Digital Processing of Speech Signals*
RABINER & JUANG *Fundamentals of Speech Recognition*
ROBINSON & TREITEL *Geophysical Signal Analysis*
STEARNS & DAVID *Signal Processing Algorithms in Fortran and C*
THERRIEN *Discrete Random Signals and Statistical Signal Processing*
TRIBOLET *Seismic Applications of Homomorphic Signal Processing*
VAIDYANATHAN *Multirate Systems and Filter Banks*
VETTERLI & KOVACEVIC *Wavelets and Subband Coding*
WIDROW & STEARNS *Adaptive Signal Processing*

Wavelets and Subband Coding

Martin Vetterli
University of California at Berkeley

Jelena Kovačević
AT&T Bell Laboratories

Prentice Hall PTR, Englewood Cliffs, New Jersey 07632

Library of Congress Cataloging-in-Publication Data

Vetterli, Martin.
Wavelets and subband coding / Martin Vetterli, Jelena Kovačević.
p. cm.
ISBN 0-13-097080-8
1. Signal processing--Digital techniques. 2. Wavelets (Mathematics) I. Kovačević, Jelena. II. Title.
TK5102.9.V48 1995
621.382'2--dc20 95-5967
CIP

Editorial production: *bookworks*
Acquisitions editor: *Bernard Goodwin*
Manufacturing manager: *Alexis R. Heydt*
Cover designer: *Anthony Gemmellaro*
Cover illustration: *M. Vetterli*
About the cover: The three dimensional plot represents the first four iterations of a four-tap filter. This filter converges to a scaling function used in the construction of a compactly supported Daubechies' wavelet. The background is a tiling of the plane with twin dragons. This is a two-dimensional generalization of the Haar scaling function.

Prentice-Hall, Inc.
A Simon & Schuster Company
Englewood Cliffs, New Jersey 07632

The publisher offers discounts on this book when ordered in bulk quantities. For more information, contact: Corporate Sales Department, Prentice Hall P T R, 113 Sylvan Avenue, Englewood Cliffs, New Jersey 07632, Phone: (800) 382-3419, FAX: (201) 592-2249, E-mail: dan_rush@prenhall.com

Printed in the United States of America

10 9 8 7 6 5 4 3 2 1

ISBN 0-13-097080-8

Prentice-Hall International (UK) Limited, *London*
Prentice-Hall of Australia Pty. Limited, *Sydney*
Prentice-Hall Canada Inc., *Toronto*
Prentice-Hall Hispanoamericana, S.A., *Mexico*
Prentice-Hall of India Private Limited, *New Delhi*
Prentice-Hall of Japan, Inc., *Tokyo*
Simon & Schuster of Southeast Asia Pte. Ltd., *Singapore*
Editora Prentice-Hall do Brasil, Ltda., *Rio de Janeiro*

Für meine Eltern.
A Marie-Laure.
— MV

A Giovanni.
Mojoj zvedzdici, mami i tati.
— JK

Contents

Preface

A central goal of signal processing is to describe real life signals, be it for computation, compression, or understanding. In that context, transforms or linear expansions have always played a key role. Linear expansions are present in Fourier's original work and in Haar's construction of the first wavelet, as well as in Gabor's work on time-frequency analysis. Today, transforms are central in fast algorithms such as the FFT as well as in applications such as image and video compression.

Over the years, depending on open problems or specific applications, theoreticians and practitioners have added more and more tools to the toolbox called signal processing. Two of the newest additions have been wavelets and their discrete-time cousins, filter banks or subband coding. From work in harmonic analysis and mathematical physics, and from applications such as speech/image compression and computer vision, various disciplines built up methods and tools with a similar flavor, which can now be cast into the common framework of wavelets.

This unified view, as well as the number of applications where this framework is useful, are motivations for writing this book. The unification has given a new understanding and a fresh view of some classic signal processing problems. Another motivation is that the subject is exciting and the results are cute!

The aim of the book is to present this unified view of wavelets and subband coding. It will be done from a signal processing perspective, but with sufficient background material such that people without signal processing knowledge will find it useful as well. The level is that of a first year graduate engineering book (typically electrical engineering and computer sciences), but elementary Fourier analysis and some knowledge of linear systems in discrete time are enough to follow most of the book.

After the introduction (Chapter 1) and a review of the basics of vector spaces, linear algebra, Fourier theory and signal processing (Chapter 2), the book covers the five main topics in as many chapters. The discrete-time case, or filter banks, is thoroughly developed in Chapter 3. This is the basis for most applications, as well as for some of the wavelet constructions. The concept of wavelets is developed in Chapter 4, both with direct approaches and based on filter banks. This chapter describes wavelet series and their computation, as well as the construction of modified local Fourier transforms. Chapter 5 discusses continuous wavelet and local Fourier transforms, which are used in signal analysis, while Chapter 6 addresses efficient algorithms for filter banks and wavelet computations. Finally, Chapter 7 describes signal compression, where filter banks and wavelets play an important role. Speech/audio, image and video compression using transforms, quantization and entropy coding are discussed in detail. Throughout the book we give examples to illustrate the concepts, and more technical parts are left to appendices.

This book evolved from class notes used at Columbia University and the University of California at Berkeley. Parts of the manuscript have also been used at the University of Illinois at Urbana-Champaign and the University of Southern California. The material was covered in a semester, but it would also be easy to carve out a subset or skip some of the more mathematical subparts when developing a curriculum. For example, Chapters 3, 4 and 7 can form a good core for a course in Wavelets and Subband Coding. Homework problems are included in all chapters, complemented with project suggestions in Chapter 7. Since there is a detailed review chapter that makes the material as self-contained as possible, we think that the book is useful for self-study as well.

The subjects covered in this book have recently been the focus of books, special issues of journals, special conference proceedings, numerous articles and even new journals! To us, the book by I. Daubechies [73] has been invaluable, and Chapters 4 and 5 have been substantially influenced by it. Like the standard book by Meyer [194] and a recent book by Chui [49], it is a more mathematically oriented book than the present text. Another, more recent, tutorial book by Meyer gives an excellent overview of the history of the subject, its mathematical implications and current applications [195]. On the engineering side, the book by Vaidyanathan [308] is an excellent reference on filter banks, as is Malvar's book [188] for lapped orthogonal transforms and compression. Several other texts, including edited books, have appeared on wavelets [27, 51, 251], as well as on subband coding [335] and multiresolution signal decompositions [3]. Recent tutorials on wavelets can be found in [128, 140, 247, 281], and on filter banks in [305, 307].

From the above, it is obvious that there is no lack of literature, yet we hope to provide a text with a broad coverage of theory and applications and a different perspective based on signal processing. We enjoyed preparing this material, and simply hope that the reader will find some pleasure in this exciting subject, and share some of our enthusiasm!

ACKNOWLEDGEMENTS

Some of the work described in this book resulted from research supported by the National Science Foundation, whose support is gratefully acknowledged. We would like also to thank Columbia University, in particular the Center for Telecommunications Research, the University of California at Berkeley and AT&T Bell Laboratories for providing support and a pleasant work environment. We take this opportunity to thank A. Oppenheim for his support and for including this book in his distinguished series. We thank K. Gettman and S. Papanikolau of Prentice-Hall for their patience and help, and K. Fortgang of bookworks for her expert help in the production stage of the book.

To us, one of the attractions of the topic of *Wavelets and Subband Coding* is its interdisciplinary nature. This allowed us to interact with people from many different disciplines, and this was an enrichment in itself. The present book is the result of this interaction and the help of many people.

Our gratitude goes to I. Daubechies, whose work and help has been invaluable, to C. Herley, whose research, collaboration and help has directly influenced this book, and O. Rioul, who first taught us about wavelets and has always been helpful.

We would like to thank M.J.T. Smith and P.P. Vaidyanathan for a continuing and fruitful interaction on the topic of filter banks, and S. Mallat for his insights and interaction on the topic of wavelets.

Over the years, discussions and interactions with many experts have contributed to our understanding of the various fields relevant to this book, and we would like to acknowledge in particular the contributions of E. Adelson, T. Barnwell, P. Burt, A. Cohen, R. Coifman, R. Crochiere, P. Duhamel, C. Galand, W. Lawton, D. LeGall, Y. Meyer, T. Ramstad, G. Strang, M. Unser and V. Wickerhauser.

Many people have commented on several versions of the present text.

We thank I. Daubechies, P. Heller, M. Unser, P.P. Vaidyanathan, and G. Wornell for going through a complete draft and making many helpful suggestions. Comments on parts of the manuscript were provided by C. Chan, G. Chang, Z. Cvetković, V. Goyal, C. Herley, T. Kalker, M. Khansari, M. Kobayashi, H. Malvar, P. Moulin, A. Ortega, A. Park, J. Princen, K. Ramchandran, J. Shapiro and G. Strang, and are acknowledged with many thanks.

Coding experiments and associated figures were prepared by S. Levine (audio compression) and J. Smith (image compression), with guidance from A. Ortega and K. Ramchandran, and we thank them for their expert work. The images used in the experiments were made available by the Independent Broadcasting Association (UK).

The preparation of the manuscript relied on the help of many people. D. Heap is thanked for his invaluable contributions in the overall process, and in preparing the final version, and we thank C. Colbert, S. Elby, T. Judson, M. Karabatur, B. Lim, S. McCanne and T. Sharp for help at various stages of the manuscript.

The first author would like to acknowledge, with many thanks, the fruitful collaborations with current and former graduate students whose research has influenced this text, in particular Z. Cvetković, M. Garrett, C. Herley, J. Hong, G. Karlsson, E. Linzer, A. Ortega, H. Radha, K. Ramchandran, I. Shah, N.T. Thao and K.M. Uz. The early guidance by H.J. Nussbaumer, and the support of M. Kunt and G. Moschytz is gratefully acknowledged.

The second author would like to acknowledge friends and colleagues who contributed to the book, in particular C. Herley, G. Karlsson, A. Ortega and K. Ramchandran. Internal reviewers at Bell Labs are thanked for their efforts, in particular A. Reibman, G. Daryanani, P. Crouch, and T. Restaino.

Wavelets and Subband Coding

1

Wavelets, Filter Banks and Multiresolution Signal Processing

"It is with logic that one proves;
it is with intuition that one invents."
— Henri Poincaré

The topic of this book is very old and very new. Fourier series, or expansion of periodic functions in terms of harmonic sines and cosines, date back to the early part of the 19th century when Fourier proposed harmonic trigonometric series [100]. The first wavelet (the only example for a long time!) was found by Haar early in this century [126]. But the construction of more general wavelets to form bases for square-integrable functions was investigated in the 1980's, along with efficient algorithms to compute the expansion. At the same time, applications of these techniques in signal processing have blossomed.

While linear expansions of functions are a classic subject, the recent constructions contain interesting new features. For example, wavelets allow good resolution in time and frequency, and should thus allow one to see "the forest and the trees." This feature is important for nonstationary signal analysis. While Fourier basis functions are given in closed form, many wavelets can only be obtained through a computational procedure (and even then, only at specific rational points). While this might seem to be a drawback, it turns out that if one is interested in implementing a signal expansion on real data, then a computational procedure is better than a closed-form expression!

The recent surge of interest in the types of expansions discussed here is due to the convergence of ideas from several different fields, and the recognition that techniques developed independently in these fields could be cast into a common framework.

The name "wavelet" had been used before in the literature,[1] but its current meaning is due to J. Goupillaud, J. Morlet and A. Grossman [119, 125]. In the context of geophysical signal processing they investigated an alternative to local Fourier analysis based on a single prototype function, and its scales and shifts. The modulation by complex exponentials in the Fourier transform is replaced by a scaling operation, and the notion of scale[2] replaces that of frequency. The simplicity and elegance of the wavelet scheme was appealing and mathematicians started studying wavelet analysis as an alternative to Fourier analysis. This led to the discovery of wavelets which form orthonormal bases for square-integrable and other function spaces by Meyer [194], Daubechies [71], Battle [21, 22], Lemarié [175], and others. A formalization of such constructions by Mallat [180] and Meyer [194] created a framework for wavelet expansions called multiresolution analysis, and established links with methods used in other fields. Also, the wavelet construction by Daubechies is closely connected to filter bank methods used in digital signal processing as we shall see.

Of course, these achievements were preceded by a long-term evolution from the 1910 Haar wavelet (which, of course, was not called a wavelet back then) to work using octave division of the Fourier spectrum (Littlewood-Paley) and results in harmonic analysis (Calderon-Zygmund operators). Other constructions were not recognized as leading to wavelets initially (for example, Stromberg's work [283]).

Paralleling the advances in pure and applied mathematics were those in signal processing, but in the context of discrete-time signals. Driven by applications such as speech and image compression, a method called subband coding was proposed by Croisier, Esteban, and Galand [69] using a special class of filters called quadrature mirror filters (QMF) in the late 1970's, and by Crochiere, Webber and Flanagan [68]. This led to the study of perfect reconstruction filter banks, a problem solved in the 1980's by several people, including Smith and Barnwell [270, 271], Mintzer [196], Vetterli [315], and Vaidyanathan [306].

In a particular configuration, namely when the filter bank has octave bands, one obtains a discrete-time wavelet series. Such a configuration has been popular in signal processing less for its mathematical properties than because an octave band or logarithmic spectrum is more natural for certain appli-

[1]For example, for the impulse response of a layer in geophysical signal processing by Ricker [237] and for a causal finite-energy function by Robinson [248].

[2]For a beautiful illustration of the notion of scale, and an argument for geometric spacing of scale in natural imagery, see [197].

cations such as audio compression since it emulates the hearing process. Such an octave-band filter bank can be used, under certain conditions, to generate wavelet bases, as shown by Daubechies [71].

In computer vision, multiresolution techniques have been used for various problems, ranging from motion estimation to object recognition [249]. Images are successively approximated starting from a coarse version and going to a fine-resolution version. In particular, Burt and Adelson proposed such a scheme for image coding in the early 1980's [41], calling it pyramid coding.[3] This method turns out to be similar to subband coding. Moreover, the successive approximation view is similar to the multiresolution framework used in the analysis of wavelet schemes.

In computer graphics, a method called successive refinement iteratively interpolates curves or surfaces, and the study of such interpolators is related to wavelet constructions from filter banks [45, 92].

Finally, many computational procedures use the concept of successive approximation, sometimes alternating between fine and coarse resolutions. The multigrid methods used for the solution of partial differential equations [39] are an example.

While these interconnections are now clarified, this has not always been the case. In fact, maybe one of the biggest contributions of wavelets has been to bring people from different fields together, and from that cross fertilization and exchange of ideas and methods, progress has been achieved in various fields.

In what follows, we will take mostly a signal processing point of view of the subject. Also, most applications discussed later are from signal processing.

1.1 Series Expansions of Signals

We are considering linear expansions of signals or functions. That is, given any signal x from some space S, where S can be finite-dimensional (for example, $\mathcal{R}^n$, $\mathcal{C}^n$) or infinite-dimensional (for example, $l_2(\mathcal{Z})$, $L_2(\mathcal{R})$), we want to find a set of elementary signals $\{\varphi_i\}_{i\in\mathcal{Z}}$ for that space so that we can write x as a linear combination

$$x = \sum_i \alpha_i \, \varphi_i. \tag{1.1.1}$$

The set $\{\varphi_i\}$ is complete for the space S, if all signals $x \in S$ can be expanded as in (1.1.1). In that case, there will also exist a dual set $\{\tilde{\varphi}_i\}_{i\in\mathcal{Z}}$ such that the

[3]The importance of the pyramid algorithm was not immediately recognized. One of the reviewers of the original Burt and Adelson paper said, "I suspect that no one will ever use this algorithm again."

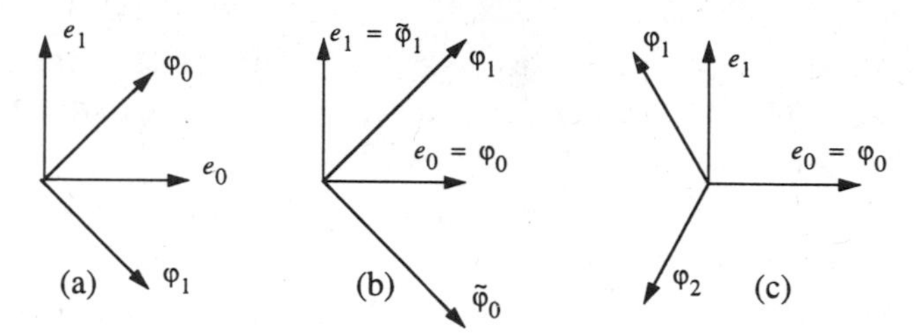

Figure 1.1 Examples of possible sets of vectors for the expansion of R^2. (a) Orthonormal case. (b) Biorthogonal case. (c) Overcomplete case.

expansion coefficients in (1.1.1) can be computed as

$$\alpha_i = \sum_n \tilde{\varphi}_i[n]\, x[n],$$

when x and $\tilde{\varphi}_i$ are real discrete-time sequences, and

$$\alpha_i = \int \tilde{\varphi}_i(t)\, x(t)\, dt,$$

when they are real continuous-time functions. The above expressions are the *inner products* of the $\tilde{\varphi}_i$'s with the signal x, denoted by $\langle \tilde{\varphi}_i, x \rangle$. An important particular case is when the set $\{\varphi_i\}$ is orthonormal and complete, since then we have an *orthonormal basis* for S and the basis and its dual are the same, that is, $\varphi_i = \tilde{\varphi}_i$. Then

$$\langle \varphi_i, \varphi_j \rangle \;=\; \delta[i-j],$$

where $\delta[i]$ equals 1 if $i = 0$, and 0 otherwise. If the set is complete and the vectors φ_i are linearly independent but not orthonormal, then we have a *biorthogonal basis,* and the basis and its dual satisfy

$$\langle \varphi_i, \tilde{\varphi}_j \rangle \;=\; \delta[i-j].$$

If the set is complete but redundant (the φ_i's are not linearly independent), then we do not have a basis but an overcomplete representation called a *frame.* To illustrate these concepts, consider the following example.

Example 1.1 *Set of Vectors for the Plane*

We show in Figure 1.1 some possible sets of vectors for the expansion of the plane, or $\mathcal{R}^2$. The standard Euclidean basis is given by e_0 and e_1. In part (a), an orthonormal basis is given by $\varphi_0 = [1,1]^T/\sqrt{2}$ and $\varphi_1 = [1,-1]^T/\sqrt{2}$. The dual basis is identical, or $\tilde{\varphi}_i = \varphi_i$. In part (b), a biorthogonal basis is given, with $\varphi_0 = e_0$ and $\varphi_1 = [1,1]^T$. The dual basis is now $\tilde{\varphi}_0 = [1,-1]^T$ and $\tilde{\varphi}_1 = [0,1]^T$. Finally, in part (c), an overcomplete set is given, namely $\varphi_0 = [1,0]^T$, $\varphi_1 = [-1/2, \sqrt{3}/2]^T$ and $\varphi_2 = [-1/2, -\sqrt{3}/2]^T$. Then, it can be verified that a possible reconstruction basis is identical (up to a scale factor), namely, $\tilde{\varphi}_i = 2/3\; \varphi_i$ (the reconstruction basis is not unique). This set behaves as an orthonormal basis, even though the vectors are linearly dependent.

The representation in (1.1.1) is a change of basis, or, conceptually, a change of point of view. The obvious question is, what is a good basis $\{\varphi_i\}$ for S? The answer depends on the class of signals we want to represent, and on the choice of a criterion for quality. However, in general, a good basis is one that allows compact representation or less complex processing. For example, the Karhunen-Loève transform concentrates as much energy in as few coefficients as possible, and is thus good for compression, while, for the implementation of convolution, the Fourier basis is computationally more efficient than the standard basis.

We will be interested mostly in expansions with some structure, that is, expansions where the various basis vectors are related to each other by some elementary operations such as shifting in time, scaling, and modulation (which is shifting in frequency). Because we are concerned with expansions for very high-dimensional spaces (possibly infinite), bases without such structure are useless for complexity reasons.

Historically, the Fourier series for periodic signals is the first example of a signal expansion. The basis functions are harmonic sines and cosines. Is this a good set of basis functions for signal processing? Besides its obvious limitation to periodic signals, it has very useful properties, such as the convolution property which comes from the fact that the basis functions are eigenfunctions of linear time-invariant systems. The extension of the scheme to nonperiodic signals,[4] by segmentation and piecewise Fourier series expansion of each segment, suffers from artificial boundary effects and poor convergence at these boundaries (due to the Gibbs phenomenon).

An attempt to create local Fourier bases is the Gabor transform or short-time Fourier transform (STFT). A smooth window is applied to the signal centered around $t = nT_0$ (where T_0 is some basic time step), and a Fourier expansion is applied to the windowed signal. This leads to a time-frequency representation since we get an approximate information about the frequency content of the signal around the location nT_0. Usually, frequency points spaced $2\pi/T_0$ apart are used and we get a sampling of the time-frequency plane on a rectangular grid. The spectrogram is related to such a time-frequency analysis. Note that the functions used in the expansion are related to each other by shift in time and modulation, and that we obtain a linear frequency analysis. While the STFT has proven useful in signal analysis, there are no good orthonormal bases based on this construction. Also, a logarithmic frequency scale, or constant relative bandwidth, is often preferable to the linear frequency scale obtained with the

[4]The Fourier transform of nonperiodic signals is also possible. It is an integral transform rather than a series expansion and lacks any time locality.

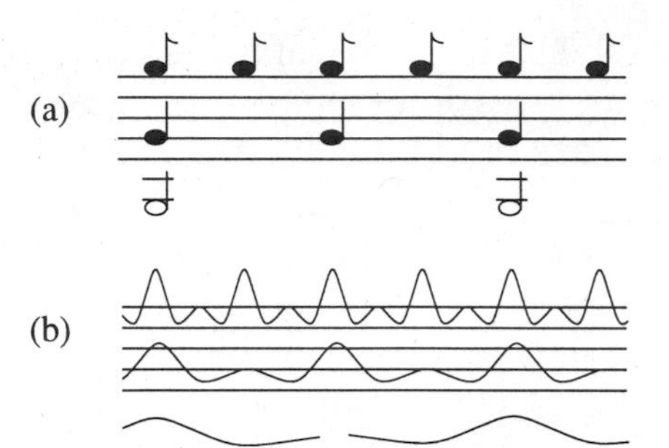

Figure 1.2 Musical notation and orthonormal wavelet bases. (a) The western musical notation uses a logarithmic frequency scale with twelve halftones per octave. In this example, notes are chosen as in an orthonormal wavelet basis, with long low-pitched notes, and short high-pitched ones. (b) Corresponding time-domain functions.

STFT. For example, the human auditory system uses constant relative bandwidth channels (critical bands), and therefore, audio compression systems use a similar decomposition.

A popular alternative to the STFT is the wavelet transform. Using scales and shifts of a prototype wavelet, a linear expansion of a signal is obtained. Because the scales used are powers of an elementary scale factor (typically 2), the analysis uses a constant relative bandwidth (or, the frequency axis is logarithmic). The sampling of the time-frequency plane is now very different from the rectangular grid used in the STFT. Lower frequencies, where the bandwidth is narrow (that is, the basis functions are stretched in time) are sampled with a large time step, while high frequencies (which correspond to short basis functions) are sampled more often. In Figure 1.2, we give an intuitive illustration of this time-frequency trade-off, and relate it to musical notation which also uses a logarithmic frequency scale.[5] What is particularly interesting is that such a wavelet scheme allows good orthonormal bases whereas the STFT does not.

In the discussions above, we implicitly assumed continuous-time signals. Of course there are discrete-time equivalents to all these results. A local analysis can be achieved using a block transform, where the sequence is segmented into adjacent blocks of N samples, and each block is individually transformed. As is to be expected, such a scheme is plagued by boundary effects, also called blocking effects. A more general expansion relies on filter banks, and can achieve both STFT-like analysis (rectangular sampling of the time-frequency plane) or wavelet-like analysis (constant relative bandwidth in frequency). Discrete-time expansions based on filter banks are not arbitrary, rather they

[5] This is the standard western musical notation based on J.S. Bach's "Well Tempered Piano". Thus one could argue that wavelets were actually invented by J.S. Bach!

are structured expansions. Again, for complexity reasons, it is useful to impose such a structure on the basis chosen for the expansion. For example, filter banks correspond to basis sequences which satisfy a block shift invariance property. Sometimes, a modulation constraint can also be added, in particular in STFT-like discrete-time bases. Because we are in discrete time, scaling cannot be done exactly (unlike in continuous time), but an approximate scaling property between basis functions holds for the discrete-time wavelet series.

Interestingly, the relationship between continuous- and discrete-time bases runs deeper than just these conceptual similarities. One of the most interesting constructions of wavelets is the one by Daubechies [71]. It relies on the iteration of a discrete-time filter bank so that, under certain conditions, it converges to a continuous-time wavelet basis. Furthermore, the multiresolution framework used in the analysis of wavelet decompositions automatically associates a discrete-time perfect reconstruction filter bank to any wavelet decomposition. Finally, the wavelet series decomposition can be computed with a filter bank algorithm. Therefore, especially in the wavelet type of a signal expansion, there is a very close interaction between discrete and continuous time.

It is to be noted that we have focused on STFT and wavelet type of expansions mainly because they are now quite standard. However, there are many alternatives, for example the wavelet packet expansion introduced by Coifman and coworkers [62, 64], and generalizations thereof. The main ingredients remain the same: they are structured bases in discrete or continuous time, and they permit different time versus frequency resolution trade-offs. An easy way to interpret such expansions is in terms of their time-frequency tiling: each basis function has a region in the time-frequency plane where most of its energy is concentrated. Then, given a basis and the expansion coefficients of a signal, one can draw a tiling where the shading corresponds to the value of the expansion coefficient.[6]

Example 1.2 *Different Time-Frequency Tilings*

Figure 1.3 shows schematically different possible expansions of a very simple discrete-time signal, namely a sine wave plus an impulse (see part (a)). It would be desirable to have an expansion that captures both the isolated impulse (or Dirac in time) and the isolated frequency component (or Dirac in frequency). The first two expansions, namely the identity transform in part (b) and the discrete-time Fourier series[7] in part (c), isolate the time and frequency impulse, respectively, but not both. The local discrete-time Fourier series in part (d) achieves a compromise, by locating both impulses to a certain degree. The discrete-time

[6]Such tiling diagrams were used by Gabor [102], and he called an elementary tile a "logon."

[7]Discrete-time series expansions are often called discrete-time transforms, both in the Fourier and in the wavelet case.

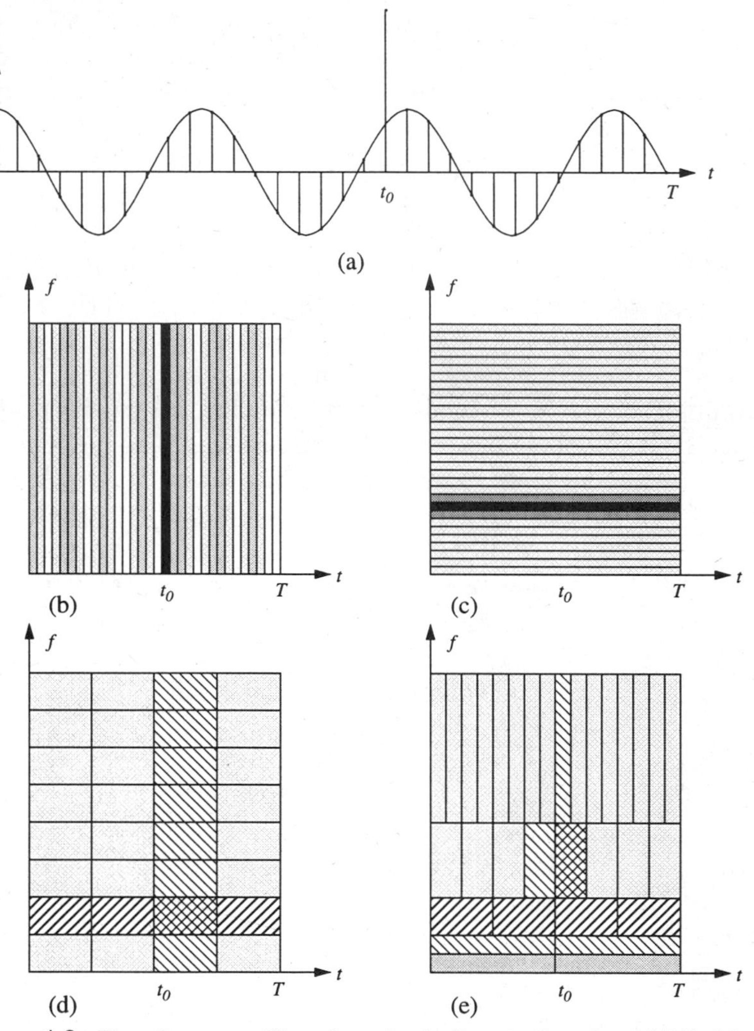

Figure 1.3 Time-frequency tilings for a simple discrete-time signal [130]. (a) Sine wave plus impulse. (b) Expansion onto the identity basis. (c) Discrete-time Fourier series. (d) Local discrete-time Fourier series. (e) Discrete-time wavelet series.

wavelet series in part (e) achieves better localization of the time-domain impulse, without sacrificing too much of the frequency localization. However, a high-frequency sinusoid would not be well localized. This simple example indicates some of the trade-offs involved.

Note that the local Fourier transform and the wavelet transform can be used for signal analysis purposes. In that case, the goal is not to obtain orthonormal bases, but rather to characterize the signal from the transform. The local Fourier transform retains many of the characteristics of the usual Fourier transform with a localization given by the window function, which is thus constant at all frequencies (this phenomenon can be seen already in Figure 1.3(d)). The wavelet, on the other hand, acts as a microscope, focusing on smaller time phenomenons as the scale becomes small (see Figure 1.3(e) to see how the impulse gets better localized at high frequencies). This behavior permits a local characterization of functions, which the Fourier transform does not.[8]

1.2 Multiresolution Concept

A slightly different expansion is obtained with multiresolution pyramids since the expansion is actually redundant (the number of samples in the expansion is bigger than in the original signal). However, conceptually, it is intimately related to subband and wavelet decompositions. The basic idea is successive approximation. A signal is written as a coarse approximation (typically a lowpass, subsampled version) plus a prediction error which is the difference between the original signal and a prediction based on the coarse version. Reconstruction is immediate: simply add back the prediction to the prediction error. The scheme can be iterated on the coarse version. It can be shown that if the lowpass filter meets certain constraints of orthogonality, then this scheme is identical to an oversampled discrete-time wavelet series. Otherwise, the successive approximation approach is still at least conceptually identical to the wavelet decomposition since it performs a multiresolution analysis of the signal.

A schematic diagram of a pyramid decomposition, with attached resulting images, is shown in Figure 1.4. After the encoding, we have a coarse resolution image of half size, as well as an error image of full size (thus the redundancy). For applications, the decomposition into a coarse resolution which gives an approximate but adequate version of the full image, plus a difference or detail image, is conceptually very important.

Example 1.3 *Multiresolution Image Database*

Let us consider the following practical problem: Users want to access and retrieve electronic images from an image database using a computer network with limited bandwidth. Because the users have an approximate idea of which image they want, they will first browse through some images before settling on a target image [214]. Given the limited bandwidth, browsing is best done on coarse versions of the images which can be transmitted faster. Once an image is chosen, the residual can be sent. Thus, the scheme shown in Figure 1.4 can be used, where the coarse and residual images are further compressed to diminish the transmission time.

[8]For example, in [137], this mathematical microscope is used to analyze some famous lacunary Fourier series that was proposed over a century ago.

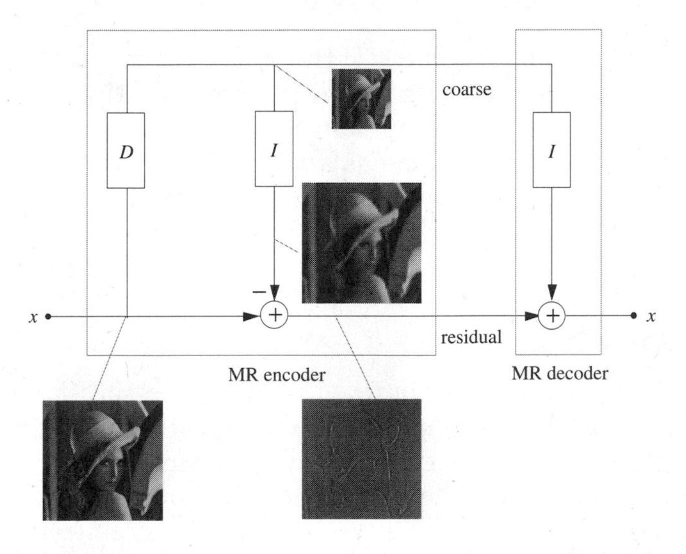

Figure 1.4 Pyramid decomposition of an image where encoding is shown on the left and decoding is shown on the right. The operators D and I correspond to decimation and interpolation operators, respectively. For example, D produces an $N/2 \times N/2$ image from an $N \times N$ original, while I interpolates an $N \times N$ image based on an $N/2 \times N/2$ original.

The above example is just one among many schemes where multiresolution decompositions are useful in communications problems. Others include transmission over error-prone channels, where the coarse resolution can be better protected to guarantee some minimum level of quality.

Multiresolution decompositions are also important for computer vision tasks such as image segmentation or object recognition: the task is performed in a successive approximation manner, starting on the coarse version and then using this result as an initial guess for the full task. However, this is a greedy approach which is sometimes suboptimal. Figure 1.5 shows a famous counter-example, where a multiresolution approach would be seriously misleading ...

Interestingly, the multiresolution concept, besides being intuitive and useful in practice, forms the basis of a mathematical framework for wavelets [181, 194]. As in the pyramid example shown in Figure 1.4, one can decompose a function into a coarse version plus a residual, and then iterate this to infinity. If properly done, this can be used to analyze wavelet schemes and derive wavelet bases.

1.3 Overview of the Book

We start with a review of fundamentals in Chapter 2. This chapter should make the book as self-contained as possible. It reviews Hilbert spaces at an el-

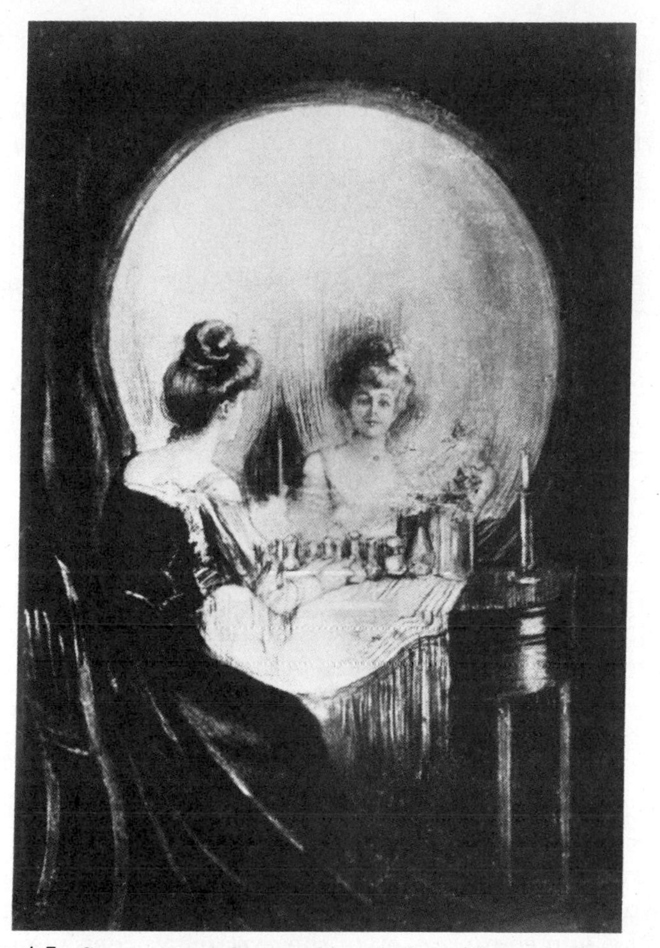

Figure 1.5 Counter-example to multiresolution technique. The coarse approximation is unrelated to the full-resolution image (Comet Photo AG).

ementary but sufficient level, linear algebra (including matrix polynomials) and Fourier theory, with material on sampling and discrete-time Fourier transforms in particular. The review of continuous-time and discrete-time signal processing is followed by a discussion of multirate signal processing, which is a topic central to later chapters. Finally, a short introduction to time-frequency distributions discusses the local Fourier transform and the wavelet transform, and shows the uncertainty principle. The appendix gives factorizations of unitary matrices, and reviews results on convergence and regularity of functions.

Chapter 3 focuses on discrete-time bases and filter banks. This topic is important for several later chapters as well as for applications. We start with two simple expansions which will reappear throughout the book as a recurring theme: the Haar and the sinc bases. They are limit cases of orthonormal expansions with good time localization (Haar) and good frequency localization (sinc). This naturally leads to an in-depth study of two-channel filter banks, including analytical tools for their analysis as well as design methods. The construction of orthonormal and linear phase filter banks is described. Multichannel filter banks are developed next, first through tree structures and then in the general case. Modulated filter banks, corresponding conceptually to a discrete-time local Fourier analysis, are addressed as well. Next, pyramid schemes and overcomplete representations are explored. Such schemes, while not critically sampled, have some other attractive features, such as time invariance. Then, the multidimensional case is discussed both for simple separable systems, as well as for general nonseparable ones. The latter systems involve lattice sampling which is detailed in an appendix. Finally, filter banks for telecommunications, namely transmultiplexers and adaptive subband filtering, are presented briefly. The appendix details factorizations of orthonormal filter banks (corresponding to paraunitary matrices).

Chapter 4 is devoted to the construction of bases for continuous-time signals, in particular wavelets and local cosine bases. Again, the Haar and sinc cases play illustrative roles as extremes of wavelet constructions. After an introduction to series expansions, we develop multiresolution analysis as a framework for wavelet constructions. This naturally leads to the classic wavelets of Meyer and Battle-Lemarié or Stromberg. These are based on Fourier-domain analysis. This is followed by Daubechies' construction of wavelets from iterated filter banks. This is a time-domain construction based on the iteration of a multirate filter. Study of the iteration leads to the notion of regularity of the discrete-time filter. Then, the wavelet series expansion is considered both in terms of properties and computation of the expansion coefficients. Some generalizations of wavelet constructions are considered next, first in one dimension (including biorthogonal and multichannel wavelets) and then in multiple dimensions, where nonseparable wavelets are shown. Finally, local cosine bases are derived and they can be seen as a real-valued local Fourier transform.

Chapter 5 is concerned with continuous wavelet and Fourier transforms. Unlike the series expansions in Chapters 3 and 4, these are very redundant representations useful for signal analysis. Both transforms are analyzed, inverses are derived, and their main properties are given. These transforms can be sampled, that is, scale/frequency and time shift can be discretized. This leads to redundant series representations called frames. In particular, reconstruction or inversion is discussed, and the case of wavelet and local Fourier frames is considered in some detail.

Chapter 6 treats algorithmic and computational aspects of series expansions. First, a review of classic fast algorithms for signal processing is given since they form the ingredients used in subsequent algorithms. The key role of the fast Fourier transform (FFT) is pointed out. The complexity of computing filter banks, that is, discrete-time expansions, is studied in detail. Important cases include the discrete-time wavelet series or transform and modulated filter banks. The latter corresponds to a local discrete-time Fourier series or transform, and uses FFT's for efficient computation. These filter bank algorithms have direct applications in the computation of wavelet series. Overcomplete expansions are considered next, in particular for the computation of a sampled continuous wavelet transform. The chapter concludes with a discussion of special topics related to efficient convolution algorithms and also application of wavelet ideas to numerical algorithms.

The last chapter is devoted to one of the main applications of wavelets and filter banks in signal processing, namely signal compression. The technique is often called subband coding because signals are considered in spectral bands for compression purposes. First comes a review of transform based compression, including quantization and entropy coding. Then follow specific discussions of one-, two- and three-dimensional signal compression methods based on transforms. Speech and audio compression, where subband coding was first invented, is discussed. The success of subband coding in current audio coding algorithms is shown on specific examples such as the MUSICAM standard. A thorough discussion of image compression follows. While current standards such as JPEG are block transform based, some innovative subband or wavelet schemes are very promising and are described in detail. Video compression is considered next. Besides expansions, motion estimation/compensation methods play a key role and are discussed. The multiresolution feature inherent in pyramid and subband coding is pointed out as an attractive feature for video compression, just as it is for image coding. The final section discusses the interaction of source coding, particularly the multiresolution type, and channel coding or transmission. This joint source-channel coding is key to new applications of image and video compression, as in transmission over packet networks. An appendix gives a brief review of statistical signal processing which underlies coding methods.

2

Fundamentals of Signal Decompositions

"A journey of a thousand miles must begin with a single step."
— Lao-Tzu, *Tao Te Ching*

The mathematical framework necessary for our later developments is established in this chapter. While we review standard material, we also cover the broad spectrum from Hilbert spaces and Fourier theory to signal processing and time-frequency distributions. Furthermore, the review is done from the point of view of the chapters to come, namely, signal expansions. This chapter attempts to make the book as self-contained as possible.

We tried to keep the level of formalism reasonable, and refer to standard texts for many proofs. While this chapter may seem dry, basic mathematics is the foundation on which the rest of the concepts are built, and therefore, some solid groundwork is justified.

After defining notations, we discuss Hilbert spaces. In their finite-dimensional form, Hilbert spaces are familiar to everyone. Their infinite-dimensional counterparts, in particular $L_2(\mathcal{R})$ and $l_2(\mathcal{Z})$, are derived, since they are fundamental to signal processing in general and to our developments in particular. Linear operators on Hilbert spaces and (in finite dimensions) linear algebra are discussed briefly. The key ideas of orthonormal bases, orthogonal projection and best approximation are detailed, as well as general bases and overcomplete expansions, or, frames.

We then turn to a review of Fourier theory which starts with the Fourier transform and series. The expansion of bandlimited signals and sampling naturally lead to the discrete-time Fourier transform and series.

Next comes a brief review of continuous-time and discrete-time signal processing, followed by a discussion of multirate discrete-time signal processing. It should be emphasized that this last topic is central to the rest of the book, but not often treated in standard signal processing books.

Finally, we review time-frequency representations, in particular short-time Fourier or Gabor expansions as well as the newer wavelet expansion. We also discuss the uncertainty relation, which is a fundamental limit in linear time-frequency representations. A bilinear expansion, the Wigner-Ville transform, is also introduced.

2.1 NOTATIONS

Let $\mathcal{C}$, $\mathcal{R}$, $\mathcal{Z}$ and $\mathcal{N}$ denote the sets of complex, real, integer and natural numbers, respectively. Then, $\mathcal{C}^n$, and $\mathcal{R}^n$ will be the sets of all n-tuples $(x_1, \ldots, x_n)$ of complex and real numbers, respectively.

The superscript * denotes complex conjugation, or, $(a+jb)^* = (a-jb)$, where the symbol j is used for the square root of -1 and $a, b \in \mathcal{R}$. The subscript $_*$ is used to denote complex conjugation of the constants but not the complex variable, for example, $(az)_* = a^* z$ where z is a complex variable. The superscript ^{T} denotes the transposition of a vector or a matrix, while the superscript * on a vector or matrix denotes hermitian transpose, or transposition and complex conjugation. $Re(z)$ and $Im(z)$ denote the real and imaginary parts of the complex number z.

We define the Nth root of unity as $W_N = e^{-j2\pi/N}$. It satisfies the following:

$$W_N^N = 1, \tag{2.1.1}$$

$$W_N^{kN+i} = W_N^i, \text{ with } k, i \text{ in } \mathcal{Z}, \tag{2.1.2}$$

$$\sum_{k=0}^{N-1} W_N^{k \cdot n} = \begin{cases} N & n = lN, \ l \in \mathcal{Z}, \\ 0 & \text{otherwise.} \end{cases} \tag{2.1.3}$$

The last relation is often referred to as *orthogonality of the roots of unity.*

Often we deal with functions of a continuous variable, and a related sequence indexed by an integer (typically, the latter is a sampled version of the former). To avoid confusion, and in keeping with the tradition of the signal processing literature [211], we use parentheses around a continuous variable and brackets around a discrete one, for example, $f(t)$ and $x[n]$, where

$$x[n] = f(nT), \qquad n \in \mathcal{Z}, \ T \in \mathcal{R}.$$

In particular, $\delta(t)$ and $\delta[n]$ denote continuous-time and discrete-time Dirac functions, which are very different indeed. The former is a generalized function (see Section 2.4.4) while the latter is the sequence which is 1 for $n = 0$ and 0 otherwise (the Dirac functions are also called delta or impulse functions).

In discrete-time signal processing, we will often encounter 2π-periodic functions (namely, discrete-time Fourier transforms of sequences, see Section 2.4.6), and we will write, for example, $H(e^{j\omega})$ to make the periodicity explicit.

2.2 HILBERT SPACES

Finite-dimensional vector spaces, as studied in linear algebra [106, 280], involve vectors over $\mathcal{R}$ or $\mathcal{C}$ that are of finite dimension n. Such spaces are denoted by $\mathcal{R}^n$ and $\mathcal{C}^n$, respectively. Given a set of vectors, $\{v_k\}$, in $\mathcal{R}^n$ or $\mathcal{C}^n$, important questions include:

(a) Does the set $\{v_k\}$ span the space $\mathcal{R}^n$ or $\mathcal{C}^n$, that is, can every vector in $\mathcal{R}^n$ or $\mathcal{C}^n$ be written as a linear combination of vectors from $\{v_k\}$?

(b) Are the vectors linearly independent, that is, is it true that no vector from $\{v_k\}$ can be written as a linear combination of the others?

(c) How can we find bases for the space to be spanned, in particular, orthonormal bases?

(d) Given a subspace of $\mathcal{R}^n$ or $\mathcal{C}^n$ and a general vector, how can we find an approximation in the least-squares sense, (see below) that lies in the subspace?

Two key notions used in addressing these questions include:

(a) The length, or norm,[1] of a vector (we take $\mathcal{R}^n$ as an example),

$$\|x\| = \left(\sum_{i=1}^{n} x_i^2\right)^{1/2}.$$

(b) The orthogonality of a vector with respect to another vector (or set of vectors), for example,

$$\langle x, y \rangle = 0,$$

with an appropriately defined scalar product,

$$\langle x, y \rangle = \sum_{i=1}^{n} x_i y_i.$$

[1]Unless otherwise specified, we will assume a squared norm.

So far, we relied on the fact that the spaces were finite-dimensional. Now, the idea is to generalize our familiar notion of a vector space to infinite dimensions. It is necessary to restrict the vectors to have finite length or norm (even though they are infinite-dimensional). This leads naturally to Hilbert spaces. For example, the space of square-summable sequences, denoted by $l_2(\mathcal{Z})$, is the vector space "$\mathcal{C}^\infty$" with a norm constraint. An example of a set of vectors spanning $l_2(\mathcal{Z})$ is the set $\{\delta[n-k]\}$, $k \in \mathcal{Z}$. A further extension with respect to linear algebra is that vectors can be generalized from n-tuples of real or complex values to include functions of a continuous variable. The notions of norm and orthogonality can be extended to functions using a suitable inner product between functions, which are thus viewed as vectors. A classic example of such orthogonal vectors is the set of harmonic sine and cosine functions, $\sin(nt)$ and $\cos(nt)$, $n = 0, 1, \ldots$, on the interval $[-\pi, \pi]$.

The classic questions from linear algebra apply here as well. In particular, the question of completeness, that is, whether the span of the set of vectors $\{v_k\}$ covers the whole space, becomes more involved than in the finite-dimensional case. The norm plays a central role, since any vector in the space must be expressed by a linear combination of v_k's such that the norm of the difference between the vector and the linear combination of v_k's is zero. For $l_2(\mathcal{Z})$, $\{\delta[n-k]\}$, $k \in \mathcal{Z}$, constitute a complete set which is actually an orthonormal basis. For the space of square-integrable functions over the interval $[-\pi, \pi]$, denoted by $L_2([-\pi, \pi])$, the harmonic sines and cosines are complete since they form the basis used in the Fourier series expansion.

If only a subset of the complete set of vectors $\{v_k\}$ is used, one is interested in the best approximation of a general element of the space by an element from the subspace spanned by the vectors in the subset. This question has a particularly easy answer when the set $\{v_k\}$ is orthonormal and the goal is least-squares approximation (that is, the norm of the difference is minimized). Because the geometry of Hilbert spaces is similar to Euclidean geometry, the solution is the orthogonal projection onto the approximation subspace, since this minimizes the distance or approximation error.

In the following, we formally introduce vector spaces and in particular Hilbert spaces. We discuss orthogonal and general bases and their properties. We often use the finite-dimensional case for intuition and examples. The treatment is not very detailed, but sufficient for the remainder of the book. For a thorough treatment, we refer the reader to [113].

2.2.1 Vector Spaces and Inner Products

Let us start with a formal definition of a vector space.

DEFINITION 2.1

A *vector space* over the set of complex or real numbers, $\mathcal{C}$ or $\mathcal{R}$, is a set of vectors, E, together with *addition* and *scalar multiplication*, which, for general x, y in E, and α, β in $\mathcal{C}$ or $\mathcal{R}$, satisfy the following:

(a) *Commutativity:* $x + y = y + x$.

(b) *Associativity:* $(x + y) + z = x + (y + z)$, $(\alpha\beta)x = \alpha(\beta x)$.

(c) *Distributivity:* $\alpha(x + y) = \alpha x + \alpha y$, $(\alpha + \beta)x = \alpha x + \beta x$.

(d) *Additive identity:* there exists 0 in E, such that $x + 0 = x$, for all x in E.

(e) *Additive inverse:* for all x in E, there exists a $(-x)$ in E, such that $x + (-x) = 0$.

(f) *Multiplicative identity:* $1 \cdot x = x$ for all x in E.

Often, x, y in E will be n-tuples or sequences, and then we define

$$x + y = (x_1, x_2, \ldots) + (y_1, y_2, \ldots) = (x_1 + y_1, x_2 + y_2, \ldots)$$
$$\alpha x = \alpha(x_1, x_2, \ldots) = (\alpha x_1, \alpha x_2, \ldots).$$

While the scalars are from $\mathcal{C}$ or $\mathcal{R}$, the vectors can be arbitrary, and apart from n-tuples and infinite sequences, we could also take functions over the real line.

A subset M of E is a *subspace* of E if

(a) For all x and y in M, $x + y$ is in M.

(b) For all x in M and α in $\mathcal{C}$ or $\mathcal{R}$, αx is in M.

Given $S \subset E$, the *span* of S is the subspace of E consisting of all linear combinations of vectors in S, for example, in finite dimensions,

$$\text{span}(S) = \left\{ \sum_{i=1}^{n} \alpha_i x_i \mid \alpha_i \in \mathcal{C} \text{ or } \mathcal{R},\ x_i \in S \right\}.$$

Vectors $x_1, \ldots, x_n$ are called *linearly independent*, if $\sum_{i=1}^{n} \alpha_i x_i = 0$ is true only if $\alpha_i = 0$, for all i. Otherwise, these vectors are *linearly dependent*. If there are infinitely many vectors $x_1, x_2, \ldots$, they are linearly independent if for each k, $x_1, x_2, \ldots, x_k$ are linearly independent.

A subset $\{x_1, \ldots, x_n\}$ of a vector space E is called a *basis* for E, when $E = \text{span}(x_1, \ldots, x_n)$ and $x_1, \ldots, x_n$ are linearly independent. Then, we say that E has dimension n. E is *infinite-dimensional* if it contains an infinite linearly independent set of vectors. As an example, the space of infinite sequences is spanned by the infinite set $\{\delta[n-k]\}_{k\in\mathcal{Z}}$. Since they are linearly independent, the space is infinite-dimensional.

Next, we equip the vector space with an inner product that is a complex function fundamental for defining norms and orthogonality.

DEFINITION 2.2

An *inner product* on a vector space E over $\mathcal{C}$ (or $\mathcal{R}$), is a comple-valued function $\langle\cdot,\cdot\rangle$, defined on $E \times E$ with the following properties:

(a) $\langle x+y, z\rangle \;=\; \langle x,z\rangle + \langle y,z\rangle$.

(b) $\langle x, \alpha y\rangle \;=\; \alpha\langle x,y\rangle$.

(c) $\langle x,y\rangle^* \;=\; \langle y,x\rangle$.

(d) $\langle x,x\rangle \;\geq\; 0$, and $\langle x,x\rangle \;=\; 0$ *if and only if* $x \;\equiv\; 0$.

Note that (b) and (c) imply $\langle ax, y\rangle \;=\; a^*\langle x,y\rangle$. From (a) and (b), it is clear that the inner product is linear. Note that we choose the definition of the inner product which takes the complex conjugate of the first vector (follows from (b)). For illustration, the standard inner product for complex-valued functions over $\mathcal{R}$ and sequences over $\mathcal{Z}$ are

$$\langle f,g\rangle \;=\; \int_{-\infty}^{\infty} f^*(t)\; g(t)dt,$$

and

$$\langle x,y\rangle \;=\; \sum_{n=-\infty}^{\infty} x^*[n]\; y[n],$$

respectively (if they exist). The norm of a vector is defined from the inner product as

$$\|x\| \;=\; \sqrt{\langle x,x\rangle}, \tag{2.2.1}$$

and the distance between two vectors x and y is simply the norm of their difference $\|x-y\|$. Note that other norms can be defined (see (2.2.16)), but since we will only use the usual Euclidean or square norm as defined in (2.2.1), we use the symbol $\|\,.\,\|$ without a particular subscript.

The following hold for inner products over a vector space:

(a) *Cauchy-Schwarz inequality*

$$|\langle x,y\rangle| \;\leq\; \|x\|\;\|y\|, \tag{2.2.2}$$

with equality *if and only if* $x = \alpha y$.

(b) *Triangle inequality*

$$\|x + y\| \;\leq\; \|x\| + \|y\|,$$

with equality *if and only if* $x = \alpha y$, where α is a positive real constant.

(c) *Parallelogram law*

$$\|x + y\|^2 + \|x - y\|^2 \;=\; 2(\|x\|^2 + \|y\|^2).$$

Finally, the inner product can be used to define orthogonality of two vectors x and y, that is, vectors x and y are orthogonal *if and only if*

$$\langle x, y\rangle \;=\; 0.$$

If two vectors are orthogonal, which is denoted by $x \perp y$, then they satisfy the *Pythagorean theorem*,

$$\|x + y\|^2 \;=\; \|x\|^2 + \|y\|^2,$$

since $\|x + y\|^2 = \langle x + y, x + y\rangle = \|x\|^2 + \langle x, y\rangle + \langle y, x\rangle + \|y\|^2$.

A vector x is said to be orthogonal to a set of vectors $S = \{y_i\}$ if $\langle x, y_i\rangle = 0$ for all i. We denote this by $x \perp S$. More generally, two subspaces S_1 and S_2 are called orthogonal if all vectors in S_1 are orthogonal to all of the vectors in S_2, and this is written $S_1 \perp S_2$. A set of vectors $\{x_1, x_2, \ldots\}$ is called *orthogonal* if $x_i \perp x_j$ when $i \neq j$. If the vectors are normalized to have unit norm, we have an *orthonormal system*, which therefore satisfies

$$\langle x_i, x_j\rangle \;=\; \delta[i - j].$$

Vectors in an orthonormal system are linearly independent, since $\sum \alpha_i x_i = 0$ implies $0 = \langle x_j, \sum \alpha_i x_i\rangle = \sum \alpha_i \langle x_j, x_i\rangle = \alpha_j$. An orthonormal system in a vector space E is an *orthonormal basis* if it spans E.

2.2.2 Complete Inner Product Spaces

A vector space equipped with an inner product is called an *inner product space*. One more notion is needed in order to obtain a Hilbert space, *completeness*. To this end, we consider sequences of vectors $\{x_n\}$ in E, which are said to *converge* to x in E if $\|x_n - x\| \to 0$ as $n \to \infty$. A sequence of vectors $\{x_n\}$ is called a *Cauchy* sequence, if $\|x_n - x_m\| \to 0$, when $n,\ m \to \infty$. If every Cauchy sequence in E, converges to a vector in E, then E is called *complete*. This leads to the following definition:

DEFINITION 2.3

A complete inner product space is called a *Hilbert space*.

We are particularly interested in those Hilbert spaces which are *separable* because a Hilbert space contains a countable orthonormal basis *if and only if* it is separable. Since all Hilbert spaces with which we are going to deal are separable, we implicitly assume that this property is satisfied (refer to [113] for details on separability). Note that a closed subspace of a separable Hilbert space is separable, that is, it also contains a countable orthonormal basis.

Given a Hilbert space E and a subspace S, we call the orthogonal complement of S in E, denoted $S^{\perp}$, the set $\{x \in E \mid x \perp S\}$. Assume further that S is closed, that is, it contains all limits of sequences of vectors in S. Then, given a vector y in E, there exists a unique v in S and a unique w in $S^{\perp}$ such that $y = v + w$. We can thus write

$$E = S \oplus S^{\perp},$$

or, E is the direct sum of the subspace and its orthogonal complement.

Let us consider a few examples of Hilbert spaces.

Complex/Real Spaces The complex space $\mathcal{C}^n$ is the set of all n-tuples $x = (x_1, \ldots, x_n)$, with finite x_i in $\mathcal{C}$. The inner product is defined as

$$\langle x, y \rangle = \sum_{i=1}^{n} x_i^* y_i,$$

and the norm is

$$\|x\| = \sqrt{\langle x, x \rangle} = \sqrt{\sum_{i=1}^{n} |x_i|^2}.$$

The above holds for the real space $\mathcal{R}^n$ as well (note that then $y_i^* = y_i$). For example, vectors $e_i = (0, \ldots, 0, 1, 0, \ldots, 0)$, where 1 is in the ith position, form an orthonormal basis both for $\mathcal{R}^n$ and $\mathcal{C}^n$. Note that these are the usual spaces considered in linear algebra.

Space of Square-Summable Sequences In discrete-time signal processing we will be dealing almost exclusively with sequences $x[n]$ having finite square sum or finite energy,[2] where $x[n]$ is, in general, complex-valued and n belongs to $\mathcal{Z}$. Such a sequence $x[n]$ is a vector in the Hilbert space $l_2(\mathcal{Z})$. The inner product is

$$\langle x, y \rangle = \sum_{n=-\infty}^{\infty} x[n]^* y[n],$$

and the norm is

$$\|x\| = \sqrt{\langle x, x \rangle} = \sqrt{\sum_{n \in \mathcal{Z}} |x[n]|^2}.$$

[2]In physical systems, the sum or integral of a squared function often corresponds to energy.

Thus, $l_2(\mathcal{Z})$ is the space of all sequences such that $\|x\| < \infty$. This is obviously an infinite-dimensional space, and a possible orthonormal basis is $\{\delta[n-k]\}_{k\in\mathcal{Z}}$.

For the completeness of $l_2(\mathcal{Z})$, one has to show that if $x_n[k]$ is a sequence of vectors in $l_2(\mathcal{Z})$ such that $\|x_n - x_m\| \to 0$ as $n, m \to \infty$ (that is, a Cauchy sequence), then there exists a limit x in $l_2(\mathcal{Z})$ such that $\|x_n - x\| \to 0$. The proof can be found, for example, in [113].

Space of Square-Integrable Functions A function $f(t)$ defined on $\mathcal{R}$ is said to be in the Hilbert space $L_2(\mathcal{R})$, if $|f(t)|^2$ is integrable,[3] that is, if

$$\sqrt{\int_{t\in\mathcal{R}} |f(t)|^2 dt} < \infty.$$

The inner product on $L_2(\mathcal{R})$ is given by

$$\langle f, g \rangle = \int_{t\in\mathcal{R}} f(t)^* g(t) dt,$$

and the norm is

$$\|f\| = \sqrt{\langle f, f \rangle} = \sqrt{\int_{t\in\mathcal{R}} |f(t)|^2 dt}.$$

This space is infinite-dimensional (for example, $e^{-t^2}, te^{-t^2}, t^2e^{-t^2} \ldots$ are linearly independent).

2.2.3 Orthonormal Bases

Among all possible bases in a Hilbert space, orthonormal bases play a very important role. We start by recalling the standard linear algebra procedure which can be used to orthogonalize an arbitrary basis.

Gram-Schmidt Orthogonalization Given a set of linearly independent vectors $\{x_i\}$ in E, we can construct an orthonormal set $\{y_i\}$ with the same span as $\{x_i\}$ as follows: Start with

$$y_1 = \frac{x_1}{\|x_1\|}.$$

Then, recursively set

$$y_k = \frac{x_k - v_k}{\|x_k - v_k\|}, \qquad k = 2, 3, \ldots$$

where

$$v_k = \sum_{i=1}^{k-1} \langle y_i, x_k \rangle y_i.$$

[3]Actually, $|f|^2$ has to be Lebesgue integrable.

As will be seen shortly, the vector v_k is the orthogonal projection of x_k onto the subspace spanned by the previous orthogonalized vectors and this is subtracted from x_k, followed by normalization.

A standard example of such an orthogonalization procedure is the Legendre polynomials over the interval $[-1, 1]$. Start with $x_k(t) = t^k$, $k = 0, 1, \ldots$ and apply the Gram-Schmidt procedure to get $y_k(t)$, of degree k, norm 1 and orthogonal to $y_i(t)$, $i < k$ (see Problem 2.1).

Bessel's Inequality If we have an orthonormal system of vectors $\{x_k\}$ in E, then for every y in E the following inequality, known as Bessel's inequality, holds:

$$\|y\|^2 \;\geq\; \sum_k |\langle x_k, y\rangle|^2.$$

If we have an orthonormal system that is complete in E, then we have an orthonormal basis for E, and Bessel's relation becomes an equality, often called Parseval's equality (see Theorem 2.4).

Orthonormal Bases For a set of vectors $S = \{x_i\}$ to be an orthonormal basis, we first have to check that the set of vectors S is orthonormal and then that it is complete, that is, that every vector from the space to be represented can be expressed as a linear combination of the vectors from S. In other words, an orthonormal system $\{x_i\}$ is called an orthonormal basis for E, if for every y in E,

$$y \;=\; \sum_k \alpha_k x_k. \tag{2.2.3}$$

The coefficients α_k of the expansion are called the *Fourier coefficients* of y (with respect to $\{x_i\}$) and are given by

$$\alpha_k \;=\; \langle x_k, y\rangle. \tag{2.2.4}$$

This can be shown by using the continuity of the inner product (that is, if $x_n \to x$, and $y_n \to y$, then $\langle x_n, y_n\rangle \to \langle x, y\rangle$) as well as the orthogonality of the x_k's. Given that y is expressed as (2.2.3), we can write

$$\langle x_k, y\rangle \;=\; \lim_{n\to\infty} \langle x_k, \sum_{i=0}^{n} \alpha_i x_i\rangle \;=\; \alpha_k,$$

where we used the linearity of the inner product.

In finite dimensions (that is, $\mathcal{R}^n$ or $\mathcal{C}^n$), having an orthonormal set of size n is sufficient to have an orthonormal basis. As expected, this is more delicate in infinite dimensions (that is, it is not sufficient to have an infinite orthonormal set). The following theorem gives several equivalent statements which permit us to check if an orthonormal system is also a basis:

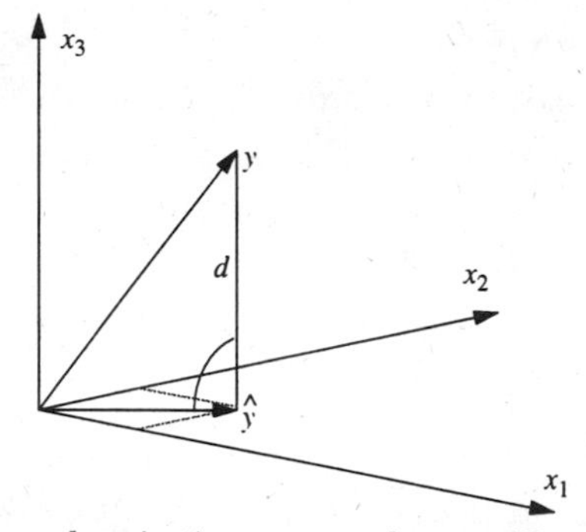

Figure 2.1 Orthogonal projection onto a subspace. Here, $y \in \mathcal{R}^3$ and $\hat{y}$ is its projection onto the span of $\{x_1, x_2\}$. Note that $y - \hat{y}$ is orthogonal to the span $\{x_1, x_2\}$.

THEOREM 2.4

Given an orthonormal system $\{x_1, x_2, \ldots\}$ in E, the following are equivalent:

(a) The set of vectors $\{x_1, x_2, \ldots\}$ is an orthonormal basis for E.

(b) If $\langle x_i, y\rangle = 0$ for $i = 1, 2, \ldots$, then $y = 0$.

(c) span($\{x_i\}$) is dense in E, that is, every vector in E is a limit of a sequence of vectors in span($\{x_i\}$).

(d) For every y in E,

$$\|y\|^2 = \sum_i |\langle x_i, y\rangle|^2, \qquad (2.2.5)$$

which is called *Parseval's equality.*

(e) For every y_1 and y_2 in E,

$$\langle y_1, y_2\rangle = \sum_i \langle x_i, y_1\rangle^* \langle x_i, y_2\rangle, \qquad (2.2.6)$$

which is often called the *generalized Parseval's equality.*

For a proof, see [113].

Orthogonal Projection and Least-Squares Approximation Often, a vector from a Hilbert space E has to be approximated by a vector lying in a (closed) subspace S. We assume that E is separable, thus, S contains an orthonormal basis $\{x_1, x_2, \ldots\}$. Then, the orthogonal projection of $y \in E$ onto S is given by

$$\hat{y} = \sum_i \langle x_i, y\rangle x_i.$$

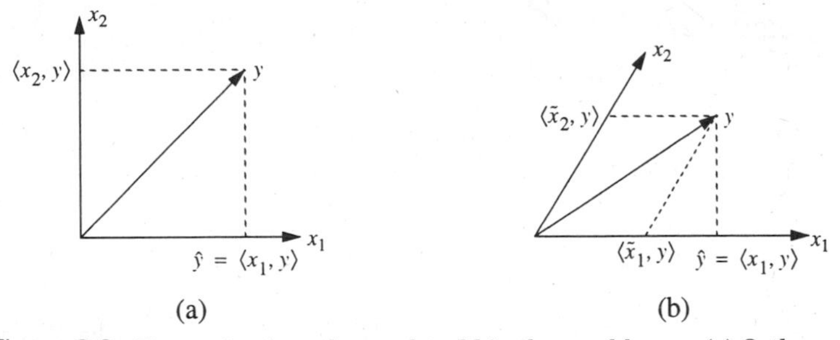

Figure 2.2 Expansion in orthogonal and biorthogonal bases. (a) Orthogonal case: The successive approximation property holds. (b) Biorthogonal case: The first approximation cannot be used in the full expansion.

Note that the difference $d = y - \hat{y}$ satisfies

$$d \perp S$$

and, in particular, $d \perp \hat{y}$, as well as

$$\|y\|^2 = \|\hat{y}\|^2 + \|d\|^2.$$

This is shown pictorially in Figure 2.1. An important property of such an approximation is that it is best in the least-squares sense, that is,

$$\min \|y - x\|$$

for x in S is attained for $x = \sum_i \alpha_i x_i$ with

$$\alpha_i = \langle x_i, y\rangle,$$

that is, the Fourier coefficients. An immediate consequence of this result is the successive approximation property of orthogonal expansions. Call $\hat{y}^{(k)}$ the best approximation of y on the subspace spanned by $\{x_1, x_2, \ldots, x_k\}$ and given by the coefficients $\{\alpha_1, \alpha_2, \ldots, \alpha_k\}$ where $\alpha_i = \langle x_i, y\rangle$. Then, the approximation $\hat{y}^{(k+1)}$ is given by

$$\hat{y}^{(k+1)} = \hat{y}^{(k)} + \langle x_{k+1}, y\rangle x_{k+1},$$

that is, the previous approximation plus the projection along the added vector x_{k+1}. While this is obvious, it is worth pointing out that this successive approximation property does not hold for nonorthogonal bases. When calculating the approximation $\hat{y}^{(k+1)}$, one cannot simply add one term to the previous approximation, but has to recalculate the whole approximation (see Figure 2.2). For a further discussion of projection operators, see Appendix 2.A.

2.2.4 General Bases

While orthonormal bases are very convenient, the more general case of nonorthogonal or biorthogonal bases is important as well. In particular, biorthogonal bases will be constructed in Chapters 3 and 4. A system $\{x_i, \tilde{x}_i\}$ constitutes a *pair of biorthogonal bases* of a Hilbert space E *if and only if* [56, 73]

(a) For all i, j in $\mathcal{Z}$

$$\langle x_i, \tilde{x}_j \rangle = \delta[i-j]. \tag{2.2.7}$$

(b) There exist strictly positive constants A, B, $\tilde{A}$, $\tilde{B}$ such that, for all y in E

$$A\,\|y\|^2 \leq \sum_k |\langle x_k, y\rangle|^2 \leq B\,\|y\|^2, \tag{2.2.8}$$

$$\tilde{A}\,\|y\|^2 \leq \sum_k |\langle \tilde{x}_k, y\rangle|^2 \leq \tilde{B}\,\|y\|^2. \tag{2.2.9}$$

Compare these inequalities with (2.2.5) in the orthonormal case. Bases which satisfy (2.2.8) or (2.2.9) are called Riesz bases [73]. Then, the signal expansion formula becomes

$$y = \sum_k \langle x_k, y\rangle\, \tilde{x}_k = \sum_k \langle \tilde{x}_k, y\rangle\, x_k. \tag{2.2.10}$$

It is clear why the term biorthogonal is used, since to the (nonorthogonal) basis $\{x_i\}$ corresponds a dual basis $\{\tilde{x}_i\}$ which satisfies the biorthogonality constraint (2.2.7). If the basis $\{x_i\}$ is orthogonal, then it is its own dual, and the expansion formula (2.2.10) becomes the usual orthogonal expansion given by (2.2.3–2.2.4).

Equivalences similar to Theorem 2.4 hold in the biorthogonal case as well, and we give the Parseval's relations which become

$$\|y\|^2 = \sum_i \langle x_i, y\rangle^* \langle \tilde{x}_i, y\rangle, \tag{2.2.11}$$

and

$$\langle y_1, y_2\rangle = \sum_i \langle x_i, y_1\rangle^* \langle \tilde{x}_i, y_2\rangle, \tag{2.2.12}$$

$$= \sum_i \langle \tilde{x}_i, y_1\rangle^* \langle x_i, y_2\rangle. \tag{2.2.13}$$

For a proof, see [213] and Problem 2.8.

2.2.5 Overcomplete Expansions

So far, we have considered signal expansion onto bases, that is, the vectors used in the expansion were linearly independent. However, one can also write signals in terms of a linear combination of an overcomplete set of vectors, where the vectors are not independent anymore. A more detailed treatment of such overcomplete sets of vectors, called *frames*, can be found in Chapter 5 and in [73, 89]. We will only discuss a few basic notions here.

A family of functions $\{x_k\}$ in a Hilbert space H is called a *frame* if there exist two constants $A > 0$, $B < \infty$, such that for all y in H

$$A\;\|y\|^2 \;\leq\; \sum_k |\langle x_k, y\rangle|^2 \;\leq\; B\;\|y\|^2.$$

A, B are called *frame bounds*, and when they are equal, we call the frame *tight*. In a tight frame we have

$$\sum_k |\langle x_k, y\rangle|^2 \;=\; A\;\|y\|^2,$$

and the signal can be expanded as follows:

$$y \;=\; A^{-1}\sum_k \langle x_k, y\rangle x_k. \tag{2.2.14}$$

While this last equation resembles the expansion formula in the case of an orthonormal basis, a frame does not constitute an orthonormal basis in general. In particular, the vectors may be linearly dependent and thus not form a basis. If all the vectors in a tight frame have unit norm, then the constant A gives the redundancy ratio (for example, $A = 2$ means there are twice as many vectors as needed to cover the space). Note that if $A - B = 1$, and $\|x_k\| = 1$ for all k, then $\{x_k\}$ constitutes an orthonormal basis.

Because of the linear dependence which exists among the vectors used in the expansion, the expansion is not unique anymore. Consider the set $\{x_1, x_2, \ldots\}$ where $\sum_i \beta_i x_i = 0$ (where not all β_i's are zero) because of linear dependence. If y can be written as

$$y \;=\; \sum_i \alpha_i x_i, \tag{2.2.15}$$

then one can add β_i to each α_i without changing the validity of the expansion (2.2.15). The expansion (2.2.14) is unique in the sense that it minimizes the norm of the expansion among all valid expansions. Similarly, for general frames, there exists a unique dual frame which is discussed in Section 5.3.2 (in the tight frame case, the frame and its dual are equal).

This concludes for now our brief introduction of signal expansions. Later, more specific expansions will be discussed, such as Fourier and wavelet expansions. The fundamental properties seen above will reappear in more specialized forms (for example, Parseval's equality).

While we have only discussed Hilbert spaces, there are of course many other spaces of functions which are of interest. For example, $L_p(\mathcal{R})$ spaces are those containing functions f for which $|f|^p$ is integrable [113]. The norm on these spaces is defined as

$$\|f\|_p \;=\; (\int_{-\infty}^{\infty} |f(t)|^p dt)^{1/p}, \tag{2.2.16}$$

which for $p = 2$ is the usual L_2 norm.[4] Two L_p spaces which will be useful later are $L_1(\mathcal{R})$, the space of functions $f(t)$ satisfying $\int_{-\infty}^{\infty} |f(t)|dt < \infty$, and $L_\infty(\mathcal{R})$, the space of functions $f(t)$ such that $\sup |f(t)| < \infty$. Their discrete-time equivalents are $l_1(\mathcal{Z})$ (space of sequences $x[n]$ such that $\sum_n |x[n]| < \infty$) and $l_\infty(\mathcal{Z})$ (space of sequences $x[n]$ such that $\sup |x[n]| < \infty$). Associated with these spaces are the corresponding norms. However, many of the intuitive geometric interpretations we have seen so far for $L_2(\mathcal{R})$ and $l_2(\mathcal{Z})$ do not hold in these spaces (see Problem 2.3). Recall that in the following, since we use mostly L_2 and l_2, we use $\| \, . \, \|$ to mean $\| \, . \, \|_2$.

2.3 Elements of Linear Algebra

The finite-dimensional cases of Hilbert spaces, namely $\mathcal{R}^n$ and $\mathcal{C}^n$, are very important, and linear operators on such spaces are studied in linear algebra. Many good reference texts exist on the subject, see [106, 280]. Good reviews can also be found in [150] and [308]. We give only a brief account here, focusing on basic concepts and topics which are needed later, such as polynomial matrices.

2.3.1 Basic Definitions and Properties

We can view matrices as representations of bounded linear operators (see Appendix 2.A). The familiar system of equations

$$\begin{array}{ccccc} A_{11}x_1 & + \cdots + & A_{1n}x_n & = & y_1, \\ \vdots & & \vdots & \vdots & \vdots \\ A_{m1}x_1 & + \cdots + & A_{mn}x_n & = & y_m, \end{array}$$

can be compactly represented as

$$\boldsymbol{A}\boldsymbol{x} = \boldsymbol{y}. \tag{2.3.1}$$

Therefore, any finite matrix, or a rectangular (m rows and n columns) array of numbers, can be interpreted as an operator $\boldsymbol{A}$

$$\boldsymbol{A} = \begin{pmatrix} A_{11} & \cdots & A_{1m} \\ \vdots & \ddots & \vdots \\ A_{m1} & \cdots & A_{mn} \end{pmatrix}.$$

An $m \times 1$ matrix is called a *column vector*, while a $1 \times n$ matrix is a *row vector*. As seen in (2.3.1), we write matrices as bold capital letters, and column vectors as lower-case bold letters. A row vector would then be written as $\boldsymbol{v}^T$, where ^{T} denotes transposition (interchange of rows and columns, that is, if $\boldsymbol{A}$ has

[4]For $p \neq 2$, the norm $\| \, . \, \|_p$ cannot be derived from an inner product as in Definition 2.2.

elements A_{ij}, $\boldsymbol{A}^T$ has elements A_{ji}). If the entries are complex, one often uses *hermitian transposition*, which is complex conjugation followed by usual transposition, and is denoted by a superscript $*$.

When $m = n$, the matrix is called *square*, otherwise it is called *rectangular*. A 1×1 matrix is called *scalar*. We denote by $\mathbf{0}$ the *null matrix* (all elements are zero) and by $\boldsymbol{I}$ the *identity* ($A_{ii} = 1$, and 0 otherwise). The identity matrix is a special case of a diagonal matrix. The *antidiagonal* matrix $\boldsymbol{J}$ has all the elements on the other diagonal equal to 1, while the rest are 0, that is, $A_{ij} = 1$, for $j = n + 1 - i$, and $A_{ij} = 0$ otherwise. A *lower* (or *upper*) *triangular* matrix is a square matrix with all of its elements above (or below) the main diagonal equal to zero.

Beside addition/subtraction of same-size matrices (by adding/subtracting the corresponding elements), one can multiply matrices $\boldsymbol{A}$ and $\boldsymbol{B}$ with sizes $m \times n$ and $n \times p$ respectively, yielding a matrix $\boldsymbol{C}$ whose elements are given by

$$C_{ij} = \sum_{k=1}^{n} A_{ik} B_{kj}.$$

Note that the matrix product is not commutative in general, that is, $\boldsymbol{A}\,\boldsymbol{B} \neq \boldsymbol{B}\,\boldsymbol{A}$.[5] It can be shown that $(\boldsymbol{A}\,\boldsymbol{B})^T = \boldsymbol{B}^T\,\boldsymbol{A}^T$.

The *inner product* of two (column) vectors from $\mathcal{R}^N$ is $\langle \boldsymbol{v}_1, \boldsymbol{v}_2 \rangle = \boldsymbol{v}_1^T \cdot \boldsymbol{v}_2$, and if the vectors are from $\mathcal{C}^n$, then $\langle \boldsymbol{v}_1, \boldsymbol{v}_2 \rangle = \boldsymbol{v}_1^* \cdot \boldsymbol{v}_2$. The *outer product* of two vectors from $\mathcal{R}^n$ and $\mathcal{R}^m$ is an $n \times m$ matrix given by $\boldsymbol{v}_1 \cdot \boldsymbol{v}_2^T$.

To define the notion of a determinant, we first need to define a minor. A *minor* $\boldsymbol{M}_{ij}$ is a submatrix of the matrix $\boldsymbol{A}$ obtained by deleting its ith row and jth column. More generally, a minor can be any submatrix of the matrix $\boldsymbol{A}$ obtained by deleting some of its rows and columns. Then the *determinant* of an $n \times n$ matrix can be defined recursively as

$$\det(\boldsymbol{A}) = \sum_{i=1}^{n} A_{ij}(-1)^{i+j} \det(\boldsymbol{M}_{ij})$$

where j is fixed and belongs to $\{1, \ldots, n\}$. The *cofactor* $\boldsymbol{C}_{ij}$ is $(-1)^{i+j} \det(\boldsymbol{M}_{ij})$. A square matrix is said to be *singular* if $\det(\boldsymbol{A}) = 0$. The product of two matrices is nonsingular only if both matrices are nonsingular. Some properties of interest include the following:

(a) If $\boldsymbol{C} = \boldsymbol{A}\,\boldsymbol{B}$, then $\det(\boldsymbol{C}) = \det(\boldsymbol{A}) \det(\boldsymbol{B})$.

[5] When there is possible confusion, we will denote a matrix product by $\boldsymbol{A} \cdot \boldsymbol{B}$; otherwise we will simply write $\boldsymbol{AB}$.

(b) If $\boldsymbol{B}$ is obtained by interchanging two rows/columns of $\boldsymbol{A}$, then $\det(\boldsymbol{B}) = -\det(\boldsymbol{A})$.

(c) $\det(\boldsymbol{A}^T) = \det(\boldsymbol{A})$.

(d) For an $n \times n$ matrix $\boldsymbol{A}$, $\det(c\boldsymbol{A}) = c^n \det(\boldsymbol{A})$.

(e) The determinant of a triangular, and in particular, of a diagonal matrix is the product of the elements on the main diagonal.

An important interpretation of the determinant is that it corresponds to the volume of the parallelepiped obtained when taking the column vectors of the matrix as its edges (one can take the row vectors as well, leading to a different parallelepiped, but the volume remains the same). Thus, a zero determinant indicates linear dependence of the row and column vectors of the matrix, since the parallelepiped is not of full dimension.

The *rank* of a matrix is the size of its largest nonsingular minor (possibly the matrix itself). In a rectangular $m \times n$ matrix, the column rank equals the row rank, that is, the number of linearly independent rows equals the number of linearly independent columns. In other words, the dimension of span(columns) is equal to the dimension of span(rows). For an $n \times n$ matrix to be nonsingular, its rank should equal n. Also $\text{rank}(\boldsymbol{AB}) \leq \min(\text{rank}(\boldsymbol{A}), \text{rank}(\boldsymbol{B}))$.

For a square nonsingular matrix $\boldsymbol{A}$, the *inverse matrix* $\boldsymbol{A}^{-1}$ can be computed using Cramer's formula

$$\boldsymbol{A}^{-1} = \frac{\text{adjugate}(\boldsymbol{A})}{\det(\boldsymbol{A})},$$

where the elements of adjugate($\boldsymbol{A}$) are $(\text{adjugate}(\boldsymbol{A}))_{ji} = \text{cofactor of } A_{ji} = \boldsymbol{C}_{ji}$. For a square matrix, $\boldsymbol{AA}^{-1} = \boldsymbol{A}^{-1}\boldsymbol{A} = \boldsymbol{I}$. Also, $(\boldsymbol{AB})^{-1} = \boldsymbol{B}^{-1}\boldsymbol{A}^{-1}$. Note that Cramer's formula is not actually used to compute the inverse in practice; rather, it serves as a tool in proofs.

For an $m \times n$ rectangular matrix $\boldsymbol{A}$, an $n \times m$ matrix $\boldsymbol{L}$ is its left inverse if $\boldsymbol{LA} = \boldsymbol{I}$. Similarly, an $n \times m$ matrix $\boldsymbol{R}$ is a right inverse of $\boldsymbol{A}$ if $\boldsymbol{AR} = \boldsymbol{I}$. These inverses are not unique and may not even exist. However, if the matrix $\boldsymbol{A}$ is square and has full rank, then its right inverse equals its left inverse, and we can apply Cramer's formula to find that inverse.

The Kronecker product of two matrices is defined as (we show a 2×2 matrix as an example)

$$\begin{bmatrix} a & b \\ c & d \end{bmatrix} \otimes \boldsymbol{M} = \begin{bmatrix} a\boldsymbol{M} & b\boldsymbol{M} \\ c\boldsymbol{M} & d\boldsymbol{M} \end{bmatrix}, \tag{2.3.2}$$

where a, b, c and d are scalars and $\boldsymbol{M}$ is a matrix (neither matrix need be square). See Problem 2.19 for an application of Kronecker products. The Kronecker product has the following useful property with respect to the usual matrix product [32]:

$$(\boldsymbol{A} \otimes \boldsymbol{B})(\boldsymbol{C} \otimes \boldsymbol{D}) = (\boldsymbol{AC}) \otimes (\boldsymbol{BD}) \tag{2.3.3}$$

where all the matrix products have to be well-defined.

2.3.2 Linear Systems of Equations and Least Squares

Going back to the equation $\boldsymbol{A}\,\boldsymbol{x} = \boldsymbol{y}$, one can say that the system has a unique solution provided $\boldsymbol{A}$ is nonsingular, and this solution is given by $\boldsymbol{x} = \boldsymbol{A}^{-1}\,\boldsymbol{y}$. Note that one would rarely compute the inverse matrix in order to solve a linear system of equations; rather Gaussian elimination would be used, since it is much more efficient. In the following, the column space of $\boldsymbol{A}$ denotes the linear span of the columns of $\boldsymbol{A}$, and similarly, the row space is the linear span of the rows of $\boldsymbol{A}$.

Let us give an interpretation of solving the problem $\boldsymbol{Ax} = \boldsymbol{y}$. The product $\boldsymbol{Ax}$ constitutes a linear combination of the columns of $\boldsymbol{A}$ weighted by the entries of $\boldsymbol{x}$. Thus, if $\boldsymbol{y}$ belongs to the column space of $\boldsymbol{A}$, also called the *range* of $\boldsymbol{A}$, there will be a solution. If the columns are linearly independent, the solution is unique, if they are not, there are infinitely many solutions. The *null space* of $\boldsymbol{A}$ is spanned by the vectors orthogonal to the row space, or $\boldsymbol{Av} = \boldsymbol{0}$. If $\boldsymbol{A}$ is of size $m \times n$ (the system of equations has m equations in n unknowns), then the dimension of the range (which equals the rank ρ) plus the dimension of the null space is equal to m. A similar relation holds for row spaces (which are column spaces of $\boldsymbol{A}^T$) and the sum is then equal to n. If $\boldsymbol{y}$ is not in the range of $\boldsymbol{A}$ there is no exact solution and only approximations are possible, such as the orthogonal projection of $\boldsymbol{y}$ onto the span of the columns of $\boldsymbol{A}$, which results in a least-squares solution. Then, the error between $\boldsymbol{y}$ and its projection $\hat{\boldsymbol{y}}$ (see Figure 2.1) is orthogonal to the column space of $\boldsymbol{A}$. That is, any linear combination of the columns of $\boldsymbol{A}$, for example $\boldsymbol{A}\alpha$, is orthogonal to $\boldsymbol{y} - \hat{\boldsymbol{y}} = \boldsymbol{y} - \boldsymbol{A}\hat{\boldsymbol{x}}$ where $\hat{\boldsymbol{x}}$ is the least-squares solution. Thus

$$(\boldsymbol{A}\alpha)^T(\boldsymbol{y} - \boldsymbol{A}\hat{\boldsymbol{x}}) = \boldsymbol{0}$$

or

$$\boldsymbol{A}^T\boldsymbol{A}\hat{\boldsymbol{x}} = \boldsymbol{A}^T\boldsymbol{y},$$

which are called the *normal equations* of the least-squares problem. If the columns of $\boldsymbol{A}$ are linearly independent, then $\boldsymbol{A}^T\boldsymbol{A}$ is invertible. The unique least-squares solution is

$$\hat{\boldsymbol{x}} = (\boldsymbol{A}^T\boldsymbol{A})^{-1}\boldsymbol{A}^T\boldsymbol{y} \tag{2.3.4}$$

(recall that A is either rectangular or rank deficient, and does not have a proper inverse) and the orthogonal projection $\hat{y}$ is equal to

$$\hat{\boldsymbol{y}} = \boldsymbol{A}(\boldsymbol{A}^T\boldsymbol{A})^{-1}\boldsymbol{A}^T\boldsymbol{y}. \qquad (2.3.5)$$

Note that the matrix $\boldsymbol{P} = \boldsymbol{A}(\boldsymbol{A}^T\boldsymbol{A})^{-1}\boldsymbol{A}^T$ satisfies $\boldsymbol{P}^2 = \boldsymbol{P}$ and is symmetric $\boldsymbol{P} = \boldsymbol{P}^T$, thus satisfying the condition for an orthogonal projection operator (see Appendix 2.A). Also, it can be verified that the partial derivatives of the squared error with respect to the components of $\hat{\boldsymbol{x}}$ are zero for the above choice (see Problem 2.6).

2.3.3 Eigenvectors and Eigenvalues

The *characteristic polynomial* for a matrix $\boldsymbol{A}$ is $D(x) = \det(x\boldsymbol{I} - \boldsymbol{A})$, whose roots are called *eigenvalues* λ_i. In particular, a vector $\boldsymbol{p} \neq \boldsymbol{0}$ for which

$$\boldsymbol{A}\boldsymbol{p} = \lambda\boldsymbol{p},$$

is an *eigenvector* associated with the eigenvalue λ. If a matrix of size $n \times n$ has n linearly independent eigenvectors, then it can be *diagonalized*, that is, it can be written as

$$\boldsymbol{A} = \boldsymbol{T}\boldsymbol{\Lambda}\boldsymbol{T}^{-1},$$

where $\boldsymbol{\Lambda}$ is a diagonal matrix containing the eigenvalues of $\boldsymbol{A}$ along the diagonal and $\boldsymbol{T}$ contains its eigenvectors as its columns. An important case is when $\boldsymbol{A}$ is symmetric or, in the complex case, hermitian symmetric, $\boldsymbol{A}^* = \boldsymbol{A}$. Then, the eigenvalues are real, and a full set of orthogonal eigenvectors exists. Taking them as columns of a matrix $\boldsymbol{U}$ after normalizing them to have unit norm so that $\boldsymbol{U}^* \cdot \boldsymbol{U} = \boldsymbol{I}$, we can write a hermitian symmetric matrix as

$$\boldsymbol{A} = \boldsymbol{U}\boldsymbol{\Lambda}\boldsymbol{U}^*.$$

This result constitutes the spectral theorem for hermitian matrices. Hermitian symmetric matrices commute with their hermitian transpose. More generally, a matrix $\boldsymbol{N}$ that commutes with its hermitian transpose is called *normal*, that is, it satisfies $\boldsymbol{N}^*\boldsymbol{N} = \boldsymbol{N}\boldsymbol{N}^*$. Normal matrices are exactly those that have a complete set of orthogonal eigenvectors.

The importance of eigenvectors in the study of linear operators comes from the following fact: Assuming a full set of eigenvectors, a vector $\boldsymbol{x}$ can be written as a linear combination of eigenvectors $\boldsymbol{x} = \sum \alpha_i \boldsymbol{v}_i$. Then,

$$\boldsymbol{A}\boldsymbol{x} = \boldsymbol{A}\left(\sum_i \alpha_i \boldsymbol{v}_i\right) = \sum_i \alpha_i (\boldsymbol{A}\boldsymbol{v}_i) = \sum_i \alpha_i \lambda_i \boldsymbol{v}_i.$$

The concept of eigenvectors generalizes to eigenfunctions for continuous operators, which are functions $f_\omega(t)$ such that $Af_\omega(t) = \lambda(\omega)f_\omega(t)$. A classic example is the complex sinusoid, which is an eigenfunction of the convolution operator, as will be shown in Section 2.4.

2.3.4 Unitary Matrices

We just explained an instance of a square unitary matrix, that is, an $m \times m$ matrix $\boldsymbol{U}$ which satisfies

$$\boldsymbol{U}^*\boldsymbol{U} = \boldsymbol{U}\boldsymbol{U}^* = \boldsymbol{I}, \tag{2.3.6}$$

or, its inverse is its (hermitian) transpose. When the matrix has real entries, it is often called orthogonal or orthonormal, and sometimes, a scale factor is allowed on the left of (2.3.6). Rectangular unitary matrices are also possible, that is, an $m \times n$ matrix $\boldsymbol{U}$ with $m < n$ is unitary if

$$\|\boldsymbol{U}\boldsymbol{x}\| = \|\boldsymbol{x}\|, \qquad \forall \boldsymbol{x} \in \mathcal{C}^n,$$

as well as

$$\langle \boldsymbol{U}\boldsymbol{x}, \boldsymbol{U}\boldsymbol{y} \rangle = \langle \boldsymbol{x}, \boldsymbol{y} \rangle, \qquad \forall \boldsymbol{x}, \boldsymbol{y} \in \mathcal{C}^n,$$

which are the usual Parseval's relations. Then it follows that

$$\boldsymbol{U}\boldsymbol{U}^* = \boldsymbol{I},$$

where $\boldsymbol{I}$ is of size $m \times m$ (and the product does not commute). Unitary matrices have eigenvalues of unit modulus and a complete set of orthogonal eigenvectors. Note that a unitary matrix performs a rotation, thus, the l_2 norm is preserved.

When a square $m \times m$ matrix $\boldsymbol{A}$ has full rank its columns (or rows) form a basis for $\mathcal{R}^m$ and we recall that the Gram-Schmidt orthogonalization procedure can be used to get an orthogonal basis. Gathering the steps of the Gram-Schmidt procedure into a matrix form, we can write $\boldsymbol{A}$ as

$$\boldsymbol{A} = \boldsymbol{Q}\boldsymbol{R},$$

where the columns of $\boldsymbol{Q}$ form the orthonormal basis and $\boldsymbol{R}$ is upper triangular.

Unitary matrices form an important but restricted class of matrices, which can be parametrized in various forms. For example, an $n \times n$ real orthogonal matrix has $n(n-1)/2$ degrees of freedom (up to a permutation of its rows or columns and a sign change in each vector). If we want to find an orthonormal basis for $\mathcal{R}^n$, start with an arbitrary vector and normalize it to have unit norm. This gives $n-1$ degrees of freedom. Next, choose a norm-1 vector in the orthogonal complement with respect to the first vector, which is of dimension $n-1$, giving another $n-2$ degrees of freedom. Iterate until the nth vector is chosen, which is unique up to a sign. We have $\sum_{i=0}^{n-1} i = n(n-1)/2$ degrees of freedom. These degrees of freedom can be used in various parametrizations, based either on planar or Givens rotations or, on Householder building blocks (see Appendix 2.B).

2.3.5 Special Matrices

A (right) *circulant* matrix is a matrix where each row is obtained by a (right) circular shift of the previous row, or

$$C = \begin{pmatrix} c_0 & c_1 & \cdots & & c_{n-1} \\ c_{n-1} & c_0 & c_1 & \cdots & c_{n-2} \\ \vdots & & & & \vdots \\ c_1 & c_2 & \cdots & & c_0 \end{pmatrix}.$$

A *Toeplitz* matrix is a matrix whose (i, j)th entry depends only on the value of $i - j$ and thus it is constant along the diagonals, or

$$T = \begin{pmatrix} t_0 & t_1 & t_2 & \cdots & t_{n-1} \\ t_{-1} & t_0 & t_1 & \cdots & t_{n-2} \\ t_{-2} & t_{-1} & t_0 & \cdots & t_{n-3} \\ \vdots & \vdots & \vdots & \ddots & \vdots \\ t_{-n+1} & t_{-n+2} & t_{-n+3} & \cdots & t_0 \end{pmatrix}.$$

Sometimes, the elements t_i are matrices themselves, in which case the matrix is called *block Toeplitz*. Another important matrix is the *DFT* (Discrete Fourier Transform) matrix. The (i, k)th element of the DFT matrix of size $n \times n$ is $W_n^{ik} = e^{-j2\pi ik/n}$. The DFT matrix diagonalizes circulant matrices, that is, its columns and rows are the eigenvectors of circulant matrices (see Section 2.4.8 and Problem 2.18).

A real symmetric matrix $\boldsymbol{A}$ is called *positive definite* if all its eigenvalues are greater than 0. Equivalently, for all nonzero vectors $\boldsymbol{x}$, the following is satisfied:

$$\boldsymbol{x}^T \boldsymbol{A} \boldsymbol{x} > 0.$$

Finally, for a positive definite matrix $\boldsymbol{A}$, there exists a nonsingular matrix $\boldsymbol{W}$ such that

$$\boldsymbol{A} = \boldsymbol{W}^T \boldsymbol{W},$$

where $\boldsymbol{W}$ is intuitively a "square root" of $\boldsymbol{A}$. One possible way to choose such a square root is to diagonalize $\boldsymbol{A}$ as $\boldsymbol{A} = \boldsymbol{Q}\boldsymbol{\Lambda}\boldsymbol{Q}^T$ and then, since all the eigenvalues are positive, choose $\boldsymbol{W}^T = \boldsymbol{Q}\sqrt{\boldsymbol{\Lambda}}$ (the square root is applied on each eigenvalue in the diagonal matrix $\boldsymbol{\Lambda}$). The above discussion carries over to hermitian symmetric matrices by using hermitian transposes.

2.3.6 Polynomial Matrices

Since a fair amount of the results given in Chapter 3 will make use of polynomial matrices, we will present a brief overview of this subject. For more details, the reader is referred to [106], while self-contained presentations on polynomial matrices can be found in [150, 308].

A *polynomial matrix* (or a *matrix polynomial*) is a matrix whose entries are polynomials. The fact that the above two names can be used interchangeably is due to the following forms of a polynomial matrix $\boldsymbol{H}(x)$:

$$\boldsymbol{H}(x) = \begin{pmatrix} \sum a_i x^i & \cdots & \sum b_i x^i \\ \vdots & \ddots & \vdots \\ \sum c_i x^i & \cdots & \sum d_i x^i \end{pmatrix} = \sum_i \boldsymbol{H}_i\, x^i,$$

that is, it can be written either as a matrix containing polynomials as its entries, or a polynomial having matrices as its coefficients.

The question of the rank in polynomial matrices is more subtle. For example, the matrix

$$\begin{pmatrix} a+bx & 3(a+bx) \\ c+dx & \lambda(c+dx) \end{pmatrix},$$

with $\lambda = 3$, always has rank less than 2, since the two columns are proportional to each other. On the other hand, if $\lambda = 2$, then the matrix would have the rank less than 2 only if $x = -a/b$ or $x = -c/d$. This leads to the notion of *normal rank*. First, note that $\boldsymbol{H}(x)$ is nonsingular only if $\det(\boldsymbol{H}(x))$ is different from 0 for some x. Then, the normal rank of $\boldsymbol{H}(x)$ is the largest of the orders of minors that have a determinant not identically zero. In the above example, for $\lambda = 3$, the normal rank is 1, while for $\lambda = 2$, the normal rank is 2.

An important class of polynomial matrices are *unimodular* matrices, whose determinant is not a function of x. An example is the following matrix:

$$\boldsymbol{H}(x) = \begin{pmatrix} 1+x & x \\ 2+x & 1+x \end{pmatrix},$$

whose determinant is equal to 1. There are several useful properties pertaining to unimodular matrices. For example, the product of two unimodular matrices is again unimodular. The inverse of a unimodular matrix is unimodular as well. Also, one can prove that a polynomial matrix $\boldsymbol{H}(x)$ is unimodular, *if and only if* its inverse is a polynomial matrix. All these facts can be proven using properties of determinants (see, for example, [308]).

The extension of the concept of unitary matrices to polynomial matrices leads to *paraunitary* matrices [308] as studied in circuit theory. In fact, these matrices are unitary on the unit circle or the imaginary axis, depending if they correspond to discrete-time or continuous-time linear operators (z-transforms or Laplace transforms). Consider the discrete-time case and $x = e^{j\omega}$. Then, a square matrix $\boldsymbol{U}(x)$ is unitary on the unit circle if

$$[\boldsymbol{U}(e^{j\omega})]^* \boldsymbol{U}(e^{j\omega}) = \boldsymbol{U}(e^{j\omega})[\boldsymbol{U}(e^{j\omega})]^* = \boldsymbol{I}.$$

Extending this beyond the unit circle leads to

$$[\boldsymbol{U}(x^{-1})]^T\boldsymbol{U}(x) = \boldsymbol{U}(x)[\boldsymbol{U}(x^{-1})]^T = \boldsymbol{I}, \qquad (2.3.7)$$

since $(e^{j\omega})^* = e^{-j\omega}$. If the coefficients of the polynomials are complex, the coefficients need to be conjugated in (2.3.7), which is usually written $[\boldsymbol{U}_*(x^{-1})]^T$. This will be studied in Chapter 3.

As a generalization of polynomial matrices, one can consider the case of *rational* matrices. In that case, each entry is a ratio of two polynomials. As will be shown in Chapter 3, polynomial matrices in z correspond to finite impulse response (FIR) discrete-time filters, while rational matrices can be associated with infinite impulse response (IIR) filters. Unimodular and unitary matrices can be defined in the rational case, as in the polynomial case.

2.4 Fourier Theory and Sampling

This section reviews the Fourier transform and its variations when signals have particular properties (such as periodicity). Sampling, which establishes the link between continuous- and discrete-time signal processing, is discussed in detail. Then, discrete versions of the Fourier transform are examined. The recurring theme is that complex exponentials form an orthonormal basis on which many classes of signals can be expanded. Also, such complex exponentials are eigenfunctions of convolution operators, leading to convolution theorems. The material in this section can be found in many sources, and we refer to [37, 91, 108, 215, 326] for details and proofs.

2.4.1 Signal Expansions and Nomenclature

Let us start by discussing some naming conventions. First, the signal to be expanded is either continuous or discrete in time. Then, the expansion involves an integral (a transform) or a summation (a series). This leads to four possible combinations of continuous/discrete time and integral/series expansions. Note that in the integral case, strictly speaking, we do not have an expansion, but a transform. We use lower case and capital letters for the signal and its expansion (or transform) and denote by ψ_ω and ψ_i a continuous and discrete set of basis functions. In general, there is a basis $\{\psi\}$ and its dual $\{\tilde{\psi}\}$, which are equal in the orthogonal case. Thus, we have

(a) Continuous-time integral expansion, or transform

$$x(t) = \int X_\omega \psi_\omega(t) d\omega \text{ with } X_\omega = \langle \tilde{\psi}_\omega(t), x(t) \rangle.$$

(b) Continuous-time series expansion

$$x(t) = \sum_i X_i \psi_i(t) \text{ with } X_i = \langle \tilde{\psi}_i(t), x(t) \rangle.$$

(c) Discrete-time integral expansion

$$x[n] = \int X_\omega \psi_\omega[n] d\omega \text{ with } X_\omega = \langle \tilde{\psi}_\omega[n], x[n] \rangle.$$

(d) Discrete-time series expansion

$$x[n] = \sum_i X_i \psi_i[n] \text{ with } X_i = \langle \tilde{\psi}_i[n], x[n] \rangle.$$

In the classic Fourier cases, this leads to

(a) The continuous-time Fourier transform (CTFT), often simply called the Fourier transform.

(b) The continuous-time Fourier series (CTFS), or simply Fourier series.

(c) The discrete-time Fourier transform (DTFT).

(d) The discrete-time Fourier series (DTFS).

In all the Fourier cases, $\{\psi\} = \{\tilde{\psi}\}$. The above transforms and series will be discussed in this section. Later, more general expansions will be introduced, in particular, series expansions of discrete-time signals using filter banks in Chapter 3, series expansions of continuous-time signals using wavelets in Chapter 4, and integral expansions of continuous-time signals using wavelets and short-time Fourier bases in Chapter 5.

2.4.2 Fourier Transform

Given an absolutely integrable function $f(t)$, its *Fourier transform* is defined by

$$F(\omega) = \int_{-\infty}^{\infty} f(t) e^{-j\omega t} dt = \langle e^{j\omega t}, f(t) \rangle, \tag{2.4.1}$$

which is called the Fourier analysis formula. The inverse Fourier transform is given by

$$f(t) = \frac{1}{2\pi} \int_{-\infty}^{\infty} F(\omega) e^{j\omega t} d\omega, \tag{2.4.2}$$

or, the Fourier synthesis formula. Note that $e^{j\omega t}$ is not in $L_2(\mathcal{R})$, and that the set $\{e^{j\omega t}\}$ is not countable. The exact conditions under which (2.4.2) is the inverse of (2.4.1) depend on the behavior of $f(t)$ and are discussed in standard texts on Fourier theory [46, 326]. For example, the inversion is exact if $f(t)$ is continuous (or if $f(t)$ is defined as $(f(t^+) + f(t^-))/2$ at a point of discontinuity).[6]

When $f(t)$ is square-integrable, then the formulas above hold in the L_2 sense (see Appendix 2.C), that is, calling $\hat{f}(t)$ the result of the analysis followed by the synthesis formula,

$$\|f(t) - \hat{f}(t)\| = 0.$$

Assuming that the Fourier transform and its inverse exist, we will denote by

$$f(t) \longleftrightarrow F(\omega)$$

a Fourier transform pair. The Fourier transform satisfies a number of properties, some of which we briefly review below. For proofs, see [215].

Linearity Since the Fourier transform is an inner product (see (2.4.1)), it follows immediately from the linearity of the inner product that

$$\alpha f(t) + \beta g(t) \longleftrightarrow \alpha F(\omega) + \beta G(\omega).$$

Symmetry If $F(\omega)$ is the Fourier transform of $f(t)$, then

$$F(t) \longleftrightarrow 2\pi f(-\omega), \tag{2.4.3}$$

which indicates the essential symmetry of the Fourier analysis and synthesis formulas.

Shifting A shift in time by t_0 results in multiplication by a phase factor in the Fourier domain,

$$f(t - t_0) \longleftrightarrow e^{-j\omega t_0} F(\omega). \tag{2.4.4}$$

Conversely, a shift in frequency results in a phase factor, or modulation by a complex exponential, in the time domain,

$$e^{j\omega_0 t} f(t) \longleftrightarrow F(\omega - \omega_0).$$

Scaling Scaling in time results in inverse scaling in frequency as given by the following transform pair (a is a real constant):

$$f(at) \longleftrightarrow \frac{1}{|a|} F\left(\frac{\omega}{a}\right). \tag{2.4.5}$$

[6] We assume that $f(t)$ is of bounded variation. That is, for $f(t)$ defined on a closed interval $[a, b]$, there exists a constant A such that $\sum_{n=1}^{N} |f(t_n) - f(t_{n-1})| < A$ for any finite set $\{t_i\}$ satisfying $a \leq t_0 < t_1 < \ldots < t_N \leq b$. Roughly speaking, the graph of $f(t)$ cannot oscillate over an infinite distance as t goes over a finite interval.

Differentiation/Integration Derivatives in time lead to multiplication by $(j\omega)$ in frequency,

$$\frac{\partial^n f(t)}{\partial t^n} \longleftrightarrow (j\omega)^n F(\omega), \tag{2.4.6}$$

if the transform actually exists. Conversely, if $F(0) = 0$, we have

$$\int_{-\infty}^{t} f(\tau) d\tau \longleftrightarrow \frac{F(\omega)}{j\omega}.$$

Differentiation in frequency leads to

$$(-jt)^n f(t) \longleftrightarrow \frac{\partial^n F(\omega)}{\partial \omega^n}.$$

Moments Calling m_n the nth moment of $f(t)$,

$$m_n = \int_{-\infty}^{\infty} t^n f(t) dt, \quad n = 0, 1, 2, \ldots, \tag{2.4.7}$$

the moment theorem of the Fourier transform states that

$$(-j)^n m_n = \frac{\partial^n F(\omega)}{\partial \omega^n}\Big|_{\omega=0}, \qquad n = 0, 1, 2, \ldots. \tag{2.4.8}$$

Convolution The convolution of two functions $f(t)$ and $g(t)$ is given by

$$h(t) = \int_{-\infty}^{\infty} f(\tau) g(t-\tau) d\tau, \tag{2.4.9}$$

and is denoted $h(t) = f(t) * g(t) = g(t) * f(t)$ since (2.4.9) is symmetric in $f(t)$ and $g(t)$. Denoting by $F(\omega)$ and $G(\omega)$ the Fourier transforms of $f(t)$ and $g(t)$, respectively, the *convolution theorem* states that

$$f(t) * g(t) \longleftrightarrow F(\omega)\, G(\omega).$$

This result is fundamental, and we will prove it for $f(t)$ and $g(t)$ being in $L_1(\mathcal{R})$. Taking the Fourier transform of $f(t) * g(t)$,

$$\int_{-\infty}^{\infty} \left[\int_{-\infty}^{\infty} f(\tau) g(t-\tau) d\tau \right] e^{-j\omega t} dt,$$

changing the order of integration (which is allowed when $f(t)$ and $g(t)$ are in $L_1(\mathcal{R})$; see Fubini's theorem in [73, 250]) and using the shift property, we get

$$\int_{-\infty}^{\infty} f(\tau) \left[\int_{-\infty}^{\infty} g(t-\tau) e^{-j\omega t} dt \right] d\tau = \int_{-\infty}^{\infty} f(\tau) e^{-j\omega\tau} G(\omega) d\tau = F(\omega)\, G(\omega).$$

The result holds as well when $f(t)$ and $g(t)$ are square-integrable, but requires a different proof [108].

An alternative view of the convolution theorem is to identify the complex exponentials $e^{j\omega t}$ as the eigenfunctions of the convolution operator, since

$$\int_{-\infty}^{\infty} e^{j\omega(t-\tau)} g(\tau) d\tau \;=\; e^{j\omega t} \int_{-\infty}^{\infty} e^{-j\omega\tau} g(\tau) d\tau \;=\; e^{j\omega t} G(\omega).$$

The associated eigenvalue $G(\omega)$ is simply the Fourier transform of the impulse response $g(\tau)$ at frequency ω.

By symmetry, the product of time-domain functions leads to the convolution of their Fourier transforms,

$$f(t)\, g(t) \;\longleftrightarrow\; \frac{1}{2\pi} F(\omega) * G(\omega). \tag{2.4.10}$$

This is known as the *modulation theorem* of the Fourier transform.

As an application of both the convolution theorem and the derivative property, consider taking the derivative of a convolution,

$$h'(t) \;=\; \frac{\partial[f(t) * g(t)]}{dt}.$$

The Fourier transform of $h'(t)$, following (2.4.6), is equal to

$$j\omega\,(F(\omega)G(\omega)) \;=\; (j\omega F(\omega))\; G(\omega) \;=\; F(\omega)\; (j\omega G(\omega)),$$

that is,

$$h'(t) \;=\; f'(t) * g(t) \;=\; f(t) * g'(t).$$

This is useful when convolving a signal with a filter which is known to be the derivative of a given function such as a Gaussian, since one can think of the result as being the convolution of the derivative of the signal with a Gaussian.

Parseval's Formula Because the Fourier transform is an orthogonal transform, it satisfies an energy conservation relation known as Parseval's formula. See also Section 2.2.3 where we proved Parseval's formula for orthonormal bases. Here, we need a different proof because the Fourier transform does not correspond to an orthonormal basis expansion (first, exponentials are not in $L_2(\mathcal{R})$ and also the complex exponentials are uncountable, whereas we considered countable orthonormal bases [113]). The general form of Parseval's formula for the Fourier transform is given by

$$\int_{-\infty}^{\infty} f^*(t)\; g(t)\; dt \;=\; \frac{1}{2\pi} \int_{-\infty}^{\infty} F^*(\omega)\; G(\omega)\; d\omega, \tag{2.4.11}$$

which reduces, when $g(t) = f(t)$, to

$$\int_{-\infty}^{\infty} |f(t)|^2\; dt \;=\; \frac{1}{2\pi} \int_{-\infty}^{\infty} |F(\omega)|^2\; d\omega. \tag{2.4.12}$$

Note that the factor $1/2\pi$ comes from our definition of the Fourier transform (2.4.1–2.4.2). A symmetric definition, with a factor $1/\sqrt{2\pi}$ in both the analysis and synthesis formulas (see, for example, [73]), would remove the scale factor in (2.4.12).

The proof of (2.4.11) uses the fact that

$$f^*(t) \longleftrightarrow F^*(-\omega)$$

and the frequency-domain convolution relation (2.4.10). That is, since $f^*(t) \cdot g(t)$ has Fourier transform $(1/2\pi)(F^*(-\omega) * G(\omega))$, we have

$$\int_{-\infty}^{\infty} f^*(t)\, g(t)\, e^{-j\omega t}\, dt \;=\; \frac{1}{2\pi} \int_{-\infty}^{\infty} F^*(-\Omega)\, G(\omega - \Omega)\, d\Omega,$$

where (2.4.11) follows by setting $\omega = 0$.

2.4.3 Fourier Series

A periodic function $f(t)$ with period T,

$$f(t+T) \;=\; f(t),$$

can be expressed as a linear combination of complex exponentials with frequencies $n\omega_0$ where $\omega_0 = 2\pi/T$. In other words,

$$f(t) \;=\; \sum_{k=-\infty}^{\infty} F[k] e^{jk\omega_0 t}, \tag{2.4.13}$$

with

$$F[k] \;=\; \frac{1}{T} \int_{-T/2}^{T/2} f(t)\, e^{-jk\omega_0 t}\, dt. \tag{2.4.14}$$

If $f(t)$ is continuous, then the series converges uniformly to $f(t)$. If a period of $f(t)$ is square-integrable but not necessarily continuous, then the series converges to $f(t)$ in the L_2 sense; that is, calling $\hat{f}_N(t)$ the truncated series with k going from $-N$ to N, the error $\|f(t) - \hat{f}_N(t)\|$ goes to zero as $N \to \infty$. At points of discontinuity, the infinite sum (2.4.13) equals the average $(f(t^+) + f(t^-))/2$. However, convergence is not uniform anymore but plagued by the Gibbs phenomenon. That is, $\hat{f}_N(t)$ will overshoot or undershoot near the point of discontinuity. The amount of over/undershooting is independent of the number of terms N used in the approximation. Only the width diminishes as N is increased.[7] For further discussions on the convergence of Fourier series, see Appendix 2.C and [46, 326].

[7] Again, we consider nonpathological functions (that is, of bounded variation).

Of course, underlying the Fourier series construction is the fact that the set of functions used in the expansion (2.4.13) is a complete orthonormal system for the interval $[-T/2, T/2]$ (up to a scale factor). That is, defining $\varphi_k(t) = (1/\sqrt{T})\, e^{jk\omega_0 t}$ for t in $[-T/2, T/2]$ and k in $\mathcal{Z}$, we can verify that

$$\langle \varphi_k(t), \varphi_l(t) \rangle_{[-\frac{T}{2}, \frac{T}{2}]} = \delta[k-l].$$

When $k = l$, the inner product equals 1. If $k \neq l$, we have

$$\frac{1}{T} \int_{-T/2}^{T/2} e^{j\frac{2\pi}{T}(l-k)t} dt = \frac{1}{\pi(l-k)} \sin(\pi(l-k)) = 0.$$

That the set $\{\varphi_k\}$ is complete is shown in [326] and means that there exists no periodic function $f(t)$ with L_2 norm greater than zero that has all its Fourier series coefficients equal to zero. Actually, there is equivalence between norms, as shown below.

Parseval's Relation With the Fourier series coefficients as defined in (2.4.14), and the inner product of periodic functions taken over one period, we have

$$\langle f(t), g(t) \rangle_{[-\frac{T}{2}, \frac{T}{2}]} = T \langle F[k], G[k] \rangle,$$

where the factor T is due to the normalization chosen in (2.4.13–2.4.14). In particular, for $g(t) = f(t)$,

$$\| f(t) \|^2_{[-\frac{T}{2}, \frac{T}{2}]} = T \| F[k] \|^2.$$

This is an example of Theorem 2.4, up to the scaling factor T.

Best Approximation Property While the following result is true in a more general setting (see Section 2.2.3), it is sufficiently important to be restated for Fourier series, namely

$$\left\| f(t) - \sum_{k=-N}^{N} \langle \varphi_k, f \rangle \varphi_k(t) \right\| \leq \left\| f(t) - \sum_{k=-N}^{N} a_k \varphi_k(t) \right\|,$$

where $\{a_k\}$ is an arbitrary set of coefficients. That is, the Fourier series coefficients are the best ones for an approximation in the span of $\{\varphi_k(t)\}$, $k = -N, \ldots, N$. Moreover, if N is increased, new coefficients are added without affecting the previous ones.

Fourier series, beside their obvious use for characterizing periodic signals, are useful for problems of finite size through periodization. The immediate concern, however, is the introduction of a discontinuity at the boundary, since periodization of a continuous signal on an interval results, in general, in a discontinuous periodic signal.

Fourier series can be related to the Fourier transform seen earlier by using sequences of Dirac functions which are also used in sampling. We will turn our attention to these functions next.

2.4.4 Dirac Function, Impulse Trains and Poisson Sum Formula

The *Dirac function* [215], which is a generalized function or distribution, is defined as a limit of rectangular functions. For example, if

$$\delta_\varepsilon(t) = \begin{cases} 1/\varepsilon & 0 \le t < \varepsilon, \\ 0 & \text{otherwise}, \end{cases} \tag{2.4.15}$$

then $\delta(t) = \lim_{\varepsilon \to 0} \delta_\varepsilon(t)$. More generally, one can use any smooth function $\psi(t)$ with integral 1 and define [278]

$$\delta(t) = \lim_{\epsilon \to 0} \frac{1}{\epsilon} \psi\left(\frac{t}{\epsilon}\right).$$

Any operation involving a Dirac function requires a limiting operation. Since we are reviewing standard results, and for notational convenience, we will skip the limiting process. However, let us emphasize that Dirac functions have to be handled with care in order to get meaningful results. When in doubt, it is best to go back to the definition and the limiting process. For details see, for example, [215]. It follows from (2.4.15) that

$$\int_{-\infty}^{\infty} \delta(t)\, dt = 1, \tag{2.4.16}$$

as well as[8]

$$\int_{-\infty}^{\infty} f(t - t_0)\, \delta(t)\, dt = \int_{-\infty}^{\infty} f(t)\, \delta(t - t_0)\, dt = f(t_0). \tag{2.4.17}$$

Actually, the preceding two relations can be used as an alternative definition of the Dirac function. That is, the Dirac function is a linear operator over a class of functions satisfying (2.4.16–2.4.17). From the above, it follows that

$$f(t) * \delta(t - t_0) = f(t - t_0). \tag{2.4.18}$$

One more standard relation useful for the Dirac function is [215]

$$f(t)\, \delta(t) = f(0)\, \delta(t).$$

The Fourier transform of $\delta(t - t_0)$ is, from (2.4.1) and (2.4.17), equal to

$$\delta(t - t_0) \longleftrightarrow e^{-j\omega t_0}.$$

[8]Note that this holds only for points of continuity.

Using the symmetry property (2.4.3) and the previous results, we see that

$$e^{j\omega_0 t} \longleftrightarrow 2\pi\delta(\omega - \omega_0). \tag{2.4.19}$$

According to the above and using the modulation theorem (2.4.10), $f(t)\ e^{j\omega_0 t}$ has Fourier transform $F(\omega - \omega_0)$.

Next, we introduce the train of Dirac functions spaced $T > 0$ apart, denoted $s_T(t)$ and given by

$$s_T(t) = \sum_{n=-\infty}^{\infty} \delta(t - nT). \tag{2.4.20}$$

Before getting its Fourier transform, we derive the Poisson sum formula. Note that, given a function $f(t)$ and using (2.4.18),

$$\int_{-\infty}^{\infty} f(\tau)\ s_T(t-\tau)\ d\tau = \sum_{n=-\infty}^{\infty} f(t - nT). \tag{2.4.21}$$

Call the above T-periodic function $f_0(t)$. Further assume that $f(t)$ is sufficiently smooth and decaying rapidly such that the above series converges uniformly to $f_0(t)$. We can then expand $f_0(t)$ into a uniformly convergent Fourier series

$$f_0(t) = \sum_{k=-\infty}^{\infty} \left[\frac{1}{T} \int_{-T/2}^{T/2} f_0(\tau) e^{-j2\pi k\tau/T}\ d\tau \right] e^{j2\pi kt/T}.$$

Consider the Fourier series coefficient in the above formula, using the expression for $f_0(t)$ in (2.4.21)

$$\begin{aligned}\int_{-T/2}^{T/2} f_0(\tau) e^{-j2\pi k\tau/T}\ d\tau &= \sum_{n=-\infty}^{\infty} \int_{(2n-1)T/2}^{(2n+1)T/2} f(\tau)\ e^{-j2\pi k\tau/T}\ d\tau \\ &= F\left(\frac{2\pi k}{T}\right).\end{aligned}$$

This leads to the Poisson sum formula.

THEOREM 2.5 *Poisson Sum Formula*

For a function $f(t)$ with sufficient smoothness and decay,

$$\sum_{n=-\infty}^{\infty} f(t - nT) = \frac{1}{T} \sum_{k=-\infty}^{\infty} F\left(\frac{2\pi k}{T}\right) e^{j2\pi kt/T}. \tag{2.4.22}$$

In particular, taking $T = 1$ and $t = 0$,

$$\sum_{n=-\infty}^{\infty} f(n) = \sum_{k=-\infty}^{\infty} F(2\pi k).$$

One can use the Poisson formula to derive the Fourier transform of the impulse train $s_T(t)$ in (2.4.20). It can be shown that

$$S_T(\omega) \;=\; \frac{2\pi}{T}\sum_{k=-\infty}^{\infty}\delta(\omega-\frac{2\pi k}{T}). \tag{2.4.23}$$

We have explained that sampling the spectrum and periodizing the time-domain function are equivalent. We will see the dual situation, when sampling the time-domain function leads to a periodized spectrum. This is also an immediate application of the Poisson formula.

2.4.5 Sampling

The process of sampling is central to discrete-time signal processing, since it provides the link with the continuous-time domain. Call $f_T(t)$ the sampled version of $f(t)$, obtained as

$$f_T(t) \;=\; f(t)\,s_T(t) \;=\; \sum_{n=-\infty}^{\infty} f(nT)\,\delta(t-nT). \tag{2.4.24}$$

Using the modulation theorem of the Fourier transform (2.4.10) and the transform of $s_T(t)$ given in (2.4.23), we get

$$F_T(\omega) \;=\; F(\omega) * \frac{1}{T}\sum_{k=-\infty}^{\infty}\delta\left(\omega-k\frac{2\pi}{T}\right) \;=\; \frac{1}{T}\sum_{k=-\infty}^{\infty}F\left(\omega-k\frac{2\pi}{T}\right), \tag{2.4.25}$$

where we used (2.4.18). Thus, $F_T(\omega)$ is periodic with period $2\pi/T$, and is obtained by overlapping copies of $F(\omega)$ at every multiple of $2\pi/T$. Another way to prove (2.4.25) is to use the Poisson formula. Taking the Fourier transform of (2.4.24) results in

$$F_T(\omega) \;=\; \sum_{n=-\infty}^{\infty} f(nT)\,e^{-jnT\omega},$$

since $f_T(t)$ is a weighted sequence of Dirac functions with weights $f(nT)$ and shifts of nT. To use the Poisson formula, consider the function $g_\Omega(t) = f(t)\,e^{-jt\Omega}$, which has Fourier transform $G_\Omega(\omega) = F(\omega+\Omega)$ according to (2.4.19). Now, applying (2.4.22) to $g_\Omega(t)$, we find

$$\sum_{n=-\infty}^{\infty} g_\Omega(nT) \;=\; \frac{1}{T}\sum_{k=-\infty}^{\infty} G_\Omega\left(\frac{2\pi k}{T}\right)$$

or changing Ω to ω and switching the sign of k,

$$\sum_{n=-\infty}^{\infty} f(nT)\,e^{-jnT\omega} \;=\; \frac{1}{T}\sum_{k=-\infty}^{\infty}F\left(\omega-k\frac{2\pi}{T}\right), \tag{2.4.26}$$

which is the desired result (2.4.25).

Equation (2.4.25) leads immediately to the famous sampling theorem of Whittaker, Kotelnikov and Shannon. If the sampling frequency $\omega_s = 2\pi/T_s$ is larger than $2\omega_m$ (where $F(\omega)$ is bandlimited[9] to ω_m), then we can extract one instance of the spectrum without overlap. If this were not true, then, for example for $k = 0$ and $k = 1$, $F(\omega)$ and $F(\omega - 2\pi/T)$ would overlap and reconstruction would not be possible.

THEOREM 2.6 *Sampling Theorem*

If $f(t)$ is continuous and bandlimited to ω_m, then $f(t)$ is uniquely defined by its samples taken at twice ω_m or $f(n\pi/\omega_m)$. The minimum sampling frequency is $\omega_s = 2\omega_m$ and $T = \pi/\omega_m$ is the maximum sampling period. Then $f(t)$ can be recovered by the interpolation formula

$$f(t) = \sum_{n=-\infty}^{\infty} f(nT) \operatorname{sinc}_T(t - nT), \tag{2.4.27}$$

where

$$\operatorname{sinc}_T(t) = \frac{\sin(\pi t/T)}{\pi t/T}.$$

Note that $\operatorname{sinc}_T(nT) = \delta[n]$, that is, it has the interpolation property since it is 1 at the origin but 0 at nonzero multiples of T. It follows immediately that (2.4.27) holds at the sampling instants $t = nT$.

PROOF

The proof that (2.4.27) is valid for all t goes as follows: Consider the sampled version of $f(t)$, $f_T(t)$, consisting of weighted Dirac functions (2.4.24). We showed that its Fourier transform is given by (2.4.25). The sampling frequency ω_s equals $2\omega_m$, where ω_m is the bandlimiting frequency of $F(\omega)$. Thus, $F(\omega - k\omega_s)$ and $F(\omega - l\omega_s)$ do not overlap for $k \neq l$. To recover $F(\omega)$, it suffices to keep the term with $k = 0$ in (2.4.25) and normalize it by T. This is accomplished with a function that has a Fourier transform which is equal to T from $-\omega_m$ to ω_m and 0 elsewhere. This is called an ideal lowpass filter. Its time-domain impulse response, denoted $\operatorname{sinc}_T(t)$ where $T = \pi/\omega_m$, is equal to (taking the inverse Fourier transform)

$$\operatorname{sinc}_T(t) = \frac{1}{2\pi}\int_{-\omega_m}^{\omega_m} T\, e^{-j\omega t} d\omega = \frac{T}{2\pi j t}\left[e^{j\pi t/T} - e^{-j\pi t/T}\right] = \frac{\sin(\pi t/T)}{\pi t/T}. \tag{2.4.28}$$

Convolving $f_T(t)$ with $\operatorname{sinc}_T(t)$ filters out the repeated spectrums (terms with $k \neq 0$ in (2.4.25)) and recovers $f(t)$, as is clear in frequency domain. Because $f_T(t)$ is a sequence

[9]We will say that a function $f(t)$ is bandlimited to ω_m if its Fourier transform $F(\omega) = 0$ for $|\omega| \geq \omega_m$.

of Dirac functions of weights $f(nT)$, the convolution results in a weighted sum of shifted impulse responses,

$$\left[\sum_{n=-\infty}^{\infty} f(nT)\delta(t-nT)\right] * \text{sinc}_T(t) = \sum_{n=-\infty}^{\infty} f(nT)\,\text{sinc}_T(t-nT),$$

proving (2.4.27)

An alternative interpretation of the sampling theorem is as a series expansion on an orthonormal basis for bandlimited signals. Define

$$\varphi_{n,T}(t) = \frac{1}{\sqrt{T}}\,\text{sinc}_T(t-nT), \tag{2.4.29}$$

whose Fourier transform magnitude is $\sqrt{T}$ from $-\omega_m$ to ω_m, and 0 otherwise. One can verify that $\varphi_{n,T}(t)$ form an orthonormal set using Parseval's relation. The Fourier transform of (2.4.29) is (from (2.4.28) and the shift property (2.4.4))

$$\Phi_{n,T}(\omega) \longleftrightarrow \begin{cases} \sqrt{\pi/\omega_m}\, e^{-j\omega n\pi/\omega_m} & -\omega_m \le \omega \le \omega_m, \\ 0 & \text{otherwise,} \end{cases}$$

where $T = \pi/\omega_m$. From (2.4.11), we find

$$\langle \varphi_{n,T}, \varphi_{k,T} \rangle = \frac{1}{2\omega_m}\int_{-\omega_m}^{\omega_m} e^{j\omega(n-k)\pi/\omega_m}\, d\omega = \delta[n-k].$$

Now, assume a bandlimited signal $f(t)$ and consider the inner product $\langle \varphi_{n,T}, f\rangle$. Again using Parseval's relation,

$$\langle \varphi_{n,T}, f \rangle = \frac{\sqrt{T}}{2\pi}\int_{-\omega_m}^{\omega_m} e^{j\omega nT}\, F(\omega)\, d\omega = \sqrt{T} f(nT),$$

because the integral is recognized as the inverse Fourier transform of $F(\omega)$ at $t = nT$ (the bounds $[-\omega_m, \omega_m]$ do not alter the computation of $F(\omega)$ because it is bandlimited to ω_m). Therefore, another way to write the interpolation formula (2.4.27) is

$$f(t) = \sum_{n=-\infty}^{\infty} \langle \varphi_{n,T}, f \rangle\, \varphi_{n,T}(t) \tag{2.4.30}$$

(the only change is that we normalized the sinc basis functions to have unit norm).

What happens if $f(t)$ is not bandlimited? Because $\{\varphi_{n,T}\}$ is an orthogonal set, the interpolation formula (2.4.30) represents the orthogonal projection of the input signal onto the subspace of bandlimited signals. Another way to write the inner product in (2.4.30) is

$$\langle \varphi_{n,T}, f \rangle = \int_{-\infty}^{\infty} \varphi_{0,T}(\tau - nT)\, f(\tau)\, d\tau = \varphi_{0,T}(-t) * f(t)|_{t=nT},$$

which equals $\varphi_{0,T}(t) * f(t)$ since $\varphi_{0,T}(t)$ is real and symmetric in t. That is, the inner products, or coefficients, in the interpolation formula are simply the outputs of an ideal lowpass filter with cutoff π/T sampled at multiples of T. This is the usual view of the sampling theorem as a bandlimiting convolution followed by sampling and reinterpolation.

To conclude this section, we will demonstrate a fact that will be used in Chapter 4. It states that the following can be seen as a Fourier transform pair:

$$\langle f(t), f(t+n)\rangle \;=\; \delta[n] \;\longleftrightarrow\; \sum_{k\in\mathcal{Z}} |F(\omega+2k\pi)|^2 \;=\; 1. \tag{2.4.31}$$

The left side of the equation is simply the deterministic autocorrelation[10] of $f(t)$ evaluated at integers, that is, sampled autocorrelation. If we denote the autocorrelation of $f(t)$ as $p(\tau) = \langle f(t), f(t+\tau)\rangle$, then the left side of (2.4.31) is $p_1(\tau) = p(\tau)s_1(\tau)$, where $s_1(\tau)$ is as defined in (2.4.20) with $T = 1$. The Fourier transform of $p_1(\tau)$ is (apply (2.4.25))

$$P_1(\omega) \;=\; \sum_{k\in\mathcal{Z}} P(\omega - 2k\pi).$$

Since the Fourier transform of $p(t)$ is $P(\omega) = |F(\omega)|^2$, we get that the Fourier transform of the right side of (2.4.31) is the left side of (2.4.31).

2.4.6 Discrete-Time Fourier Transform

Given a sequence $\{f[n]\}_{n\in\mathcal{Z}}$, its *discrete-time Fourier transform* (DTFT) is defined by

$$F(e^{j\omega}) \;=\; \sum_{n=-\infty}^{\infty} f[n]\, e^{-j\omega n}, \tag{2.4.32}$$

which is 2π-periodic. Its inverse is given by

$$f[n] \;=\; \frac{1}{2\pi}\int_{-\pi}^{\pi} F(e^{j\omega})\, e^{j\omega n}\, d\omega. \tag{2.4.33}$$

A sufficient condition for the convergence of (2.4.32) is that the sequence $f[n]$ be absolutely summable. Then, convergence is uniform to a continuous function of ω [211]. If the sequence is square-summable, then we have mean square convergence of the series in (2.4.32) (that is, the energy of the error goes to zero as the summation limits go to infinity). By using distributions, one can define discrete-time transforms of more general sequences as well, for example [211]

$$e^{j\omega_0 n} \;\longleftrightarrow\; 2\pi \sum_{k=-\infty}^{\infty} \delta(\omega - \omega_0 + 2\pi k).$$

[10]The deterministic autocorrelation of a real function $f(t)$ is $f(t) * f(-t) \;=\; \int f(\tau)\, f(\tau+t)\, d\tau$.

Comparing (2.4.32–2.4.33) with the equivalent expressions for Fourier series (2.4.13–2.4.14), one can see that they are duals of each other (within scale factors). Furthermore, if the sequence $f[n]$ is obtained by sampling a continuous-time function $f(t)$ at instants nT,

$$f[n] = f(nT), \tag{2.4.34}$$

then the discrete-time Fourier transform is related to the Fourier transform of $f(t)$. Denoting the latter by $F_c(\omega)$, the Fourier transform of its sampled version is equal to (see (2.4.26))

$$F_T(\omega) = \sum_{n=-\infty}^{\infty} f(nT)\, e^{-jnT\omega} = \frac{1}{T} \sum_{k=-\infty}^{\infty} F_c\left(\omega - k\frac{2\pi}{T}\right). \tag{2.4.35}$$

Now consider (2.4.32) at ωT and use (2.4.34), thus

$$F(e^{j\omega T}) = \sum_{n=-\infty}^{\infty} f(nT)\, e^{-jn\omega T}$$

and, using (2.4.35),

$$F(e^{j\omega T}) = \frac{1}{T} \sum_{k=-\infty}^{\infty} F_c\left(\omega - k\frac{2\pi}{T}\right). \tag{2.4.36}$$

Because of these close relationships with the Fourier transform and Fourier series, it follows that all properties seen earlier carry over and we will only repeat two of the most important ones (for others, see [211]).

Convolution Given two sequences $f[n]$ and $g[n]$ and their discrete-time Fourier transforms $F(e^{j\omega})$ and $G(e^{j\omega})$, then

$$f[n] * g[n] = \sum_{l=-\infty}^{\infty} f[n-l]\, g[l] = \sum_{l=-\infty}^{\infty} f[l]\, g[n-l] \longleftrightarrow F(e^{j\omega})\, G(e^{j\omega}).$$

Parseval's Equality With the same notations as above, we have

$$\sum_{n=-\infty}^{\infty} f^*[n]\, g[n] = \frac{1}{2\pi} \int_{-\pi}^{\pi} F^*(e^{j\omega})\, G(e^{j\omega})\, d\omega, \tag{2.4.37}$$

and in particular, when $g[n] = f[n]$,

$$\sum_{n=-\infty}^{\infty} |f[n]|^2 = \frac{1}{2\pi} \int_{-\pi}^{\pi} |F(e^{j\omega})|^2\, d\omega.$$

2.4.7 Discrete-Time Fourier Series

If a discrete-time sequence is periodic with period N, that is, $f[n] = f[n + lN]$, $l \in \mathcal{Z}$, then its *discrete-time Fourier series* representation is given by

$$F[k] = \sum_{n=0}^{N-1} f[n]\, W_N^{nk}, \qquad k \in \mathcal{Z}, \tag{2.4.38}$$

$$f[n] = \frac{1}{N} \sum_{k=0}^{N-1} F[k]\, W_N^{-nk}, \qquad n \in \mathcal{Z}, \tag{2.4.39}$$

where W_N is the Nth root of unity. That this is an analysis-synthesis pair is easily verified by using the orthogonality of the roots of unity (see (2.1.3)). Again, all the familiar properties of Fourier transforms hold, taking periodicity into account. For example, convolution is now periodic convolution, that is,

$$f[n] * g[n] = \sum_{l=0}^{N-1} f[n-l]\, g[l] = \sum_{l=0}^{N-1} f_0[(n-l) \bmod N]\, g_0[l], \tag{2.4.40}$$

where $f_0[\cdot]$ and $g_0[\cdot]$ are equal to one period of $f[\cdot]$ and $g[\cdot]$ respectively. That is, $f_0[n] = f[n]$, $n = 0, \ldots, N-1$, and 0 otherwise, and similarly for $g_0[n]$. Then, the convolution property is given by

$$f[n] * g[n] = f_0[n] *_p g_0[n] \longleftrightarrow F[k]\, G[k], \tag{2.4.41}$$

where $*_p$ denotes periodic convolution. Parseval's formula then follows as

$$\sum_{n=0}^{N-1} f^*[n]\, g[n] = \frac{1}{N} \sum_{k=0}^{N-1} F^*[k]\, G[k].$$

Just as the Fourier series coefficients were related to the Fourier transform of one period (see (2.4.14)), the coefficients of the discrete-time Fourier series can be obtained from the discrete-time Fourier transform of one period. If we call $F_0(e^{j\omega})$ the discrete-time Fourier transform of $f_0[n]$, (2.4.32) and (2.4.38) imply that

$$F_0(e^{j\omega}) = \sum_{n=-\infty}^{\infty} f_0[n]\, e^{-j\omega n} = \sum_{n=0}^{N-1} f[n]\, e^{-j\omega n},$$

leading to

$$F[k] = F_0(e^{j\omega})|_{\omega = k2\pi/N}.$$

The sampling of $F_0(e^{j\omega})$ simply repeats copies of $f_0[n]$ at integer multiples of N, and thus we have

$$f[n] = \sum_{l=-\infty}^{\infty} f_0[n - lN] = \frac{1}{N} \sum_{k=0}^{N-1} F[k]\, e^{jnk2\pi/N} = \frac{1}{N} \sum_{k=0}^{N-1} F_0\left[e^{jk2\pi/N}\right] e^{jnk2\pi/N}, \tag{2.4.42}$$

which is the discrete-time version of the Poisson sum formula. It actually holds for $f_0[\cdot]$ with support larger than $0, \ldots, N-1$, as long as the first sum in (2.4.42) converges. For $n = 0$, (2.4.42) yields

$$\sum_{l=-\infty}^{\infty} f_0[lN] = \frac{1}{N} \sum_{k=0}^{N-1} F_0 \left[e^{jk2\pi/N} \right].$$

2.4.8 Discrete Fourier Transform

The importance of the discrete-time Fourier transform of a finite-length sequence (which can be one period of a periodic sequence) leads to the definition of the *discrete Fourier transform* (DFT). This transform is very important for computational reasons, since it can be implemented using the fast Fourier transform (FFT) algorithm (see Chapter 6). The DFT is defined as

$$F[k] = \sum_{n=0}^{N-1} f[n] \, W_N^{nk}, \tag{2.4.43}$$

and its inverse as

$$f[n] = \frac{1}{N} \sum_{k=0}^{N-1} F[k] \, W_N^{-nk}, \tag{2.4.44}$$

where $W_N = e^{-j2\pi/N}$. These are the same formulas as (2.4.38–2.4.39), except that $f[n]$ and $F[k]$ are not defined for $n, k \notin \{0, \ldots, N-1\}$. Recall that the discrete-time Fourier transform of a finite-length sequence can be sampled at $\omega = 2\pi/N$ (which periodizes the sequence). Therefore, it is useful to think of the DFT as the transform of one period of a periodic signal, or a sampling of the DTFT of a finite-length signal. In both cases, there is an underlying periodic signal. Therefore, all properties are with respect to this inherent periodicity. For example, the convolution property of the DFT leads to periodic convolution (see (2.4.40)). Because of the finite-length signals involved, the DFT is a mapping on $\mathcal{C}^N$ and can thus be best represented as a matrix-vector product. Calling $\boldsymbol{F}$ the Fourier matrix with entries

$$F_{n,k} = W_N^{nk}, \qquad n, k = 0, \ldots, N-1,$$

then its inverse is equal to (following (2.4.44))

$$\boldsymbol{F}^{-1} = \frac{1}{N} \boldsymbol{F}^*. \tag{2.4.45}$$

Given a sequence $\{f[0], f[1], \ldots, f[N-1]\}$, we can define a circular convolution matrix $\boldsymbol{C}$ with a first line equal to $\{f[0], f[N-1], \ldots, f[1]\}$ and each subsequent line being a right circular shift of the previous one. Then, circular convolution of $\{f[n]\}$ with a sequence $\{g[n]\}$ can be written as

$$f *_p g = \boldsymbol{C}\boldsymbol{g} = \boldsymbol{F}^{-1} \boldsymbol{\Lambda} \boldsymbol{F} \boldsymbol{g},$$

according to the convolution property (2.4.40–2.4.41), where $\boldsymbol{\Lambda}$ is a diagonal matrix with $F[k]$ on its diagonal. Conversely, this means that $\boldsymbol{C}$ is diagonalized by $\boldsymbol{F}$ or that the complex exponential sequences $\{e^{j(2\pi/N)nk}\} = W_N^{-nk}$ are eigenvectors of the convolution matrix $\boldsymbol{C}$, with eigenvalues $F[k]$. Note that the time reversal associated with convolution is taken into account in the definition of the circulant matrix $\boldsymbol{C}$.

Using matrix notation, Parseval's formula for the DFT follows easily. Call $\hat{\boldsymbol{f}}$ the Fourier transform of the vector $\boldsymbol{f} = (\, f[0] \quad f[1] \quad \cdots \quad f[N-1]\,)^T$, that is

$$\hat{\boldsymbol{f}} \;=\; \boldsymbol{F}\boldsymbol{f},$$

and a similar definition for $\hat{\boldsymbol{g}}$ as the Fourier transform of $\boldsymbol{g}$. Then

$$\hat{\boldsymbol{f}}^* \hat{\boldsymbol{g}} \;=\; (\boldsymbol{F}\boldsymbol{f})^*(\boldsymbol{F}\boldsymbol{g}) \;=\; \boldsymbol{f}^*\boldsymbol{F}^*\boldsymbol{F}\boldsymbol{g} \;=\; N\boldsymbol{f}^*\boldsymbol{g},$$

where we used (2.4.45), that is, the fact that $\boldsymbol{F}^*$ is the inverse of $\boldsymbol{F}$ up to a scale factor of N.

Other properties of the DFT follow from their counterparts for the discrete-time Fourier transform, bearing in mind the underlying circular structure implied by the discrete-time Fourier series (for example, a shift is a circular shift).

2.4.9 Summary of Various Flavors of Fourier Transforms

Between the Fourier transform, where both time and frequency variables are continuous, and the discrete-time Fourier series (DTFS), where both variables are discrete, there are a number of intermediate cases.

First, in Table 2.1 and Figure 2.3, we compare the Fourier transform, Fourier series, discrete-time Fourier transform and discrete-time Fourier series. The table shows four combinations of continuous versus discrete variables in time and frequency. As defined in Section 2.4.1, we use a short-hand CT or DT for continuous- versus discrete-time variable, and we call it a Fourier transform or series if the synthesis formula involves an integral or a summation.

Then, in Table 2.2 and Figure 2.4, we consider the same transforms but when the signal satisfies some additional restrictions, that is, when it is limited either in time or in frequency. In that case, the continuous function (of time or frequency) can be sampled without loss of information.

2.5 Signal Processing

This section briefly covers some fundamental notions of continuous and discrete-time signal processing. Our focus is on linear time-invariant or periodically time-varying systems. For these, weighted complex exponentials play a special role, leading to the Laplace and z-transform as useful generalizations

Table 2.1 Fourier transforms with various combinations of continuous/discrete time and frequency variables. CT and DT stand for continuous and discrete time, while FT and FS stand for Fourier transform (integral synthesis) and Fourier series (summation synthesis). P stands for a periodic signal. The relation between sampling period T and sampling frequency ω_s is $\omega_s = 2\pi/T$. Note that in the DTFT case, ω_s is usually equal to 2π $(T = 1)$.

Transform	Time	Freq.	Analysis Synthesis	Duality
(a) Fourier transform CTFT	C	C	$F(\omega) = \int_t f(t)\, e^{-j\omega t} dt$ $f(t) = 1/2\pi \int_\omega F(\omega)\, e^{j\omega t} d\omega$	self- dual
(b) Fourier series CTFS	C P	D	$F[k] = 1/T \int_{-T/2}^{T/2} f(t)\, e^{-j2\pi kt/T}\, dt$ $f(t) = \sum_k F[k]\, e^{j2\pi kt/T}$	dual with DTFT
(c) Discrete-time Fourier transform DTFT	D	C P	$F(e^{j\omega}) = \sum_n f[n]\, e^{-j2\pi\omega n/\omega_s}$ $f[n] = 1/\omega_s \int_{-\omega_s/2}^{\omega_s/2} F(e^{j\omega})\, e^{j2\pi\omega n/\omega_s}\, d\omega$	dual with CTFS
(d) Discrete-time Fourier series DTFS	D P	D P	$F[k] = \sum_{n=0}^{N-1} f[n]\, e^{-j2\pi nk/N}$ $f[n] = 1/N \sum_{n=0}^{N-1} F[k]\, e^{j2\pi nk/N}$	self -dual

Table 2.2 Various Fourier transforms with restrictions on the signals involved. Either the signal is of finite length (FL) or the Fourier transform is bandlimited (BL).

Transform	Time	Frequency	Equivalence	Duality
(a) Fourier transform of bandlimited signal BL-CTFT	Can be sampled	$(-\frac{\omega_s}{2}, \frac{\omega_s}{2})$	Sample time. Periodize frequency.	Dual with FL-CTFT
(b) Fourier transform of finite-length signal FL-CTFT	$(0, T)$	Can be sampled	Periodize time. Sample frequency.	Dual with BL-CTFT
(c) Fourier series of band-limited periodic signal BL-CTFS	Periodic can be sampled	Finite number of Fourier coefficients	Sample time. Finite Fourier series in time.	Dual with FL-DTFT
(d) Discrete-time Fourier transform of finite-length sequence FL-DTFT	Finite number of samples	Periodic can be sampled	Sample frequency. Finite Fourier series in frequency.	Dual with BL-CTFS

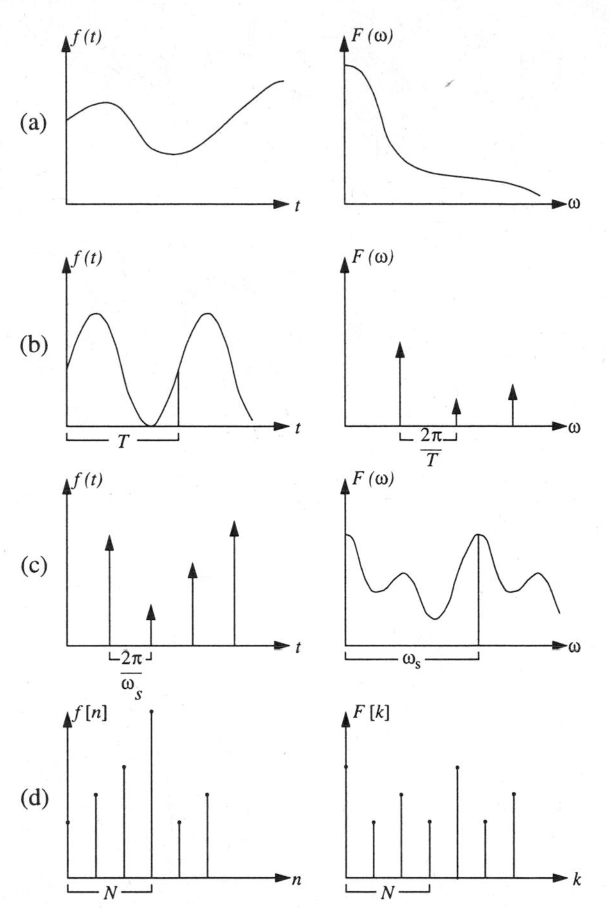

Figure 2.3 Fourier transforms with various combinations of continuous/discrete time and frequency variables (see also Table 2.1). (a) Continuous-time Fourier transform. (b) Continuous-time Fourier series (note that the frequency-domain function is discrete in frequency, appearing at multiples of $2\pi/T$, with weights $F[k]$). (c) Discrete-time Fourier transform (note that the time-domain function is discrete in time, appearing at multiples of $2\pi/\omega_s$, with weights $f[n]$). (d) Discrete-time Fourier series.

of the continuous and discrete-time Fourier transforms. Within this class of systems, we are particularly interested in those having finite-complexity realizations or finite-order differential/difference equations. These will have rational Laplace or z-transforms, which we assume in what follows. For further details, see [211, 212]. We also discuss the basics of multirate signal processing which is at the heart of the material on discrete-time bases in Chapter 3. More material on multirate signal processing can be found in [67, 308].

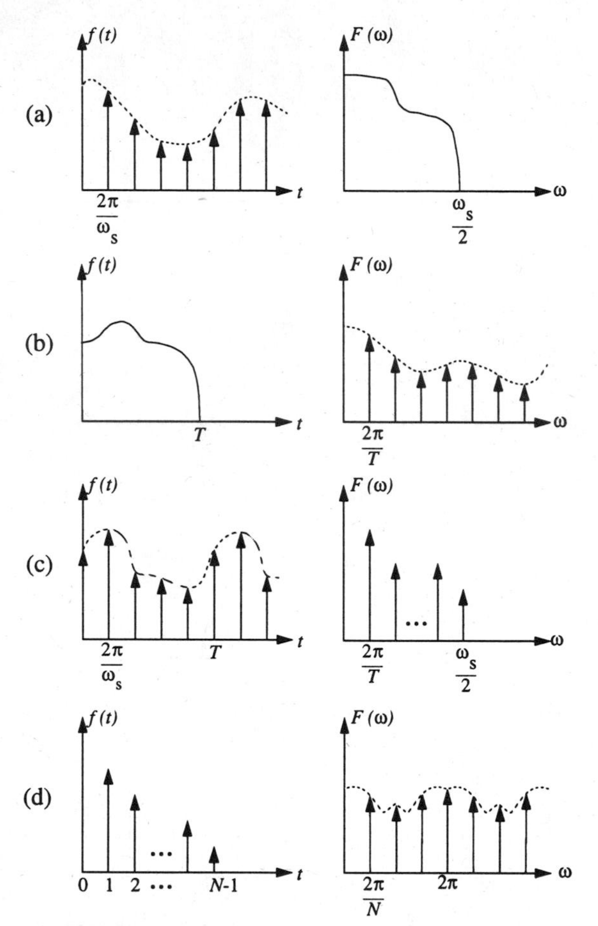

Figure 2.4 Fourier transform with length and bandwidth restrictions on the signal (see also Table 2.2). (a) Fourier transform of bandlimited signals, where the time-domain signal can be sampled. Note that the function in frequency domain has support on $(-\omega_{s/2}, \omega_{s/2})$. (b) Fourier transform of finite-length signals, where the frequency-domain signal can be sampled. (c) Fourier series of bandlimited periodic signals (it has a finite number of Fourier components). (d) Discrete-time Fourier transform of finite-length sequences.

2.5.1 Continuous-Time Signal Processing

Signal processing, which is based on Fourier theory, is concerned with actually implementing algorithms. So, for example, the study of filter structures and their associated properties is central to the subject.

The Laplace Transform An extension of the Fourier transform to the complex plane (instead of just the frequency axis) is the following:

$$F(s) = \int_{-\infty}^{\infty} f(t) e^{-st}\, dt,$$

where $s = \sigma + j\omega$. This is equivalent, for a given σ, to the Fourier transform of $f(t) \cdot e^{-\sigma t}$, that is, the transform of an exponentially weighted signal. Now, the above transform does not in general converge for all s, that is, associated with it is a region of convergence (ROC). The ROC has the following important properties [212]: The ROC is made up of strips in the complex plane parallel to the $j\omega$-axis. If the $j\omega$-axis is contained in the ROC, then the Fourier transform converges. Note that if the Laplace transform is rational, then the ROC cannot contain any poles. If a signal is right-sided (that is, zero for $t < T_0$) or left-sided (zero for $t > T_1$), then the ROC is right- or left-sided, respectively, in the sense that it extends from some vertical line (corresponding to the limit value of $Re(s)$ up to where the Laplace transform converges) all the way to $Re(s)$ becoming plus or minus infinity. It follows that a finite-length signal has the whole complex plane as its ROC (assuming it converges anywhere), since it is both left- and right-sided and connected.

If a signal is two-sided, that is, neither left- nor right-sided, then its ROC is the intersection of the ROC's of its left- and right-sided parts. This ROC is therefore either empty or of the form of a vertical strip.

Given a Laplace transform (such as a rational expression), different ROC's lead to different time-domain signals. Let us illustrate this with an example.

Example 2.1

Assume $F(s) = 1/((s+1)(s+2))$. The ROC $\{Re(s) < -2\}$ corresponds to a left-sided signal

$$f(t) = -(e^{-t} - e^{-2t})\, u(-t).$$

The ROC $\{Re(s) > -1\}$ corresponds to a right-sided signal

$$f(t) = (e^{-t} - e^{-2t})\, u(t).$$

Finally, the ROC $\{-2 < Re(s) < -1\}$ corresponds to a two-sided signal

$$f(t) = -e^{-t}\, u(-t) - e^{-2t}\, u(t).$$

Note that only the right-sided signal would also have a Fourier transform (since its ROC includes the $j\omega$-axis).

For the inversion of the Laplace transform, recall its relation to the Fourier transform of an exponentially weighted signal. Then, it can be shown that its inverse is

$$f(t) = \frac{1}{2\pi j} \int_{\sigma - j\infty}^{\sigma + j\infty} F(s)\, e^{st}\, ds,$$

where σ is chosen inside the ROC. We will denote a Laplace transform pair by

$$f(t) \longleftrightarrow F(s), \qquad s \in \text{ROC}.$$

For a review of Laplace transform properties, see [212]. Next, we will concentrate on filtering only.

Linear Time-Invariant Systems The convolution theorem of the Laplace transform follows immediately from the fact that exponentials are eigenfunctions of the convolution operator. For, if $f(t) = h(t) * g(t)$ and $h(t) = e^{st}$, then

$$f(t) = \int h(t-\tau)\, g(\tau)\, d\tau = \int e^{s(t-\tau)}\, g(\tau)\, d\tau = e^{st} \int e^{-s\tau}\, g(\tau)\, d\tau = e^{st}\, G(s).$$

The eigenvalue attached to e^{st} is the Laplace transform of $g(t)$ at s. Thus,

$$f(t) = h(t) * g(t) \longleftrightarrow F(s) = H(s)\, G(s),$$

with an ROC containing the intersection of the ROC's of $H(s)$ and $G(s)$.

The differentiation property of the Laplace transform says that

$$\frac{\partial f(t)}{\partial t} \longleftrightarrow s\, F(s),$$

with ROC containing the ROC of $F(s)$. Then, it follows that *linear constant-coefficient differential equations* can be characterized by a Laplace transform called the *transfer function* $H(s)$. Linear, time-invariant differential equations, given by

$$\sum_{k=0}^{N} a_k \frac{\partial^k y(t)}{\partial t^k} = \sum_{k=0}^{M} b_k \frac{\partial^k x(t)}{\partial t^k}, \tag{2.5.1}$$

lead, after taking the Laplace transform, to the following ratio:

$$H(s) = \frac{Y(s)}{X(s)} = \frac{\sum_{k=0}^{M} b_k s^k}{\sum_{k=0}^{N} a_k s^k},$$

that is, the input and the output are related by a convolution with a filter having impulse response $h(t)$, where $h(t)$ is the inverse Laplace transform of $H(s)$.

To take this inverse Laplace transform, we need to specify the ROC. Typically, we look for a *causal* solution, where we solve the differential equation forward in time. Then, the ROC extends to the right of the vertical line which passes through the rightmost pole. Stability[11] of the filter corresponding to the transfer function requires that the ROC include the $j\omega$-axis. This leads to the well-known requirement that a causal system with rational transfer function is stable *if and only if* all the poles are in the left half-plane (the real part of the

[11] Stability of a filter means that a bounded input produces a bounded output.

pole location is smaller than zero). In the above discussion, we have assumed initial rest conditions, that is, the homogeneous solution of differential Equation (2.5.1) is zero (otherwise, the system is neither linear nor time-invariant).

Example 2.2 *Butterworth Filters*

Among various classes of continuous-time filters we will briefly describe the Butterworth filters, both because they are simple and because they will reappear later as useful filters in the context of wavelets. The magnitude squared of the Fourier transform of an Nth-order Butterworth filter is given by

$$|H_N(j\omega)|^2 = \frac{1}{1+(j\omega/j\omega_c)^{2N}}, \tag{2.5.2}$$

where ω_c is a parameter which will specify the cutoff frequency beyond which sinusoids are substantially attenuated. Thus, ω_c defines the bandwidth of the lowpass Butterworth filter. Since $|H_N(j\omega)|^2 = H(j\omega)H^*(j\omega) = H(j\omega)H(-j\omega)$ when the filter is real, and noting that (2.5.2) is the Laplace transform for $s = j\omega$, we get

$$H(s)\,H(-s) = \frac{1}{1+(s/j\omega_c)^{2N}}. \tag{2.5.3}$$

The poles of $H(s)H(-s)$ are thus at $(-1)^{1/2N}(j\omega_c)$, or

$$|s_k| = \omega_c, \qquad \arg[s_k] = \frac{\pi(2k+1)}{2N} + \frac{\pi}{2},$$

and $k = 0, \ldots, 2N-1$. The poles thus lie on a circle, and they appear in pairs at $\pm s_k$. To get a stable and causal filter, one simply chooses the N poles which lie on the left-hand side half-circle. Since pole locations specify the filter only up to a scale factor, set $s = 0$ in (2.5.3) which leads to $H(0) = 1$. For example, a second-order Butterworth filter has the following Laplace transform:

$$H_2(s) = \frac{\omega_c^2}{(s+\omega_c e^{j\pi/4})(s+\omega_c e^{-j\pi/4})}. \tag{2.5.4}$$

One can find its "physical" implementation by going back, through the inverse Laplace transform, to the equivalent linear constant-coefficient differential equation. See also Example 3.6 in Chapter 3, for discrete-time Butterworth filters.

2.5.2 Discrete-Time Signal Processing

Just as the Laplace transform was a generalization of the Fourier transform, the z-transform will be introduced as a generalization of the discrete-time Fourier transform [149]. Again, it will be most useful for the study of difference equations (the discrete-time equivalent of differential equations) and the associated discrete-time filters.

The z-Transform The forward z-transform is defined as

$$F(z) = \sum_{n=-\infty}^{\infty} f[n]\, z^{-n}, \tag{2.5.5}$$

where $z \in \mathcal{C}$. On the unit circle $z = e^{j\omega}$, this is the discrete-time Fourier transform (2.4.32), and for $z = \rho e^{j\omega}$, it is the discrete-time Fourier transform of the sequence $f[n] \cdot \rho^n$. Similarly to the Laplace transform, there is a region of convergence (ROC) associated with the z-transform $F(z)$, namely a region of the complex plane where $F(z)$ converges. Consider the case where the z-transform is rational and the sequence is bounded in amplitude. The ROC does not contain any pole. If the sequence is right-sided (left-sided), the ROC extends outward (inward) from a circle with the radius corresponding to the modulus of the outermost (innermost) pole. If the sequence is two-sided, the ROC is a ring. The discrete-time Fourier transform converges absolutely *if and only if* the ROC contains the unit circle. From the above discussion, it is clear that the unit circle in the z-plane of the z-transform and the $j\omega$-axis in the s-plane of the Laplace transform play equivalent roles.

Also, just as in the Laplace transform, a given z-transform corresponds to different signals, depending on the ROC attached to it.

The inverse z-transform involves contour integration in the ROC and Cauchy's integral theorem [211]. If the contour of integration is the unit circle, the inversion formula reduces to the discrete-time Fourier transform inversion (2.4.33). On circles centered at the origin but of radius ρ different from 1, one can think of forward and inverse z-transforms as the Fourier analysis and synthesis of a sequence $f'[n] = \rho^n f[n]$. Thus, convergence properties are as for the Fourier transform of the exponentially weighted sequence. In the ROC, we can write formally a z-transform pair as

$$f[n] \longleftrightarrow F(z), \qquad z \in \text{ROC}.$$

When z-transforms are rational functions, the inversion is best done by partial fraction expansion followed by term-wise inversion. Then, the z-transform pairs,

$$a^n u[n] \longleftrightarrow \frac{1}{1 - az^{-1}} \qquad |z| > |a|, \tag{2.5.6}$$

and

$$-a^n u[-n-1] \longleftrightarrow \frac{1}{1 - az^{-1}} \qquad |z| < |a|, \tag{2.5.7}$$

are useful, where $u[n]$ is the unit-step function ($u[n] = 1, n \geq 0$, and 0 otherwise). The above transforms follow from the definition (2.5.5) and the sum of geometric series, and they are a good example of identical z-transforms with different ROC's corresponding to different signals.

As a simple example, consider the sequence

$$f[n] = a^{|n|}$$

which, following (2.5.6–2.5.7), has a z-transform

$$F(z) = \frac{1}{1-az^{-1}} - \frac{1}{1-1/az^{-1}}, \qquad \text{ROC } |a| < |z| < \left|\frac{1}{a}\right|,$$

that is, a nonempty ROC only if $|a| < 1$. For more z-transform properties, see [211].

Convolutions, Difference Equations and Discrete-Time Filters Just as in continuous time, complex exponentials are eigenfunctions of the convolution operator. That is, if $f[n] = h[n] * g[n]$ and $h[n] = z^n$, $z \in \mathcal{C}$, then

$$f[n] = \sum_k h[n-k]\, g[k] = \sum_k z^{(n-k)} g[k] = z^n \sum_k z^{-k} g[k] = z^n G(z).$$

The z-transform $G(z)$ is thus the eigenvalue of the convolution operator for that particular value of z. The convolution theorem follows as

$$f[n] = h[n] * g[n] \longleftrightarrow F(z) = H(z)\, G(z),$$

with an ROC containing the intersection of the ROC's of $H(z)$ and $G(z)$. Convolution with a time-reversed filter can be expressed as an inner product,

$$f[n] = \sum_k x[k]\, h[n-k] = \sum_k x[k]\, \tilde{h}[k-n] = \langle x[k], \tilde{h}[k-n] \rangle,$$

where " ~ " denotes time reversal, $\tilde{h}[n] = h[-n]$.

It is easy to verify that the "delay by one" operator, that is, a discrete-time filter with impulse response $\delta[n-1]$ has a z-transform z^{-1}. That is why z^{-1} is often called a delay, or z^{-1} is used in block diagrams to denote a delay. Then, given $x[n]$ with the z-transform $X(z)$, $x[n-k]$ has a z-transform

$$x[n-k] \longleftrightarrow z^{-k} X(z).$$

Thus, a linear constant-coefficient difference equation can be analyzed with the z-transform, leading to the notion of a transfer function. We assume initial rest conditions in the following, that is, all delay operators are set to zero initially. Then, the homogeneous solution to the difference equation is zero. Assume a linear, time-invariant difference equation given by

$$\sum_{k=0}^{N} a_k y[n-k] = \sum_{k=0}^{M} b_k\, x[n-k], \tag{2.5.8}$$

and taking its z-transform using the delay property, we get the transfer function as the ratio of the output and input z-transforms,

$$H(z) = \frac{Y(z)}{X(z)} = \frac{\sum_{k=0}^{M} b_k z^{-1}}{\sum_{k=0}^{N} a_k z^{-1}}.$$

The output is related to the input by a convolution with a discrete-time filter having as impulse response $h[n]$, the inverse z-transform of $H(z)$. Again, the ROC depends on whether we wish a causal[12] or an anticausal solution, and the system is stable *if and only if* the ROC includes the unit circle. This leads to the conclusion that a causal system with rational transfer function is stable *if and only if* all poles are inside the unit circle (their modulus is smaller than one).

Note, however, that a system with poles inside and outside the unit circle can still correspond to a stable system (but not a causal one). Simply gather poles inside the unit circle into a causal impulse response, while poles outside correspond to an anticausal impulse response, and thus, the stable impulse response is two-sided.

From a transfer function given by a z-transform it is always possible to get a difference equation and thus a possible hardware implementation. However, many different realizations have the same transfer function and depending on the application, certain realizations will be vastly superior to others (for example, in finite-precision implementation). Let us just mention that the most obvious implementation which realizes the difference equation (2.5.8), called the direct-form implementation is poor as far as coefficient quantization is concerned. A better solution is obtained by factoring $H(z)$ into single and/or complex conjugate roots and implementing a cascade of such factors. For a detailed discussion of numerical behavior of filter structures see [211].

Autocorrelation and Spectral Factorization An important concept which we will use later in the book, is that of *deterministic autocorrelation* (autocorrelation in the statistical sense will be discussed in Chapter 7, Appendix 7.A). We will say that

$$p[m] \;=\; \langle h[n], h[n+m] \rangle,$$

is the deterministic autocorrelation (or, simply autocorrelation from now on) of the sequence $h[n]$. In Fourier domain, we have that

$$\begin{aligned} P(e^{j\omega}) \;&=\; \sum_{n=-\infty}^{\infty} p[n]\, e^{-j\omega n} \;=\; \sum_{n=-\infty}^{\infty} \sum_{k=-\infty}^{\infty} h^*[k]\, h[k+n]\, e^{-j\omega n}, \\ &=\; H^*(e^{j\omega})\, H(e^{j\omega}) \;=\; |H(e^{j\omega})|^2, \end{aligned}$$

that is, $P(e^{j\omega})$ is a nonnegative function on the unit circle. In other words, the following is a Fourier-transform pair:

$$p[m] \;=\; \langle h[n], h[n+m] \rangle \;\longleftrightarrow\; P(e^{j\omega}) \;=\; |H(e^{j\omega})|^2.$$

[12]A discrete-time sequence $x[n]$ is said to be causal if $x[n] = 0$ for $n < 0$.

Similarly, in z-domain, the following is a transform pair:

$$p[m] = \langle h[n], h[n+m] \rangle \longleftrightarrow P(z) = H(z)\, H_*(1/z)$$

(recall that the subscript * implies conjugation of the coefficients but not of z). Note that from the above, it is obvious that if z_k is a zero of $P(z)$, so is $1/z_k^*$ (that also means that zeros on the unit circle are of even multiplicity). When $h[n]$ is real, and z_k is a zero of $H(z)$, then $z_k^*, 1/z_k, 1/z_k^*$ are zeros as well (they are not necessarily different).

Suppose now that we are given an autocorrelation function $P(z)$ and we want to find $H(z)$. Here, $H(z)$ is called a *spectral factor* of $P(z)$ and the technique of extracting it, *spectral factorization*. These spectral factors are not unique, and are obtained by assigning one zero out of each zero pair to $H(z)$ (we assume here that $p[m]$ is FIR, otherwise allpass functions (2.5.10) can be involved). The choice of which zeros to assign to $H(z)$ leads to different spectral factors. To obtain a spectral factor, first factor $P(z)$ into its zeros as follows:

$$P(z) = \alpha \prod_{i=1}^{N_u} ((1 - z_{1_i}\, z^{-1})\, (1 - z_{1_i}\, z)) \prod_{i=1}^{N} (1 - z_{2_i}\, z^{-1}) \prod_{i=1}^{N} (1 - z_{2_i}^*\, z),$$

where the first product contains the zeros on the unit circle, and thus $|z_{1_i}| = 1$, and the last two contain pairs of zeros inside/outside the unit circle, respectively. In that case, $|z_{2_i}| < 1$. To obtain various $H(z)$, one has to take one zero out of each zero pair on the unit circle, as well as one of two zeros inside/outside the unit circle. Note that all these solutions have the same magnitude response but different phase behavior. An important case is the minimum phase solution which is the one, among all causal spectral factors, that has the smallest phase term. To get a minimum phase solution, we will consistently choose the zeros inside the unit circle. Thus, $H(z)$ would be of the form

$$H(z) = \sqrt{\alpha} \prod_{i=1}^{N_u} (1 - z_{1_i}\, z^{-1}) \prod_{i=1}^{N} (1 - z_{2_i}\, z^{-1}).$$

Examples of Discrete-Time Filters Discrete-time filters come in two major classes. The first class consists of infinite impulse response (IIR) filters, which correspond to difference equations where the present output depends on past outputs (that is, $N \geq 1$ in (2.5.8)). IIR filters often depend on a finite number of past outputs ($N < \infty$) in which case the transfer function is a ratio of polynomials in z^{-1}. Often, by abuse of language, we will call an IIR filter a filter with a rational transfer function. The second class corresponds to nonrecursive, or finite impulse response (FIR) filters, where the output only depends on the inputs (or $N = 0$ in (2.5.8)). The z-transform is thus a polynomial in z^{-1}. An important class of FIR filters are those which have symmetric or antisymmetric impulse responses because this leads to a linear phase behavior of their Fourier transform. Consider causal FIR filters of length L. When the impulse response is symmetric, one can write

$$H(e^{j\omega}) = e^{-j\omega(L-1)/2} A(\omega),$$

where L is the length of the filter, and $A(\omega)$ is a real function of ω. Thus, the phase is a linear function of ω. Similarly, when the impulse response is antisymmetric, one can write

$$H(e^{j\omega}) = je^{-j\omega(L-1)/2} B(\omega),$$

where $B(\omega)$ is a real function of ω. Here, the phase is an affine function of ω (but usually called linear phase).

One way to design discrete-time filters is by transformation of an analog filter. For example, one can sample the impulse response of the analog filter if its magnitude frequency response is close enough to being bandlimited. Another approach consists of mapping the s-plane of the Laplace transform into the z-plane. From our previous discussion of the relationship between the two planes, it is clear that the $j\omega$-axis should map into the unit circle and the left half-plane should become the inside of the unit circle in order to preserve stability. Such a mapping is given by the bilinear transformation [211]

$$B(z) = \beta\frac{1-z^{-1}}{1+z^{-1}}.$$

Then, the discrete-time filter H_d is obtained from a continuous-time filter H_c by setting

$$H_d(z) = H_c(B(z)).$$

Considering what happens on the $j\omega$-axis and the unit circle, it can be verified that the bilinear transform warps the frequency axis as $\omega = 2\arctan(\omega_c/\beta)$, where ω and ω_c are the discrete and continuous frequency variables, respectively.

As an example, the discrete-time Butterworth filter has a magnitude frequency response equal to

$$|H(e^{j\omega})|^2 = \frac{1}{1+\left(\tan(\omega/2)/\tan(\omega_0/2)\right)^{2N}}. \tag{2.5.9}$$

This squared magnitude is flat at the origin, in the sense that its first $2N-1$ derivatives are zero at $\omega = 0$. Note that since we have a closed-form factorization of the continuous-time Butterworth filter (see (2.5.4)), it is best to apply the bilinear transform to the factored form rather than factoring (2.5.9) in order to obtain $H(e^{j\omega})$ in its cascade form.

Instead of the above indirect construction, one can design discrete-time filters directly. This leads to better designs at a given complexity of the filter or, conversely, to lower-complexity filters for a given filtering performance.

In the particular case of FIR linear phase filters (that is, a finite-length symmetric or antisymmetric impulse response), a powerful design method called the Parks-McClellan algorithm [211] leads to optimal filters in the minimax sense (the maximum deviation from the desired Fourier transform magnitude is minimized). The resulting approximation of the desired frequency response becomes equiripple both in the passband and stopband (the approximation error is evenly spread out). It is thus very different from a monotonically decreasing approximation as achieved by a Butterworth filter.

Finally, we discuss the allpass filter, which is an example of what could be called a unitary filter. An allpass filter has the property that

$$|H_{ap}(e^{j\omega})| \; = \; 1, \tag{2.5.10}$$

for all ω. Calling $y[n]$ the output of the allpass when $x[n]$ is input, we have

$$\|y\|^2 \; = \; \frac{1}{2\pi}\|Y(e^{j\omega})\|^2 \; = \; \frac{1}{2\pi}\|H_{ap}(e^{j\omega})\, X(e^{j\omega})\|^2 \; = \; \frac{1}{2\pi}\|X(e^{j\omega})\|^2 \; = \; \|x\|^2,$$

which means it conserves the energy of the signal it filters. An elementary single-pole/zero allpass filter is of the following form (see also Appendix 3.A in Chapter 3):

$$H_{ap}(z) \; = \; \frac{z^{-1} - a^*}{1 - az^{-1}}. \tag{2.5.11}$$

Writing the pole location as $a = \rho e^{j\theta}$, the zero is at $1/a^* = (1/\rho)e^{j\theta}$. A general allpass filter is made up of elementary sections as in (2.5.11)

$$H_{ap}(z) \; = \; \prod_{i=1}^{N} \frac{z^{-1} - a_i^*}{1 - a_i\, z^{-1}} \; = \; \frac{\tilde{P}(z)}{P(z)}, \tag{2.5.12}$$

where $\tilde{P}(z) = z^{-N} P_*(z^{-1})$ is the time-reversed and coefficient-conjugated version of $P(z)$ (recall that the subscript $_*$ stands for conjugation of the coefficients of the polynomial, but not of z). On the unit circle,

$$H_{ap}(e^{j\omega}) \; = \; e^{-j\omega N}\, \frac{P^*(e^{j\omega})}{P(e^{j\omega})},$$

and property (2.5.10) follows easily. That all rational functions satisfying (2.5.10) can be factored as in (2.5.12) is shown in [308].

2.5.3 Multirate Discrete-Time Signal Processing

As implied by its name, multirate signal processing deals with discrete-time sequences taken at different rates. While one can always go back to an underlying continuous-time signal and resample it at a different rate, most often, the rate changes are being done in the discrete-time domain. We review some of the key results. For further details, see [67] and [308].

Sampling Rate Changes *Downsampling* or *subsampling*[13] a sequence $x[n]$ by an integer factor N results in a sequence $y[n]$ given by

$$y[n] = x[nN],$$

that is, all samples with indexes modulo N different from zero are discarded. In the Fourier domain, we get

$$Y(e^{j\omega}) = \frac{1}{N}\sum_{k=0}^{N-1} X\left(e^{j(\omega-2\pi k)/N}\right), \tag{2.5.13}$$

that is, the spectrum is stretched by N, and $(N-1)$ aliased versions at multiples of 2π are added. They are called aliased because they are copies of the original spectrum (up to a stretch) but shifted in frequency. That is, low-frequency components will be replicated at the aliasing frequencies $\omega_i = 2\pi i/N$, as will high frequencies (with an appropriate shift). Thus, some high-frequency sinusoid might have a low-frequency alias. Note that the aliased components are non-harmonically related to the original frequency component; a fact that can be very disturbing in applications such as audio. Sometimes, it is useful to extend the above relation to the z-transform domain;

$$Y(z) = \frac{1}{N}\sum_{k=0}^{N-1} X\left(W_N^k\, z^{1/N}\right), \tag{2.5.14}$$

where $W_N = e^{-j2\pi/N}$ as usual. To prove (2.5.14), consider first a signal $x'[n]$ which equals $x[n]$ at multiples of N, and 0 elsewhere. If $x[n]$ has z-transform $X(z)$, then $X'(z)$ equals

$$X'(z) = \frac{1}{N}\sum_{k=0}^{N-1} X(W_N^k\, z) \tag{2.5.15}$$

as can be shown by using the orthogonality of the roots of unity (2.1.3). To obtain $y[n]$ from $x'[n]$, one has to drop the extra zeros between the nonzero terms or contract the signal by a factor of N. This is obtained by substituting $z^{1/N}$ for z in (2.5.15), leading to (2.5.14). Note that (2.5.15) contains the signal X as well as its $N-1$ modulated versions (on the unit circle, $X(W_N^k z) = X(e^{j(\omega-k2\pi/N)})$). This is the reason why in Chapter 3, we will call the analysis dealing with $X(W_N^k\, z)$, modulation-domain analysis.

[13] Sometimes, the term decimation is used even though it historically stands for "keep 9 out of 10" in reference to a Roman practice of killing every tenth soldier of a defeated army.

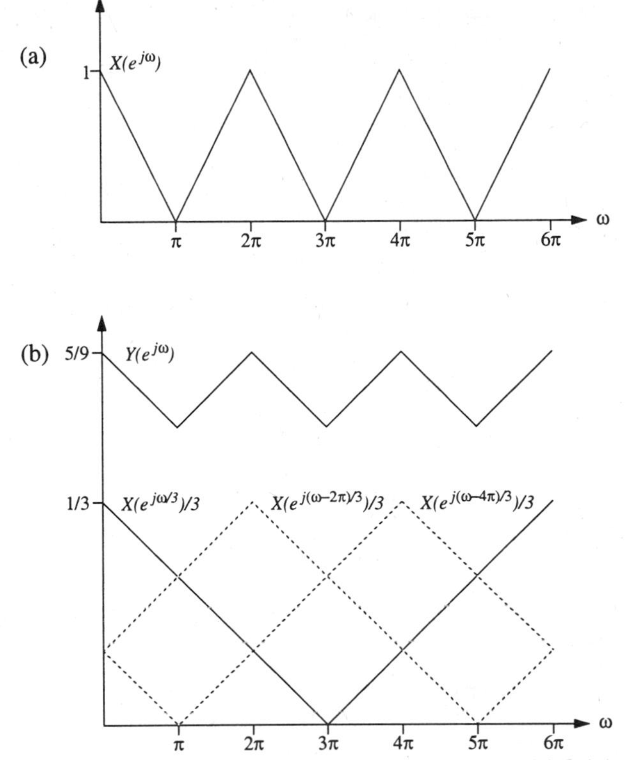

Figure 2.5 Downsampling by 3 in the frequency domain. (a) Original spectrum (we assume a real spectrum for simplicity). (b) The three stretched replicas and the sum $Y(e^{j\omega})$.

An alternative proof of (2.5.13) (which is (2.5.14) on the unit circle) consists of going back to the underlying continuous-time signal and resampling with an N-times larger sampling period. This is considered in Problem 2.10.

By way of an example, we show the case $N = 3$ in Figure 2.5. It is obvious that in order to avoid aliasing, downsampling by N should be preceded by an ideal lowpass filter with cutoff frequency π/N (see Figure 2.6(a)). Its impulse response $h[n]$ is given by

$$h[n] = \frac{1}{2\pi}\int_{-\pi/N}^{\pi/N} e^{j\omega n}\, d\omega = \frac{\sin \pi n/N}{\pi n}. \tag{2.5.16}$$

The converse of downsampling is *upsampling* by an integer M. That is, to obtain a new sequence, one simply inserts $M-1$ zeros between consecutive

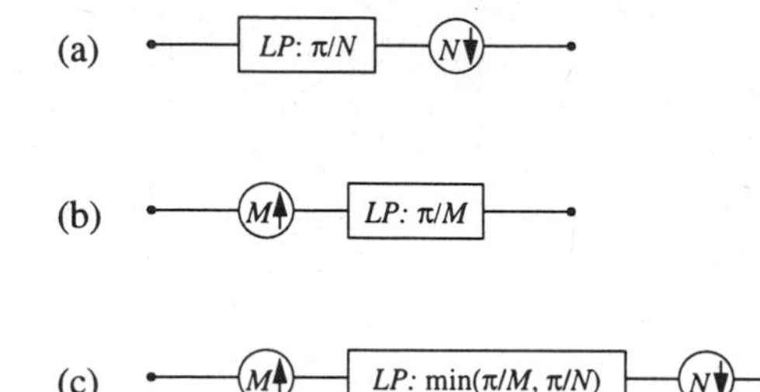

Figure 2.6 Sampling rate changes. (a) Downsampling by N preceded by ideal lowpass filtering with cutoff frequency π/N. (b) Upsampling by M followed by interpolation with an ideal lowpass filter with cutoff frequency π/M. (c) Sampling rate change by a rational factor M/N, with an interpolation filter in between. The cutoff frequency is the lesser of π/M and π/N.

samples of the input sequence, or

$$y[n] = \begin{cases} x[n/M] & n = kM, k \in \mathcal{Z} \\ 0 & \text{otherwise.} \end{cases}$$

In Fourier domain, this amounts to

$$Y(e^{j\omega}) = X(e^{jM\omega}), \tag{2.5.17}$$

and similarly, in z-transform domain

$$Y(z) = X(z^M). \tag{2.5.18}$$

Due to upsampling, the spectrum contracts by M. Besides the "base spectrum" at multiples of 2π, there are spectral images in between which are due to the interleaving of zeros in the upsampling. To get rid of these spectral images, a perfect interpolator or a lowpass filter with cutoff frequency π/M has to be used, as shown in Figure 2.6(b). Its impulse response is as given in (2.5.16), but with a different scale factor,

$$h[n] = \frac{\sin \pi n/M}{\pi n/M}.$$

It is easy to see that $h[nM] = \delta[n]$. Therefore, calling $u[n]$ the result of the interpolation, or $u[n] = y[n] * h[n]$, it follows that $u[nM] = x[n]$. Thus, $u[n]$ is a perfect interpolation of $x[n]$ in the sense that the missing samples have been filled in without disturbing the original ones.

A rational sampling rate change by M/N is obtained by cascading upsampling and downsampling with an interpolation filter in the middle, as shown in Figure 2.6(c). The interpolation filter is the cascade of the ideal lowpass for the upsampling and for the downsampling, that is, the narrower of the two in the ideal filter case.

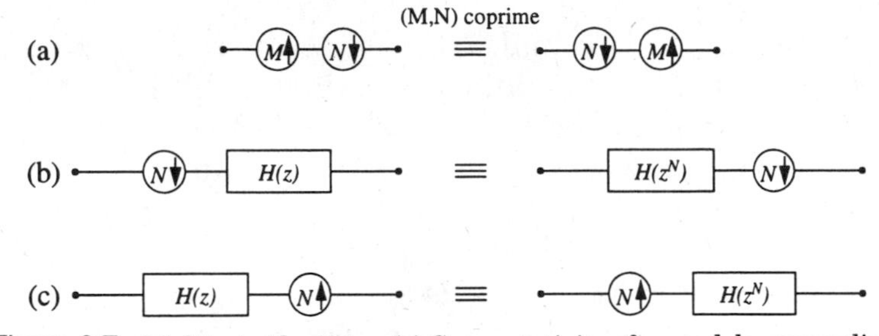

Figure 2.7 Multirate identities. (a) Commutativity of up and downsampling. (b) Interchange of downsampling and filtering. (c) Interchange of filtering and upsampling.

Finally, we demonstrate a fact that will be extensively used in Chapter 3. It can be seen as an application of downsampling followed by upsampling to the deterministic autocorrelation of $g[n]$. This is the discrete-time equivalent of (2.4.31). We want to show that the following holds:

$$\langle g[n], g[n+Nl]\rangle = \delta[l] \longleftrightarrow \sum_{k=0}^{N-1} G(W_N^k z)\, G(W_N^{-k} z^{-1}) = N. \tag{2.5.19}$$

The left side of the above equation is simply the autocorrelation of $g[n]$ evaluated at every Nth index $m = Nl$. If we denote the autocorrelation of $g[n]$ as $p[n]$, then the left side of (2.5.19) is $p'[n] = p[Nn]$. The z-transform of $p'[n]$ is (apply (2.5.14))

$$P'(z) = \frac{1}{N}\sum_{k=0}^{N-1} P(W_N^k\, z^{1/N}).$$

Replace now $z^{1/N}$ by z and since the z-transform of $p[n]$ is $P(z) = G(z)G(z^{-1})$, we get that the z-transform of the left side of (2.5.19) is the right side of (2.5.19).

Multirate Identities

Commutativity of Sampling Rate Changes Upsampling by M and downsampling by N commute *if and only if* M and N are coprime.

The relation is shown pictorially in Figure 2.7(a). Using (2.5.14) and (2.5.18) for down and upsampling in z-domain, we find that upsampling by M followed by downsampling by N leads to

$$Y_{u/d}(z) = \sum_{k=0}^{N-1} X(W_N^k\, z^{M/N}),$$

while the reverse order leads to

$$Y_{d/u}(z) = \sum_{k=0}^{N-1} X(W_N^{kM}\, z^{M/N}).$$

For the two expressions to be equal, $kM \bmod N$ has to be a permutation, that is, $kM \bmod N = l$ has to have a unique solution for all $l \in \{0, \ldots, N-1\}$. If M and N have a common factor $L > 1$, then $M = M'\,L$ and $N = N'\,L$. Note that $(kM \bmod N) \bmod L$ is zero, or $kM \bmod N$ is a multiple of L and thus not a permutation. If M and N are coprime, then Bezout's identity [209] guarantees that there exist two integers m and n such that $mM + nN = 1$. It follows that $mM \bmod N = 1$ thus, $k = ml \bmod N$ is the desired solution to the equation $k\,M \bmod N = l$. This property has an interesting generalization in multiple dimensions (see for example [152]).

Interchange of Filtering and Downsampling Downsampling by N followed by filtering with a filter having z-transform $H(z)$ is equivalent to filtering with the upsampled filter $H(z^N)$ before the downsampling.

Using (2.5.14), it follows that downsampling the filtered signal with the z-transform $X(z)H(z^N)$ results in

$$\sum_{k=0}^{N-1} X(W_N^K\, z^{1/N})\, H\left((W_N^k\, z^{1/N})^N\right) = H(z) \sum_{k=0}^{N-1} X(W_N^k\, z^{1/N}),$$

which is equal to filtering a downsampled version of $X(z)$.

Interchange of Filtering and Upsampling Filtering with a filter having the z-transform $H(z)$, followed by upsampling by N, is equivalent to upsampling followed by filtering with $H(z^N)$.

Using (2.5.18), it is immediate that both systems lead to an output with z-transform $X(z^N)H(z^N)$ when the input is $X(z)$.

In short, the last two properties simply say that filtering in the downsampled domain can always be realized by filtering in the upsampled domain, but then with the upsampled filter (down and upsampled stand for low versus high sampling rate domain). The last two relations are shown in Figures 2.7(b) and (c).

Polyphase Transform Recall that in a time-invariant system, if input $x[n]$ produces output $y[n]$, then input $x[n+m]$ will produce output $y[n+m]$. In a time-varying system this is not true. However, there exist *periodically time-varying* systems for which if input $x[n]$ produces output $y[n]$, then $x[n+Nm]$ produces output $y[n+mN]$. These systems are periodically time-varying with period N. For example, a downsampler by N followed by an upsampler by N is such a system. A downsampler alone is also periodically time-varying, but with a time-scale change. Then, if $x[n]$ produces $y[n]$, $x[n+mN]$ produces $y[n+m]$ (note that $x[n]$ and $y[n]$ do not live on the same time-scale). Such periodically time-varying systems can be analyzed with a simple but useful transform

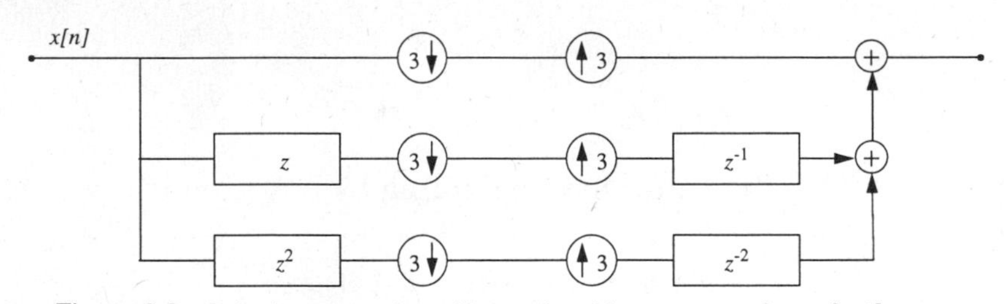

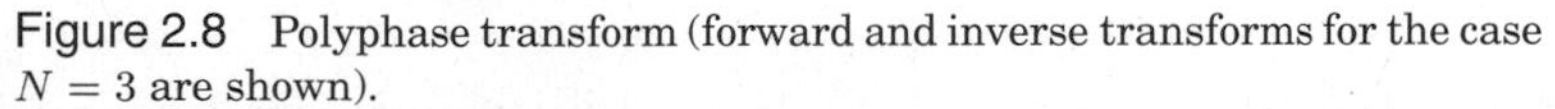

Figure 2.8 Polyphase transform (forward and inverse transforms for the case $N = 3$ are shown).

where a sequence is mapped into N sequences with each being a shifted and downsampled version of the original sequence. Obviously, the original sequence can be recovered by simply interleaving the subsequences. Such a transform is called a *polyphase transform* of size N since each subsequence has a different phase and there are N of them. The simplest example is the case $N = 2$, where a sequence is subdivided into samples of even and odd indexes, respectively. In general, we define the size-N polyphase transform of a sequence $x[n]$ as a vector of sequences $(\, x_0[n] \quad x_1[n] \quad \cdots \quad x_{N-1}[n]\,)^T$, where

$$x_i[n] \;=\; x[nN + i].$$

These are called *signal polyphase components*. In z-transform domain, we can write $X(z)$ as the sum of shifted and upsampled polyphase components. That is,

$$X(z) \;=\; \sum_{i=0}^{N-1} z^{-i} X_i(z^N), \tag{2.5.20}$$

where

$$X_i(z) \;=\; \sum_{n=-\infty}^{\infty} x[nN + i]\; z^{-n}. \tag{2.5.21}$$

Figure 2.8 shows the signal polyphase transform and its inverse (for the case $N = 3$). Because the forward shift requires advance operators which are noncausal, a causal version would produce a total delay of $N - 1$ samples between forward and inverse polyphase transform. Such a causal version is obtained by multiplying the noncausal forward polyphase transform by z^{-N+1}.

Later we will need to express the output of filtering with H followed by downsampling in terms of the polyphase components of the input signal. That is, we need the 0th polyphase component of $H(z)X(z)$. This is easiest if we define a polyphase decomposition of the filter to have the reverse phase of the one used for the signal, or

$$H(z) \;=\; \sum_{i=0}^{N-1} z^{i} H_i(z^N), \tag{2.5.22}$$

with

$$H_i(z) = \sum_{n=-\infty}^{\infty} h[Nn - i]z^{-n}, \qquad i = 0, \ldots, N-1. \tag{2.5.23}$$

Then the product $H(z)X(z)$ after downsampling by N becomes

$$Y(z) = \sum_{i=0}^{N-1} H_i(z)\, X_i(z).$$

The same operation (filtering by $h[n]$ followed by downsampling by N) can be expressed in matrix notation as

$$\begin{pmatrix} \vdots \\ y[0] \\ y[1] \\ \vdots \end{pmatrix} = \begin{pmatrix} & \vdots & & \vdots & \vdots & & \\ \cdots & h[L-1] & \cdots & h[L-N] & h[L-N-1] & \cdots \\ \cdots & 0 & \cdots & 0 & h[L-1] & \cdots \\ & \vdots & & \vdots & \vdots & & \end{pmatrix} \begin{pmatrix} \vdots \\ x[0] \\ x[1] \\ \vdots \end{pmatrix},$$

where L is the filter length, and the matrix operator will be denoted by $\boldsymbol{H}$. Similarly, upsampling by N followed by filtering by $g[n]$ can be expressed as

$$\begin{pmatrix} \vdots \\ x[0] \\ x[1] \\ \vdots \end{pmatrix} = \begin{pmatrix} & \vdots & \vdots & \\ \cdots & g[0] & 0 & \cdots \\ \cdots & \vdots & \vdots & \cdots \\ \cdots & g[N-1] & 0 & \cdots \\ \cdots & g[N] & g[0] & \cdots \\ & \vdots & \vdots & \end{pmatrix} \begin{pmatrix} \vdots \\ y[0] \\ y[1] \\ \vdots \end{pmatrix}.$$

Here the matrix operator is denoted by $\boldsymbol{G}$. Note that if $h[n] = g[-n]$, then $\boldsymbol{H} = \boldsymbol{G}^T$, a fact that will be important when analyzing orthonormal filter banks in Chapter 3.

2.6 Time-Frequency Representations

While the Fourier transform and its variations are very useful mathematical tools, practical applications require basic modifications. These modifications aim at "localizing" the analysis, so that it is not necessary to have the signal over $(-\infty, \infty)$ to perform the transform (as required with the Fourier integral) and so that local effects (transients) can be captured with some accuracy. The classic example is the short-time Fourier [204], or Gabor transform[14] [102],

[14] Gabor's original paper proposed synthesis of signals using complex sinusoids windowed by a Gaussian, and is thus a synthesis rather than an analysis tool. However, it is closely related to

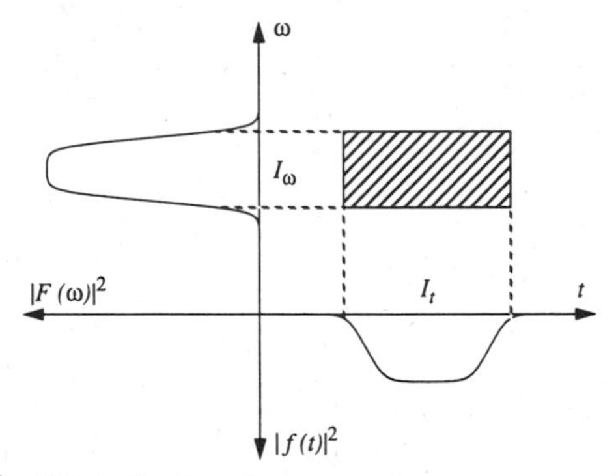

Figure 2.9 Tile in the time-frequency plane as an approximation of the time-frequency localization of $f(t)$. Intervals I_t and I_ω contain 90% of the energy of the time- and frequency-domain functions, respectively.

which uses windowed complex exponentials and their translates as expansion functions. We therefore discuss the localization properties of basis functions and derive the uncertainty principle which gives a lower bound on the joint time and frequency resolutions. We then review the short-time Fourier transform and its associated energy distribution called the spectrogram and introduce the wavelet transform. Block transforms are also discussed. Finally, an example of a bilinear expansion, namely the Wigner-Ville distribution, is also discussed.

2.6.1 Frequency, Scale and Resolution

When calculating a signal expansion, a primary concern is the localization of a given basis function in time and frequency. For example, in the Fourier transform, the functions used in the analysis are infinitely sharp in their frequency localization (they exist at one precise frequency) but have no time localization because of their infinite extent.

There are various ways to define the localization of a particular basis function, but they are all related to the "spread" of the function in time and frequency. For example, one can define intervals I_t and I_ω which contain 90% of the energy of the time- and frequency-domain functions, respectively, and are centered around the center of gravity of $|f(t)|^2$ and $|F(\omega)|^2$ (see Figure 2.9). This defines what we call a *tile* in the time-frequency domain, as shown in Figure 2.9. For simplicity, we assumed a complex basis function. A real basis function would be represented by two mirror tiles at positive and negative frequencies.

Consider now elementary operations on a basis function and their effects on the tile. Obviously, a shift in time by τ results in shifting of the tile by

the short-time Fourier transform, and we call Gabor transform a short-time Fourier transform using a Gaussian window.

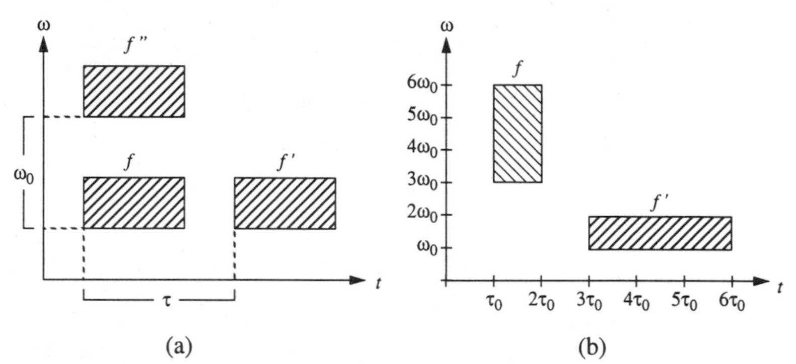

Figure 2.10 Elementary operations on a basis function f and effect on the time-frequency tile. (a) Shift in time by τ producing f' and modulation by ω_0 producing f''. (b) Scaling $f'(t) = f(at)$ ($a = 1/3$ is shown).

τ. Similarly, modulation by $e^{j\omega_0 t}$ shifts the tile by ω_0 in frequency (vertically). This is shown in Figure 2.10(a). Finally, scaling by a, or $f'(t) = f(at)$, results in $I'_t = (1/a)I_t$ and $I'_\omega = aI_\omega$, following the scaling property of the Fourier transform (2.4.5). That is, both the shape and localization of the tile have been affected, as shown in Figure 2.10(b). Note that all elementary operations conserve the surface of the time-frequency tile. In the scaling case, resolution in frequency was traded for resolution in time.

Since scaling is a fundamental operation used in the wavelet transform, we need to define it properly. While frequency has a natural ordering, the notion of scale is defined differently by different authors. The analysis functions for the wavelet transform will be defined as

$$\psi_{a,b}(t) = \frac{1}{\sqrt{a}}\psi\left(\frac{t-b}{a}\right), \qquad a \in \mathcal{R}^+$$

where the function $\psi(t)$ is usually a bandpass filter. Thus, large a's ($a \gg 1$) correspond to long basis functions, and will identify long-term trends in the signal to be analyzed. Small a's ($0 < a < 1$) lead to short basis functions, which will follow short-term behavior of the signal. This leads to the following: *Scale* is proportional to the duration of the basis functions used in the signal expansion.

Because of this, and assuming that a basis function is a bandpass filter as in wavelet analysis, high-frequency basis functions are obtained by going to small scales, and therefore, scale is loosely related to inverse frequency. This is only a qualitative statement, since scaling and modulation are fundamentally different operations as was seen in Figure 2.10. The discussed scale is similar to those in geographical maps, where large means a coarse, global view, and small corresponds to a fine, detailed view.

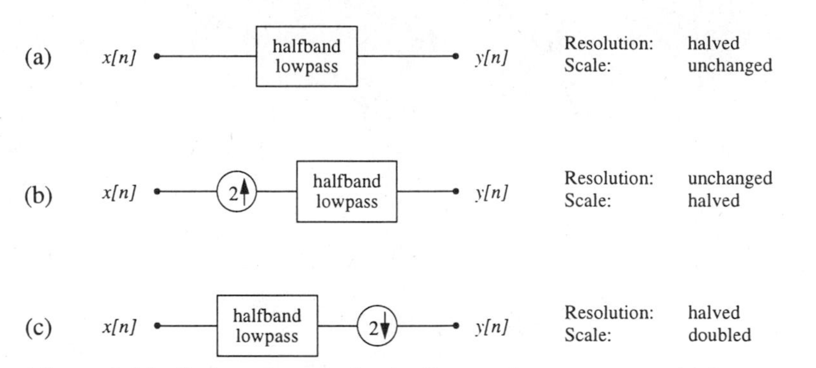

Figure 2.11 Scale and resolution in discrete-time sequences. (a) Lowpass filtering reduces the resolution. (b) Upsampling and interpolation change the scale but not the resolution. (c) Lowpass filtering and downsampling increase scale and reduces resolution.

Scale changes can be inverted if the function is continuous-time. In discrete time, the situation is more complicated. From the discussion of multirate signal processing in Section 2.5.3, we can see that upsampling (that is, a stretching of the sequence) can be undone by downsampling by the same factor, and this with no loss of information if done properly. Downsampling (or contraction of a sequence) involves loss of information in general, since either a bandlimitation precedes the downsampling, or aliasing occurs. This naturally leads to the notion of *resolution* of a signal. We will thus say that the *resolution* of a finite-length signal is the minimum number of samples required to represent it. It is thus related to the *information content* of the signal. For infinite-length signals having finite energy and sufficient decay, one can define the length as the essential support (for example, where 99% of the energy is).

In continuous time, scaling does not change the resolution, since a scale change affects both the sampling rate and the length of the signal, thus keeping the number of samples constant. In discrete time, upsampling followed by interpolation does not affect the resolution, since the interpolated samples are redundant. Downsampling by N decreases the resolution by N, and cannot be undone. Figure 2.11 shows the interplay of scale and resolution on simple discrete-time examples. Note that the notion of resolution is central to multiresolution analysis developed in Chapters 3 and 4. There, the key idea is to split a signal into several lower-resolution components, from which the original, full-resolution signal can be recovered.

2.6.2 Uncertainty Principle

As indicated in the discussion of scaling in the previous section, sharpness of the time analysis can be traded off for sharpness in frequency, and vice versa. But there is no way to get arbitrarily sharp analysis in both domains simultaneously, as shown below [37, 102, 215]. Note that the sharpness is also called resolution in time and frequency (but is different from the resolution discussed just above, which was related to information content). Consider a unit energy signal $f(t)$ with Fourier transform $F(\omega)$ centered around the origin in time as well as in frequency, that is, satisfying $\int t|f(t)|^2\,dt = 0$ and $\int \omega|F(\omega)|^2\,d\omega = 0$ (this can always be obtained by appropriate translation and modulation). Define the time width Δt of $f(t)$ by

$$\Delta_t^2 = \int_{-\infty}^{\infty} t^2|f(t)|^2 dt, \tag{2.6.1}$$

and its frequency width Δ_ω by

$$\Delta_\omega^2 = \int_{-\infty}^{\infty} \omega^2|F(\omega)|^2 d\omega.$$

THEOREM 2.7 *Uncertainty Principle*

If $f(t)$ vanishes faster than $1/\sqrt{t}$ as $t \to \pm\infty$, then

$$\Delta_t^2\,\Delta_\omega^2 \geq \frac{\pi}{2}, \tag{2.6.2}$$

where equality holds only for Gaussian signals

$$f(t) = \sqrt{\frac{\alpha}{\pi}}\, e^{-\alpha t^2}. \tag{2.6.3}$$

PROOF

Consider the integral of $t\,f(t)\,f'(t)$. Using Cauchy-Schwarz inequality (2.2.2),

$$\left|\int_{\mathcal{R}} tf(t)\,f'(t)\,dt\right|^2 \leq \int_{\mathcal{R}} |tf(t)|^2 dt \int_{\mathcal{R}} |f'(t)|^2\,dt. \tag{2.6.4}$$

The first integral on the right side is equal to Δ_t^2. Because $f'(t)$ has Fourier transform $j\omega F(\omega)$, and using Parseval's formula, we find that the second integral is equal to $(1/(2\pi))\Delta_\omega^2$. Thus, the integral on the left side of (2.6.4) is bounded from above by $(1/(2\pi))\Delta_t^2\Delta_\omega^2$. Using integration by parts, and noting that $f(t)f'(t) = (1/2)(\partial f^2(t))/(\partial t)$,

$$\int_{\mathcal{R}} tf(t)\,f'(t)\,dt = \frac{1}{2}\int_{\mathcal{R}} t\frac{\partial f^2(t)}{\partial t}dt = \frac{1}{2}\,t\,f^2(t)\Big|_{-\infty}^{\infty} - \frac{1}{2}\int_{\mathcal{R}} f^2(t)\,dt.$$

By assumption, the limit of $tf^2(t)$ is zero at infinity, and, because the function is of unit norm, the above equals $-1/2$. Replacing this into (2.6.4), we obtain

$$\frac{1}{4} \leq \frac{1}{2\pi}\Delta_t^2\,\Delta_\omega^2,$$

or (2.6.2). To find a function that meets the lower bound note that Cauchy-Schwarz inequality is an equality when the two functions involved are equal within a multiplicative factor, that is, from (2.6.4),

$$f'(t) = ktf(t).$$

Thus, $f(t)$ is of the form

$$f(t) = ce^{kt^2/2} \tag{2.6.5}$$

and (2.6.3) follows for $k = -2\alpha$ and $c = \sqrt{\alpha/\pi}$.

The uncertainty principle is fundamental since it sets a bound on the maximum joint sharpness or resolution in time and frequency of any linear transform. It is easy to check that scaling does not change the time-bandwidth product, it only exchanges one resolution for the other, similarly to what was shown in Figure 2.10.

Example 2.3 *Prolate Spheroidal Wave Functions*

A related problem is that of finding bandlimited functions which are maximally concentrated around the origin in time (recall that there exist no functions that are both bandlimited and of finite duration). That is, find a function $f(t)$ of unit norm and bandlimited to ω_0 ($F(\omega) = 0, |\omega| > \omega_0$) such that, for a given $T \in (0, \infty)$

$$\alpha = \int_{-T}^{T} |f(t)|^2 \, dt$$

is maximized. It can be shown [216, 268] that the solution $f(t)$ is the eigenfunction with the largest eigenvalue satisfying

$$\int_{-T}^{T} f(\tau) \frac{\sin \omega_0 (t-\tau)}{\pi(t-\tau)} \, d\tau = \lambda f(t). \tag{2.6.6}$$

An interpretation of the above formula is the following. If $T \to \infty$, then we have the usual convolution with an ideal lowpass filter, and thus, any bandlimited function is an eigenfunction with eigenvalue 1. For finite T, because of the truncation, the eigenvalues will be strictly smaller than one. Actually, it turns out that the eigenvalues belong to $(0, 1)$ and are all different, or

$$1 > \lambda_0 > \lambda_1 > \cdots > \lambda_n \to 0, \qquad n \to \infty.$$

Call $f_n(t)$ the eigenfunction of (2.6.6) with eigenvalue λ_n. Then (i) each $f_n(t)$ is unique (up to a scale factor), (ii) $f_n(t)$ and $f_m(t)$ are orthogonal for $n \neq m$, and (iii) with proper normalization the set $\{f_n(t)\}$ forms an orthonormal basis for functions bandlimited to $(-\omega_0, \omega_0)$ [216]. These functions are called prolate spheroidal wave functions. Note that while (2.6.6) seems to depend on both T and ω_0, the solution depends only on the product $T \cdot \omega_0$.

2.6.3 Short-Time Fourier Transform

To achieve a "local" Fourier transform, one can define a windowed Fourier transform. The signal is first multiplied by a window function $w(t-\tau)$ and then the usual Fourier transform is taken. This results in a two-indexed transform, $STFT_f(\omega, \tau)$, given by

$$STFT_f(\omega, \tau) = \int_{-\infty}^{\infty} w^*(t-\tau) \, f(t) e^{-j\omega t} \, dt.$$

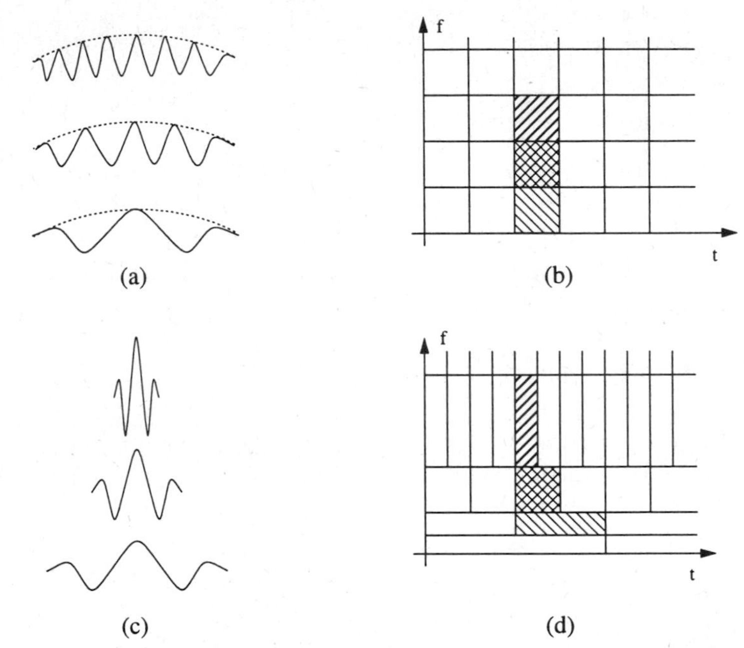

Figure 2.12 The short-time Fourier and wavelet transforms. (a) Modulates and shifts of a Gaussian window used in the expansion. (b) Tiling of the time-frequency plane. (c) Shifts and scales of the prototype bandpass wavelet. (d) Tiling of the time-frequency plane.

That is, one measures the similarity between the signal and shifts and modulates of an elementary window, or

$$STFT_f(\omega, \tau) = \langle g_{\omega,\tau}(t), f(t) \rangle,$$

where

$$g_{\omega,\tau}(t) = w(t-\tau)e^{j\omega t}.$$

Thus, each elementary function used in the expansion has the same time and frequency resolution, simply a different location in the time-frequency plane. It is thus natural to discretize the STFT on a rectangular grid $(m\omega_0, n\tau_0)$. If the window function is a lowpass filter with a cutoff frequency of ω_b, or a bandwidth of $2\omega_b$, then ω_0 is chosen smaller than $2\omega_b$ and τ_0 smaller than π/ω_b in order to get an adequate sampling. Typically, the STFT is actually oversampled. A more detailed discussion of the sampling of the STFT is given in Section 5.2, where the inversion formula is also given. A real-valued version of the STFT, using cosine modulation and an appropriate window, leads to orthonormal bases, which are discussed in Section 4.8.

Examples of STFT basis functions and the tiling of the time-frequency plane are given in Figures 2.12(a) and (b). To achieve good time-frequency resolution, a Gaussian window (see (2.6.5)) can be used, as originally proposed by Gabor [102]. Thus, the STFT is often called *Gabor transform* as well.

The *spectrogram* is the energy distribution associated with the STFT, that is,

$$S(\omega, \tau) = |STFT(\omega, \tau)|^2. \tag{2.6.7}$$

Because the STFT can be thought of as a bank of filters with impulse responses $g_{\omega,\tau}(-t) = w(-t-\tau)\, e^{-j\omega\tau}$, the spectrogram is the magnitude squared of the filter outputs.

2.6.4 Wavelet Transform

Instead of shifts and modulates of a prototype function, one can choose shifts and scales, and obtain a constant relative bandwidth analysis known as the wavelet transform. To achieve this, take a real bandpass filter with impulse response $\psi(t)$ and zero mean

$$\int_{-\infty}^{\infty} \psi(t)\, dt = \Psi(0) = 0.$$

Then, define the continuous wavelet transform as

$$CWT_f(a, b) = \frac{1}{\sqrt{a}} \int_{\mathcal{R}} \psi^* \left(\frac{t-b}{a}\right) f(t)\, dt, \tag{2.6.8}$$

where $a \in \mathcal{R}^+$ and $b \in \mathcal{R}$. That is, we measure the similarity between the signal $f(t)$ and shifts and scales of an elementary function, since

$$CWT_f(a, b) = \langle \psi_{a,b}(t), f(t) \rangle,$$

where

$$\psi_{a,b}(t) = \frac{1}{\sqrt{a}}\, \psi\left(\frac{t-b}{a}\right)$$

and the factor $1/\sqrt{a}$ is used to conserve the norm. Now, the functions used in the expansion have changing time-frequency tiles because of the scaling. For small a ($a < 1$), $\psi_{a,b}(t)$ will be short and of high frequency, while for large a ($a > 1$), $\psi_{a,b}(t)$ will be long and of low frequency. Thus, a natural discretization will use large time steps for large a, and conversely, choose fine time steps for small a. The discretization of (a, b) is then of the form $(a_0^n, a_0^n \cdot \tau_0)$, and leads to functions for the expansion as shown in Figure 2.12(c). The resulting tiling of the time-frequency plane is shown in Figure 2.12(d) (the case $a = 2$ is shown). Special choices for $\psi(t)$ and the discretization lead to orthonormal bases or wavelet series as studied in Chapter 4, while the overcomplete, continuous wavelet transform in (2.6.8) is discussed in Section 5.1.

2.6.5 Block Transforms

An easy way to obtain a time-frequency representation is to slice the signal into nonoverlapping adjacent blocks and expand each block independently. For example, this can be done using a window function on the signal which is the indicator function of the interval $[nT, (n+1)T)$, periodizing each windowed signal with period T and applying an expansion such as the Fourier series on each periodized signal (see Section 4.1.2). Of course, the arbitrary segmentation at points nT creates artificial boundary problems. Yet, such transforms are used due to their simplicity. For example, in discrete time, block transforms such as the Karhunen-Loève transform (see Section 7.1.1) and its approximations are quite popular.

2.6.6 Wigner-Ville Distribution

An alternative to linear expansions of signals are *bilinear expansions*, of which the Wigner-Ville distribution is the most well-known [53, 59, 135].

Bilinear or quadratic time-frequency representations are motivated by the idea of an "instantaneous power spectrum", of which the spectrogram (see (2.6.7)) is a possible example. In addition, the time-frequency distribution $TFD_f(\omega, \tau)$ of a signal $f(t)$ with Fourier transform $F(\omega)$ should satisfy the following marginal properties: Its integral along τ given ω should equal $|F(\omega)|^2$, and its integral along ω given τ should equal $|f(\tau)|^2$. Also, time-frequency shift invariance is desirable, that is, if $g(t) = f(t - \tau_0)e^{j\omega_0 t}$, then

$$TFD_g(\omega, \tau) = TDF_f(\omega - \omega_0, \tau - \tau_0).$$

The Wigner-Ville distribution satisfies the above requirements, as well as several other desirable ones [135]. It is defined, for a signal $f(t)$, as

$$WD_f(\omega, \tau) = \int_{-\infty}^{\infty} f\left(\tau + t/2\right)\, f^*\left(\tau - t/2\right) e^{-j\omega t}\, dt. \qquad (2.6.9)$$

A related distribution is the ambiguity function [216], which is dual to (2.6.9) through a two-dimensional Fourier transform.

The attractive feature of time-frequency distributions such as the Wigner-Ville distribution above is the possible improved time-frequency resolution. For signals with a single time-frequency component (such as a linear chirp signal), the Wigner-Ville distribution gives a very clear and concentrated energy ridge in the time-frequency plane.

However, the increased resolution for single component signals comes at a price for multicomponent signals, with the appearance of cross terms or interferences. If there are N components in the signal, there will be N signal terms and one cross term for each pair of components, that is, $\binom{N}{2}$ or $N(N-1)/2$

cross terms. While these interferences can be smoothed, this smoothing will come at the price of some resolution loss. In any case, the interference patterns make it difficult to visually interpret quadratic time-frequency distributions of complex signals.

APPENDIX 2.A BOUNDED LINEAR OPERATORS ON HILBERT SPACES

DEFINITION 2.8

An operator A which maps one Hilbert space H_1 into another Hilbert space H_2 (which may be the same) is called a *linear operator* if for all x, y in H_1 and α in $\mathcal{C}$

(a) $A(x+y) = Ax + Ay.$

(b) $A(\alpha x) = \alpha Ax.$

The norm of A, denoted by $\|A\|$, is given by

$$\|A\| = \sup_{\|x\|=1} \|Ax\|.$$

A linear operator $A : H_1 \to H_2$ is called *bounded* if

$$\sup_{\|x\|\leq 1} \|Ax\| < \infty.$$

An important property of bounded linear operators is that they are continuous, that is, if $x_n \to x$ then $Ax_n \to Ax$. An example of a bounded operator is the multiplication operator in $l_2(\mathcal{Z})$, defined as

$$Ax[n] = m[n]\, x[n],$$

where $m[n] \in l_\infty(\mathcal{Z})$. Because

$$\|Ax\|^2 = \sum_n (m[n])^2\, (x[n])^2 \leq \max(m[n])^2\, \|x\|^2,$$

the operator is bounded. A bounded linear operator $A : H_1 \to H_2$ is called *invertible* if there exists a bounded linear operator $A^{-1} : H_2 \to H_1$ such that

$$A^{-1}Ax = x, \text{ for every } x \text{ in } H_1,$$
$$AA^{-1}y = y, \text{ for every } y \text{ in } H_2.$$

The operator A^{-1} is called the *inverse* of A. An important result is the following: Suppose A is a bounded linear operator mapping H onto itself, and $\|A\| < 1$. Then $I - A$ is invertible, and for every y in H,

$$(I-A)^{-1}y = \sum_{k=0}^{\infty} A^k y. \tag{2.A.1}$$

Note that although the above expansion has the same form for a scalar as well as an operator, one should not forget the distinction between the two. Another important notion is that of an *adjoint* operator.[15] It can be shown that for every x in H_1 and y in H_2, there exists a unique y^* from H_1, such that

$$\langle Ax, y\rangle_{H_2} \;=\; \langle x, y^*\rangle_{H_1} \;=\; \langle x, A^*y\rangle_{H_1}. \tag{2.A.2}$$

The operator $A^* : H_2 \rightarrow H_1$ defined by $A^*y = y^*$, is the adjoint of A. Note that A^* is also linear and bounded, and that $\|A\| = \|A^*\|$. If $H_2 = H_1$ and $A = A^*$, then A is called a *self-adjoint* or *hermitian* operator.

Finally, an important type of operators are *projection operators*. Given a closed subspace S of a Hilbert space E, an operator P is called an *orthogonal projection* onto S if

$$P(v+w) \;=\; v \quad \text{for all } v \in S \text{ and } w \in S^\perp.$$

It can be shown that an operator is an orthogonal projection *if and only if* $P^2 = P$ and P is self-adjoint.

Let us now show how we can associate a possibly infinite matrix[16] with a given bounded linear operator on a Hilbert space. Given is a bounded linear operator $\boldsymbol{A}$ on a Hilbert space H with the orthonormal basis $\{\boldsymbol{x}_i\}$. Then any $\boldsymbol{x}$ from H can be written as $\boldsymbol{x} = \sum_i \langle \boldsymbol{x}_i, \boldsymbol{x}\rangle \boldsymbol{x}_i$, and

$$\boldsymbol{A}\boldsymbol{x} \;=\; \sum_i \langle \boldsymbol{x}_i, \boldsymbol{x}\rangle \boldsymbol{A}\boldsymbol{x}_i, \quad \boldsymbol{A}\boldsymbol{x}_i \;=\; \sum_k \langle \boldsymbol{x}_k, \boldsymbol{A}\boldsymbol{x}_i\rangle \boldsymbol{x}_k.$$

Similarly, writing $\boldsymbol{y} = \sum_i \langle \boldsymbol{x}_i, \boldsymbol{y}\rangle \boldsymbol{x}_i$, we can write $\boldsymbol{A}\boldsymbol{x} = \boldsymbol{y}$ as

$$\begin{pmatrix} \langle \boldsymbol{x}_1, \boldsymbol{A}\boldsymbol{x}_1\rangle & \langle \boldsymbol{x}_1, \boldsymbol{A}\boldsymbol{x}_2\rangle & \cdots \\ \langle \boldsymbol{x}_2, \boldsymbol{A}\boldsymbol{x}_1\rangle & \langle \boldsymbol{x}_2, \boldsymbol{A}\boldsymbol{x}_2\rangle & \cdots \\ \vdots & \vdots & \end{pmatrix} \begin{pmatrix} \langle \boldsymbol{x}_1, \boldsymbol{x}\rangle \\ \langle \boldsymbol{x}_2, \boldsymbol{x}\rangle \\ \vdots \end{pmatrix} = \begin{pmatrix} \langle \boldsymbol{x}_1, \boldsymbol{y}\rangle \\ \langle \boldsymbol{x}_2, \boldsymbol{y}\rangle \\ \vdots \end{pmatrix},$$

or, in other words, the matrix $\{a_{ij}\}$ corresponding to the operator $\boldsymbol{A}$ expressed with respect to the basis $\{\boldsymbol{x}_i\}$ is defined by $a_{ij} = \langle \boldsymbol{x}_i, \boldsymbol{A}\boldsymbol{x}_j\rangle$.

APPENDIX 2.B PARAMETRIZATION OF UNITARY MATRICES

Our aim in this appendix is to show two ways of factoring real, $n \times n$, unitary matrices, namely using Givens rotations and Householder building blocks. We concentrate here on real, square matrices, since these are the ones we will be using in Chapter 3. The treatment here is fairly brisk; for a more detailed, yet succinct account of these two factorizations, see [308].

[15] In the case of matrices, the adjoint is the hermitian transpose.

[16] To be consistent with our notation throughout the book, in this context, matrices will be denoted by capital bold letters, while vectors will be denoted by lower-case bold letters.

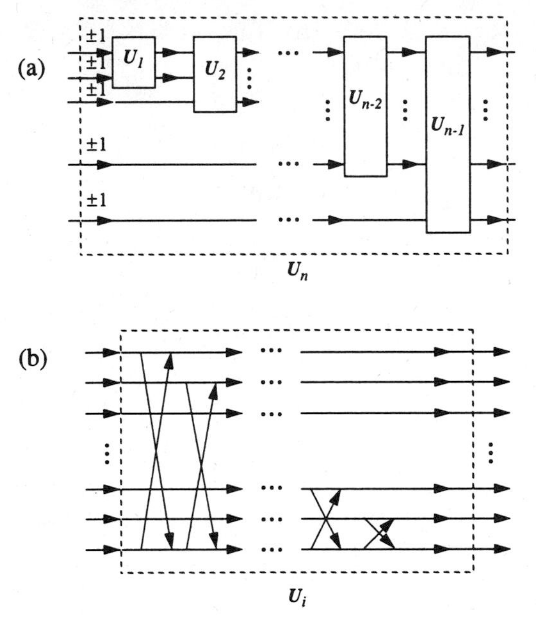

Figure 2.13 Unitary matrices. (a) Factorization of a real, unitary, $n \times n$ matrix. (b) The structure of the block $\boldsymbol{U}_i$.

2.B.1 Givens Rotations

Recall that a real, $n \times n$, unitary matrix $\boldsymbol{U}$ satisfies (2.3.6). We want to show that such a matrix can be factored as in Figure 2.13, where each cross in part (b) represents a Givens (planar) rotation

$$\boldsymbol{G}_\alpha = \begin{pmatrix} \cos\alpha & -\sin\alpha \\ \sin\alpha & \cos\alpha \end{pmatrix}. \tag{2.B.1}$$

The way to demonstrate this is to show that any real, unitary $n \times n$ matrix $\boldsymbol{U}_n$ can be expressed as

$$\boldsymbol{U}_n = \boldsymbol{R}_{n-2} \cdots \boldsymbol{R}_0 \begin{pmatrix} \boldsymbol{U}_{n-1} & 0 \\ 0 & \pm 1 \end{pmatrix}, \tag{2.B.2}$$

where $\boldsymbol{U}_{n-1}$ is an $(n-1)\times(n-1)$, real, unitary matrix, and $\boldsymbol{R}_i$ is of the following form:

$$\boldsymbol{R}_i = \begin{pmatrix} 1 & \dots & 0 & 0 & 0 & \dots & 0 & 0 \\ \vdots & \vdots & \vdots & \vdots & \vdots & \vdots & \vdots & \vdots \\ 0 & \dots & 1 & 0 & 0 & \dots & 0 & 0 \\ 0 & \dots & 0 & \cos\alpha_i & 0 & \dots & 0 & -\sin\alpha_i \\ 0 & \dots & 0 & 0 & 1 & \dots & 0 & 0 \\ \vdots & \vdots & \vdots & \vdots & \vdots & \vdots & \vdots & \vdots \\ 0 & \dots & 0 & 0 & 0 & \dots & 1 & 0 \\ 0 & \dots & 0 & \sin\alpha_i & 0 & \dots & 0 & \cos\alpha_i \end{pmatrix},$$

that is, we have a planar rotation in rows $(i-1)$ and n. By repeating the process on the matrix $\boldsymbol{U}_{n-1}$, we obtain the factorization as in Figure 2.13. The proof that any real, unitary matrix can be written as in (2.B.2) can be found in [308]. Note that the number of free variables (angles in Givens rotations) is $n(n-1)/2$.

2.B.2 Householder Building Blocks

A unitary matrix can be factored in terms of Householder building blocks, where each block has the form $\boldsymbol{I} - 2\cdot\boldsymbol{u}\boldsymbol{u}^T$, and $\boldsymbol{u}$ is a unitary vector. Thus, an $n\times n$ unitary matrix $\boldsymbol{U}$ can be written as

$$\boldsymbol{U} = \sqrt{c}\,\boldsymbol{H}_1\cdots\boldsymbol{H}_{n-1}\cdot\boldsymbol{D}, \tag{2.B.3}$$

where $\boldsymbol{D}$ is diagonal with $d_{ii} = e^{j\theta_i}$, and $\boldsymbol{H}_i$ are Householder blocks $\boldsymbol{I} - 2\boldsymbol{u}_i\,\boldsymbol{u}_i^T$.

The fact that we mention the Householder factorization here is because we will use its polynomial version to factor lossless matrices in Chapter 3.

Note that the Householder building block is unitary, and that the factorization in (2.B.3) can be proved similarly to the factorization using Givens rotations. That is, we can first show that

$$\frac{1}{\sqrt{c}}\boldsymbol{H}_1\boldsymbol{U} = \begin{pmatrix} e^{j\alpha_0} & 0 \\ 0 & \boldsymbol{U}_1 \end{pmatrix},$$

where $\boldsymbol{U}_1$ is an $(n-1)\times(n-1)$ unitary matrix. Repeating the process on $\boldsymbol{U}_1, \boldsymbol{U}_2, \dots$, we finally obtain

$$\frac{1}{\sqrt{c}}\boldsymbol{H}_{n-1}\dots\boldsymbol{H}_1\boldsymbol{U} = \boldsymbol{D},$$

but since $\boldsymbol{H}_i = \boldsymbol{H}_i^{-1}$, we obtain (2.B.3).

APPENDIX 2.C CONVERGENCE AND REGULARITY OF FUNCTIONS

In Section 2.4.3, when discussing Fourier series, we pointed out possible convergence problems such as the Gibbs phenomenon. In this appendix, we first review different types of convergence and then discuss briefly some convergence properties of Fourier series and transforms. Then, we discuss regularity of functions and the associated decay of the Fourier series and transforms. More details on these topics can be found for example in [46, 326].

2.C.1 Convergence

Pointwise Convergence Given an infinite sequence of functions $\{f_n\}_{n=1}^{\infty}$, we say that it converges pointwise to a limit function $f = \lim_{n\to\infty} f_n$ if for each value of t we have

$$\lim_{n\to\infty} f_n(t) = f(t).$$

This is a relatively weak form of convergence, since certain properties of $f_n(t)$, such as continuity, are not passed on to the limit. Consider the truncated Fourier series, that is (from (2.4.13))

$$f_n(t) = \sum_{k=-n}^{n} F[k]\, e^{jkw_o t}. \tag{2.C.1}$$

This Fourier series converges pointwise for all t when $F[k]$ are the Fourier coefficients (see (2.4.14)) of a piecewise smooth[17] function $f(t)$. Note that while each $f_n(t)$ is continuous, the limit need not be.

Uniform Convergence An infinite sequence of functions $\{f_n\}_{n=1}^{\infty}$ converges uniformly to a limit $f(t)$ on a closed interval $[a, b]$ if (i) the sequence converges pointwise on $[a, b]$ and (ii) given any $\epsilon > 0$, there exists an integer N such that for $n > N$, $f_n(t)$ satisfies $|f(t) - f_n(t)| < \epsilon$ for all t in $[a, b]$.

Uniform convergence is obviously stronger than pointwise convergence. For example, uniform convergence of the truncated Fourier series (2.C.1) implies continuity of the limit, and conversely, continuous piecewise smooth functions have uniformly convergent Fourier series [326]. An example of pointwise convergence without uniform convergence is the Fourier series of piecewise smooth but discontinuous functions and the associated Gibbs phenomenon around discontinuities.

[17] A piecewise smooth function on an interval is piecewise continuous (finite number of discontinuities) and its derivative is also piecewise continuous.

Mean Square Convergence An infinite sequence of functions $\{f_n\}_{n=1}^{\infty}$ converges in the mean square sense to a limit $f(t)$ if

$$\lim_{n\to\infty} \|f - f_n\|_2 \;=\; 0.$$

Note that this does not mean that $\lim_{n\to\infty} f_n = f$ for all t, but only almost everywhere. For example, the truncated Fourier series (2.C.1) of a piecewise smooth function converges in the mean square sense to $f(t)$ when $F[k]$ are the Fourier series coefficients of $f(t)$, even though at a point of discontinuity t_0, $f(t_0)$ might be different from $\lim_{n\to\infty} f_n(t_0)$ which equals the mean of the right and left limits.

In the case of the Fourier transform, the concept analogous to the truncated Fourier series (2.C.1) is the truncated integral defined from the Fourier inversion formula (2.4.2) as

$$f_c(t) \;=\; \frac{1}{2\pi}\int_{-c}^{c} F(\omega)\, e^{j\omega t}\, d\omega$$

where $F(\omega)$ is the Fourier transform of $f(t)$ (see (2.4.1)). The convergence of the above integral as $c \to \infty$ is an important question, since the limit $\lim_{c\to\infty} f_c(t)$ might not equal $f(t)$. Under suitable restrictions on $f(t)$, equality will hold. As an example, if $f(t)$ is piecewise smooth and absolutely integrable, then $\lim_{c\to\infty} f_c(t_0) = f(t_0)$ at each point of continuity and is equal to the mean of the left and right limits at discontinuity points [326].

2.C.2 Regularity

So far, we have mostly discussed functions satisfying some integral conditions (absolutely or square-integrable functions for example). Instead, regularity is concerned with differentiability. The space of continuous functions is called C^0, and similarly, C^n is the space of functions having n continuous derivatives.

A finer analysis is obtained using Lipschitz (or Hölder) exponents. A function f is called Lipschitz of order α, $0 < \alpha \leq 1$, if for any t and some small ϵ, we have

$$|f(t) - f(t+\epsilon)| \;\leq\; c|\epsilon|^{\alpha}. \tag{2.C.2}$$

Higher orders $r = n + \alpha$ can be obtained by replacing f with its nth derivative. This defines Hölder spaces of order r. Note that condition (2.C.2) for $\alpha = 1$ is weaker than differentiability. For example, the triangle function or linear spline $f(t) = 1 - |t|, t \in [0, 1]$, and 0 otherwise is Lipschitz of order 1 but only C^0.

How does regularity manifest itself in the Fourier domain? Since differentiation amounts to a multiplication by $(j\omega)$ in Fourier domain (see (2.4.6)), existence of derivatives is related to sufficient decay of the Fourier spectrum.

It can be shown (see [216]) that if a function $f(t)$ and all its derivatives up to order n exist and are of bounded variation, then the Fourier transform can be bounded by

$$F(\omega) \leq \frac{c}{1 + |\omega|^{n+1}}, \tag{2.C.3}$$

that is, it decays as $O(1/|\omega|^{n+1})$ for large ω. Conversely, if $F(\omega)$ has a decay as in (2.C.3), then $f(t)$ has $n-1$ continuous derivatives, and the nth derivative exists but might be discontinuous. A finer analysis of regularity and associated localization in Fourier domain can be found in [241], in particular for functions in Hölder spaces and using different norms in Fourier domain.

PROBLEMS

2.1 *Legendre polynomials:* Consider the interval $[-1, 1]$ and the vectors $1, t, t^2, t^3, \ldots$. Using Gram-Schmidt orthogonalization, find an equivalent orthonormal set.

2.2 Prove Theorem 2.4, parts (a), (b), (d), (e), for finite-dimensional Hilbert spaces, $\mathcal{R}^n$ or $\mathcal{C}^n$.

2.3 *Orthogonal transforms and l_∞ norm:* Orthogonal transforms conserve the l_2 norm, but not others, in general. The l_∞ norm of a vector is defined as (assume $v \in R^n$):

$$l_\infty[v] = \max_{i=0,\ldots,n-1} |v_i|.$$

(a) Consider $n = 2$ and the set of real orthogonal transforms T_2, that is, plane rotations. Given the set of vectors v with unit l_2 norm (that is, vectors on the unit circle), give lower and upper bounds such that

$$a_2 \leq l_\infty[T_2 \cdot v] \leq b_2.$$

(b) Give the lower and upper bounds for the general case $n > 2$, that is, a_n and b_n.

2.4 *Norm of operators:* Consider operators that map $l_2(\mathcal{Z})$ to itself, and indicate their norm, or bounds on their norm.

(a) $(\boldsymbol{A}\boldsymbol{x})[n] = \boldsymbol{m}[n] \cdot \boldsymbol{x}[n]$, $\boldsymbol{m}[n] = e^{j\Theta_n}$, $n \in \mathcal{Z}$.

(b) $(\boldsymbol{A}\boldsymbol{x})[2n] = \boldsymbol{x}[2n] + \boldsymbol{x}[2n+1]$, $(\boldsymbol{A}\boldsymbol{x})[2n+1] = \boldsymbol{x}[2n] - \boldsymbol{x}[2n+1], n \in \mathcal{Z}$.

2.5 Assume a finite-dimensional space $\mathcal{R}^N$ and an orthonormal basis $\{\boldsymbol{x}_1, \boldsymbol{x}_2, \ldots, \boldsymbol{x}_N\}$. Any vector $\boldsymbol{y}$ can thus be written as $\boldsymbol{y} = \sum_i \alpha_i \boldsymbol{x}_i$ where $\alpha_i = \langle \boldsymbol{x}_i, \boldsymbol{y} \rangle$. Consider the best approximation to $\boldsymbol{y}$ in the least-squares sense and living on the subspace spanned by the first K vectors, $\{\boldsymbol{x}_1, \boldsymbol{x}_2, \ldots, \boldsymbol{x}_K\}$, or $\hat{\boldsymbol{y}} = \sum_{i=1}^{K} \beta_i \boldsymbol{x}_i$. Prove that $\beta_i = \alpha_i$ for $i = 1, \ldots, K$, by showing that it minimizes $\|\boldsymbol{y} - \hat{\boldsymbol{y}}\|$. *Hint:* Use Parseval's equality.

2.6 *Least-squares solution:* Show that for the least-squares solution obtained in Section 2.3.2, the partial derivatives $\partial(|\boldsymbol{y} - \hat{\boldsymbol{y}}|^2)/\partial \hat{x}_i$ are all zero.

2.7 *Least-squares solution to a linear system of equations:* The general solution was given in Equation (2.3.4–2.3.5).

(a) Show that if $\boldsymbol{y}$ belongs to the column space of $\boldsymbol{A}$, then $\hat{\boldsymbol{y}} = \boldsymbol{y}$.

(b) Show that if $\boldsymbol{y}$ is orthogonal to the column space of $\boldsymbol{A}$, then $\hat{\boldsymbol{y}} = \boldsymbol{0}$.

2.8 Parseval's formulas can be proven by using orthogonality and biorthogonality relations of the basis vectors.

(a) Show relations (2.2.5–2.2.6) using the orthogonality of the basis vectors.

(b) Show relations (2.2.11–2.2.13) using the biorthogonality of the basis vectors.

2.9 Consider the space of square-integrable real functions on the interval $[-\pi, \pi]$, $L_2([-\pi, \pi])$, and the associated orthonormal basis given by

$$\left\{ \frac{1}{\sqrt{2\pi}}, \frac{\cos nx}{\sqrt{\pi}}, \frac{\sin nx}{\sqrt{\pi}} \right\}, \qquad n = 1, 2, \ldots$$

Consider the following two subspaces: S – space of symmetric functions, that is, $f(x) = f(-x)$, on $[-\pi, \pi]$, and A – space of antisymmetric functions, $f(x) = -f(-x)$, on $[-\pi, \pi]$.

(a) Show how any function $f(x)$ from $L_2([-\pi, \pi])$ can be written as $f(x) = f_s(x) + f_a(x)$, where $f_s(x) \in S$ and $f_a(x) \in A$.

(b) Give orthonormal bases for S and A.

(c) Verify that $L_2([-\pi, \pi]) = S \oplus A$.

2.10 *Downsampling by N:* Prove (2.5.13) by going back to the underlying time-domain signal and resampling it with an N-times longer sampling period. That is, consider $x[n]$ and $y[n] = x[nN]$ as two sampled versions of the same continuous-time signal, with sampling periods T and NT, respectively. *Hint:* Recall that the discrete-time Fourier transform $X(e^{j\omega})$ of $x[n]$ is (see (2.4.36))

$$X(e^{j\omega}) = X_T(\frac{\omega}{T}) = \frac{1}{T} \sum_{k=-\infty}^{\infty} X_C\left(\frac{\omega}{T} - k\frac{2\pi}{T}\right),$$

where T is the sampling period. Then $Y(e^{j\omega}) = X_{NT}(\omega/NT)$ (since the sampling period is now NT), where $X_{NT}(\omega/NT)$ can be written similarly to the above equation. Finally, split the sum involved in $X_{NT}(\omega/NT)$ into $k = nN + l$, and gathering terms, (2.5.13) will follow.

2.11 *Downsampling and aliasing:* If an arbitrary discrete-time sequence $x[n]$ is input to a filter followed by downsampling by 2, we know that an ideal half-band lowpass filter (that is, $|H(e^{j\omega})| = 1$, $|\omega| < \pi/2$, and $H(e^{j\omega}) = 0$, $\pi/2 \leq |\omega| \leq \pi$) will avoid aliasing.
(a) Show that $H'(e^{j\omega}) = H(e^{j2\omega})$ will also avoid aliasing.
(b) Same for $H''(e^{j\omega}) = H(e^{j(2\omega-\pi)})$.
(c) A two-channel system using $H(e^{j\omega})$ and $H(e^{j(\omega-\pi)})$ followed by downsampling by 2 will keep all parts of the input spectrum untouched in either channel (except at $\omega = \pi/2$). Show that this is also true if $H'(e^{j\omega})$ and $H''(e^{j\omega})$ are used instead.

2.12 In pattern recognition, it is sometimes useful to expand a signal using the desired pattern, or template, and its shifts, as basis functions. For simplicity, consider a signal of length N, $x[n], n = 0, \ldots, N-1$, and a pattern $p[n], n = 0, \ldots, N-1$. Then, choose as basis functions

$$\varphi_k[n] = p[(n-k) \bmod N], \qquad k = 0, \ldots, N-1,$$

that is, circular shifts of $p[n]$.
(a) Derive a simple condition on $p[n]$, so that any $x[n]$ can be written as a linear combination of $\{\varphi_k\}$.
(b) Assuming the previous condition is met, give the coefficients α_k of the expansion

$$x[n] = \sum_{k=0}^{N-1} \alpha_k \, \varphi_k[n].$$

2.13 Show that a linear, periodically time-varying system of period N can be implemented with a polyphase transform followed by upsampling by N, N filter operations and a summation.

2.14 *Interpolation of oversampled signals:* Assume a function $f(t)$ bandlimited to $\omega_m = \pi$. If the sampling frequency is chosen at the Nyquist rate, $\omega_s = 2\pi$, the interpolation filter is the usual sinc filter with slow decay ($\sim 1/t$). If $f(t)$ is oversampled, for example, with $\omega_s = 3\pi$, then filters with faster decay can be used for interpolating $f(t)$ from its samples. Such filters are obtained by convolving (in frequency) elementary rectangular filters (two for $H_2(\omega)$, three for $H_3(\omega)$, while $H_1(\omega)$ would be the usual sinc filter).
(a) Give the expression for $h_2(t)$, and verify that it decays as $1/t^2$.
(b) Same for $h_3(t)$, which decays as $1/t^3$. Show that $H_3(\omega)$ has a continuous derivative.
(c) By generalizing the construction above of $H_2(\omega)$ and $H_3(\omega)$, show that one can obtain $h_i(t)$ with decay $1/t^i$. Also, show that $H_i(\omega)$ has a continuous $(i-2)$th derivative. However, the filters involved become spread out in time, and the result is only interesting asymptotically.

2.15 *Uncertainty relation:* Consider the uncertainty relation $\Delta_\omega^2\,\Delta_t^2 \geq \pi/2$.
(a) Show that scaling does not change $\Delta_\omega^2 \cdot \Delta_t^2$. Either use scaling that conserves the L_2 norm ($f'(t) = \sqrt{a}f(at)$) or be sure to renormalize Δ_ω^2, Δ_t^2.
(b) Can you give the time-bandwidth product of a rectangular pulse, $p(t) = 1$, $-1/2 \leq t \leq 1/2$, and 0 otherwise?
(c) Same as above, but for a triangular pulse.
(d) What can you say about the time-bandwidth product as the time-domain function is obtained from convolving more and more rectangular pulse with themselves?

2.16 Consider allpass filters where

$$H(z) = \prod_i \frac{a_i^* + z^{-1}}{1 + a_i z^{-1}}.$$

(a) Assume the filter has real coefficients. Show pole-zero locations, and that numerator and denominator polynomials are mirrors of each other.
(b) Given $h[n]$, the causal, real-coefficient impulse response of a stable allpass filter, give its autocorrelation $a[k] = \sum_n h[n]h[n-k]$. Show that the set $\{h[n-k]\}, k \in \mathcal{Z}$, is an orthonormal basis for $l_2(\mathcal{Z})$. *Hint:* Use Theorem 2.4.
(c) Show that the set $\{h[n-2k]\}$ is an orthonormal set but not a basis for $l_2(\mathcal{Z})$.

2.17 *Parseval's relation for nonorthogonal bases:* Consider the space $V = \mathcal{R}^n$ and a biorthogonal basis, that is, two sets $\{\alpha_i\}$ and $\{\beta_i\}$ such that

$$\langle \alpha_i, \beta_i \rangle \;=\; \delta[i-j] \qquad i, j = 0, \ldots, n-1$$

(a) Show that any vector $v \in V$ can be written in the following two ways:

$$v \;=\; \sum_{i=0}^{n-1} \langle \alpha_i, v \rangle\, \beta_i \;=\; \sum_{i=0}^{n-1} \langle \beta_i, v \rangle\, \alpha_i$$

(b) Call v_α the vector with entries $\langle \alpha_i, v \rangle$ and similarly v_β with entries $\langle \beta_i, v \rangle$. Given $\|v\|$, what can you say about $\|v_\alpha\|$ and $\|v_\beta\|$?
(c) Show that the generalization of Parseval's identity to biorthogonal systems is

$$\|v\|^2 \;=\; \langle v, v \rangle \;=\; \langle v_\alpha, v_\beta \rangle$$

and

$$\langle v, g \rangle \;=\; \langle v_\alpha, g_\beta \rangle.$$

2.18 *Circulant matrices:* An $N \times N$ circulant matrix $\boldsymbol{C}$ is defined by its first line, since subsequent lines are obtained by a right circular shift. Denote the first line by $\{c_0, c_{N-1}, \ldots, c_1\}$ so that $\boldsymbol{C}$ corresponds to a circular convolution with a filter having impulse response $\{c_0, c_1, c_2, \ldots, c_{N-1}\}$.
(a) Give a simple test for the singularity of $\boldsymbol{C}$.
(b) Give a formula for $\det(\boldsymbol{C})$.
(c) Prove that $\boldsymbol{C}^{-1}$ is circulant.
(d) Show that $\boldsymbol{C}_1\,\boldsymbol{C}_2 = \boldsymbol{C}_2\,\boldsymbol{C}_1$ and that the result is circulant.

2.19 *Walsh basis:* To define the Walsh basis, we need the Kronecker product of matrices defined in (2.3.2). Then, the matrix $\boldsymbol{W}_k$, of size $2^k \times 2^k$, is

$$\boldsymbol{W}_k = \begin{bmatrix} 1 & 1 \\ 1 & -1 \end{bmatrix} \otimes \boldsymbol{W}_{k-1}, \quad \boldsymbol{W}_0 = [1], \quad \boldsymbol{W}_1 = \begin{bmatrix} 1 & 1 \\ 1 & -1 \end{bmatrix}.$$

(a) Give W_2, W_3 and W_4 (last one only partially).
(b) Show that W_k is orthonormal (within a scale factor you should indicate).
(c) Create a block matrix T

$$T = \begin{bmatrix} W_0 & & & & \\ & 1/\sqrt{2}W_1 & & & \\ & & 1/2W_2 & & \\ & & & 1/2^{3/2}W_3 & \\ & & & & \ddots \end{bmatrix},$$

and show that T is unitary. Sketch the upper left corner of T.
(d) Consider the rows of T as basis functions in an orthonormal expansion of $l_2(\mathcal{Z}^+)$ (right-sided sequences). Sketch the tiling of the time-frequency plane achieved by this expansion.

3

Discrete-Time Bases and Filter Banks

"What is more beautiful than the Quincunx,
which, from whatever direction you look,
is correct?"
— Quintilian

Our focus in this chapter will be directed to series expansions of discrete-time sequences. The reasons for expanding signals, discussed in Chapter 1, are linked to signal analysis, approximation and compression, as well as algorithms and implementations. Thus, given an arbitrary sequence $x[n]$, we would like to write it as

$$x[n] = \sum_{k \in \mathcal{Z}} \langle \varphi_k, x \rangle \, \varphi_k[n], \qquad n \in \mathcal{Z}.$$

Therefore, we would like to construct orthonormal sets of basis functions, $\{\varphi_k[n]\}$, which are complete in the space of square-summable sequences, $l_2(\mathcal{Z})$. More general, biorthogonal and overcomplete sets, will be considered as well.

The discrete-time Fourier series, seen in Chapter 2, is an example of such an orthogonal series expansion, but it has a number of shortcomings. Discrete-time bases better suited for signal processing tasks will try to satisfy two conflicting requirements, namely to achieve good frequency resolution while keeping good time locality as well. Additionally, for both practical and computational reasons, the set of basis functions has to be structured. Typically, the infinite set of basis functions $\{\varphi_k\}$ is obtained from a finite number of prototype sequences and their shifted versions in time. This leads to discrete-time *filter banks* for the implementation of such structured expansions. This filter bank point of view has been central to the developments in the digital signal processing community, and to the design of good basis functions or filters in particular. While the expansion is not time-invariant, it will at least be *periodically time-invariant*. Also, the expansions will often have a *successive approximation* property. This means that a reconstruction based on an appropriate subset of the basis functions leads to a good approximation of the signal, which is an important feature for applications such as signal compression.

Linear signal expansions have been used in digital signal processing since at least the 1960's, mainly as block transforms, such as piecewise Fourier series and Karhunen-Loève transforms [143]. They have also been used as overcomplete expansions, such as the short-time Fourier transform (STFT) for signal analysis and synthesis [8, 226] and in transmultiplexers [25]. Increased interest in the subject, especially in orthogonal and biorthogonal bases, arose with work on compression, where redundancy of the expansion such as in the STFT is avoided. In particular, subband coding of speech [68, 69] spurred a detailed study of critically sampled filter banks. The discovery of *quadrature mirror filters* (QMF) by Croisier, Esteban and Galand in 1976 [69], which allows a signal to be split into two downsampled subband signals and then reconstructed without aliasing (spectral foldbacks) even though nonideal filters are used, was a key step forward.

Perfect reconstruction filter banks, that is, subband decompositions, where the signal is a perfect replica of the input, followed soon. The first orthogonal solution was discovered by Smith and Barnwell [270, 271] and Mintzer [196] for the two-channel case. Fettweiss and coworkers [98] gave an orthogonal solution related to wave digital filters [97]. Vaidyanathan, who established the relation between these results and certain unitary operators (paraunitary matrices of polynomials) studied in circuit theory [23], gave more general orthogonal solutions [305, 306] as well as lattice factorizations for orthogonal filter banks [308, 310]. Biorthogonal solutions were given by Vetterli [315], as well as multidimensional quadrature mirror filters [314]. Biorthogonal filter banks, in particular with linear phase filters, were investigated in [208, 321] and multidimensional filter banks were further studied in [155, 164, 257, 264, 325]. Recent work includes filter banks with rational sampling factors [160, 206] and filter banks with block sampling [158]. Additional work on the design of filter banks has been done in [144, 205] among others.

In parallel to this work on filter banks, a generalization of block transforms called lapped orthogonal transforms (LOT's) was derived by Cassereau [43] and Malvar [186, 188, 189]. An attractive feature of a subclass of LOT's is the existence of fast algorithms for their implementation since they are modulated filter banks (similar to a "real" STFT). The connection of LOT's with filter banks was shown, in [321].

Another development, which happened independently of filter banks but turns out to be closely related, is the pyramid decomposition of Burt and Adelson [41]. While it is oversampled (overcomplete), it clearly uses multiresolution concepts, by decomposing a signal into a coarse approximation plus added details. This framework is central to wavelet decompositions and establishes conceptually the link between filter banks and wavelets, as shown by Mallat [179, 180, 181] and Daubechies [71, 73]. This connection has led to a renewed interest in filter banks, especially with the work of Daubechies who first constructed wavelets from filter banks [71] and Mallat who showed that a wavelet series expansion could be implemented with filter banks [181]. Recent work on this topic includes [117, 240, 319].

As can be seen from the above short historical discussion, there are two different points of view on the subject, namely, expansion of signals in terms of structured bases, and perfect reconstruction filter banks. While the two are equivalent, the former is more in tune with Fourier and wavelet theory, while the latter is central to the construction of implementable systems. In what follows, we use both points of view, using whichever is more appropriate to explain the material.

The outline of the chapter is as follows: First, we review discrete-time series expansions, and consider two cases in some detail, namely the Haar and the sinc bases. They are two extreme cases of two-channel filter banks. The general two-channel filter bank is studied in detail in Section 3.2, where both the expansion and the more traditional filter bank point of view are given. The orthogonal case with finite-length basis functions or finite impulse response (FIR) filters is thoroughly studied. The biorthogonal FIR case, in particular with linear phase filters (symmetric or antisymmetric basis functions), is considered, and the infinite impulse response (IIR) filter case (which corresponds to basis functions with exponential decay) is given as well.

In Section 3.3, the study of filter banks with more than two channels starts with tree-structured filter banks. In particular, a constant relative bandwidth (or constant-Q) tree is shown to compute a discrete-time wavelet series. Such a transform has a multiresolution property that provides an important framework for wavelet transforms. More general filter bank trees, also known as wavelet packets, are presented as well.

Filter banks with N channels are treated next. The two particular cases of block transforms and lapped orthogonal transforms are discussed first, leading to the analysis of general N-channel filter banks. An important case, namely modulated filter banks, is studied in detail, both because of its relation to short-time Fourier-like expansions, and because of its computational efficiency.

Overcomplete discrete-time expansions are discussed in Section 3.5. The pyramid decomposition is studied, as well as the classic overlap-add/save algorithm for convolution computation which is a filter bank algorithm.

Multidimensional expansions and filter banks are derived in Section 3.6. Both separable and nonseparable systems are considered. In the nonseparable case, the focus is mostly on two-channel decompositions, while more general cases are indicated as well.

Section 3.7 discusses a scheme that has received less attention in the filter bank literature, but is nonetheless very important in applications, and is called a transmultiplexer. It is dual to the analysis/synthesis scheme used in compression applications, and is used in telecommunications.

The two appendices contain more details on orthogonal solutions and their factorizations as well as on multidimensional sampling.

The material in this chapter covers filter banks at a level of detail which is adequate for the remainder of the book. For a more exhaustive treatment of filter banks, we refer the reader to the text by Vaidyanathan [308]. Discussions of filter banks and multiresolution signal processing are also contained in the book by Akansu and Haddad [3].

3.1 Series Expansions of Discrete-Time Signals

We start by recalling some general properties of discrete-time expansions. Then, we discuss a very simple structured expansion called the Haar expansion, and give its filter bank implementation. The dual of the Haar expansion — the sinc expansion — is examined as well. These two examples are extreme cases of filter bank expansions and set the stage for solutions that lie in between.

Discrete-time series expansions come in various flavors, which we briefly review (see also Sections 2.2.3–2.2.5). As usual, $x[n]$ is an arbitrary square-summable sequence, or $x[n] \in l_2(\mathcal{Z})$. First, *orthonormal expansions* of signals $x[n]$ from $l_2(\mathcal{Z})$ are of the form

$$x[n] \;=\; \sum_{k\in\mathcal{Z}} \langle \varphi_k[l], x[l] \rangle \, \varphi_k[n] \;=\; \sum_{k\in\mathcal{Z}} X[k] \, \varphi_k[n], \tag{3.1.1}$$

where

$$X[k] \;=\; \langle \varphi_k[l], x[l] \rangle \;=\; \sum_{l} \varphi_k^*[l] \, x[l], \tag{3.1.2}$$

is the *transform* of $x[n]$. The basis functions φ_k satisfy the orthonormality[1] constraint

$$\langle \varphi_k[n], \varphi_l[n] \rangle \;=\; \delta[k-l]$$

and the set of basis functions is complete, so that every signal from $l_2(\mathcal{Z})$ can

[1] The first constraint is orthogonality between basis vectors. Then, normalization leads to orthonormality. The terms "orthogonal" and "orthonormal" will often be used interchangeably, unless we want to insist on the normalization and then use the latter.

be expressed using (3.1.1). An important property of orthonormal expansions is conservation of energy,

$$\|x\|^2 = \|X\|^2.$$

Biorthogonal expansions, on the other hand, are given as

$$\begin{aligned} x[n] &= \sum_{k\in\mathcal{Z}} \langle \varphi_k[l], x[l]\rangle \, \tilde{\varphi}_k[n] = \sum_{k\in\mathcal{Z}} \tilde{X}[k] \, \tilde{\varphi}_k[n], \\ &= \sum_{k\in\mathcal{Z}} \langle \tilde{\varphi}_k[l], x[l]\rangle \, \varphi_k[n] = \sum_{k\in\mathcal{Z}} X[k] \, \varphi_k[n], \end{aligned} \tag{3.1.3}$$

where

$$\tilde{X}[k] = \langle \varphi_k[l], x[l]\rangle \quad \text{and} \quad X[k] = \langle \tilde{\varphi}_k[l], x[l]\rangle$$

are the transform coefficients of $x[n]$ with respect to $\{\tilde{\varphi}_k\}$ and $\{\varphi_k\}$. The dual bases $\{\varphi_k\}$ and $\{\tilde{\varphi}_k\}$ satisfy the biorthogonality constraint

$$\langle \varphi_k[n], \tilde{\varphi}_l[n]\rangle = \delta[k-l].$$

Note that in this case, conservation of energy does not hold. For stability of the expansion, the transform coefficients have to satisfy

$$A\sum_k |X[k]|^2 \le \|x\|^2 \le B\sum_k |X[k]|^2$$

with a similar relation for the coefficients $\tilde{X}[k]$. In the biorthogonal case, conservation of energy can be expressed as

$$\|x\|^2 = \langle X[k], \tilde{X}[k]\rangle.$$

Finally, *overcomplete expansions* can be of the form (3.1.1) or (3.1.3), but with redundant sets of functions, that is, the functions $\varphi_k[n]$ used in the expansions are not linearly independent.

3.1.1 Discrete-Time Fourier Series

The discrete-time Fourier transform (see also Section 2.4.6) is given by

$$x[n] = \frac{1}{2\pi}\int_{-\pi}^{\pi} X(\omega)\, e^{j\omega n}\, dw \tag{3.1.4}$$

$$X(\omega) = \sum_{n=-\infty}^{\infty} x[n]\, e^{-j\omega n}. \tag{3.1.5}$$

It is a series expansion of the 2π-periodic function $X(\omega)$ as given by (3.1.5), while $x[n]$ is written in terms of an integral of the continuous-time function $X(\omega)$. While this is an important tool in the analysis of discrete-time signals and systems [211], the fact that the synthesis of $x[n]$ given by (3.1.4) involves integration rather than series expansion, makes it of limited practical use. An example of a series expansion is the discrete-time Fourier series

$$x[n] = \frac{1}{N}\sum_{k=0}^{N-1} X[k]\, e^{j2\pi kn/N}, \tag{3.1.6}$$
$$X[k] = \sum_{n=0}^{N-1} x[n]\, e^{-j2\pi kn/N},$$

where $x[n]$ is either periodic ($n \in \mathcal{Z}$) or of finite length ($n = 0, 1, \ldots, N-1$). In the latter case, the above is often called the discrete Fourier transform (DFT).

Because it only applies to such restricted types of signals, the Fourier series is somewhat limited in its applications. Since the basis functions are complex exponentials

$$\varphi_k[n] = \begin{cases} \frac{1}{N}e^{j2\pi kn/N} & n = 0, 1, \ldots, N-1, \\ 0 & \text{otherwise}, \end{cases}$$

for the finite-length case (or the periodic extension in the periodic case), there is no decay of the basis function over the length-N window, that is, no time localization (note that $\|\varphi_k\| = 1/\sqrt{N}$ in the above definition).

In order to expand arbitrary sequences we can segment the signal, and obtain a piecewise Fourier series (one for each segment). Simply segment the sequence $x[n]$ into subsequences $x^{(i)}[n]$ such that

$$x^{(i)}[n] = \begin{cases} x[n] & n = i\,N + l, \quad l = 0, 1, \ldots, N-1, \quad i \in \mathcal{Z}, \\ 0 & \text{otherwise}, \end{cases} \tag{3.1.7}$$

and take the discrete Fourier transform of each subsequence independently,

$$X^{(i)}[k] = \sum_{l=0}^{N-1} x^{(i)}[iN + l]\, e^{-j2\pi kl/N} \qquad k = 0, 1, \ldots, N-1. \tag{3.1.8}$$

Reconstruction of $x[n]$ from $X^{(i)}[k]$ is obvious. Recover $x^{(i)}[n]$ by inverting (3.1.8) (see also (3.1.6)) and then get $x[n]$ following (3.1.7) by juxtaposing the various $x^{(i)}[n]$. This leads to

$$x[n] = \sum_{i=-\infty}^{\infty}\sum_{k=0}^{N-1} X^{(i)}[k]\, \varphi_k^{(i)}[n],$$

where

$$\varphi_k^{(i)}[n] = \begin{cases} \frac{1}{N}e^{j2\pi kn/N} & n = iN + l, \quad l = 0, 1, \ldots, N-1, \\ 0 & \text{otherwise}. \end{cases}$$

The $\varphi_k^{(i)}[n]$ are simply the basis functions of the DFT shifted to the appropriate interval $[iN, \ldots, (i+1)N - 1]$.

The above expansion is called a *block discrete-time Fourier series*, since the signal is divided into blocks of size N, which are then Fourier transformed. In matrix notation, the overall expansion of the transform is given by a block diagonal matrix, where each block is an $N \times N$ Fourier matrix $\boldsymbol{F}_N$,

$$\begin{pmatrix} \vdots \\ \boldsymbol{X}^{(-1)} \\ \boldsymbol{X}^{(0)} \\ \boldsymbol{X}^{(1)} \\ \vdots \end{pmatrix} = \begin{pmatrix} \ddots & & & \\ & \boldsymbol{F}_N & & \\ & & \boldsymbol{F}_N & \\ & & & \boldsymbol{F}_N \\ & & & & \ddots \end{pmatrix} \begin{pmatrix} \vdots \\ \boldsymbol{x}^{(-1)} \\ \boldsymbol{x}^{(0)} \\ \boldsymbol{x}^{(1)} \\ \vdots \end{pmatrix},$$

and $\boldsymbol{X}^{(i)}$, $\boldsymbol{x}^{(i)}$ are size-N vectors. Up to a scale factor of $1/\sqrt{N}$ (see (3.1.6)), this is a unitary transform. This transform is not shift-invariant in general, that is, if $x[n]$ has transform $X[k]$, then $x[n-l]$ does not necessarily have the transform $X[k-l]$. However, it can be seen that

$$x[n - l\,N] \longleftrightarrow X[k - l\,N]. \tag{3.1.9}$$

That is, the transform is periodically time-varying with period N.[2] Note that we have achieved a certain time locality. Components of the signal that exist only in an interval $[iN \ldots (i+1)N-1]$ will only influence transform coefficients in the same interval. Finally, the basis functions in this block transform are naturally divided into size-N subsets, with no overlaps between subsets, that is

$$\langle \varphi_k^{(i)}[n], \varphi_l^{(m)}[n] \rangle = 0, \qquad i \neq m,$$

simply because the supports of the basis functions are disjoint. This abrupt change between intervals, and the fact that the interval length and position are arbitrary, are the drawbacks of this block DTFS.

In this chapter, we will extend the idea of block transforms in order to address these drawbacks, and this will be done using filter banks. But first, we turn our attention to the simplest block transform case, when $N = 2$. This is followed by the simplest filter bank case, when the filters are ideal sinc filters. The general case, to which these are a prelude, lies between these extremes.

3.1.2 Haar Expansion of Discrete-Time Signals

The Haar basis, while very simple, should nonetheless highlight key features such as periodic time variance and the relation with filter bank implementations. The basic unit is a two-point average and difference operation. While this is a 2×2 unitary transform that could be called a DFT just as well, we

[2]Another way to say this is that the "shift by N" and the size-N block transform operators commute.

refer to it as the elementary Haar basis because we will see that its suitable iteration will lead to both the discrete-time Haar decomposition (in Section 3.3) as well as the continuous-time Haar wavelet (in Chapter 4).

The basis functions in the Haar case are given by

$$\varphi_{2k}[n] = \begin{cases} \frac{1}{\sqrt{2}} & n = 2k,\ 2k+1, \\ 0 & \text{otherwise}, \end{cases} \qquad \varphi_{2k+1}[n] = \begin{cases} \frac{1}{\sqrt{2}} & n = 2k, \\ -\frac{1}{\sqrt{2}} & n = 2k+1, \\ 0 & \text{otherwise}. \end{cases} \tag{3.1.10}$$

It follows that the even-indexed basis functions are translates of each other, and so are the odd-indexed ones, or

$$\varphi_{2k}[n] = \varphi_0[n-2k], \qquad \varphi_{2k+1}[n] = \varphi_1[n-2k]. \tag{3.1.11}$$

The transform is

$$X[2k] = \langle \varphi_{2k}, x \rangle = \frac{1}{\sqrt{2}}\left(x[2k] + x[2k+1]\right), \tag{3.1.12}$$

$$X[2k+1] = \langle \varphi_{2k+1}, x \rangle = \frac{1}{\sqrt{2}}\left(x[2k] - x[2k+1]\right). \tag{3.1.13}$$

The reconstruction is obtained from

$$x[n] = \sum_{k \in \mathcal{Z}} X[k]\, \varphi_k[n], \tag{3.1.14}$$

as usual for an orthonormal basis. Let us prove that the set $\varphi_k[n]$ given in (3.1.10) is an orthonormal basis for $l_2(\mathcal{Z})$. While the proof is straightforward in this simple case, we indicate it for two reasons. First, it is easy to extend it to any block transform, and second, the method of the proof can be used in more general cases as well.

PROPOSITION 3.1

The set of functions as given in (3.1.10) is an orthonormal basis for signals from $l_2(\mathcal{Z})$.

PROOF

To check that the set of basis functions $\{\varphi_k\}_{k\in\mathcal{Z}}$ indeed constitutes an orthonormal basis for signals from $l_2(\mathcal{Z})$, we have to verify that:

(a) $\{\varphi_k\}_{k\in\mathcal{Z}}$ is an orthonormal family.

(b) $\{\varphi_k\}_{k\in\mathcal{Z}}$ is complete.

Consider (a). We want to show that $\langle \varphi_k, \varphi_l \rangle = \delta[k-l]$. Take k even, $k = 2i$. Then, for l smaller than $2i$ or larger than $2i+1$, the inner product is automatically zero since the basis functions do not overlap. For $l = 2i$, we have

$$\langle \varphi_{2i}, \varphi_{2i} \rangle = \varphi_{2i}^2[2i] + \varphi_{2i}^2[2i+1] = \frac{1}{2} + \frac{1}{2} = 1.$$

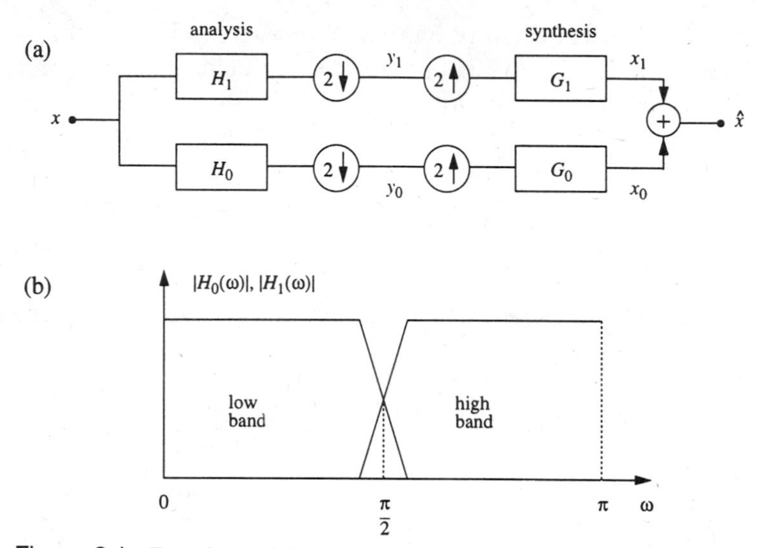

Figure 3.1 Two-channel filter bank with analysis filters $h_0[n]$, $h_1[n]$ and synthesis filters $g_0[n]$, $g_1[n]$. If the filter bank implements an orthonormal transform, then $g_0[n] = h_0[-n]$ and $g_1[n] = h_1[-n]$. (a) Block diagram. (b) Spectrum splitting performed by the filter bank.

For $l = 2i + 1$, we get

$$\langle \varphi_{2i}, \varphi_{2i+1} \rangle = \varphi_{2i}[2i] \cdot \varphi_{2i+1}[2i] + \varphi_{2i}[2i+1] \cdot \varphi_{2i+1}[2i+1] = 0.$$

A similar argument can be followed for odd l's, and thus, orthonormality is proven. Now consider (b). We have to demonstrate that any signal belonging to $l_2(\mathcal{Z})$ can be expanded using (3.1.14). This is equivalent to showing that there exists no $x[n]$ with $\|x\| > 0$, such that it has a zero expansion, that is, such that $\|\langle \varphi_k, x \rangle\| = 0$, for all k. To prove this, suppose it is not true, that is, suppose that there exists an $x[n]$ with $\|x\| > 0$, such that $\|\langle \varphi_k, x \rangle\| = 0$, for all k. Thus

$$\|\langle \varphi_k, x \rangle\| = 0 \iff \|\langle \varphi_k, x \rangle\|^2 = 0 \iff \sum_{k \in \mathcal{Z}} |\langle \varphi_k[n], x[n] \rangle|^2 = 0. \tag{3.1.15}$$

Since the last sum consists of strictly nonnegative terms, (3.1.15) is possible *if and only if*

$$X[k] = \langle \varphi_k[n], x[n] \rangle = 0, \qquad \text{for all } k.$$

First, take k even, and consider $X[2k] = 0$. Because of (3.1.12), it means that $x[2k] = -x[2k+1]$ for all k. Now take the odd k's, and look at $X[2k+1] = 0$. From (3.1.13), it follows that $x[2k] = x[2k+1]$ for all k. Thus, the only solution to the above two requirements is $x[2k] = x[2k+1] = 0$, or a contradiction with our assumption. This shows that there is no sequence $x[n]$, $\|x\| > 0$ such that $\|X\| = 0$, and proves completeness.

Now, we would like to show how the expansion (3.1.12–3.1.14) can be implemented using convolutions, thus leading to filter banks. Consider the filter

$h_0[n]$ with the following impulse response:

$$h_0[n] = \begin{cases} \frac{1}{\sqrt{2}} & n = -1,\ 0, \\ 0 & \text{otherwise.} \end{cases} \tag{3.1.16}$$

Note that this is a noncausal filter. Then, $X[2k]$ in (3.1.12) is the result of the convolution of $h_0[n]$ with $x[n]$ at instant $2k$ since

$$h_0[n] * x[n]\ |_{n=2k} = \sum_{l \in \mathcal{Z}} h_0[2k-l]\ x[l] = \frac{1}{\sqrt{2}} x[2k] + \frac{1}{\sqrt{2}} x[2k+1] = X[2k].$$

Similarly, by defining the filter $h_1[n]$ with the impulse response

$$h_1[n] = \begin{cases} \frac{1}{\sqrt{2}} & n = 0, \\ -\frac{1}{\sqrt{2}} & n = -1, \\ 0 & \text{otherwise,} \end{cases} \tag{3.1.17}$$

we obtain that $X[2k+1]$ in (3.1.13) follows from

$$\begin{aligned} h_1[n] * x[n]\ |_{n=2k} &= \sum_{l \in \mathcal{Z}} h_1[2k-l]\ x[l] \\ &= \frac{1}{\sqrt{2}} x[2k] - \frac{1}{\sqrt{2}} x[2k+1] = X[2k+1]. \end{aligned}$$

We recall (from Section 2.5.3) that evaluating a convolution at even indexes corresponds to a filter followed by downsampling by 2. Therefore, $X[2k]$ and $X[2k+1]$ can be obtained from a two-channel filter bank, with filters $h_0[n]$ and $h_1[n]$, followed by downsampling by 2, as shown in the left half of Figure 3.1(a). This is called an *analysis filter bank*. Often, we will specifically label the channel signals as y_0 and y_1, where

$$y_0[k] = X[2k], \qquad y_1[k] = X[2k+1].$$

It is important to note that the impulse responses of the analysis filters are time-reversed versions of the basis functions,

$$h_0[n] = \varphi_0[-n], \qquad h_1[n] = \varphi_1[-n],$$

since convolution is an inner product involving time reversal. Also, the filters we defined in (3.1.16) and (3.1.17) are noncausal, which is to be expected since, for example, the computation of $X[2k]$ in (3.1.12) involves $x[2k+1]$, that is, a future sample. To summarize this discussion, it is easiest to visualize the analysis in matrix notation as

$$\begin{pmatrix} \vdots \\ y_0[0] \\ y_1[0] \\ y_0[1] \\ y_1[1] \\ \vdots \end{pmatrix} = \begin{pmatrix} \vdots \\ X[0] \\ X[1] \\ X[2] \\ X[3] \\ \vdots \end{pmatrix} = \begin{pmatrix} \ddots & & & & & \\ & \overbrace{\begin{matrix} h_0[0] & h_0[-1] \end{matrix}}^{\varphi_0[n]} & & & \\ & \underbrace{\begin{matrix} h_1[0] & h_1[-1] \end{matrix}}_{\varphi_1[n]} & & & \\ & & & \overbrace{\begin{matrix} h_0[0] & h_0[-1] \end{matrix}}^{\varphi_2[n]} & \\ & & & \underbrace{\begin{matrix} h_1[0] & h_1[-1] \end{matrix}}_{\varphi_3[n]} & \\ & & & & \ddots \end{pmatrix} \begin{pmatrix} \vdots \\ x[0] \\ x[1] \\ x[2] \\ x[3] \\ \vdots \end{pmatrix}, \tag{3.1.18}$$

where we again see the shift property of the basis functions (see (3.1.11)). We can verify the shift invariance of the analysis with respect to even shifts. If $x'[n] = x[n-2l]$, then

$$\begin{aligned} X'[2k] &= \frac{1}{\sqrt{2}}(x'[2k] + x'[2k+1]) = \frac{1}{\sqrt{2}}(x[2k-2l] + x[2k+1-2l]) \\ &= X[2k-2l] \end{aligned}$$

and similarly for $X'[2k+1]$ which equals $X[2k+1-2l]$, thus verifying (3.1.9). This does not hold for odd shifts, however. For example, $\delta[n]$ has the transform $(\delta[n] + \delta[n-1])/\sqrt{2}$ while $\delta[n-1]$ leads to $(\delta[n] - \delta[n-1])/\sqrt{2}$.

What about the synthesis or reconstruction given by (3.1.14)? Define two filters g_0 and g_1 with impulse responses equal to the basis functions φ_0 and φ_1

$$g_0[n] = \varphi_0[n], \qquad g_1[n] = \varphi_1[n]. \tag{3.1.19}$$

Therefore

$$\varphi_{2k}[n] = g_0[n-2k], \qquad \varphi_{2k+1}[n] = g_1[n-2k], \tag{3.1.20}$$

following (3.1.11). Then (3.1.14) becomes, using (3.1.19) and (3.1.20),

$$x[n] = \sum_{k \in \mathcal{Z}} y_0[k]\varphi_{2k}[n] + \sum_{k \in \mathcal{Z}} y_1[k]\varphi_{2k+1}[n] \tag{3.1.21}$$

$$= \sum_{k \in \mathcal{Z}} y_0[k]g_0[n-2k] + \sum_{k \in \mathcal{Z}} y_1[k]g_1[n-2k]. \tag{3.1.22}$$

That is, each sample from $y_i[k]$ adds a copy of the impulse response of $g_i[n]$ shifted by $2k$. This can be implemented by an upsampling by 2 (inserting a zero between every two samples of $y_i[k]$) followed by a convolution with $g_i[n]$ (see also Section 2.5.3). This is shown in the right side of Figure 3.1(a), and is called a *synthesis filter bank*.

What we have just explained is a way of implementing a structured orthogonal expansion by means of filter banks. We summarize two characteristics of the filters which will hold in general orthogonal cases as well.

(a) The impulse responses of the synthesis filters equal the first set of basis functions

$$g_i[n] \;=\; \varphi_i[n], \qquad i = 0,\ 1.$$

(b) The impulse responses of the analysis filters are the time-reversed versions of the synthesis ones

$$h_i[n] \;=\; g_i[-n], \qquad i = 0,\ 1.$$

What about the signal processing properties of our decomposition? From (3.1.12) and (3.1.13), we recall that one channel computes the average and the other the difference of two successive samples. While these are not the "best possible" lowpass and highpass filters (they have, however, good time localization), they lead to an important interpretation. The reconstruction from $y_0[k]$ (that is, the first sum in (3.1.21)) is the orthogonal projection of the input onto the subspace spanned by $\varphi_{2k}[n]$, that is, an *average* or *coarse* version of $x[n]$. Calling it x_0, it equals

$$x_0[2k] \;=\; x_0[2k+1] \;=\; \frac{1}{2}\left(x[2k] + x[2k+1]\right).$$

The other sum in (3.1.21), which is the reconstruction from $y_1[k]$, is the orthogonal projection onto the subspace spanned by $\varphi_{2k+1}[n]$. Denoting it by x_1, it is given by

$$x_1[2k] \;=\; \frac{1}{2}\left(x[2k] - x[2k+1]\right), \qquad x_1[2k+1] \;=\; -x_1[2k].$$

This is the *difference* or *added detail* necessary to reconstruct $x[n]$ from its coarse version $x_0[n]$. The two subspaces spanned by $\{\varphi_{2k}\}$ and $\{\varphi_{2k+1}\}$ are orthogonal and the sum of the two projections recovers $x[n]$ perfectly, since summing $(x_0[2k] + x_1[2k])$ yields $x[2k]$ and similarly $(x_0[2k+1] + x_1[2k+1])$ gives $x[2k+1]$.

3.1.3 Sinc Expansion of Discrete-Time Signals

Although remarkably simple, the Haar basis suffers from an important drawback — the frequency resolution of its basis functions (filters), is not very good. We now look at a basis which uses ideal half-band lowpass and highpass filters. The frequency selectivity is ideal (out-of-band signals are perfectly rejected), but the time localization suffers (the filter impulse response is infinite, and decays only proportionally to $1/n$).

Let us start with an ideal half-band lowpass filter $g_0[n]$, defined by its 2π-periodic discrete-time Fourier transform $G_0(e^{j\omega}) = \sqrt{2}$, $\omega \in [-\pi/2,\ \pi/2]$ and 0 for $\omega \in [\pi/2,\ 3\pi/2]$. The scale factor is so chosen that $\|G_0\| = 2\pi$ or $\|g_0\| = 1$ following Parseval's relation for the DTFT. The inverse DTFT yields

$$g_0[n] \;=\; \frac{\sqrt{2}}{2\pi}\int_{\pi/2}^{\pi/2} e^{j\omega n}\,d\omega \;=\; \frac{1}{\sqrt{2}}\frac{\sin \pi n/2}{\pi n/2}. \qquad (3.1.23)$$

Note that $g_0[2n] = 1/\sqrt{2}\cdot\delta[n]$. As the highpass filter, choose a modulated version of $g_0[n]$, with a twist, namely a time reversal and a shift by one

$$g_1[n] \;=\; (-1)^n g_0[-n+1]. \qquad (3.1.24)$$

While the time reversal is only formal here (since $g_0[n]$ is symmetric in n), the shift by one is important for the completeness of the highpass and lowpass impulse responses in the space of square-summable sequences.

Just as in the Haar case, the basis functions are obtained from the filter impulse responses and their even shifts,

$$\varphi_{2k}[n] \;=\; g_0[n-2k], \qquad \varphi_{2k+1}[n] \;=\; g_1[n-2k], \qquad (3.1.25)$$

and the coefficients of the expansion $\langle\varphi_{2k}, x\rangle$ and $\langle\varphi_{2k+1}, x\rangle$ are obtained by filtering with $h_0[n]$ and $h_1[n]$ followed by downsampling by 2, with $h_i[n] = g_i[-n]$.

PROPOSITION 3.2

The set of functions as given in (3.1.25) is an orthonormal basis for signals from $l_2(\mathcal{Z})$.

PROOF

To prove that the set of functions $\varphi_k[n]$ is indeed an orthonormal basis, again we would have to demonstrate orthonormality of the set as well as completeness. Let us demonstrate orthonormality of basis functions. We will do that only for

$$\langle\varphi_{2k}[n], \varphi_{2l}[n]\rangle \;=\; \delta[k-l], \qquad (3.1.26)$$

and leave the other two cases

$$\langle\varphi_{2k}[n], \varphi_{2l+1}[n]\rangle \;=\; 0, \qquad (3.1.27)$$

$$\langle\varphi_{2k+1}[n], \varphi_{2l+1}[n]\rangle \;=\; \delta[k-l], \qquad (3.1.28)$$

as an exercise (Problem 3.1). First, because $\varphi_{2k}[n] = \varphi_0[n-2k]$, it suffices to show (3.1.26) for $k=0$, or equivalently, to prove that

$$\langle g_0[n]\ ,\ g_0[n-2l]\rangle = \delta[l].$$

From (2.5.19) this is equivalent to showing

$$|G_0(e^{j\omega})|^2 + |G_0(e^{j(\omega+\pi)})|^2 = 2,$$

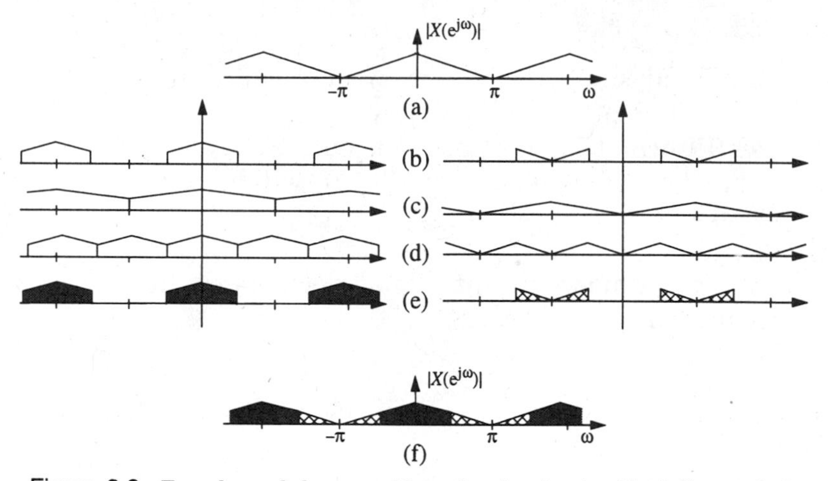

Figure 3.2 Two-channel decomposition of a signal using ideal filters. Left side depicts the process in the lowpass channel, while the right side depicts the process in the highpass channel. (a) Original spectrum. (b) Spectrums after filtering. (c) Spectrums after downsampling. (d) Spectrums after upsampling. (e) Spectrums after interpolation filtering. (f) Reconstructed spectrum.

which holds true since $G_0(e^{j\omega}) = \sqrt{2}$ between $-\pi/2$ and $\pi/2$. The proof of the other orthogonality relations is similar.

The proof of completeness, which can be made along the lines of the proof in Proposition 3.1, is left to the reader (see Problem 3.1).

As we said, the filters in this case have perfect frequency resolution. However, the decay of the filters in time is rather poor, being of the order of $1/n$. The multiresolution interpretation we gave for the Haar case holds here as well. The perfect lowpass filter h_0, followed by downsampling, upsampling and interpolation by g_0, leads to a projection of the signal onto the subspace of sequences bandlimited to $[-\pi/2, \pi/2]$, given by x_0. Similarly, the other path in Figure 3.1 leads to a projection onto the subspace of half-band highpass signals given by x_1. The two subspaces are orthogonal and their sum is $l_2(\mathcal{Z})$. It is also clear that x_0 is a coarse, lowpass approximation to x, while x_1 contains the additional frequencies necessary to reconstruct x from x_0.

An example describing the decomposition of a signal into downsampled lowpass and highpass components, with subsequent reconstruction using upsampling and interpolation, is shown in Figure 3.2. Ideal half-band filters are assumed. The reader is encouraged to verify this spectral decomposition using the downsampling and upsampling formulas (see (2.5.13) and (2.5.17)) from Section 2.5.3.

3.1.4 Discussion

In both the Haar and sinc cases above, we noticed that the expansion was not time-invariant, but periodically time-varying. We show below that time invariance in orthonormal expansions leads only to trivial solutions, and thus, any meaningful orthonormal expansion of $l_2(\mathcal{Z})$ will be time-varying.

PROPOSITION 3.3

An orthonormal time-invariant signal decomposition will have no frequency resolution.

PROOF

An expansion is time-invariant if $x[n] \longleftrightarrow X[k]$, then $x[n-m] \longleftrightarrow X[k-m]$ for all $x[n]$ in $l_2(\mathcal{Z})$. Thus, we have that

$$\langle \varphi_k[n],\, x[n-m] \rangle \;=\; \langle \varphi_{k-m}[n],\, x[n] \rangle.$$

By a change of variable, the left side is equal to $\langle \varphi_k[n+m],\, x[n] \rangle$, and then using $k' = k - m$, we find that

$$\varphi_{k'+m}[n+m] \;=\; \varphi_{k'}[n], \tag{3.1.29}$$

that is, the expansion operator is Toeplitz. Now, we want the expansion to be orthonormal, that is, using (3.1.29),

$$\langle \varphi_k[n],\, \varphi_{k+m}[n] \rangle \;=\; \langle \varphi_k[n],\, \varphi_k[n-m] \rangle \;=\; \delta[m],$$

or the autocorrelation of $\varphi_k[n]$ is a Dirac function. In Fourier domain, this leads to

$$|\Phi(e^{j\omega})|^2 \;=\; 1,$$

showing that the basis functions have no frequency selectivity since they are allpass functions.

Therefore, time variance is an inherent feature of orthonormal expansions. Note that Proposition 3.3 does not hold if the orthogonality constraint is removed (see Problem 3.3). Another consequence of Proposition 3.3 is that there are no banded[3] orthonormal Toeplitz matrices, since an allpass filter has necessarily infinite impulse response. However, in (3.1.18), we saw a banded block Toeplitz matrix (actually, block diagonal) that was orthonormal. The construction of orthonormal FIR filter banks is the study of such banded block Toeplitz matrices.

We have seen two extreme cases of structured series expansions of sequences, based on Haar and sinc filters respectively (Table 3.1 gives basis functions for both of these cases). More interesting cases exist between these extremes and they will be implemented with filter banks as shown in Figure 3.1(a). Thus, we did not consider arbitrary expansions of $l_2(\mathcal{Z})$, but rather

[3] A banded Toeplitz matrix has a finite number of nonzero diagonals.

Table 3.1 Basis functions (synthesis filters) in Haar and sinc cases.

	Haar	Sinc
$g_0[n]$	$(\delta[n] + \delta[n-1])/\sqrt{2}$	$\frac{1}{\sqrt{2}} \frac{\sin(\pi/2)n}{(\pi/2)n}$
$g_1[n]$	$(\delta[n] - \delta[n-1])/\sqrt{2}$	$(-1)^n g_0[-n+1]$
$G_0(e^{j\omega})$	$\sqrt{2}e^{-j(\omega/2)}\cos(\omega/2)$	$\begin{cases} \sqrt{2} & \text{for } \omega \in [-\pi/2,\ \pi/2], \\ 0 & \text{otherwise.} \end{cases}$
$G_1(e^{j\omega})$	$\sqrt{2}je^{-j(\omega/2)}\sin(\omega/2)$	$-e^{-j\omega}G_0(-e^{-j\omega})$

a structured subclass. These expansions will have the multiresolution characteristic already built in, which will be shown to be a framework for a large body of work on filter banks that appeared in the literature of the last decade.

3.2 Two-Channel Filter Banks

We saw in the last section how Haar and sinc expansions of discrete-time signals could be implemented using a two-channel filter bank (see Figure 3.1(a)). The aim in this section is to examine two-channel filter banks in more detail. The main idea is that perfect reconstruction filter banks implement series expansions of discrete-time signals as in the Haar and sinc cases. Recall that in both of these cases, the expansion is orthonormal and the basis functions are actually the impulse responses of the synthesis filters and their even shifts. In addition to the orthonormal case, we will consider biorthogonal (or general) expansions (filter banks) as well.

The present section serves as a core for the remainder of the chapter; all important notions and concepts will be introduced here. For the sake of simplicity, we concentrate on the two-channel case. More general solutions are given later in the chapter. We start with tools for analyzing general filter banks. Then, we examine orthonormal and linear phase two-channel filter banks in more detail. We then present results valid for general two-channel filter banks and examine some special cases, such as IIR solutions.

3.2.1 Analysis of Filter Banks

Consider Figure 3.1(a). We saw in the Haar and sinc cases, that such a two-channel filter bank implements an orthonormal series expansion of discrete-time signals with synthesis filters being the time-reversed version of the analysis filters, that is $g_i[n] = h_i[-n]$. Here, we relax the assumption of orthonormality and consider a general filter bank, with analysis filters $h_0[n]$, $h_1[n]$ and synthesis filters $g_0[n]$, $g_1[n]$. Our only requirement will be that such a filter bank implements an expansion of discrete-time signals (not necessarily orthonormal). Such an expansion will be termed *biorthogonal*. In the filter bank literature, such a system is called a *perfect reconstruction filter bank*.

Looking at Figure 3.1, besides filtering, the key elements in the filter bank computation of an expansion are downsamplers and upsamplers. These perform the sampling rate changes and the downsampler creates a periodically time-varying linear system. As discussed in Section 2.5.3, special analysis techniques are needed for such systems. We will present three ways to look at periodically time-varying systems, namely in time, modulation, and polyphase domains. The first approach was already used in our discussion of the Haar case. The two other approaches are based on the Fourier or z-transform and aim at decomposing the periodically time-varying system into several time-invariant subsystems.

Time-Domain Analysis Recall that in the Haar case (see (3.1.18)), in order to visualize block time invariance, we expressed the transform coefficients via an infinite matrix, that is

$$\underbrace{\begin{pmatrix} \vdots \\ y_0[0] \\ y_1[0] \\ y_0[1] \\ y_1[1] \\ \vdots \end{pmatrix}}_{\boldsymbol{y}} = \underbrace{\begin{pmatrix} \vdots \\ X[0] \\ X[1] \\ X[2] \\ X[3] \\ \vdots \end{pmatrix}}_{\boldsymbol{X}} = \boldsymbol{T}_a \cdot \underbrace{\begin{pmatrix} \vdots \\ x[0] \\ x[1] \\ x[2] \\ x[3] \\ \vdots \end{pmatrix}}_{\boldsymbol{x}}. \tag{3.2.1}$$

Here, the transform coefficients $X[k]$ are expressed in another form as well. In the filter bank literature, it is more common to write $X[k]$ as outputs of the two branches in Figure 3.1(a), that is, as two *subband* outputs denoted by $y_0[k] = X[2k]$, and $y_1[k] = X[2k+1]$. Also, in (3.2.1), $\boldsymbol{T}_a \cdot \boldsymbol{x}$ represents the inner products, where $\boldsymbol{T}_a$ is the *analysis* matrix and can be expressed as

$$\boldsymbol{T}_a = \begin{pmatrix} \vdots & \vdots & \vdots & & \vdots & \vdots & \vdots \\ h_0[L-1] & h_0[L-2] & h_0[L-3] & \cdots & h_0[0] & 0 & 0 \\ h_1[L-1] & h_1[L-2] & h_1[L-3] & \cdots & h_1[0] & 0 & 0 \\ 0 & 0 & h_0[L-1] & \cdots & h_0[2] & h_0[1] & h_0[0] \\ 0 & 0 & h_1[L-1] & \cdots & h_1[2] & h_1[1] & h_1[0] \\ \vdots & \vdots & \vdots & & \vdots & \vdots & \vdots \end{pmatrix},$$

where we assume that the analysis filters $h_i[n]$ are finite impulse response (FIR) filters of length $L = 2K$. To make the block Toeplitz structure of $\boldsymbol{T}_a$ more explicit, we can write

$$\boldsymbol{T}_a = \begin{pmatrix} & \vdots & \vdots & & \vdots & \vdots & \\ \cdots & \boldsymbol{A}_0 & \boldsymbol{A}_1 & \cdots & \boldsymbol{A}_{K-1} & \boldsymbol{0} & \cdots \\ \cdots & \boldsymbol{0} & \boldsymbol{A}_0 & \cdots & \boldsymbol{A}_{K-2} & \boldsymbol{A}_{K-1} & \cdots \\ & \vdots & \vdots & & \vdots & \vdots & \end{pmatrix}. \tag{3.2.2}$$

The block $\boldsymbol{A}_i$ is given by

$$\boldsymbol{A}_i = \begin{pmatrix} h_0[2K-1-2i] & h_0[2K-2-2i] \\ h_1[2K-1-2i] & h_1[2K-2-2i] \end{pmatrix}. \tag{3.2.3}$$

The transform coefficient

$$X[k] = \langle \varphi_k[n], x[n] \rangle,$$

equals (in the case $k = 2k'$)

$$y_0[k'] = \langle h_0[2k'-n], x[n] \rangle,$$

and (in the case $k = 2k'+1$)

$$y_1[k'] = \langle h_1[2k'-n], x[n] \rangle.$$

The analysis basis functions are thus

$$\varphi_{2k}[n] = h_0[2k-n], \tag{3.2.4}$$

$$\varphi_{2k+1}[n] = h_1[2k-n]. \tag{3.2.5}$$

To resynthesize the signal, we use the dual-basis, *synthesis*, matrix $\boldsymbol{T}_s$

$$\boldsymbol{x} = \boldsymbol{T}_s\, \boldsymbol{y} = \boldsymbol{T}_s\, \boldsymbol{X} = \boldsymbol{T}_s\, \boldsymbol{T}_a\, \boldsymbol{x}. \tag{3.2.6}$$

Similarly to $\boldsymbol{T}_a$, $\boldsymbol{T}_s$ can be expressed as

$$\boldsymbol{T}_s^T = \begin{pmatrix} \vdots & \vdots & \vdots & & \vdots & \vdots & \vdots \\ g_0[0] & g_0[1] & g_0[2] & \cdots & g_0[L'-1] & 0 & 0 \\ g_1[0] & g_1[1] & g_1[2] & \cdots & g_1[L'-1] & 0 & 0 \\ 0 & 0 & g_0[0] & \cdots & g_0[L'-3] & g_0[L'-2] & g_0[L'-1] \\ 0 & 0 & g_1[0] & \cdots & g_1[L'-3] & g_1[L'-2] & g_1[L'-1] \\ \vdots & \vdots & \vdots & \vdots & \vdots & \vdots & \end{pmatrix}$$

$$= \begin{pmatrix} & \vdots & \vdots & & \vdots & \vdots & \\ \cdots & \boldsymbol{S}_0^T & \boldsymbol{S}_1^T & \cdots & \boldsymbol{S}_{K'-1}^T & \boldsymbol{0} & \cdots \\ \cdots & \boldsymbol{0} & \boldsymbol{S}_0^T & \cdots & \boldsymbol{S}_{K'-2}^T & \boldsymbol{S}_{K'-1}^T & \cdots \\ & \vdots & \vdots & & \vdots & \vdots & \end{pmatrix}, \tag{3.2.7}$$

where the block $\boldsymbol{S}_i$ is of size 2×2 and FIR filters are of length $L' = 2K'$. The block $\boldsymbol{S}_i$ is

$$\boldsymbol{S}_i = \begin{pmatrix} g_0[2i] & g_1[2i] \\ g_0[2i+1] & g_1[2i+1] \end{pmatrix},$$

where $g_0[n]$ and $g_1[n]$ are the synthesis filters. The dual synthesis basis functions are

$$\begin{aligned} \tilde{\varphi}_{2k}[n] &= g_0[n-2k], \\ \tilde{\varphi}_{2k+1}[n] &= g_1[n-2k]. \end{aligned}$$

Let us go back for a moment to (3.2.6). The requirement that $\{h_0[2k-n], h_1[2k-n]\}$ and $\{g_0[n-2k],\ g_1[n-2k]\}$ form a dual bases pair is equivalent to

$$\boldsymbol{T}_s \boldsymbol{T}_a = \boldsymbol{T}_a \boldsymbol{T}_s = \boldsymbol{I}. \tag{3.2.8}$$

This is the biorthogonality condition or, in the filter bank literature, the perfect reconstruction condition. In other words,

$$\langle \varphi_k[n], \tilde{\varphi}_l[n] \rangle = \delta[k-l],$$

or in terms of filter impulse responses

$$\langle h_i[2k-n], g_j[n-2l] \rangle = \delta[k-l]\,\delta[i-j], \qquad i,j = 0,1.$$

Consider the two branches in Figure 3.1(a) which produce y_0 and y_1. Call $\boldsymbol{H}_i$ the operator corresponding to filtering by $h_i[n]$ followed by downsampling by 2. Then the output $\boldsymbol{y}_i$ can be written as (L denotes the filter length)

$$\underbrace{\begin{pmatrix} \vdots \\ y_i[0] \\ y_i[1] \\ \vdots \end{pmatrix}}_{\boldsymbol{y}_i} = \underbrace{\begin{pmatrix} & \vdots & \vdots & \vdots & \\ \cdots & h_i[L-1] & h_i[L-2] & h_i[L-4] & \cdots \\ \cdots & 0 & 0 & h_i[L-1] & \cdots \\ & \vdots & \vdots & \vdots & \end{pmatrix}}_{\boldsymbol{H}_i} \underbrace{\begin{pmatrix} \vdots \\ x[0] \\ x[1] \\ \vdots \end{pmatrix}}_{\boldsymbol{x}}, \tag{3.2.9}$$

or, in operator notation

$$\boldsymbol{y}_i = \boldsymbol{H}_i\, \boldsymbol{x}.$$

Defining $\boldsymbol{G}_i^T$ similarly to $\boldsymbol{H}_i$ but with $g_i[n]$ in reverse order (see also the definition of $\boldsymbol{T}_s$), the output of the system can now be written as

$$(\boldsymbol{G}_0\, \boldsymbol{H}_0 + \boldsymbol{G}_1\, \boldsymbol{H}_1)\, \boldsymbol{x}.$$

Thus, to resynthesize the signal (the condition for perfect reconstruction), we have that

$$\boldsymbol{G}_0\, \boldsymbol{H}_0 + \boldsymbol{G}_1\, \boldsymbol{H}_1 = \boldsymbol{I}.$$

Of course, by interleaving the rows of $\boldsymbol{H}_0$ and $\boldsymbol{H}_1$, we get $\boldsymbol{T}_a$, and similarly, $\boldsymbol{T}_s$ corresponds to interleaving the columns of $\boldsymbol{G}_0$ and $\boldsymbol{G}_1$.

To summarize this part on time-domain analysis, let us stress once more that biorthogonal expansions of discrete-time signals, where the basis functions are obtained from two prototype functions and their even shifts (for both dual bases), is implemented using a perfect reconstruction, two-channel multirate filter bank. In other words, perfect reconstruction is equivalent to the biorthogonality condition (3.2.8).

Completeness is also automatically satisfied. To prove it, we show that there exists no $x[n]$ with $\|x\| > 0$, such that it has a zero expansion, that is, such that $\|X\| = 0$. Suppose it is not true, that is, suppose that there exists an $x[n]$ with $\|x\| > 0$, such that $\|X\| = 0$. But, since $\boldsymbol{X} = \boldsymbol{T}_a\, \boldsymbol{x}$, we have that

$$\|\boldsymbol{T}_a\, \boldsymbol{x}\| = \boldsymbol{0},$$

and this is possible *if and only if*

$$\boldsymbol{T}_a\, \boldsymbol{x} = \boldsymbol{0} \tag{3.2.10}$$

(since in a Hilbert space — $l_2(\mathcal{Z})$ in this case, $\|v\|^2 = \langle v, v\rangle = 0$, *if and only if* $v \equiv 0$). We know that (3.2.10) has a nontrivial solution *if and only if* $\boldsymbol{T}_a$ is singular. However, due to (3.2.8), $\boldsymbol{T}_a$ is nonsingular and thus (3.2.10) has only a trivial solution, $\boldsymbol{x} \equiv 0$, violating our assumption and proving completeness.

Modulation-Domain Analysis This approach is based on Fourier or more generally z-transforms. Recall from Section 2.5.3, that downsampling a signal with the z-transform $X(z)$ by 2 leads to $X'(z)$ given by

$$X'(z) = \frac{1}{2}\left[X(z^{1/2}) + X(-z^{1/2})\right]. \tag{3.2.11}$$

Then, upsampling $X'(z)$ by 2 yields $X''(z) = X'(z^2)$, or

$$X''(z) = \frac{1}{2}\left[X(z) + X(-z)\right]. \tag{3.2.12}$$

To verify (3.2.12) directly, notice that downsampling followed by upsampling by 2 simply nulls out the odd-indexed coefficients, that is, $x''[2n] = x[2n]$ and $x''[2n+1] = 0$. Then, note that $X(-z)$ is the z-transform of $(-1)^n x[n]$ by the modulation property, and therefore, (3.2.12) follows.

With this preamble, the z-transform analysis of the filter bank in Figure 3.1(a) becomes easy. Consider the lower branch. The filtered signal, which has the z-transform $H_0(z) \cdot X(z)$, goes through downsampling and upsampling, yielding (according to (3.2.12))

$$\frac{1}{2}\left[H_0(z)\, X(z) + H_0(-z)\, X(-z)\right].$$

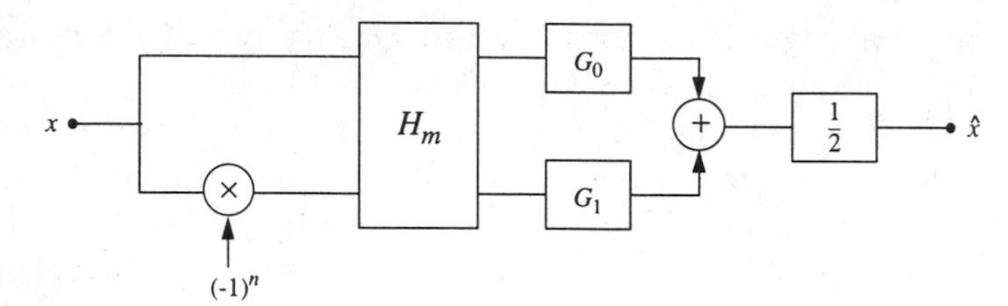

Figure 3.3 Modulation-domain analysis of the two-channel filter bank. The 2×2 matrix $\boldsymbol{H}_m(z)$ contains the z-transform of the filters and their modulated versions.

This signal is filtered with $G_0(z)$, leading to $X_0(z)$ given by

$$X_0(z) = \frac{1}{2} G_0(z) \left[H_0(z)\, X(z) + H_0(-z)\, X(-z) \right]. \tag{3.2.13}$$

The upper branch contributes $X_1(z)$, which equals to (3.2.13) up to the change of index $0 \rightarrow 1$, and the output of the analysis/synthesis filter bank is the sum of the two components $X_0(z)$ and $X_1(z)$. This is best written in matrix notation as

$$\begin{aligned} \hat{X}(z) &= X_0(z) + X_1(z) \qquad (3.2.14) \\ &= \frac{1}{2} \begin{pmatrix} G_0(z) & G_1(z) \end{pmatrix} \underbrace{\begin{pmatrix} H_0(z) & H_0(-z) \\ H_1(z) & H_1(-z) \end{pmatrix}}_{\boldsymbol{H}_m(z)} \underbrace{\begin{pmatrix} X(z) \\ X(-z) \end{pmatrix}}_{\boldsymbol{x}_m(z)}. \end{aligned}$$

In the above, $\boldsymbol{H}_m(z)$ is the *analysis modulation matrix* containing the modulated versions of the analysis filters and $\boldsymbol{x}_m(z)$ contains the modulated versions of $X(z)$. Relation (3.2.14) is illustrated in Figure 3.3, where the time-varying part is in the lower channel. If the channel signals $Y_0(z)$ and $Y_1(z)$ are desired, that is, the downsampled domain signals, it follows from (3.2.11) and (3.2.14) that

$$\begin{pmatrix} Y_0(z) \\ Y_1(z) \end{pmatrix} = \frac{1}{2} \begin{pmatrix} H_0(z^{1/2}) & H_0(-z^{1/2}) \\ H_1(z^{1/2}) & H_1(-z^{1/2}) \end{pmatrix} \begin{pmatrix} X(z^{1/2}) \\ X(-z^{1/2}) \end{pmatrix},$$

or, calling $\boldsymbol{y}(z)$ the vector $[Y_0(z) \;\; Y_1(z)]^T$,

$$\boldsymbol{y}(z) = \frac{1}{2} \boldsymbol{H}_m(z^{1/2})\, \boldsymbol{x}_m(z^{1/2}).$$

For the system to represent a valid expansion, (3.2.14) has to yield $\hat{X}(z) = X(z)$, which can be obtained when

$$G_0(z)\, H_0(z) + G_1(z)\, H_1(z) = 2, \tag{3.2.15}$$

$$G_0(z)\, H_0(-z) + G_1(z)\, H_1(-z) = 0. \tag{3.2.16}$$

The above two conditions then ensure perfect reconstruction. Expressing (3.2.15) and (3.2.16) in matrix notation, we get

$$\begin{pmatrix} G_0(z) & G_1(z) \end{pmatrix} \cdot \boldsymbol{H}_m(z) \;=\; \begin{pmatrix} 2 & 0 \end{pmatrix}. \tag{3.2.17}$$

We can solve now for $G_0(z)$ and $G_1(z)$ (transpose (3.2.17) and multiply by $(\boldsymbol{H}_m^T(z))^{-1}$ from the left)

$$\begin{pmatrix} G_0(z) \\ G_1(z) \end{pmatrix} = \frac{2}{\det(\boldsymbol{H}_m(z))} \begin{pmatrix} H_1(-z) \\ -H_0(-z) \end{pmatrix}. \tag{3.2.18}$$

In the above, we assumed that $\boldsymbol{H}_m(z)$ is nonsingular; that is, its normal rank is equal to 2. Define $P(z)$ as

$$P(z) \;=\; G_0(z)\, H_0(z) \;=\; \frac{2}{\det(\boldsymbol{H}_m(z))} H_0(z) H_1(-z), \tag{3.2.19}$$

where we used (3.2.18). Observe that $\det(\boldsymbol{H}_m(z)) = -\det(\boldsymbol{H}_m(-z))$. Then, we can express the product $G_1(z)H_1(z)$ as

$$G_1(z)\, H_1(z) \;=\; \frac{-2}{\det(\boldsymbol{H}_m(z))} H_0(-z)\, H_1(z) \;=\; P(-z).$$

It follows that (3.2.15) can be expressed in terms of $P(z)$ as

$$P(z) + P(-z) \;=\; 2. \tag{3.2.20}$$

We will show later, that the function $P(z)$ plays a crucial role in analyzing and designing filter banks. It suffices to note at this moment that, due to (3.2.20), all even-indexed coefficients of $P(z)$ equal 0, except for $p[0] = 1$. Thus, $P(z)$ is of the following form:

$$P(z) \;=\; 1 + \sum_{k \in \mathcal{Z}} p[2k+1]\; z^{-(2k+1)}.$$

A polynomial or a rational function in z satisfying (3.2.20) will be called *valid*. Following the definition of $P(z)$ in (3.2.19), we can rewrite (3.2.15) or equivalently (3.2.20) as

$$G_0(z)\, H_0(z) + G_0(-z)\, H_0(-z) \;=\; 2. \tag{3.2.21}$$

Using the modulation property, its time-domain equivalent is

$$\sum_{k \in \mathcal{Z}} g_0[k]\; h_0[n-k] + (-1)^n \sum_{k \in \mathcal{Z}} g_0[k]\; h_0[n-k] \;=\; 2\delta[n],$$

or equivalently,

$$\sum_{k \in \mathcal{Z}} g_0[k]\; h_0[2n-k] \;=\; \delta[n],$$

since odd-indexed terms are cancelled. Written as an inner product

$$\langle g_0[k], h_0[2n-k] \rangle \;=\; \delta[n],$$

this is one of the biorthogonality relations

$$\langle \tilde{\varphi}_0[k], \varphi_{2n}[k] \rangle = \delta[n].$$

Similarly, starting from (3.2.15) or (3.2.16) and expressing $G_0(z)$ and $H_0(z)$ as a function of $G_1(z)$ and $H_1(z)$ would lead to the other biorthogonality relations, namely

$$\begin{aligned} \langle \tilde{\varphi}_1[k], \varphi_{2n+1}[k] \rangle &= \delta[n], \\ \langle \tilde{\varphi}_0[k], \varphi_{2n+1}[k] \rangle &= 0, \\ \langle \tilde{\varphi}_1[k], \varphi_{2n}[k] \rangle &= 0 \end{aligned}$$

Note that we obtained these relations for $\tilde{\varphi}_0$ and $\tilde{\varphi}_1$ but they hold also for $\tilde{\varphi}_{2l}$ and $\tilde{\varphi}_{2l+1}$, respectively. This shows once again that perfect reconstruction implies the biorthogonality conditions. The converse can be shown as well, demonstrating the equivalence of the two conditions.

Polyphase-Domain Analysis Although a very natural representation, modulation-domain analysis suffers from a drawback — it is redundant. Note how in $\boldsymbol{H}_m(z)$ every filter coefficient appears twice, since both the filter $H_i(z)$ and its modulated version $H_i(-z)$ are present. A more compact way of analyzing a filter bank uses polyphase-domain analysis, which was introduced in Section 2.5.3.

Thus, what we will do is decompose both signals and filters into their polyphase components and use (2.5.23) with $N = 2$ to express the output of filtering followed by downsampling. For convenience, we introduce matrix notation to express the two channel signals Y_0 and Y_1, or

$$\underbrace{\begin{pmatrix} Y_0(z) \\ Y_1(z) \end{pmatrix}}_{\boldsymbol{y}(z)} = \underbrace{\begin{pmatrix} H_{00}(z) & H_{01}(z) \\ H_{10}(z) & H_{11}(z) \end{pmatrix}}_{\boldsymbol{H}_p(z)} \underbrace{\begin{pmatrix} X_0(z) \\ X_1(z) \end{pmatrix}}_{\boldsymbol{x}_p(z)}, \tag{3.2.22}$$

where H_{ij} is the jth polyphase component of the ith filter, or, following (2.5.22–2.5.23),

$$H_i(z) = H_{i0}(z^2) + zH_{i1}(z^2).$$

In (3.2.22) $\boldsymbol{y}(z)$ contains the signals in the middle of the system in Figure 3.1(a). $\boldsymbol{H}_p(z)$ contains the polyphase components of the analysis filters, and is consequently denoted the *analysis polyphase matrix*, while $\boldsymbol{x}_p(z)$ contains the polyphase components of the input signal or, following (2.5.20),

$$X(z) = X_0(z^2) + z^{-1}X_1(z^2).$$

It is instructive to give a block diagram of (3.2.22) as shown in Figure 3.4(b). First, the input signal X is split into its polyphase components X_0 and X_1

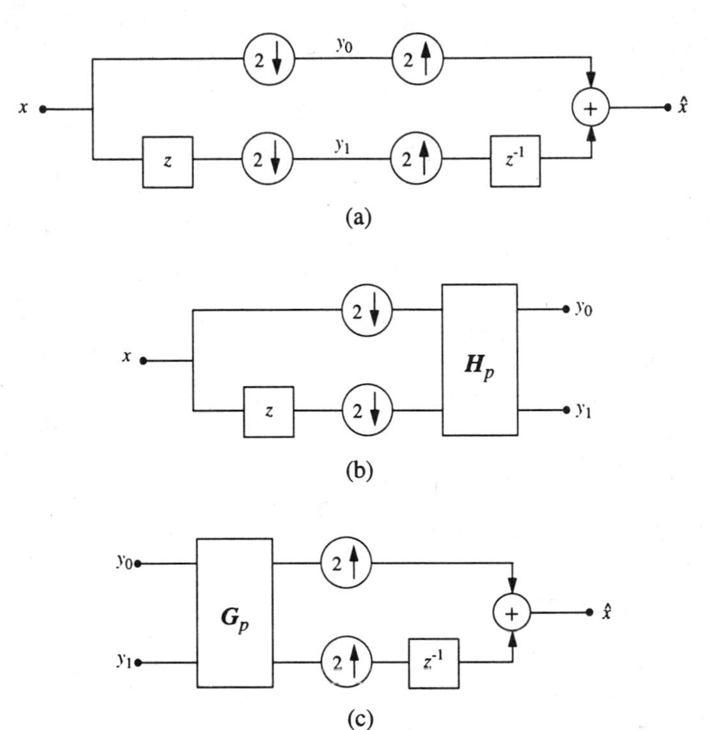

Figure 3.4 Polyphase-domain analysis. (a) Forward and inverse polyphase transform. (b) Analysis part in the polyphase domain. (c) Synthesis part in the polyphase domain.

using a forward polyphase transform. Then, a two-input, two-output system containing $\boldsymbol{H}_p(z)$ as transfer function matrix leads to the outputs y_0 and y_1.

The synthesis part of the system in Figure 3.1(a) can be analyzed in a similar fashion. It can be implemented with an inverse polyphase transform (as given on the right side of Figure 3.4(a)) preceded by a two-input two-output *synthesis polyphase matrix* $\boldsymbol{G}_p(z)$ defined by

$$\boldsymbol{G}_p(z) = \begin{pmatrix} G_{00}(z) & G_{10}(z) \\ G_{01}(z) & G_{11}(z) \end{pmatrix}, \tag{3.2.23}$$

where

$$G_i(z) = G_{i0}(z^2) + z^{-1} G_{i1}(z^2). \tag{3.2.24}$$

The synthesis filter polyphase components are defined such as those of the signal (2.5.20–2.5.21), or in reverse order of those of the analysis filters. In Figure 3.4(c), we show how the output signal is synthesized from the channel signals Y_0 and Y_1 as

$$\hat{X}(z) = (1 \quad z^{-1}) \underbrace{\begin{pmatrix} G_{00}(z^2) & G_{10}(z^2) \\ G_{01}(z^2) & G_{11}(z^2) \end{pmatrix}}_{\boldsymbol{G}_p(z^2)} \underbrace{\begin{pmatrix} Y_0(z^2) \\ Y_1(z^2) \end{pmatrix}}_{\boldsymbol{y}(z^2)}. \tag{3.2.25}$$

This equation reflects that the channel signals are first upsampled by 2 (leading to $Y_i(z^2)$) and then filtered by filters $G_i(z)$ which can be written as in (3.2.24). Note that the matrix-vector product in (3.2.25) is in z^2 and can thus be implemented before the upsampler by 2 (replacing z^2 by z) as shown in the figure.

Note the duality between the analysis and synthesis filter banks. The former uses a forward, the latter an inverse polyphase transform, and $\boldsymbol{G}_p(z)$ is a transpose of $\boldsymbol{H}_p(z)$. The phase reversal in the definition of the polyphase components in analysis and synthesis comes from the fact that z and z^{-1} are dual operators, or, on the unit circle, $e^{j\omega} = (e^{-j\omega})^*$.

Obviously the transfer function between the forward and inverse polyphase transforms defines the analysis/synthesis filter bank. This *transfer polyphase matrix* is given by

$$\boldsymbol{T}_p(z) = \boldsymbol{G}_p(z)\, \boldsymbol{H}_p(z).$$

In order to find the input-output relationship, we use (3.2.22) as input to (3.2.25), which yields

$$\begin{aligned} \hat{X}(z) &= (1 \quad z^{-1})\, \boldsymbol{G}_p(z^2)\, \boldsymbol{H}_p(z^2)\, \boldsymbol{x}_p(z^2), \\ &= (1 \quad z^{-1})\, \boldsymbol{T}_p(z^2)\, \boldsymbol{x}_p(z^2). \end{aligned} \tag{3.2.26}$$

Obviously, if $\boldsymbol{T}_p(z) = \boldsymbol{I}$, we have

$$\hat{X}(z) = (1 \quad z^{-1}) \begin{pmatrix} X_0(z^2) \\ X_1(z^2) \end{pmatrix} = X(z),$$

following (2.5.20), that is, the analysis/synthesis filter bank achieves perfect reconstruction with no delay and is equivalent to Figure 3.4(a).

Relationships Between Time, Modulation and Polyphase Representations Being different views of the same system, the representations discussed are related. A few useful formulas are given below. From (2.5.20), we can write

$$\begin{pmatrix} X_0(z^2) \\ X_1(z^2) \end{pmatrix} = \frac{1}{2} \begin{pmatrix} 1 & \\ & z \end{pmatrix} \begin{pmatrix} 1 & 1 \\ 1 & -1 \end{pmatrix} \begin{pmatrix} X(z) \\ X(-z) \end{pmatrix}, \tag{3.2.27}$$

thus relating polyphase and modulation representations of the signal, that is, $\boldsymbol{x}_p(z)$ and $\boldsymbol{x}_m(z)$. For the analysis filter bank, we have that

$$\begin{pmatrix} H_{00}(z^2) & H_{01}(z^2) \\ H_{10}(z^2) & H_{11}(z^2) \end{pmatrix} = \frac{1}{2} \begin{pmatrix} H_0(z) & H_0(-z) \\ H_1(z) & H_1(-z) \end{pmatrix} \begin{pmatrix} 1 & 1 \\ 1 & -1 \end{pmatrix} \begin{pmatrix} 1 & \\ & z^{-1} \end{pmatrix}, \tag{3.2.28}$$

establishing the relationship between $\boldsymbol{H}_p(z)$ and $\boldsymbol{H}_m(z)$. Finally, following the definition of $\boldsymbol{G}_p(z)$ in (3.2.23) and similarly to (3.2.28) we have

$$\begin{pmatrix} G_{00}(z^2) & G_{10}(z^2) \\ G_{01}(z^2) & G_{11}(z^2) \end{pmatrix} = \frac{1}{2} \begin{pmatrix} 1 & \\ & z \end{pmatrix} \begin{pmatrix} 1 & 1 \\ 1 & -1 \end{pmatrix} \begin{pmatrix} G_0(z) & G_1(z) \\ G_0(-z) & G_1(-z) \end{pmatrix}, \quad (3.2.29)$$

which relates $\boldsymbol{G}_p(z)$ with $\boldsymbol{G}_m(z)$ defined as

$$\boldsymbol{G}_m(z) = \begin{pmatrix} G_0(z) & G_1(z) \\ G_0(-z) & G_1(-z) \end{pmatrix}.$$

Again, note that (3.2.28) is the transpose of (3.2.29), with a phase change in the diagonal matrix. The change from the polyphase to the modulation representation (and vice versa) involves not only a diagonal matrix with a delay (or phase factor), but also a sum and/or a difference operation (see the middle matrix in (3.2.27–3.2.29)). This is actually a size-2 Fourier transform, as will become clear in cases of higher dimension.

The relation between time domain and polyphase domain is most obvious for the synthesis filters g_i, since their impulse responses correspond to the first basis functions φ_i. Consider the time-domain synthesis matrix, and create a matrix $\boldsymbol{T}_s(z)$

$$\boldsymbol{T}_s(z) = \sum_{i=0}^{K'-1} \boldsymbol{S}_i \, z^{-i},$$

where $\boldsymbol{S}_i$ are the successive 2×2 blocks along a column of the block Toeplitz matrix (there are K' of them for length $2K'$ filters), or

$$\boldsymbol{S}_i = \begin{pmatrix} g_0[2i] & g_1[2i] \\ g_0[2i+1] & g_1[2i+1] \end{pmatrix}.$$

Then, by inspection, it can be seen that $\boldsymbol{T}_s(z)$ is identical to $\boldsymbol{G}_p(z)$. A similar relation holds between $\boldsymbol{H}_p(z)$ and the time-domain analysis matrix. It is a bit more involved since time reversal has to be taken into account, and is given by

$$\boldsymbol{T}_a(z) = z^{-K+1} \boldsymbol{H}_p(z^{-1}) \begin{pmatrix} 0 & 1 \\ z^{-1} & 0 \end{pmatrix},$$

where

$$\boldsymbol{T}_a(z) = \sum_{i=0}^{K-1} \boldsymbol{A}_i \, z^{-i},$$

and

$$\boldsymbol{A}_i = \begin{pmatrix} h_0[2(K-i)-1] & h_0[2(K-i)-2] \\ h_1[2(K-i)-1] & h_1[2(K-i)-2] \end{pmatrix},$$

K being the number of 2×2 blocks in a row of the block Toeplitz matrix. The above relations can be used to establish equivalences between results in the various representations (see also Theorem 3.7 below).

3.2.2 Results on Filter Banks

We now use the tools just established to review several classic results from the filter bank literature. These have a slightly different flavor than the expansion results which are concerned with the existence of orthogonal or biorthogonal bases. Here, approximate reconstruction is considered, and issues of realizability of the filters involved are very important.

In the filter bank language, *perfect reconstruction* means that the output is a delayed and possibly scaled version of the input,

$$\hat{X}(z) = cz^{-k}X(z).$$

This is equivalent to saying that, up to a shift and scale, the impulse responses of the analysis filters (with time reversal) and of the synthesis filters form a biorthogonal basis.

Among approximate reconstructions, the most important one is *alias-free* reconstruction. Remember that because of the periodic time-variance of analysis/synthesis filter banks, the output is both a function of $x[n]$ and its modulated version $(-1)^n x[n]$, or $X(z)$ and $X(-z)$ in the z-transform domain. The aliased component $X(-z)$ can be very disturbing in applications and thus cancellation of aliasing is of prime importance. In particular, aliasing represents a nonharmonic distortion (new sinusoidal components appear which are not harmonically related to the input) and this is particularly disturbing in audio applications.

What follows now, are results on alias cancellation and perfect reconstruction for the two-channel case. Note that all the results are valid for a general, N-channel case as well (substitute N for 2 in statements and proofs).

For the first result, we need to introduce *pseudocirculant matrices* [311]. These are $N \times N$ circulant matrices with elements $F_{ij}(z)$, except that the lower triangular elements are multiplied by z, that is

$$F_{ij}(z) = \begin{cases} F_{0,j-i}(z) & j \geq i, \\ z \cdot F_{0,N+j-i}(z) & j < i. \end{cases}$$

Then, the following holds:

PROPOSITION 3.4

> Aliasing in a one-dimensional subband coding system will be cancelled *if and only if* the transfer polyphase matrix $\boldsymbol{T}_p$ is pseudocirculant [311].

PROOF

Consider a 2×2 pseudocirculant matrix

$$\boldsymbol{T}_p(z) = \begin{pmatrix} F_0(z) & F_1(z) \\ zF_1(z) & F_0(z) \end{pmatrix},$$

and substitute it into (3.2.26)

$$\hat{X}(z) = (1 \quad z^{-1})\, \boldsymbol{T}_p(z^2) \begin{pmatrix} X_0(z^2) \\ X_1(z^2) \end{pmatrix},$$

yielding (use $F(z) = F_0(z^2) + zF_1(z^2)$)

$$\begin{aligned} \hat{X}(z) &= (F(z) \quad z^{-1}F(z)) \cdot \begin{pmatrix} X_0(z^2) \\ X_1(z^2) \end{pmatrix}, \\ &= F(z) \cdot (X_0(z^2) + z^{-1}X_1(z^2)), \\ &= F(z) \cdot X(z), \end{aligned}$$

that is, it results in a time-invariant system or aliasing is cancelled. Given a time-invariant system, defined by a transfer function $F(z)$, it can be shown (see [311]) that its polyphase implementation is pseudocirculant.

A corollary to Proposition 3.4, is that for perfect reconstruction, the transfer function matrix has to be a pseudocirculant delay, that is, for an even delay $2k$

$$\boldsymbol{T}_p(z) = z^{-k} \begin{pmatrix} 1 & 0 \\ 0 & 1 \end{pmatrix},$$

while for an odd delay $2k+1$

$$\boldsymbol{T}_p(z) = z^{-k-1} \begin{pmatrix} 0 & 1 \\ z & 0 \end{pmatrix}.$$

The next result indicates when aliasing can be cancelled for a given analysis filter bank. Since the analysis and synthesis filter banks play dual roles, the result that we will discuss holds for synthesis filter banks as well.

PROPOSITION 3.5

Given a two-channel filter bank downsampled by 2 with the polyphase matrix $\boldsymbol{H}_p(z)$, then alias-free reconstruction is possible *if and only if* the determinant of $\boldsymbol{H}_p(z)$ is not identically zero, that is, $\boldsymbol{H}_p(z)$ has normal rank 2.

PROOF

Choose the synthesis matrix as

$$\boldsymbol{G}_p(z) = \text{cofactor}\,(\boldsymbol{H}_p(z)),$$

resulting in

$$\boldsymbol{T}_p(z) = \boldsymbol{G}_p(z)\, \boldsymbol{H}_p(z) = \det\,(\boldsymbol{H}_p(z)) \cdot \boldsymbol{I}$$

which is pseudocirculant, and thus cancels aliasing. If, on the other hand, the system is alias-free, then we know (see Proposition 3.4) that $\boldsymbol{T}_p(z)$ is pseudocirculant and therefore has full rank 2. Since the rank of a matrix product is bounded above by the ranks of its terms, $\boldsymbol{H}_p(z)$ has rank 2.[4]

[4]Note that we excluded the case of zero reconstruction, even if technically it is also aliasing free (but of zero interest!).

Often, one is interested in perfect reconstruction filter banks where all filters involved have a finite impulse response (FIR). Again, analysis and synthesis filter banks play the same role.

PROPOSITION 3.6

Given a critically sampled FIR analysis filter bank, perfect reconstruction with FIR filters is possible *if and only if* $\det(\boldsymbol{H}_p(z))$ is a pure delay.

PROOF

Suppose that the determinant of $\boldsymbol{H}_p(z)$ is a pure delay, and choose

$$\boldsymbol{G}_p(z) \;=\; \text{cofactor}\,(\boldsymbol{H}_p(z))\,.$$

It is obvious that the above choice leads to perfect reconstruction with FIR filters. Suppose, on the other hand, that we have perfect reconstruction with FIR filters. Then, $\boldsymbol{T}_p(z)$ has to be a pseudocirculant shift (corollary below Proposition 3.4), or

$$\det(\boldsymbol{T}_p(z)) \;=\; \det(\boldsymbol{G}_p(z)) \cdot \det(\boldsymbol{H}_p(z)) \;=\; z^{-l},$$

meaning that it has l poles at $z=0$. Since the synthesis has to be FIR as well, $\det(\boldsymbol{G}_p(z))$ has only zeros (or poles at the origin). Therefore, $\det(\boldsymbol{H}_p(z))$ cannot have any zeros (except possibly at the origin or ∞).

If $\det(\boldsymbol{H}_p(z))$ has no zeros, neither does $\det(\boldsymbol{H}_m(z))$ (because of (3.2.28) and assuming FIR filters). Since $\det(\boldsymbol{H}_m(z))$ is an odd function of z, it is of the form

$$\det(\boldsymbol{H}_m(z)) \;=\; \alpha z^{-2k-1},$$

(typically, $\alpha = 2$) and following (3.2.18)

$$G_0(z) \;=\; \frac{2}{\alpha} z^{2k+1}\, H_1(-z), \tag{3.2.30}$$

$$G_1(z) \;=\; -\frac{2}{\alpha} z^{2k+1}\, H_0(-z). \tag{3.2.31}$$

These filters give perfect reconstruction with zero delay but they are noncausal if the analysis filters are causal. Multiplying them by z^{-2k-1} gives a causal version with perfect reconstruction and a delay of $2k+1$ samples (note that the shift can be arbitrary, since it only changes the overall delay).

In the above results, we used the polyphase decomposition of filter banks. All these results can be translated to the other representation as well. In particular, aliasing cancellation can be studied in the modulation domain. Then, a necessary and sufficient condition for alias cancellation is that (see (3.2.14))

$$(\, G_0(z) \quad G_1(z)\,) \cdot \boldsymbol{H}_m(z)$$

be a row-vector with only the first component different from zero. One could expand $(G_0(z) \quad G_1(z))$ into a matrix $\boldsymbol{G}_m(z)$ by modulation, that is

$$\boldsymbol{G}_m(z) = \begin{pmatrix} G_0(z) & G_1(z) \\ G_0(-z) & G_1(-z) \end{pmatrix}. \tag{3.2.32}$$

It is easy to see then that for the system to be alias-free

$$\boldsymbol{T}_m(z) = \boldsymbol{G}_m(z)\, \boldsymbol{H}_m(z) = \begin{pmatrix} F(z) & \\ & F(-z) \end{pmatrix}.$$

The matrix $\boldsymbol{T}_m(z)$ is sometimes called the *aliasing cancellation matrix* [272].

Let us for a moment return to (3.2.14). As we said, $X(-z)$ is the aliased version of the signal. A necessary and sufficient condition for aliasing cancellation is that

$$G_0(z)\, H_0(-z) + G_1(z)\, H_1(-z) = 0. \tag{3.2.33}$$

The solution proposed by Croisier, Esteban, Galand [69] is known under the name QMF (quadrature mirror filters), which cancels aliasing in a two-channel filter bank:

$$H_1(z) = H_0(-z), \tag{3.2.34}$$

$$G_0(z) = H_0(z),$$

$$G_1(z) = -H_1(z) = -H_0(-z). \tag{3.2.35}$$

Substituting the above into (3.2.33) leads to $H_0(z)H_0(-z) - H_0(-z)H_0(z) = 0$, and aliasing is indeed cancelled. In order to achieve perfect reconstruction, the following has to be satisfied:

$$G_0(z)\, H_0(z) + G_1(z)\, H_1(z) = 2z^{-l}. \tag{3.2.36}$$

For the QMF solution, (3.2.36) becomes

$$H_0^2(z) - H_0^2(-z) = 2z^{-l}. \tag{3.2.37}$$

Note that the left side is an odd function of z, and thus, l has to be odd. The above relation explains the name QMF. On the unit circle $H_0(-z) = H(e^{j(\omega+\pi)})$ is the mirror image of $H_0(z)$ and both the filter and its mirror image are squared. For FIR filters, the condition (3.2.37) cannot be satisfied exactly except for the Haar filters introduced in Section 3.1. Taking a causal Haar filter, or $H_0(z) = (1 + z^{-1})/\sqrt{2}$, (3.2.37) becomes

$$\frac{1}{2}(1 + 2z^{-1} + z^{-2}) - \frac{1}{2}(1 - 2z^{-1} + z^{-2}) = 2z^{-1}.$$

For larger, linear phase filters, (3.2.37) can only be approximated (see Section 3.2.4).

Summary of Biorthogonality Relations Let us summarize our findings on biorthogonal filter banks.

THEOREM 3.7

In a two-channel, biorthogonal, real-coefficient filter bank, the following are equivalent:

(a) $\langle h_i[-n], g_j[n-2m]\rangle = \delta[i-j]\delta[m],\ i = 0,1.$

(b) $G_0(z)H_0(z) + G_1(z)H_1(z) = 2$, and $G_0(z)H_0(-z) + G_1(z)H_1(-z) = 0.$

(c) $\boldsymbol{T}_s \cdot \boldsymbol{T}_a = \boldsymbol{T}_a \cdot \boldsymbol{T}_s = \boldsymbol{I}.$

(d) $\boldsymbol{G}_m(z)\boldsymbol{H}_m(z) = \boldsymbol{H}_m(z)\boldsymbol{G}_m(z) = 2\boldsymbol{I}.$

(e) $\boldsymbol{G}_p(z)\boldsymbol{H}_p(z) = \boldsymbol{H}_p(z)\boldsymbol{G}_p(z) = \boldsymbol{I}.$

The proof follows from the equivalences between the various representations introduced in this section and is left as an exercise (see Problem 3.4). Note that we are assuming a critically sampled filter bank. Thus, the matrices in points (c)–(e) are square, and left inverses are also right inverses.

3.2.3 Analysis and Design of Orthogonal FIR Filter Banks

Assume now that we impose two constraints on our filter bank: First, it should implement an orthonormal expansion[5] of discrete-time signals and second, the filters used should be FIR.

Let us first concentrate on the orthonormality requirement. We saw in the Haar and sinc cases (both orthonormal expansions), that the expansion was of the form

$$x[n] = \sum_{k\in\mathcal{Z}} \langle \varphi_k[l], x[l]\rangle\ \varphi_k[n] = \sum_{k\in\mathcal{Z}} X[k]\ \varphi_k[n], \tag{3.2.38}$$

with the basis functions being

$$\varphi_{2k}[n] = h_0[2k-n] = g_0[n-2k], \tag{3.2.39}$$

$$\varphi_{2k+1}[n] = h_1[2k-n] = g_1[n-2k], \tag{3.2.40}$$

or, the even shifts of synthesis filters (even shifts of time-reversed analysis filters). We will show here that (3.2.38–3.2.40) describe orthonormal expansions, in the general case.

[5]The term *orthogonal* is often used, especially for the associated filters or filter banks. For filter banks, the term *unitary* or *paraunitary* is also often used, as well as the notion of *losslessness* (see Appendix 3.A).

Orthonormality in Time Domain Start with a general filter bank as given in Figure 3.1(a). Impose orthonormality on the expansion, that is, the dual basis $\{\tilde{\varphi}_k[n]\}$ becomes identical to $\{\varphi_k[n]\}$. In filter bank terms, the dual basis — synthesis filters — now becomes

$$\{g_0[n-2k], g_1[n-2k]\} = \{\tilde{\varphi}_k[n]\} = \{\varphi_k[n]\} = \{h_0[2k-n], h_1[2k-n]\}, \quad (3.2.41)$$

or,

$$g_i[n] = h_i[-n], \qquad i = 0, 1. \quad (3.2.42)$$

Thus, we have encountered the first important consequence of orthonormality: The synthesis filters are the time-reversed versions of the analysis filters. Also, since (3.2.41) holds and φ_k is an orthonormal set, the following are the orthogonality relations for the synthesis filters:

$$\langle g_i[n-2k], g_j[n-2l]\rangle = \delta[i-j]\,\delta[k-l], \quad (3.2.43)$$

with a similar relation for the analysis filters. We call this an *orthonormal filter bank*.

Let us now see how orthonormality can be expressed using matrix notation. First, substituting the expression for $g_i[n]$ given by (3.2.42) into the synthesis matrix $\boldsymbol{T}_s$ given in (3.2.7), we see that

$$\boldsymbol{T}_s = \boldsymbol{T}_a^T,$$

or, the perfect reconstruction condition is

$$\boldsymbol{T}_s\,\boldsymbol{T}_a = \boldsymbol{T}_a^T\,\boldsymbol{T}_a = \boldsymbol{I}. \quad (3.2.44)$$

That is, the above condition means that the matrix $\boldsymbol{T}_a$ is unitary. Because it is full rank, the product commutes and we have also $\boldsymbol{T}_a\,\boldsymbol{T}_a^T = \boldsymbol{I}$. Thus, having an orthonormal basis, or perfect reconstruction with an orthonormal filter bank, is equivalent to the analysis matrix $\boldsymbol{T}_a$ being unitary.

If we separate the outputs now as was done in (3.2.9), and note that

$$\boldsymbol{G}_i = \boldsymbol{H}_i^T,$$

then the following is obtained from (3.2.43):

$$\boldsymbol{H}_i\,\boldsymbol{H}_j^T = \delta[i-j]\,\boldsymbol{I}, \qquad i, j = 0, 1.$$

Now, the output of one channel in Figure 3.1(a) (filtering, downsampling, upsampling and filtering) is equal to

$$\boldsymbol{M}_i = \boldsymbol{H}_i^T\,\boldsymbol{H}_i.$$

It is easy to verify that $\boldsymbol{M}_i$ satisfies the requirements for an orthogonal projection (see Appendix 2.A) since $\boldsymbol{M}_i^T = \boldsymbol{M}_i$ and $\boldsymbol{M}_i^2 = \boldsymbol{M}_i$. Thus, the two channels of the filter bank correspond to orthogonal projections onto spaces spanned by their respective impulse responses, and perfect reconstruction can be written as the direct sum of the projections

$$\boldsymbol{H}_0^T \boldsymbol{H}_0 + \boldsymbol{H}_1^T \boldsymbol{H}_1 = \boldsymbol{I}.$$

Note also, that sometimes in order to visualize the action of the matrix $\boldsymbol{T}_a$, it is expressed in terms of 2×2 blocks $\boldsymbol{A}_i$ (see (3.2.2–3.2.3)), which can also be used to express orthonormality as follows (see (3.2.44)):

$$\sum_{i=0}^{K-1} \boldsymbol{A}_i^T \boldsymbol{A}_i = \boldsymbol{I},$$

$$\sum_{i=0}^{K-1} \boldsymbol{A}_{i+j}^T \boldsymbol{A}_i = \boldsymbol{0}, \qquad j = 1, \ldots, K-1.$$

Orthonormality in Modulation Domain To see how orthonormality translates in the modulation domain, consider (3.2.43) and $i = j = 0$. Substitute $n' = n - 2k$. Thus, we have

$$\langle g_0[n'], g_0[n' + 2(k-l)] \rangle = \delta[k-l],$$

or

$$\langle g_0[n], g_0[n + 2m] \rangle = \delta[m]. \tag{3.2.45}$$

Recall that $p[l] = \langle g_0[n], g_0[n+l] \rangle$ is the autocorrelation of the sequence $g_0[n]$ (see Section 2.5.2). Then, (3.2.45) is simply the autocorrelation of $g_0[n]$ evaluated at even indexes $l = 2m$, or $p[l]$ downsampled by 2, that is, $p'[m] = p[2m]$. The z-transform of $p'[m]$ is (see Section 2.5.3)

$$P'(z) = \frac{1}{2}[P(z^{1/2}) + P(-z^{1/2})].$$

Replacing z by z^2 (for notational convenience) and recalling that the z-transform of the autocorrelation of $g_0[n]$ is given by $P(z) = G_0(z) \cdot G_0(z^{-1})$, the z-transform of (3.2.45) becomes

$$G_0(z)\, G_0(z^{-1}) + G_0(-z)\, G_0(-z^{-1}) = 2. \tag{3.2.46}$$

Using the same arguments for the other cases in (3.2.43), we also have that

$$G_1(z)\, G_1(z^{-1}) + G_1(-z)\, G_1(-z^{-1}) = 2, \tag{3.2.47}$$

$$G_0(z)\, G_1(z^{-1}) + G_0(-z)\, G_1(-z^{-1}) = 0. \tag{3.2.48}$$

On the unit circle, (3.2.46–3.2.47) become (use $G(e^{-j\omega}) = G^*(e^{j\omega})$ since the filter has real coefficients)

$$|G_i(e^{j\omega})|^2 + |G_i(e^{j(\omega+\pi)})|^2 = 2, \tag{3.2.49}$$

that is, the filter and its modulated version are *power complementary* (their magnitudes squared sum up to a constant). Since this condition was used in [270] for designing the first orthogonal filter banks, it is also called the *Smith-Barnwell condition.* Writing (3.2.46–3.2.48) in matrix form,

$$\begin{pmatrix} G_0(z^{-1}) & G_0(-z^{-1}) \\ G_1(z^{-1}) & G_1(-z^{-1}) \end{pmatrix} \begin{pmatrix} G_0(z) & G_1(z) \\ G_0(-z) & G_1(-z) \end{pmatrix} = \begin{pmatrix} 2 & 0 \\ 0 & 2 \end{pmatrix}, \tag{3.2.50}$$

that is, using the synthesis modulation matrix $\boldsymbol{G}_m(z)$ (see (3.2.32))

$$\boldsymbol{G}_m^T(z^{-1})\, \boldsymbol{G}_m(z) = 2\boldsymbol{I}. \tag{3.2.51}$$

Since g_i and h_i are identical up to time reversal, a similar relation holds for the analysis modulation matrix $\boldsymbol{H}_m(z)$ (up to a transpose), or $\boldsymbol{H}_m(z^{-1})\, \boldsymbol{H}_m^T(z) = 2\boldsymbol{I}$.

A matrix satisfying (3.2.51) is called *paraunitary* (note that we have assumed that the filter coefficients are real). If all its entries are stable (which they are in this case, since we assumed the filters to be FIR), then such a matrix is called *lossless*. The concept of losslessness comes from classical circuit theory [23, 308] and is discussed in more detail in Appendix 3.A. It suffices to say at this point that having a lossless transfer matrix is equivalent to the filter bank implementing an orthogonal transform. Concentrating on lossless modulation matrices, we can continue our analysis of orthogonal systems in the modulation domain. First, from (3.2.50) we can see that $(\, G_1(z^{-1}) \quad G_1(-z^{-1})\,)^T$ has to be orthogonal to $(\, G_0(z) \quad G_0(-z)\,)^T$. It will be proven in Appendix 3.A (although in polyphase domain), that this implies that the two filters $G_0(z)$ and $G_1(z)$ are related as follows:

$$G_1(z) = -z^{-2K+1}\, G_0(-z^{-1}), \tag{3.2.52}$$

or, in time domain

$$g_1[n] = (-1)^n g_0[2K-1-n].$$

Equation (3.2.52) therefore establishes an important property of an orthogonal system: *In an orthogonal two-channel filter bank, all filters are obtained from a single prototype filter.*

This single prototype filter has to satisfy the power complementary property given by (3.2.49). For filter design purposes, one can use (3.2.46) and design an autocorrelation function $P(z)$ that satisfies $P(z) + P(-z) = 2$ as will be shown below. This special form of the autocorrelation function can be used to prove that the filters in an orthogonal FIR filter bank have to be of even length (Problem 3.5).

Orthonormality in Polyphase Domain We have seen that the polyphase and modulation matrices are related as in (3.2.29). Since $\boldsymbol{G}_m$ and $\boldsymbol{G}_p$ are related by unitary operations, $\boldsymbol{G}_p$ will be lossless *if and only if* $\boldsymbol{G}_m$ is lossless. Thus, one can search or examine an orthonormal system in either modulation, or polyphase domain, since

$$\boldsymbol{G}_p^T(z^{-2})\, \boldsymbol{G}_p(z^2) = \frac{1}{4}\, \boldsymbol{G}_m^T(z^{-1}) \begin{pmatrix} 1 & 1 \\ 1 & -1 \end{pmatrix} \begin{pmatrix} 1 & 0 \\ 0 & z^{-1} \end{pmatrix}$$

$$\times \begin{pmatrix} 1 & 0 \\ 0 & z \end{pmatrix} \begin{pmatrix} 1 & 1 \\ 1 & -1 \end{pmatrix} \boldsymbol{G}_m(z)$$
$$= \frac{1}{2} \boldsymbol{G}_m^T(z^{-1})\, \boldsymbol{G}_m(z) \;=\; \boldsymbol{I}, \tag{3.2.53}$$

where we used (3.2.51). Since (3.2.53) also implies $\boldsymbol{G}_p(z)\, \boldsymbol{G}_p^T(z^{-1}) = \boldsymbol{I}$ (left inverse is also right inverse), it is clear that given a paraunitary $\boldsymbol{G}_p(z)$ corresponding to an orthogonal synthesis filter bank, we can choose the analysis filter bank with a polyphase matrix $\boldsymbol{H}_p(z) = \boldsymbol{G}_p^T(z^{-1})$ and get perfect reconstruction with no delay.

Summary of Orthonormality Relations Let us summarize our findings so far.

THEOREM 3.8

In a two-channel, orthonormal, FIR, real-coefficient filter bank, the following are equivalent:

(a) $\langle g_i[n], g_j[n+2m] \rangle \;=\; \delta[i-j]\, \delta[m], \quad i = 0, 1.$

(b) $G_0(z)\, G_0(z^{-1}) + G_0(-z)\, G_0(-z^{-1}) \;=\; 2,$
and $G_1(z) \;=\; -z^{-2K+1}\, G_0(-z^{-1}), \quad k \in \mathcal{Z}.$

(c) $\boldsymbol{T}_s^T\, \boldsymbol{T}_s \;=\; \boldsymbol{T}_s\, \boldsymbol{T}_s^T \;=\; \boldsymbol{I}, \quad \boldsymbol{T}_a \;=\; \boldsymbol{T}_s^T.$

(d) $\boldsymbol{G}_m^T(z^{-1})\, \boldsymbol{G}_m(z) \;=\; \boldsymbol{G}_m(z)\, \boldsymbol{G}_m^T(z^{-1}) \;=\; 2\boldsymbol{I}, \quad \boldsymbol{H}_m(z) \;=\; \boldsymbol{G}_m^T(z^{-1}).$

(e) $\boldsymbol{G}_p^T(z^{-1})\, \boldsymbol{G}_p(z) \;=\; \boldsymbol{G}_p(z)\, \boldsymbol{G}_p^T(z^{-1}) \;=\; \boldsymbol{I}, \quad \boldsymbol{H}_p(z) \;=\; \boldsymbol{G}_p^T(z^{-1}).$

Again, we used the fact that the left inverse is also the right inverse in a square matrix in relations (c), (d) and (e). The proof follows from the relations between the various representations, and is left as an exercise (see Problem 3.7). Note that the theorem holds in more general cases as well. In particular, the filters do not have to be restricted to be FIR, and if their coefficients are complex valued, transposes have to be hermitian transposes (in the case of $\boldsymbol{G}_m$ and $\boldsymbol{G}_p$, only the coefficients of the filters have to be conjugated, not z since z^{-1} plays that role).

Because all filters are related to a single prototype satisfying (a) or (b), the other filter in the synthesis filter bank follows by modulation, time reversal and an odd shift (see (3.2.52)). The filters in the analysis are simply time-reversed versions of the synthesis filters. In the FIR case, the length of the filters is even. Let us formalize these statements:

COROLLARY 3.9

In a two-channel, orthonormal, FIR, real-coefficient filter bank, the following hold:

(a) The filter length L is even, or $L = 2K$.

(b) The filters satisfy the power complementary or Smith-Barnwell condition.

$$|G_0(e^{j\omega})|^2 + |G_0(e^{j(\omega+\pi)})|^2 = 2, \quad |G_0(e^{j\omega})|^2 + |G_1(e^{j\omega})|^2 = 2. \quad (3.2.54)$$

(c) The highpass filter is specified (up to an even shift and a sign change) by the lowpass filter as

$$G_1(z) = -z^{-2K+1} \, G_0(z^{-1}).$$

(d) If the lowpass filter has a zero at π, that is, $G_0(-1) = 0$, then

$$G_0(1) = \sqrt{2}. \quad (3.2.55)$$

Also, an orthogonal filter bank has, as any orthogonal transform, an energy conservation property:

PROPOSITION 3.10

In an orthonormal filter bank, that is, a filter bank with a unitary polyphase or modulation matrix, the energy is conserved between the input and the channel signals,

$$\|x\|^2 = \|y_0\|^2 + \|y_1\|^2. \quad (3.2.56)$$

PROOF

The energy of the subband signals equals

$$\|y_0\|^2 + \|y_1\|^2 = \frac{1}{2\pi} \int_0^{2\pi} \left(|\boldsymbol{Y}_0(e^{j\omega})|^2 + |\boldsymbol{Y}_1(e^{j\omega})|^2 \right) d\omega,$$

by Parseval's relation (2.4.37). Using the fact that $\boldsymbol{y}(z) = \boldsymbol{H}_p(z) \, \boldsymbol{x}_p(z)$, the right side can be written as,

$$\begin{aligned} \frac{1}{2\pi} \int_0^{2\pi} \left[\boldsymbol{y}(e^{j\omega})\right]^* \cdot \boldsymbol{y}(e^{j\omega}) d\omega &= \frac{1}{2\pi} \int_0^{2\pi} \left[\boldsymbol{x}_p(e^{j\omega})\right]^* \left[\boldsymbol{H}_p(e^{j\omega})\right]^* \\ &\quad \times \boldsymbol{H}_p(e^{j\omega}) \, \boldsymbol{x}_p(e^{j\omega}) \, d\omega, \\ &= \frac{1}{2\pi} \int_0^{2\pi} \left[\boldsymbol{x}_p(e^{j\omega})\right]^* \, \boldsymbol{x}_p(e^{j\omega}) \, d\omega, \\ &= \|x_0\|^2 + \|x_1\|^2. \end{aligned}$$

We used the fact that $\boldsymbol{H}_p(e^{j\omega})$ is unitary and Parseval's relation. Finally, (3.2.56) follows from the fact that the energy of the signal is equal to the sum of the polyphase components' energy, $\|x\|^2 = \|x_0\|^2 + \|x_1\|^2$.

Designing Orthogonal Filter Banks Now, we give two design procedures: the first, based on spectral factorization, and the second, based on lattice structures. Let us just note that most of the methods in the literature design analysis filters. We will give designs for synthesis filters so as to be consistent with our approach; however, analysis filters are easily obtained by time reversing the synthesis ones.

Designs Based on Spectral Factorizations The first solution we will show is due to Smith and Barnwell [271]. The approach here is to find an autocorrelation sequence $P(z) = G_0(z)G_0(z^{-1})$ that satisfies (3.2.46) and then to perform spectral factorization as explained in Section 2.5.2. However, factorization becomes numerically ill-conditioned as the filter size grows, and thus, the resulting filters are usually only approximately orthogonal.

Example 3.1

Choose $p[n]$ as a windowed version of a perfect half-band lowpass filter,

$$p[n] = \begin{cases} w[n]\frac{\sin(\pi/2n)}{\pi/2\cdot n} & n = -2K+1, \ldots, 2K-1, \\ 0 & \text{otherwise.} \end{cases}$$

where $w[n]$ is a symmetric window function with $w[0] = 1$. Because $p[2n] = \delta[n]$, the z-transform of $p[n]$ satisfies

$$P(z) + P(-z) = 2. \tag{3.2.57}$$

Also since $P(z)$ is an approximation to a half-band lowpass filter, its spectral factor will be such an approximation as well. Now, $P(e^{j\omega})$ might not be positive everywhere, in which case it is not an autocorrelation and has to be modified. The following trick can be used to find an autocorrelation sequence $p'[n]$ close to $p[n]$ [271]. Find the minimum of $P(e^{j\omega})$, $\delta_{min} = \min_\omega[P(e^{j\omega})]$. If $\delta_{min} > 0$, we need not do anything, otherwise, subtract it from $p[0]$ to get the sequence $p'[n]$. Now,

$$P'(e^{j\omega}) = P(e^{j\omega}) - \delta_{min} \geq 0,$$

and $P'(z)$ still satisfies (3.2.57) up to a scale factor $(1 - \delta_{min})$ which can be divided out.

An example of a design for $N = 8$ by Smith and Barnwell is given in Figure 3.5(a) (magnitude responses) and Table 3.2 (impulse response coefficients) [271].

Another example based on spectral factorization is Daubechies' family of maximally flat filters [71]. Daubechies' purpose was that the filters should lead to continuous-time wavelet bases (see Section 4.4). The design procedure then amounts to finding orthogonal lowpass filters with a large number of zeros at

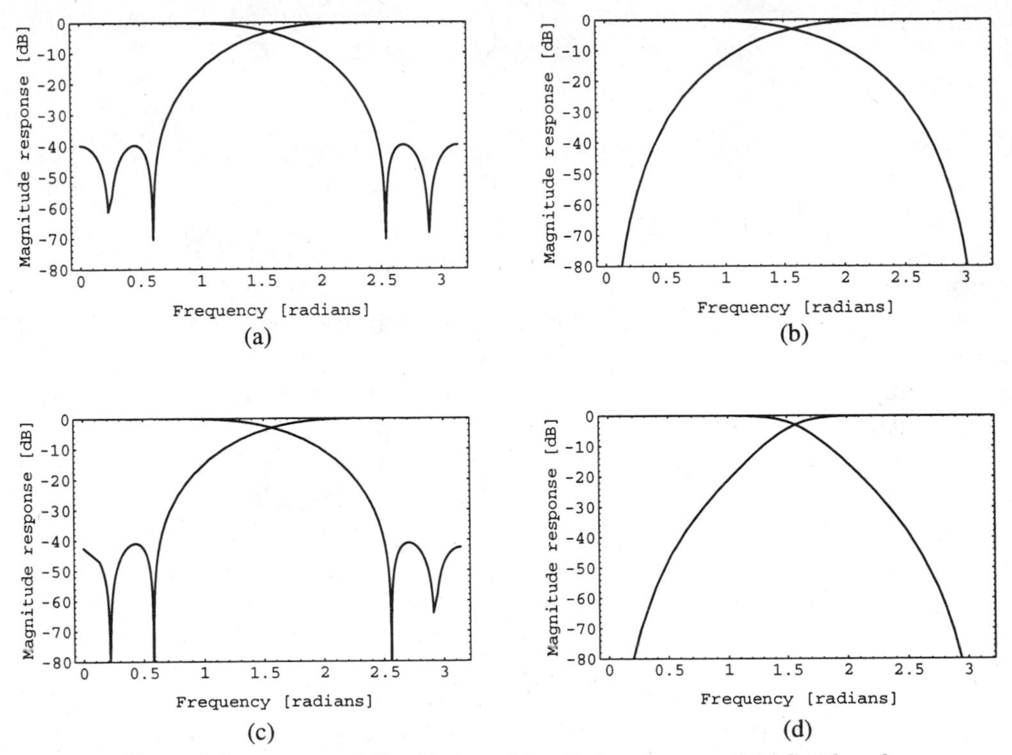

Figure 3.5 Orthogonal filter designs. Magnitude responses of: (a) Smith and Barnwell filter of length 8 [271], (b) Daubechies' filter of length 8 (D_4) [71], (c) Vaidyanathan and Hoang filter of length 8 [310], (d) Butterworth filter for $N = 4$ [133].

Table 3.2 Impulse response coefficients for Smith and Barnwell filter [271], Daubechies' filter D_4 [71] and Vaidyanathan and Hoang filter [310] (all of length 8).

n	Smith and Barnwell	Daubechies	Vaidyanathan and Hoang
0	0.04935260	0.23037781	0.27844300
1	-0.01553230	0.71484657	0.73454200
2	-0.08890390	0.63088076	0.58191000
3	0.31665300	-0.02798376	-0.05046140
4	0.78751500	-0.18703481	-0.19487100
5	0.50625500	0.03084138	0.03547370
6	-0.03380010	0.03288301	0.04692520
7	-0.10739700	-0.01059740	-0.01778800

$\omega = \pi$. Equivalently, one has to design an autocorrelation satisfying (3.2.46) and having many zeros at $\omega = \pi$. That is, we want

$$P(z) = (1+z^{-1})^k(1+z)^k R(z),$$

which satisfies (3.2.57), where $R(z)$ is symmetric ($R(z^{-1}) = R(z)$) and positive on the unit circle, $R(e^{j\omega}) \geq 0$. Of particular interest is the case when $R(z)$ is of minimal degree, which turns out to be when $R(z)$ has powers of z going from $(-k+1)$ to $(k-1)$. Once the solution to this constrained problem is found, a spectral factorization of $R(z)$ yields the desired filter $G_0(z)$, which has automatically k zeros at π. As always with spectral factorization, there is a choice of taking zeros either inside or outside the unit circle. Taking them systematically from inside the unit circle, leads to Daubechies' family of minimum-phase filters.

The function $R(z)$ which is required so that $P(z)$ satisfies (3.2.57) can be found by solving a system of linear equations or a closed form is possible in the minimum-degree case [71]. Let us indicate a straightforward approach leading to a system of linear equations. Assume the minimum-degree solution. Then $P(z)$ has powers of z going from $(-2k+1)$ to $(2k-1)$ and (3.2.57) puts $2k-1$ constraints on $P(z)$. But because $P(z)$ is symmetric, $k-1$ of them are redundant, leaving k active constraints. Because $R(z)$ is symmetric, it has k degrees of freedom (out of its $2k-1$ nonzero coefficients). Since $P(z)$ is the convolution of $(1+z^{-1})^k(1+z)^k$ with $R(z)$, it can be written as a matrix-vector product, where the matrix contains the impulse response of $(1+z^{-1})^k(1+z)^k$ and its shifts. Gathering the even terms of this matrix-vector product (which correspond to the k constraints) and expressing them in terms of the k free parameters of $R(z)$, leads to the desired $k \times k$ system of equation. It is interesting to note that the matrix involved is never singular, and the $R(z)$ obtained by solving the system of equations is positive on the unit circle. Therefore, this method automatically leads to an autocorrelation, and by spectral factorization, to an orthogonal filter bank with filters of length $2k$ having k zeros at π and 0 for the lowpass and highpass, respectively.

As an example, we will construct Daubechies' D_2 filter, that is, a length-4 orthogonal filter with two zeros at $\omega = \pi$ (the maximum number of zeros at π is equal to half the length, and indicated by the subscript).

Example 3.2

Let us choose $k = 2$ and construct length-4 filters. This means that

$$P(z) \;=\; G_0(z)G_0(z^{-1}) \;=\; (1+z^{-1})^2(1+z)^2 R(z).$$

Now, recall that since $P(z) + P(-z) = 2$, all even-indexed coefficients in $P(z)$ equal 0, except for $p[0] = 1$. To obtain a length-4 filter, the highest-degree term has to be z^{-3}, and thus $R(z)$ is of the form

$$R(z) = (az + b + az^{-1}). \qquad (3.2.58)$$

Substituting (3.2.58) into $P(z)$ we obtain

$$P(z) = az^3 + (4a+b)z^2 + (7a+4b)z + (8a+6b) + (4b+7a)z^{-1} + (b+4a)z^{-2} + az^{-3}.$$

Equating the coefficients of z^2 or z^{-2} with 0, and the one with z^0 with 1 yields

$$4a + b = 0, \quad 8a + 6b = 1.$$

The solution to this system of equations is

$$a = -\frac{1}{16}, \quad b = \frac{1}{4},$$

yielding the following $R(z)$:

$$R(z) = -\frac{1}{16}z + \frac{1}{4} - \frac{1}{16}z^{-1}.$$

We factor now $R(z)$ as

$$R(z) = \left(\frac{1}{4\sqrt{2}}\right)^2 (1+\sqrt{3}+(1-\sqrt{3})z^{-1})(1+\sqrt{3}+(1-\sqrt{3})z).$$

Taking the term with the zero inside the unit circle, that is $(1+\sqrt{3}+(1-\sqrt{3})z^{-1})$, we obtain the filter $G_0(z)$ as

$$\begin{aligned} G_0(z) &= \frac{1}{4\sqrt{2}}(1+z^{-1})^2(1+\sqrt{3}+(1-\sqrt{3})z^{-1}), \\ &= \frac{1}{4\sqrt{2}}((1+\sqrt{3}) \\ &\quad + (3+\sqrt{3})z^{-1} + (3-\sqrt{3})z^{-2} + (1-\sqrt{3})z^{-3}). \end{aligned} \qquad (3.2.59)$$

Note that this lowpass filter has a double zero at $z = -1$ (important for constructing wavelet bases, as will be seen in Section 4.4). A longer filter with four zeros at $\omega = \pi$ is shown in Figure 3.5(b) (magnitude responses of the lowpass/highpass pair) while the impulse response coefficients are given in Table 3.2 [71].

Designs Based on Vaidyanathan and Hoang Lattice Factorizations An alternative and numerically well-conditioned procedure relies on the fact that paraunitary, just like unitary matrices, possess canonical factorizations[6] into elementary paraunitary matrices [305, 310] (see also Appendix 3.A). Thus, all paraunitary filter banks with FIR filters of length $L = 2K$ can be reached by the following lattice structure (here $G_1(z) = -z^{-2K+1}G_0(-z^{-1})$):

$$\boldsymbol{G}_p(z) = \begin{pmatrix} G_{00}(z) & G_{10}(z) \\ G_{01}(z) & G_{11}(z) \end{pmatrix} = \boldsymbol{U}_0 \left[\prod_{i=1}^{K-1} \begin{pmatrix} 1 & \\ & z^{-1} \end{pmatrix} \boldsymbol{U}_i \right], \qquad (3.2.60)$$

[6]By canonical we mean complete factorizations with a minimum number of free parameters. However, such factorizations are not unique in general.

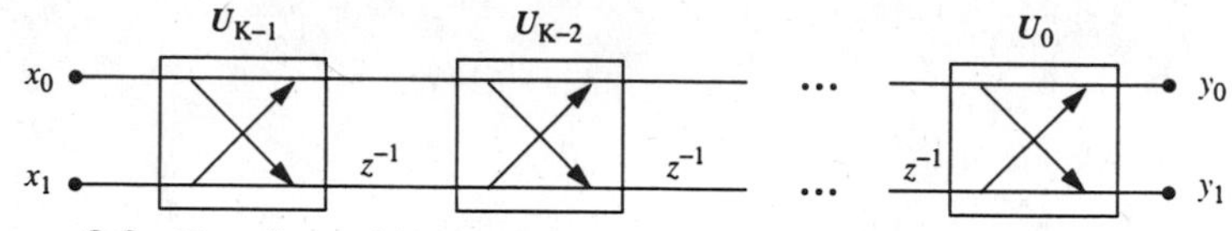

Figure 3.6 Two-channel lattice factorization of paraunitary filter banks. The 2×2 blocks $\boldsymbol{U}_i$ are rotation matrices.

where $\boldsymbol{U}_i$ is a 2×2 rotation matrix given in (2.B.1)

$$\boldsymbol{U}_i = \begin{pmatrix} \cos\alpha_i & -\sin\alpha_i \\ \sin\alpha_i & \cos\alpha_i \end{pmatrix}.$$

That the resulting structure is paraunitary is easy to check (it is the product of paraunitary elementary blocks). What is much more interesting is that all paraunitary matrices of a given degree can be written in this form [310] (see also Appendix 3.A.1). The lattice factorization is given in Figure 3.6.

As an example of this approach, we construct the D_2 filter from the previous example, using the lattice factorization.

Example 3.3

We construct the D_2 filter which is of length 4, thus $L = 2K = 4$. This means that

$$\begin{aligned} \boldsymbol{G}_p(z) &= \begin{pmatrix} \cos\alpha_0 & -\sin\alpha_0 \\ \sin\alpha_0 & \cos\alpha_0 \end{pmatrix} \begin{pmatrix} 1 & \\ & z^{-1} \end{pmatrix} \begin{pmatrix} \cos\alpha_1 & -\sin\alpha_1 \\ \sin\alpha_1 & \cos\alpha_1 \end{pmatrix}, \\ &= \begin{pmatrix} \cos\alpha_0\cos\alpha_1 - \sin\alpha_0\sin\alpha_1 z^{-1} & -\cos\alpha_0\sin\alpha_1 - \sin\alpha_0\cos\alpha_1 z^{-1} \\ \sin\alpha_0\cos\alpha_1 + \cos\alpha_0\sin\alpha_1 z^{-1} & -\sin\alpha_0\sin\alpha_1 + \cos\alpha_0\cos\alpha_1 z^{-1} \end{pmatrix}. \end{aligned} \tag{3.2.61}$$

We get the lowpass filter $G_0(z)$ as

$$\begin{aligned} G_0(z) &= G_{00}(z^2) + z^{-1}G_{01}(z^2), \\ &= \cos\alpha_0\cos\alpha_1 + \sin\alpha_0\cos\alpha_1 z^{-1} - \sin\alpha_0\sin\alpha_1 z^{-2} + \cos\alpha_0\sin\alpha_1 z^{-3}. \end{aligned}$$

We now obtain the D_2 filter by imposing a second-order zero at $z = -1$. So, we obtain the first equation as

$$G_0(-1) = \cos\alpha_1\cos\alpha_0 - \cos\alpha_1\sin\alpha_0 - \sin\alpha_1\sin\alpha_0 - \sin\alpha_1\cos\alpha_0 = 0,$$

or,

$$\cos(\alpha_0 + \alpha_1) - \sin(\alpha_0 + \alpha_1) = 0.$$

This equation implies that

$$\alpha_0 + \alpha_1 = k\pi + \frac{\pi}{4}.$$

Since we also know that $G_0(1) = \sqrt{2}$ (see (3.2.55)

$$\cos(\alpha_0 + \alpha_1) + \sin(\alpha_0 + \alpha_1) = \sqrt{2},$$

we get that

$$\alpha_0 + \alpha_1 = \frac{\pi}{4}. \tag{3.2.62}$$

Imposing now a zero at $e^{j\omega} = -1$ on the derivative of $G_0(e^{j\omega})$, we obtain

$$\left.\frac{dG_0(e^{j\omega})}{d\omega}\right|_{\omega=\pi} = \cos\alpha_1 \sin\alpha_0 + 2\sin\alpha_1 \sin\alpha_0 + 3\sin\alpha_1 \cos\alpha_0 = 0. \tag{3.2.63}$$

Solving (3.2.62) and (3.2.63), we obtain

$$\alpha_0 = \frac{\pi}{3}, \quad \alpha_1 = -\frac{\pi}{12}.$$

Substituting the angles α_0, α_1 into the expression for $G_0(z)$ (3.2.61) and comparing it to (3.2.59), we can see that we have indeed obtained the D_2 filter.

An example of a longer filter obtained by lattice factorization is given in Figure 3.5(c) (magnitude responses) and Table 3.2 (impulse response coefficients). This design example was obtained by Vaidyanathan and Hoang in [310].

3.2.4 Linear Phase FIR Filter Banks

Orthogonal filter banks have many nice features (conservation of energy, identical analysis and synthesis) but also some restrictions. In particular, there are no orthogonal linear phase solutions with real FIR filters (see Proposition 3.12) except in some trivial cases (such as the Haar filters). Since linear phase filter banks yield biorthogonal expansions, four filters are involved, namely H_0, H_1 at analysis, and G_0 and G_1 at synthesis. In our discussions, we will often concentrate on H_0 and H_1 first (that is, in this case we design the analysis part of the system, or, one of the two biorthogonal bases).

First, note that if a filter is linear phase, then it can be written as

$$H(z) = \pm z^{-L+1}\, H(z^{-1}), \tag{3.2.64}$$

where $\pm$ will mean it is a symmetric/antisymmetric filter, respectively, and L denotes the filter's length. Note that here we have assumed that $H(z)$ has the impulse response ranging from $h[0], \ldots, h[L-1]$ (otherwise, modify (3.2.64) with a phase factor). Recall from Proposition 3.6 that perfect reconstruction FIR solutions are possible *if and only if* the matrix $\boldsymbol{H}_p(z)$ (or equivalently $\boldsymbol{H}_m(z)$) has a determinant equal to a delay, that is [319]

$$H_{00}(z)\, H_{11}(z) - H_{01}(z)\, H_{10}(z) = z^{-l}, \tag{3.2.65}$$

$$H_0(z)\, H_1(-z) - H_0(-z)\, H_1(z) = 2z^{-2l-1}. \tag{3.2.66}$$

The right-hand side of (3.2.65) is the determinant of the polyphase matrix $\boldsymbol{H}_p(z)$, while the right-hand side of (3.2.66) is the determinant of the modulation matrix $\boldsymbol{H}_m(z)$. The synthesis filters are then equal to (see (3.2.30–3.2.31))

$$G_0(z) = z^{-k} H_1(-z), \quad G_1(z) = -z^{-k} H_0(-z),$$

where k is an arbitrary shift.

Of particular interest is the case when both $H_0(z)$ and $H_1(z)$ are linear phase (symmetric or antisymmetric) filters. Then, as in the paraunitary case, there are certain restrictions on possible filters [315, 319].

PROPOSITION 3.11

In a two-channel, perfect reconstruction filter bank, where all filters are linear phase, the analysis filters have one of the following forms:

(a) Both filters are symmetric and of odd lengths, differing by an odd multiple of 2.

(b) One filter is symmetric and the other is antisymmetric; both lengths are even, and are equal or differ by an even multiple of 2.

(c) One filter is of odd length, the other one of even length; both have all zeros on the unit circle. Either both filters are symmetric, or one is symmetric and the other one is antisymmetric (this is a degenerate case) .

The proof can be found in [319] and is left as an exercise (see Problem 3.8). We will discuss it briefly. The idea is to consider the product polynomial $P(z) = H_0(z)H_1(-z)$ that has to satisfy (3.2.66). Because $H_0(z)$ and $H_1(z)$ (as well as $H_1(-z)$) are linear phase, so is $P(z)$. Because of (3.2.66), when $P(z)$ has more than two nonzero coefficients, it has to be symmetric with one central coefficient at $2l-1$. Also, the end terms of $P(z)$ have to be of an even index, so they cancel in $P(z) - P(-z)$. The above two requirements lead to the symmetry and length constraints for cases (a) and (b). In addition, there is a degenerate case (c), of little practical interest, when $P(z)$ has only two nonzero coefficients,

$$P(z) = z^{-j}(1 \pm z^{2N-1-2j}),$$

which leads to zeros at odd roots at ± 1. Because these are distributed among $H_0(z)$ and $H_1(-z)$ (rather than $H_1(z)$), the resulting filters will be a poor set of lowpass and highpass filters.

Another result that we mentioned at the beginning of this section is:

PROPOSITION 3.12

There are no two-channel perfect reconstruction, orthogonal filter banks, with filters being FIR, linear phase, and with real coefficients (except for the Haar filters).

PROOF

We know from Theorem 3.8 that orthonormality implies that

$$\boldsymbol{H}_p(z)\boldsymbol{H}_p^T(z^{-1}) = \boldsymbol{I},$$

which further means that

$$H_{00}(z)H_{00}(z^{-1}) + H_{01}(z)H_{01}(z^{-1}) = 1. \tag{3.2.67}$$

We also know that in orthogonal filter banks, the filters are of even length. Therefore, following Proposition 3.11, one filter is symmetric and the other one is antisymmetric. Take the symmetric one, $H_0(z)$ for example, and use (3.2.64)

$$\begin{aligned} H_0(z) &= H_{00}(z^2) + z^{-1}H_{01}(z^2), \\ &= z^{-L+1}H_0(z^{-1}) = z^{-L+1}(H_{00}(z^{-2}) + zH_{01}(z^{-2})), \\ &= z^{-L+2}H_{01}(z^{-2}) + z^{-1}(z^{-L+2}H_{00}(z^{-2})). \end{aligned}$$

This further means that the polyphase components are related as

$$H_{00}(z) = z^{-L/2+1}H_{01}(z^{-1}), \qquad H_{01}(z) = z^{-L/2+1}H_{00}(z^{-1}). \tag{3.2.68}$$

Substituting the second equation from (3.2.68) into (3.2.67) we obtain

$$H_{00}(z)\,H_{00}(z^{-1}) = \frac{1}{2}.$$

However, the only FIR, real-coefficient polynomial satisfying the above is

$$H_{00}(z) = \frac{1}{\sqrt{2}}z^{-l}.$$

Performing a similar analysis for $H_{01}(z)$, we obtain that $H_{01}(z) = 1/\sqrt{2}z^{-k}$, which, in turn, means that

$$H_0(z) = \frac{1}{\sqrt{2}}(z^{-2l} + z^{-2k-1}), \qquad H_1(z) = H_0(-z),$$

or, the only solution yields Haar filters ($l = k = 0$) or trivial variations thereof.

We now shift our attention to design issues.

Lattice Structure for Linear Phase Filters Unlike in the paraunitary case, there are no canonical factorizations for general matrices of polynomials.[7] But there are lattice structures that will produce, for example, linear phase perfect reconstruction filters [208, 321]. To obtain it, note that $\boldsymbol{H}_p(z)$ has to satisfy (if the filters are of the same length)

$$\boldsymbol{H}_p(z) = \begin{pmatrix} 1 & 0 \\ 0 & -1 \end{pmatrix} \cdot z^{-k} \cdot \boldsymbol{H}_p(z^{-1}) \cdot \begin{pmatrix} 0 & 1 \\ 1 & 0 \end{pmatrix}. \tag{3.2.69}$$

[7]There exist factorizations of polynomial matrices based on ladder steps [151], but they are not canonical like the lattice structure in (3.2.60).

Table 3.3 Impulse response coefficients for analysis and synthesis filters in two different linear phase cases. There is a factor of $1/16$ to be distributed between $h_i[n]$ and $g_i[n]$, like $\{1/4, 1/4\}$ or $\{1/16, 1\}$ (the latter was used in the text).

n	$h_0[n]$	$h_1[n]$	$g_0[n]$	$g_1[n]$	$h_0[n]$	$h_1[n]$	$g_0[n]$	$g_1[n]$
0	1	-1	-1	-1	1	-1	-1	-1
1	3	-3	3	3	2	-2	2	2
2	3	3	3	-3	1	6	6	-1
3	1	1	-1	1		-2	2	
4						-1	-1	

Here, we assume that $H_i(z) = H_{i0}(z^2)+z^{-1}H_{i1}(z^2)$ in order to have causal filters. This is referred to as the *linear phase testing condition* (see Problem 3.9). Then, assume that $\boldsymbol{H}_p(z)$ satisfies (3.2.69) and construct $\boldsymbol{H}_p'(z)$ as

$$\boldsymbol{H}_p'(z) = \boldsymbol{H}_p(z)\begin{pmatrix} 1 & \\ & z^{-1} \end{pmatrix}\begin{pmatrix} 1 & \alpha \\ \alpha & 1 \end{pmatrix}.$$

It is then easy to show that $\boldsymbol{H}_p'(z)$ satisfies (3.2.69) as well. The lattice

$$\boldsymbol{H}_p(z) = C\begin{pmatrix} 1 & 1 \\ -1 & 1 \end{pmatrix}\left[\prod_{i=1}^{K-1}\begin{pmatrix} 1 & \\ & z^{-1} \end{pmatrix}\begin{pmatrix} 1 & \alpha_i \\ \alpha_i & 1 \end{pmatrix}\right], \tag{3.2.70}$$

with $C = -(1/2)\prod_{i=1}^{K-1}(1/(1-\alpha_i^2))$, produces length $L = 2K$ symmetric (lowpass) and antisymmetric (highpass) filters leading to perfect reconstruction filter banks. Note that the structure is incomplete [321] and that $|\alpha_i| \neq 1$. Again, just as in the paraunitary lattice, perfect reconstruction is structurally guaranteed within a scale factor (in the synthesis, replace simply α_i by $-\alpha_i$ and pick $C = 1$).

Example 3.4

Let us construct filters of length 4 where the lowpass has a maximum number of zeros at $z = -1$ (that is, the linear phase counterpart of the D_2 filter). From the cascade structure,

$$\begin{aligned}\boldsymbol{H}_p(z) &= \frac{-1}{2(1-\alpha^2)}\begin{pmatrix} 1 & 1 \\ -1 & 1 \end{pmatrix}\begin{pmatrix} 1 & \\ & z^{-1} \end{pmatrix}\begin{pmatrix} 1 & \alpha \\ \alpha & 1 \end{pmatrix} \\ &= \frac{-1}{2(1-\alpha^2)}\begin{pmatrix} 1+\alpha z^{-1} & \alpha+z^{-1} \\ -1+\alpha z^{-1} & -\alpha+z^{-1} \end{pmatrix}.\end{aligned}$$

We can now find the filter $H_0(z)$ as

$$H_0(z) = H_{00}(z^2)+z^{-1}H_{01}(z^2) = \frac{1+\alpha z^{-1}+\alpha z^{-2}+z^{-3}}{-2(1-\alpha^2)}.$$

Because $H_0(z)$ is an even-length symmetric filter, it has automatically a zero at $z = -1$, or $H_0(-1) = 0$. Take now the first derivative of $H_0(e^{j\omega})$ at $\omega = \pi$ and set it to 0 (which corresponds to imposing a double zero at $z=-1$)

$$\left.\frac{dH_0(e^{j\omega})}{d\omega}\right|_{\omega=\pi} = \frac{-1}{2(1-\alpha^2)}(\alpha - 2\alpha + 3) = 0,$$

leading to $\alpha = 3$. Substituting this into the expression for $H_0(z)$, we get

$$H_0(z) = \frac{1}{16}(1+3z^{-1}+3z^{-2}+z^{-3}) = \frac{1}{16}(1+z^{-1})^3, \tag{3.2.71}$$

which means that $H_0(z)$ has a triple zero at $z=-1$. The highpass filter is equal to

$$H_1(z) = \frac{1}{16}(-1-3z^{-1}+3z^{-2}+z^{-3}). \tag{3.2.72}$$

Note that $\det(\boldsymbol{H}_m(z)) = (1/8)\ z^{-3}$. Following (3.2.30–3.2.31), $G_0(z) = 16z^3 H_1(-z)$ and $G_1(z) = -16z^3\ H_0(-z)$. A causal version simply skips the z^3 factor. Recall that the key to perfect reconstruction is the product $P(z) = H_0(z) \cdot H_1(-z)$ in (3.2.66), which equals in this case (using (3.2.71–3.2.72))

$$\begin{aligned} P(z) &= \frac{1}{256}(-1+9z^{-1}+16z^{-3}+9z^{-4}-z^{-6}) \\ &= \frac{1}{256}(1+z^{-1})^4\ (-1+4z^{-1}-z^{-2}), \end{aligned}$$

that is, the same $P(z)$ as in Example 3.2. One can refactor this $P(z)$ into a different set of $\{H_0'(z), H_1'(-z)\}$, such as, for example,

$$\begin{aligned} P(z) &= H_0'(z)\ H_1'(-z) \\ &= \frac{1}{16}(1+2z^{-1}+z^{-2})\ \frac{1}{16}(-1+2z^{-1}+6z^{-2}+2z^{-3}-z^{-4}), \end{aligned}$$

that is, odd-length linear phase lowpass and highpass filters with impulse responses 1/16 [1, 2, 1] and 1/16 [-1, -2, 6, -2, -1], respectively. Table 3.3 gives impulse response coefficients for both analysis and synthesis filters for the two cases given above.

The above example showed again the central role played by $P(z) = H_0(z) \cdot H_1(-z)$. In some sense, designing two-channel filter banks boils down to designing $P(z)$'s with particular properties, and factoring them in a particular way.

If one relaxes the perfect reconstruction constraint, one can obtain some desirable properties at the cost of some small reconstruction error. For example, popular QMF filters have been designed by Johnston [144], which have linear phase and "almost" perfect reconstruction. The idea is to approximate perfect reconstruction in a QMF solution (see (3.2.37)) as well as possible, while obtaining a good lowpass filter (the highpass filter $H_1(z)$ being equal to $H_0(-z)$, is automatically as good as the lowpass). Therefore, define an objective function depending on two quantities: (a) stopband attenuation error of $H_0(z)$

$$S = \int_{\omega_s}^{\pi} |H_0(e^{j\omega})|^2\ d\omega,$$

and (b) reconstruction error

$$E = \int_0^{\pi} |2 - (H_0(e^{j\omega}))^2 + (H_0(e^{j(\omega+\pi)}))^2|^2 \, d\omega.$$

The objective function is

$$O = cS + (1-c)E,$$

where c assigns the relative cost to these two quantities. Then, O is minimized using the coefficients of $H_0(z)$ as free variables. Such filter designs are tabulated in [67, 144].

Complementary Filters The following question sometimes arises in the design of filter banks: given an FIR filter $H_0(z)$, is there a complementary filter $H_1(z)$ such that the filter bank allows perfect reconstruction with FIR filters? The answer is given by the following proposition which was first proven in [139]. We will follow the proof in [319]:

PROPOSITION 3.13

Given a causal FIR filter $H_0(z)$, there exists a complementary filter $H_1(z)$ *if and only if* the polyphase components of $H_0(z)$ are coprime (except for possible zeros at $z = \infty$).

PROOF

From Proposition 3.6, we know that a necessary and sufficient condition for perfect FIR reconstruction is that $\det(\boldsymbol{H}_p(z))$ be a monomial. Thus, coprimeness is obviously necessary, since if there is a common factor between $H_{00}(z)$ and $H_{01}(z)$, it will show up in the determinant. Sufficiency follows from the Euclidean algorithm or Bezout's identity: given two coprime polynomials $a(z)$ and $b(z)$, the equation $a(z)p(z)+b(z)q(z) = c(z)$ has a unique solution (see, for example, [32]). Thus, choose $c(z) = z^{-k}$ and then, the solution $\{p(z), q(z)\}$ corresponds to the two polyphase components of $H_1(z)$.

Note that the solution $H_1(z)$ is not unique [32, 319]. Also, coprimeness of $H_{00}(z), H_{01}(z)$ is equivalent with $H_0(z)$ not having any pair of zeros at locations α and $-\alpha$. This can be used to prove that the filter $H_0(z) = (1+z^{-1})^N$ always has a complementary filter (see Problem 3.12).

Example 3.5

Consider the filter $H_0(z) = (1+z^{-1})^4 = 1 + 4z^{-1} + 6z^{-2} + 4z^{-3} + z^{-4}$. It can be verified that its two polyphase components are coprime, and thus, there is a complementary filter. We will find a solution to the equation

$$\det(\boldsymbol{H}_p(z)) = H_{00}(z) \cdot H_{11}(z) - H_{01}(z) \cdot H_{10}(z) = z^{-1}, \tag{3.2.73}$$

with $H_{00}(z) = 1 + 6z^{-1} + z^{-2}$ and $H_{01}(z) = 4 + 4z^{-1}$. The right side of (3.2.73) was chosen so that there is a linear phase solution. For example,

$$H_{10}(z) = \frac{1}{16}(1+z^{-1}), \quad H_{11}(z) = \frac{1}{4},$$

is a solution to (3.2.73), that is, $H_1(z) = (1 + 4z^{-1} + z^2)/16$. This of course leads to the same $P(z)$ as in Examples 3.3 and 3.4.

3.2.5 Filter Banks with IIR Filters

We will now concentrate on orthogonal filter banks with infinite impulse response (IIR) filters. An early study of IIR filter banks was done in [313], and further developed in [234] as well as in [269] for perfect reconstruction in the context of image coding. The main advantage of such filter banks is good frequency selectivity and low computational complexity, just like in regular IIR filtering. However, this advantage comes with a cost. Recall that in orthogonal filter banks, the synthesis filter impulse response is the time-reversed version of the analysis filter. Now if the analysis uses causal filters (with impulse response going from 0 to $+\infty$), then the synthesis has anticausal filters. This is a drawback from the point of view of implementation, since in general anticausal IIR filters cannot be implemented unless their impulse responses are truncated. However, a case where anticausal IIR filters can be implemented appears when the signal to be filtered is of finite length, a case encountered in image processing [234, 269]. IIR filter banks have been less popular because of this drawback, but their attractive features justify a brief treatment as given below. For more details, the reader is referred to [133].

First, return to the lattice factorization for FIR orthogonal filter banks (see (3.2.60)). If one substitutes an allpass section[8] for the delay z^{-1} in (3.2.60), the factorization is still paraunitary. For example, instead of the diagonal matrix used in (3.2.60), take a diagonal matrix $\boldsymbol{D}(z)$ such that

$$\boldsymbol{D}(z)\,\boldsymbol{D}(z^{-1}) \;=\; \begin{pmatrix} F_0(z) & 0 \\ 0 & F_1(z) \end{pmatrix} \begin{pmatrix} F_0(z^{-1}) & 0 \\ 0 & F_1(z^{-1}) \end{pmatrix} \;=\; \boldsymbol{I},$$

where we have assumed that the coefficients are real, and have used two allpass sections (instead of 1 and z^{-1}). What is even more interesting is that such a factorization is complete [84].

Alternatively, recall that one of the ways to design orthogonal filter banks is to find an autocorrelation function $P(z)$ which is valid, that is, which satisfies

$$P(z) + P(-z) \;=\; 2, \qquad (3.2.74)$$

and then factor it into $P(z) = H_0(z)H_0(z^{-1})$. This approach is used in [133] to construct all possible orthogonal filter banks with rational filters. The method goes as follows:

[8] Remember that a filter $H(e^{j\omega})$ is allpass if $|H(e^{j\omega})| = c$, $c > 0$, for all ω. Here we choose $c = 1$.

First, one chooses an arbitrary polynomial $R(z)$ and forms $P(z)$ as

$$P(z) = \frac{2R(z)R(z^{-1})}{R(z)R(z^{-1}) + R(-z)R(-z^{-1})}. \tag{3.2.75}$$

It is easy to see that this $P(z)$ satisfies (3.2.74). Since both the numerator and the denominator are autocorrelations (the latter being the sum of two autocorrelations), $P(z)$ is as well. It can be shown that any valid autocorrelation can be written as in (3.2.75) [133]. Then factor $P(z)$ as $H(z)H(z^{-1})$ and form the filter

$$H_0(z) = A_{H_0}(z)\, H(z),$$

where $A_{H_0}(z)$ is an arbitrary allpass. Finally choose

$$H_1(z) = z^{2K-1} H_0(-z^{-1})\, A_{H_1}(z), \tag{3.2.76}$$

where $A_{H_1}(z)$ is again an arbitrary allpass. The synthesis filters are then

$$G_0(z) = H_0(z^{-1}), \quad G_1(z) = -H_1(z^{-1}). \tag{3.2.77}$$

The above construction covers the whole spectrum of possible solutions. For example, if $R(z)R(z^{-1})$ is in itself a valid function, then

$$R(z)R(z^{-1}) + R(-z)R(-z^{-1}) = 2,$$

and by choosing A_{H_0}, A_{H_1} to be pure delays, the solutions obtained by the above construction are FIR.

Example 3.6 *Butterworth Filters*

As an example, consider a family of IIR solutions constructed in [133]. It is obtained using the above construction and imposing a maximum number of zeros at $z = -1$. Choosing $R(z) = (1 + z^{-1})^N$ in (3.2.75) gives

$$P(z) = \frac{(1+z^{-1})^N (1+z)^N}{(z^{-1}+2+z)^N + (-z^{-1}+2-z)^N} = H(z)H(z^{-1}). \tag{3.2.78}$$

These filters are the IIR counterparts of the Daubechies' filters given in Example 3.2. These are, in fact, the Nth order half-band digital Butterworth filters [211] (see also Example 2.2). That these particular filters satisfy the conditions for orthogonality was also pointed out in [269]. The Butterworth filters are known to be the maximally flat IIR filters of a given order.

Choose $N = 5$, or $P(z)$ equals

$$P(z) = \frac{(1+z)^5 (1+z^{-1})^5}{10z^4 + 120z^3 + 252 + 120z^{-2} + 10z^{-4}}.$$

In this case, we can obtain a closed form spectral factorization of $P(z)$, which leads to

$$H_0(z) = \frac{1 + 5z^{-1} + 10z^{-2} + 10z^{-3} + 5z^{-4} + z^{-5}}{\sqrt{2}(1 + 10z^{-2} + 5z^{-4})}, \tag{3.2.79}$$

$$H_1(z) = z^{-1}\frac{1 - 5z + 10z^2 - 10z^3 + 5z^4 - z^5}{\sqrt{2}(1 + 10z^2 + 5z^4)}. \tag{3.2.80}$$

For the purposes of implementation, it is necessary to factor $\boldsymbol{H}_i(z)$ into stable causal (poles inside the unit circle) and anticausal (poles outside the unit circle) parts. For comparison with earlier designs, where length-8 FIR filters were designed, we show in Figure 3.5(d) the magnitude responses of $H_0(e^{j\omega})$ and $H_1(e^{j\omega})$ for $N = 4$. The form of the $P(z)$ is then

$$P(z) = \frac{z^{-4}(1+z)^4(1+z^{-1})^4}{1+28z^{-2}+70z^{-4}+28z^{-6}+z^{-8}}.$$

As we pointed out in Proposition 3.12, there are no real FIR orthogonal symmetric/antisymmetric filter banks. However, if we allow IIR filters instead, then solutions do exist. There are two cases, depending if the center of symmetry/antisymmetry is at a half integer (such as in an even-length FIR linear phase filter) or at an integer (such as in the odd-length FIR case). We will only consider the former case. For discussion of the latter case as well as further details, see [133].

It can be shown that the polyphase matrix for an orthogonal, half-integer symmetric/antisymmetric filter bank is necessarily of the form

$$\boldsymbol{H}_p(z) = \begin{pmatrix} A(z) & z^{-l}A(z^{-1}) \\ -z^{l-n}A(z) & z^{-n}A(z^{-1}) \end{pmatrix},$$

where $A(z)A(z^{-1}) = 1$, that is, $A(z)$ is an allpass filter. Choosing $l = n = 0$ gives

$$H_0(z) = A(z^2) + z^{-1}A(z^{-2}), \quad H_1(z) = -A(z^2) + z^{-1}A(z^{-2}), \tag{3.2.81}$$

which is an orthogonal, linear phase pair. For a simple example, choose

$$A(z) = \frac{1+6z^{-1}+(15/7)z^{-2}}{(15/7)+6z^{-1}+z^{-2}}. \tag{3.2.82}$$

This particular solution will prove useful in the construction of wavelets (see Section 4.6.2). Again, for the purposes of implementation, one has to implement stable causal and anticausal parts separately.

Remarks The main advantage of IIR filters is their good frequency selectivity and low computational complexity. The price one pays, however, is the fact that the filters become noncausal. For the sake of discussion, assume a finite-length signal, and a causal analysis filter, which will be followed by an anticausal synthesis filter. The output will be infinite even though the input is of finite length. One can take care of this problem in two ways. Either one stores the state of the filters after the end of the input signal and uses this as an initial state for the synthesis filters [269], or one takes advantage of the fact that the outputs of the analysis filter bank decay rapidly after the input is zero, and stores only a finite extension of these signals. While the former technique is exact, the latter is usually a good enough approximation. This short discussion indicates that the implementation of IIR filter banks is less straightforward than that of their FIR counterparts, and explains their lesser popularity.

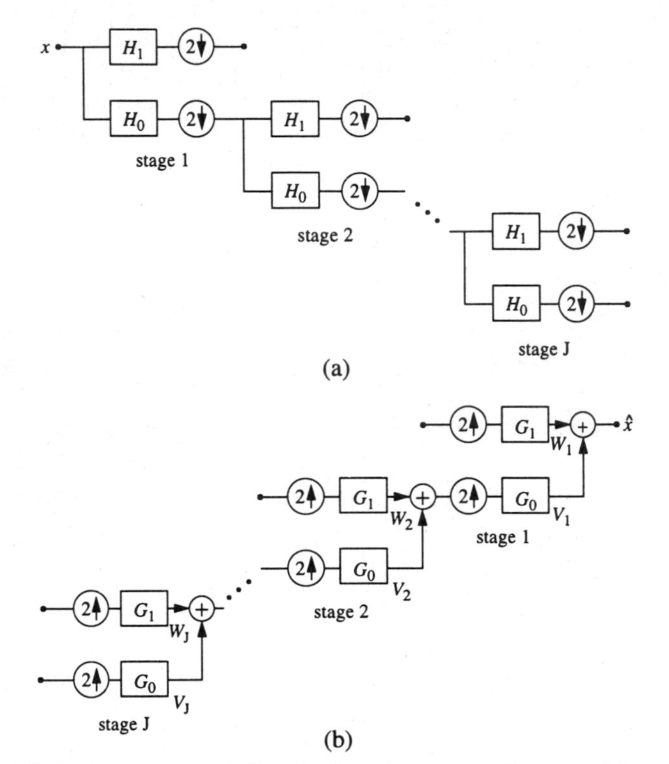

Figure 3.7 An octave-band filter bank with J stages. Decomposition spaces V_i, W_i are indicated. If $h_i[n]$ is an orthogonal filter, and $g_i[n] = h_i[-n]$, the structure implements an orthogonal discrete-time wavelet series expansion. (a) Analysis part. (b) Synthesis part.

3.3 TREE-STRUCTURED FILTER BANKS

An easy way to construct multichannel filter banks is to cascade two-channel banks appropriately. One case can be seen in Figure 3.7(a), where frequency analysis is obtained by simply iterating a two-channel division on the previous lowpass channel. This is often called a *constant-Q* or constant relative bandwidth filter bank since the bandwidth at each channel, divided by its center frequency, is constant. It is also sometimes called a logarithmic filter bank since the channels are equal bandwidth on a logarithmic scale. We will call it an *octave-band filter bank* since each successive highpass output contains an octave of the input bandwidth. Another case appears when 2^J equal bandwidth channels are desired. This can be obtained by a J-step subdivision into 2 channels, that is, the two-channel bank is now iterated on both the lowpass and highpass channels. This results in a tree with 2^J leaves, each corresponding to $(1/2^J)$th of the original bandwidth, with a downsampling by 2^J. Another possibility is building an arbitrary tree-structured filter bank, giving rise to *wavelet packets*, discussed later in this section.

3.3.1 Octave-Band Filter Bank and Discrete-Time Wavelet Series

Consider the filter bank given in Figure 3.7. We see that the signal is split first via a two-channel filter bank, then the lowpass version is split again using the same filter bank, and so on. It will be shown later that this structure implements a discrete-time biorthogonal wavelet series (we assume here that the two-channel filter banks are perfect reconstruction). If the two-channel filter bank is orthonormal, then it implements an orthonormal discrete-time wavelet series.[9]

Recall that the basis functions of the discrete-time expansion are given by the impulse responses of the synthesis filters. Therefore, we will concentrate on the synthesis filter bank (even though, in the orthogonal case, simple time reversal relates analysis and synthesis filters). Let us start with a simple example which should highlight the main features of octave-band filter bank expansions.

Example 3.7

Consider what happens if the filters $g_i[n]$ from Figure 3.7(a)-(b) are Haar filters defined in z-transform domain as

$$G_0(z) = \frac{1}{\sqrt{2}}(1+z^{-1}), \quad G_1(z) = \frac{1}{\sqrt{2}}(1-z^{-1}).$$

Take, for example, $J=3$, that is, we will use three two-channel filter banks. Then, using the multirate identity which says that $G(z)$ followed by upsampling by 2 is equivalent to upsampling by 2 followed by $G(z^2)$ (see Section 2.5.3), we can transform this filter bank into a four-channel one as given in Figure 3.8. The equivalent filters are

$$\begin{aligned}
G_1^{(1)}(z) &= G_1(z) = \frac{1}{\sqrt{2}}(1-z^{-1}),\\
G_1^{(2)}(z) &= G_0(z)\,G_1(z^2) = \frac{1}{2}(1+z^{-1}-z^{-2}-z^{-3}),\\
G_1^{(3)}(z) &= G_0(z)\,G_0(z^2)\,G_1(z^4)\\
&= \frac{1}{2\sqrt{2}}(1+z^{-1}+z^{-2}+z^{-3}-z^{-4}-z^{-5}-z^{-6}-z^{-7}),\\
G_0^{(3)}(z) &= G_0(z)\,G_0(z^2)\,G_0(z^4)\\
&= \frac{1}{2\sqrt{2}}(1+z^{-1}+z^{-2}+z^{-3}+z^{-4}+z^{-5}+z^{-6}+z^{-7}),
\end{aligned}$$

preceded by upsampling by 2, 4, 8 and 8 respectively. The impulse responses follow by

[9]This is also sometimes called a discrete-time wavelet transform in the literature.

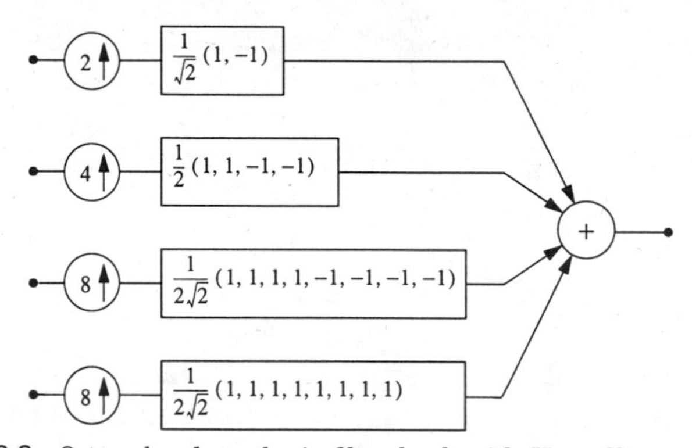

Figure 3.8 Octave-band synthesis filter bank with Haar filters and three stages. It is obtained by transforming the filter bank from Figure 3.7(b) using the multirate identity for filtering followed by upsampling.

inverse z-transform. Denote by $g_0^{(3)}[n]$ the equivalent filter obtained by going through three stages of lowpass filters $g_0[n]$ each preceded by upsampling by 2. It can be defined recursively as (we give it in z-domain for simplicity)

$$G_0^{(3)}(z) = G_0(z^{2^2})\, G_0^{(2)}(z) = \prod_{k=0}^{2} G_0(z^{2^k}).$$

Note that this implies that $G_0^{(1)}(z) = G_0(z)$. On the other hand, we denote by $g_1^{(i)}[n]$, the equivalent filter corresponding to highpass filtering followed by $(i-1)$ stages of lowpass filtering, each again preceded by upsampling by 2. It can be defined recursively as

$$G_1^{(3)}(z) = G_1(z^{2^2})\, G_0^{(2)}(z) = G_1(z^{2^2}) \prod_{k=0}^{1} G_0(z^{2^k}), \quad j = 1,2,3.$$

Since this is an orthonormal system, the time-domain matrices representing analysis and synthesis are just transposes of each other. Thus the analysis matrix $\boldsymbol{T}_a$ representing the actions of the filters $h_1^{(1)}[n]$, $h_1^{(2)}[n]$, $h_1^{(3)}[n]$, $h_0^{(3)}[n]$ contains as lines the impulse responses of $g_1^{(1)}[n]$, $g_1^{(2)}[n]$, $g_1^{(3)}[n]$, and $g_0^{(3)}[n]$ or of $h_i^{(j)}[-n]$ since analysis and synthesis filters are linked by time reversal. The matrix $\boldsymbol{T}_a$ is block-diagonal,

$$\boldsymbol{T}_a = \begin{pmatrix} \ddots & & & \\ & \boldsymbol{A}_0 & & \\ & & \boldsymbol{A}_0 & \\ & & & \ddots \end{pmatrix}, \tag{3.3.1}$$

where the block A_0 is of the following form:

$$A_0 = \frac{1}{2\sqrt{2}} \begin{pmatrix} 2 & -2 & 0 & 0 & 0 & 0 & 0 & 0 \\ 0 & 0 & 2 & -2 & 0 & 0 & 0 & 0 \\ 0 & 0 & 0 & 0 & 2 & -2 & 0 & 0 \\ 0 & 0 & 0 & 0 & 0 & 0 & 2 & -2 \\ \sqrt{2} & \sqrt{2} & -\sqrt{2} & -\sqrt{2} & 0 & 0 & 0 & 0 \\ 0 & 0 & 0 & 0 & \sqrt{2} & \sqrt{2} & -\sqrt{2} & -\sqrt{2} \\ 1 & 1 & 1 & 1 & -1 & -1 & -1 & -1 \\ 1 & 1 & 1 & 1 & 1 & 1 & 1 & 1 \end{pmatrix}. \tag{3.3.2}$$

Note how this matrix reflects the fact that the filter $g_1^{(1)}[n]$ is preceded by upsampling by 2 (the row $(2 \quad -2)$ is shifted by 2 each time and appears 4 times in the matrix). $g_1^{(2)}[n]$ is preceded by upsampling by 4 (the corresponding row is shifted by 4 and appears twice), while filters in $g_1^{(3)}[n]$, $g_0^{(3)}[n]$ are preceded by upsampling by 8 (the corresponding rows appear only once in the matrix). Note that the ordering of the rows in (3.3.2) is somewhat arbitrary; we simply gathered successive impulse responses for clarity.

Now that we have seen how it works in a simple case, we take more general filters $g_i[n]$, and a number of stages J. We concentrate on the orthonormal case (the biorthogonal one would follow similarly). In an orthonormal octave-band filter bank with J stages, the equivalent filters (basis functions) are given by (again we give them in z-domain for simplicity)

$$G_0^{(J)}(z) = G_0^{(J-1)}(z)\, G_0(z^{2^{J-1}}) = \prod_{K=0}^{J-1} G_0(z^{2^K}), \tag{3.3.3}$$

$$G_1^{(j)}(z) = G_0^{(j-1)}(z)\, G_1(z^{2^{j-1}}) = G_1(z^{2^{j-1}}) \prod_{K=0}^{j-2} G_0(z^{2^K}),$$

$$j = 1, \ldots, J. \tag{3.3.4}$$

In time domain, each of the outputs in Figure 3.7(a) can be described as

$$\boldsymbol{H}_1\, \boldsymbol{H}_0^{j-1}\, \boldsymbol{x}, \qquad j = 1, \ldots, J-1$$

except for the last, which is obtained by

$$\boldsymbol{H}_0^J\, \boldsymbol{x}.$$

Here, the time-domain matrices $\boldsymbol{H}_0$, $\boldsymbol{H}_1$ are as defined in Section 3.2.1, that is, each line is an even shift of the impulse response of $g_i[n]$, or equivalently, of $h_i[-n]$. Since each stage in the analysis bank is orthonormal and invertible, the overall scheme is as well. Thus, we get a unitary analysis matrix $\boldsymbol{T}_a$ by interleaving the rows of $\boldsymbol{H}_1$, $\boldsymbol{H}_1\boldsymbol{H}_0$, ..., $\boldsymbol{H}_1\boldsymbol{H}_0^{J-1}$, $\boldsymbol{H}_0^J$, as was done in (3.3.1–3.3.2). A formal proof of this statement will be given in Section 3.3.2 under orthogonality of basis functions.

Example 3.8

Let us go back to the Haar case and three stages. We can form matrices $\boldsymbol{H}_1$, $\boldsymbol{H}_1\boldsymbol{H}_0$, $\boldsymbol{H}_1\boldsymbol{H}_0^2$, $\boldsymbol{H}_0^3$ as

$$\boldsymbol{H}_1 = \frac{1}{\sqrt{2}}\begin{pmatrix} & \vdots & \vdots & \vdots & \vdots & \\ \cdots & 1 & -1 & 0 & 0 & \cdots \\ \cdots & 0 & 0 & 1 & -1 & \cdots \\ & \vdots & \vdots & \vdots & \vdots & \end{pmatrix}, \tag{3.3.5}$$

$$\boldsymbol{H}_0 = \frac{1}{\sqrt{2}}\begin{pmatrix} & \vdots & \vdots & \vdots & \vdots & \\ \cdots & 1 & 1 & 0 & 0 & \cdots \\ \cdots & 0 & 0 & 1 & 1 & \cdots \\ & \vdots & \vdots & \vdots & \vdots & \end{pmatrix}, \tag{3.3.6}$$

$$\boldsymbol{H}_1\boldsymbol{H}_0 = \frac{1}{2}\begin{pmatrix} & \vdots & \vdots & \vdots & \vdots & \vdots & \vdots & \vdots & \vdots & \\ \cdots & 1 & 1 & -1 & -1 & 0 & 0 & 0 & 0 & \cdots \\ \cdots & 0 & 0 & 0 & 0 & 1 & 1 & -1 & -1 & \cdots \\ & \vdots & \vdots & \vdots & \vdots & \vdots & \vdots & \vdots & \vdots & \end{pmatrix}, \tag{3.3.7}$$

$$\boldsymbol{H}_1\boldsymbol{H}_0^2 = \frac{1}{2\sqrt{2}}\begin{pmatrix} & \vdots & \vdots & \vdots & \vdots & \vdots & \vdots & \vdots & \vdots & \vdots & \vdots & \\ \cdots & 1 & 1 & 1 & 1 & -1 & -1 & -1 & -1 & 0 & 0 & \cdots \\ \cdots & 0 & 0 & 0 & 0 & 0 & 0 & 0 & 0 & 1 & 1 & \cdots \\ & \vdots & \vdots & \vdots & \vdots & \vdots & \vdots & \vdots & \vdots & \vdots & \vdots & \end{pmatrix}, \tag{3.3.8}$$

$$\boldsymbol{H}_0^3 = \frac{1}{2\sqrt{2}}\begin{pmatrix} & \vdots & \vdots & \vdots & \vdots & \vdots & \vdots & \vdots & \vdots & \vdots & \vdots & \\ \cdots & 1 & 1 & 1 & 1 & 1 & 1 & 1 & 1 & 0 & 0 & \cdots \\ \cdots & 0 & 0 & 0 & 0 & 0 & 0 & 0 & 0 & 1 & 1 & \cdots \\ & \vdots & \vdots & \vdots & \vdots & \vdots & \vdots & \vdots & \vdots & \vdots & \vdots & \end{pmatrix}. \tag{3.3.9}$$

Now, it is easy to see that by interleaving (3.3.5–3.3.9) we obtain the matrix $\boldsymbol{T}_a$ as in (3.3.1–3.3.2). To check that it is unitary, it is enough to check that $\boldsymbol{A}_0$ is unitary (which it is, just compute the product $\boldsymbol{A}_0\boldsymbol{A}_0^T$).

Until now, we have concentrated on the orthonormal case. If one would relax the orthonormality constraint, we would obtain a biorthogonal tree-structured filter bank. Now, $h_i[n]$ and $g_i[n]$ are not related by simple time reversal, but are impulse responses of a biorthogonal perfect reconstruction filter bank. We therefore have both equivalent synthesis filters $g_1^{(j)}[n - 2^j k]$, $g_0^{(J)}[n-2^J k]$ as given in (3.3.3–3.3.4) and analysis filters $h_1^{(j)}[n-2^j k]$, $h_0^{(J)}[n-2^J k]$, which are defined similarly. Therefore if the individual two-channel filter banks are biorthogonal (perfect reconstruction), then the overall scheme is as well. The proof of this statement will follow the proof for the orthonormal case (see Section 3.3.2 for the discrete-time wavelet series case), and is left as an exercise to the reader.

3.3.2 Discrete-Time Wavelet Series and Its Properties

What was obtained in the last section is called a discrete-time wavelet series. It should be noted that this is not an exact equivalent of the continuous-time wavelet transform or series discussed in Chapter 4. In continuous time, there is a single wavelet involved, whereas in the discrete-time case, there are different iterated filters.

At the risk of a slight redundancy, we go once more through the whole process leading to the discrete-time wavelet series. Consider a two-channel orthogonal filter bank with filters $h_0[n]$, $h_1[n]$, $g_0[n]$ and $g_1[n]$, where $h_i[n] = g_i[-n]$. Then, the input signal can be written as

$$x[n] = \sum_{k\in\mathcal{Z}} X^{(1)}[2k+1]\, g_1^{(1)}[n-2^1k] + \sum_{k\in\mathcal{Z}} X^{(1)}[2k]\, g_0^{(1)}[n-2^1k], \qquad (3.3.10)$$

where

$$\begin{aligned} X^{(1)}[2k] &= \langle h_0^{(1)}[2^1k-l], x[l]\rangle, \\ X^{(1)}[2k+1] &= \langle h_1^{(1)}[2^1k-l], x[l]\rangle, \end{aligned}$$

are the convolutions of the input with $h_0[n]$ and $h_1[n]$ evaluated at even indexes $2k$. In these equations $h_i^{(1)}[n] = h_i[n]$, and $g_i^{(1)}[n] = g_i[n]$. In an octave-band filter bank or discrete-time wavelet series, the lowpass channel is further split by lowpass/highpass filtering and downsampling. Then, the first term on the right side of (3.3.10) remains unchanged, while the second can be expressed as

$$\begin{aligned} \sum_{k\in\mathcal{Z}} X^{(1)}[2k]\, h_0^{(1)}[2^1k-n] &= \sum_{k\in\mathcal{Z}} X^{(2)}[2k+1]\, g_1^{(2)}[n-2^2k] \\ &\quad + \sum_{k\in\mathcal{Z}} X^{(2)}[2k]\, g_0^{(2)}[n-2^2k], \qquad (3.3.11) \end{aligned}$$

where

$$\begin{aligned} X^{(2)}[2k] &= \langle h_0^{(2)}[2^2k-l], x[l]\rangle, \\ X^{(2)}[2k+1] &= \langle h_1^{(2)}[2^2k-l], x[l]\rangle, \end{aligned}$$

that is, we applied (3.3.10) once more. In the above, basis functions $g^{(i)}[n]$ are as defined in (3.3.3) and (3.3.4). In other words, $g_0^{(2)}[n]$ is the time-domain version of

$$G_0^{(2)}(z) = G_0(z)\, G_0(z^2),$$

while $g_1^{(2)}[n]$ is the time-domain version of

$$G_1^{(2)}(z) = G_0(z)\, G_1(z^2).$$

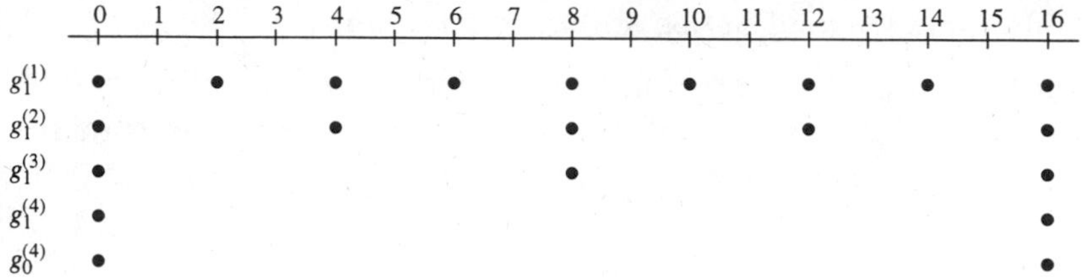

Figure 3.9 Dyadic sampling grid used in the discrete-time wavelet series. The shifts of the basis functions $g_1^{(j)}$ are shown, as well as $g_0^{(J)}$ (case $J = 4$ is shown). This corresponds to the "sampling" of the discrete-time wavelet series. Note the conservation of the number of samples between the signal and transform domains.

With (3.3.11), the input signal $x[n]$ in (3.3.10) can be written as

$$\begin{aligned} x[n] &= \sum_{k\in\mathcal{Z}} X^{(1)}[2k+1]\, g_1^{(1)}[n-2^1k] + \sum_{k\in\mathcal{Z}} X^{(2)}[2k+1]\, g_1^{(2)}[n-2^2k] \\ &\quad + \sum_{k\in\mathcal{Z}} X^{(2)}[2k]\, g_0^{(2)}[2^2k-n]. \end{aligned} \tag{3.3.12}$$

Repeating the process in (3.3.12) J times, one obtains the discrete-time wavelet series over J octaves, plus the final octave containing the lowpass version. Thus, (3.3.12) becomes

$$x[n] = \sum_{j=1}^{J}\sum_{k\in\mathcal{Z}} X^{(j)}[2k+1]\, g_1^{(j)}[n-2^jk] + \sum_{k\in\mathcal{Z}} X^{(J)}[2k]\, g_0^{(J)}[n-2^Jk], \tag{3.3.13}$$

where

$$\begin{aligned} X^{(j)}[2k+1] &= \langle h_1^{(j)}[2^jk-l], x[l]\rangle, \qquad j = 1,\ldots,J, \\ X^{(J)}[2k] &= \langle h_0^{(J)}[2^Jk-l], x[l]\rangle. \end{aligned} \tag{3.3.14}$$

In (3.3.13) the sequence $g_1^{(j)}[n]$ is the time-domain version of (3.3.4), while $g_0^{(J)}[n]$ is the time-domain version of (3.3.3) and $h_i^{(j)}[n] = g_i^{(j)}[-n]$. Because any input sequence can be decomposed as in (3.3.13), the family of functions $\{g_1^{(j)}[2^jk - n], g_0^{(J)}[2^Jk-n]\}$, $j = 1,\ldots,J$, and $k, n \in \mathcal{Z}$, is an orthonormal basis for $l_2(\mathcal{Z})$.

Note the special sampling used in the discrete-time wavelet series. Each subsequent channel is downsampled by 2 with respect to the previous one and has a bandwidth that is reduced by 2 as well. This is called a *dyadic sampling grid*, as shown in Figure 3.9.

Let us now list a few properties of the discrete-time wavelet series (orthonormal and dyadic).

Linearity Since the discrete-time wavelet series involves inner products or convolutions (which are linear operators) it is obviously linear.

Shift Recall that multirate systems are not shift-invariant in general, and two-channel filter banks downsampled by 2 are shift-invariant with respect to even shifts only. Therefore, it is intuitive that a J-octave discrete-time wavelet series will be invariant under shifts by multiples of 2^J. A visual interpretation follows from the fact that the dyadic grid in Figure 3.9, when moved by $k2^J$, will overlap with itself, whereas it will not if the shift is a noninteger multiple of 2^J.

PROPOSITION 3.14

In a discrete-time wavelet series expansion over J octaves, if

$$x[l] \longleftrightarrow X^{(j)}[2k+1], \qquad j = 1, 2, \ldots, J$$

then

$$x[l - m2^J] \longleftrightarrow X^{(j)}[2(k - m2^{J-j}) + 1].$$

PROOF

If $y[l] = x[l - m2^J]$, then its transform is, following (3.3.14),

$$\begin{aligned} Y^{(j)}[2k+1] &= \langle h_1^{(j)}[2^j k - l], x[l - m2^J] \rangle \\ &= \langle h_1^{(j)}[2^j k - l' - m2^J], x[l'] \rangle \\ &= X^{(j)}[2^j(k - m2^{J-j}) + 1]. \end{aligned}$$

Very similarly, one proves for the lowpass channel that, when $x[l]$ produces $X^{(J)}[2k]$, then $x[l - m2^J]$ leads to $X^{(J)}[2(k - m)]$.

Orthogonality We have mentioned before that $g_0^{(J)}[n]$ and $g_1^{(j)}[n]$, $j = 1, \ldots, J$, with appropriate shifts, form an orthonormal family of functions (see [274]). This stems from the fact that we have used two-channel orthogonal filter banks, for which we know that

$$\langle g_i[n - 2k], g_j[n - 2l] \rangle = \delta[i - j]\, \delta[k - l].$$

PROPOSITION 3.15

In a discrete-time wavelet series expansion, the following orthogonality relations hold:

$$\langle g_0^{(J)}[n - 2^J k], g_0^{(J)}[n - 2^J l] \rangle = \delta[k - l], \tag{3.3.15}$$

$$\langle g_1^{(j)}[n - 2^j k], g_1^{(i)}[n - 2^i l] \rangle = \delta[i - j]\, \delta[k - l], \tag{3.3.16}$$

$$\langle g_0^{(J)}[n - 2^J k], g_1^{(j)}[n - 2^j l] \rangle = 0. \tag{3.3.17}$$

PROOF

We will here prove only (3.3.15), while (3.3.16) and (3.3.17) are left as an exercise to the reader (see Problem 3.15). We prove (3.3.15) by induction.

It will be convenient to work with the z-transform of the autocorrelation of the filter $G_0^{(j)}(z)$, which we call $P^{(j)}(z)$ and equals

$$P^{(j)}(z) = G_0^{(j)}(z)\, G_0^{(j)}(z^{-1}).$$

Recall that because of the orthogonality of $g_0[n]$ with respect to even shifts, we have that

$$P^{(1)}(z) + P^{(1)}(-z) = 2,$$

or, equivalently, that the polyphase decomposition of $P^{(1)}(z)$ is of the form

$$P^{(1)}(z) = 1 + zP_1^{(1)}(z^2).$$

This is the initial step for our induction. Now, assume that $g_0^{(j)}[n]$ is orthogonal to its translates by 2^j. Therefore, the polyphase decomposition of its autocorrelation can be written as

$$P^{(j)}(z) = 1 + \sum_{i=1}^{2^j-1} z^i P_i^{(j)}(z^{2^j}).$$

Now, because of the recursion (3.3.3), the autocorrelation of $G^{(j+1)}(z)$ equals

$$P^{(j+1)}(z) = P^{(j)}(z)\; P^{(1)}(z^{2^j}).$$

Expanding both terms on the right-hand side, we get

$$P^{(j+1)}(z) = \left(1 + \sum_{i=1}^{2^j-1} z^i P_i^{(j)}(z^{2^j})\right)\left(1 + z^{2^j} P_1^{(1)}(z^{2^{j+1}})\right).$$

We need to verify that the 0th polyphase component of $P^{(j+1)}(z)$ is equal to 1, or that coefficients of z's which are raised to powers multiple of 2^{j+1} are 0. Out of the four products that appear when multiplying out the above right-hand side, only the product involving the polyphase components needs to be considered,

$$\sum_{i=1}^{2^j-1} z^i P_i^{(j)}(z^{2^j}) \cdot z^{2^j} P_1^{(1)}(z^{2^{j+1}}).$$

The powers of z appearing in the above product are of the form $l = i + k2^j + 2^j + m2^{j+1}$, where $i = 0 \cdots 2^j - 1$ and $k, m \in \mathcal{Z}$. Thus, l cannot be a multiple of 2^{j+1}, and we have shown that

$$P^{j+1}(z) = 1 + \sum_{i=1}^{2^{j+1}-1} z^i P_i^{(j+1)}(z^{2^{j+1}}),$$

thus completing the proof.

Parseval's Equality Orthogonality together with completeness (which follows from perfect reconstruction) leads to conservation of energy, also called Bessel's or Parseval's equality, that is

$$\|x[n]\|^2 = \sum_{k\in\mathcal{Z}}(|X^{(J)}[2k]|^2 + \sum_{j=1}^{J}|X^{(j)}[2k+1]|^2).$$

3.3.3 Multiresolution Interpretation of Octave-Band Filter Banks

The two-channel filter banks studied in Sections 3.1 and 3.2 have the property of splitting the signal into two lower-resolution versions. One was a lowpass or coarse resolution version, and the other was a highpass version of the input. Then, in this section, we have applied this decomposition recursively on the lowpass or coarse version. This leads to a hierarchy of resolutions, also called a *multiresolution decomposition*.

Actually, in computer vision as well as in image processing, looking at signals at various resolutions has been around for quite some time. In 1983, Burt and Adelson introduced the pyramid coding technique, that builds up a signal from its lower-resolution version plus a sequence of details (see also Section 3.5.2) [41]. In fact, one of the first links between wavelet theory and signal processing was Daubechies' [71] and Mallat's [180] recognition that the scheme of Burt and Adelson is closely related to wavelet theory and multiresolution analysis, and that filter banks or subband coding schemes can be used for the computation of wavelet decompositions. While these relations will be further explored in Chapter 4 for the continuous-time wavelet series, here we study the discrete-time wavelet series or its octave-band filter bank realization. This discrete-time multiresolution analysis was studied by Rioul [240].

Since this is a formalization of earlier concepts, we need some definitions. First we introduce the concept of embedded closed spaces. We will say that the space V_0 is the space of all square-summable sequences, that is,

$$V_0 = l_2\{\mathcal{Z}\}. \tag{3.3.18}$$

Then, a multiresolution analysis consists of a sequence of embedded closed spaces

$$V_J \subset \cdots \subset V_2 \subset V_1 \subset V_0. \tag{3.3.19}$$

It is obvious that due to (3.3.18–3.3.19)

$$\bigcup_{j=0}^{J} V_j = V_0 = l_2\{\mathcal{Z}\}.$$

The orthogonal complement of V_{j+1} in V_j will be denoted by W_{j+1}, and thus

$$V_j = V_{j+1} \oplus W_{j+1}, \tag{3.3.20}$$

with $V_{j+1} \perp W_{j+1}$, where $\oplus$ denotes the direct sum (see Section 2.2.2). Assume that there exists a sequence $g_0[n] \in V_0$ such that

$$\{g_0[n-2k]\}_{k\in\mathcal{Z}}$$

is a basis for V_1. Then, it can be shown that there exists a sequence $g_1[n] \in V$ such that

$$\{g_1[n-2k]\}_{k\in\mathcal{Z}}$$

is a basis for W_1. Such a sequence is given by

$$g_1[n] = (-1)^n g_0[-n+1]. \tag{3.3.21}$$

In other words, and having in mind (3.3.20), $\{g_0[n-2k], g_1[n-2k]\}_{k\in\mathcal{Z}}$ is an orthonormal basis for V_0. This splitting can be iterated on V_1. Therefore, one can see that V_0 can be decomposed in the following manner:

$$V_0 \;=\; W_1 \oplus W_2 \oplus \cdots \oplus W_J \oplus V_J, \tag{3.3.22}$$

by simply iterating the decomposition J times.

Now, consider the octave-band filter bank in Figure 3.7(a). The analysis filters are the time-reversed versions of $g_0[n]$ and $g_1[n]$. Therefore, the octave-band analysis filter bank computes the inner products with the basis functions for $W_1, W_2, \ldots, W_J$ and V_J.

In Figure 3.7(b), after convolution with the synthesis filters, we get the orthogonal projection of the input signal onto $W_1, W_2, \ldots, W_J$ and V_J. That is, the input is decomposed into a very coarse resolution (which exists in V_J) and added details (which exist in the spaces W_i, $i = 1, \ldots, J$). By (3.3.22), the sum of the coarse version and all the added details yields back the original signal; a result that follows from the perfect reconstruction property of the analysis/synthesis system as well.

We will call V_j's *approximation* spaces and W_j's *detail* spaces. Then, the process of building up the signal is intuitively very clear — one starts with its lower-resolution version belonging to V_J, and adds up the details until the final resolution is reached.

It will be seen in Chapter 4 that the decomposition into approximation and detail spaces is very similar to the multiresolution framework for continuous-time signals. However, there are a few important distinctions. First, in the discrete-time case, there is a "finest" resolution, associated with the space V_0, that is, one cannot refine the signal further. Then, we are considering a finite number of decomposition steps J, thus leading to a "coarsest" resolution, associated with V_J. Finally, in the continuous-time case, a simple function and its scales and translates are used, whereas here, various iterated filters are involved (which, under certain conditions, resemble scales of each other as we will see).

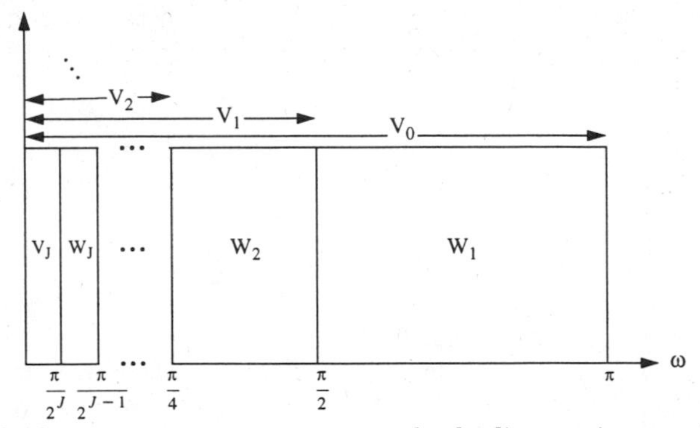

Figure 3.10 Ideal division of the spectrum by the discrete-time wavelet series using sinc filters. Note that the spectrums are symmetric around zero. Division into V_i spaces (note how $V_i \subset V_{i-1}$), and resulting W_i spaces. (Actually, V_j and W_j are of height $2^{j/2}$, so they have unit norm).

Example 3.9 *Sinc Case*

In the sinc case, introduced in Section 3.1.3, it is very easy to spot the multiresolution flavor. Since the filters used are ideal lowpass/highpass filters, respectively, at each stage the lowpass filter would halve the coarse space, while the highpass filter would take care of the difference between them. The above argument is best seen in Figure 3.10. The original signal (discrete in time and thus its spectrum occupies $(-\pi, \pi)$) is lowpass filtered using the ideal half-band filter. As a result, starting from the space V_0, we have derived a lower-resolution signal by halving V_0, resulting in V_1. Then, an even coarser version is obtained by using the same process, resulting in the space V_2. Using the above process repeatedly, one obtains the final coarse (approximation) space V_J. Along the way we have created difference spaces, W_i, as well.

For example, the space V_1 occupies the part $(-\pi/2, \pi/2)$ in the spectrum, while W_1 will occupy $(-\pi, -\pi/2) \cup (\pi/2, \pi)$. It can be seen that $g_0[n]$ as defined in (3.1.23) with its even shifts, will constitute a basis for V_1, while $g_1[n]$ following (3.3.21) constitutes a basis for W_1. In other words, $g_0[n]$, $g_1[n]$ and their even shifts would constitute a basis for the original (starting) space V_0 $(l_2(\mathcal{Z}))$.

Because we deal with ideal filters, there is an obvious frequency interpretation. However, one has to be careful with the boundaries between intervals. With our definition of $g_0[n]$ and $g_1[n]$, $\cos((\pi/2)n)$[10] belongs to V_1 while $\sin((\pi/2)n)$ belongs to W_1.

3.3.4 General Tree-Structured Filter Banks and Wavelet Packets

A major part of this section was devoted to octave-band, tree-structured filter banks. It is easy to generalize that discussion to arbitrary tree structures, starting from a single two-channel filter bank, all the way through the full

[10]To be precise, since $\cos((\pi/2)n)$ is not of finite energy and does not belong to $l_2(\mathcal{Z})$, one needs to define windowed versions of unit norm and take appropriate limits.

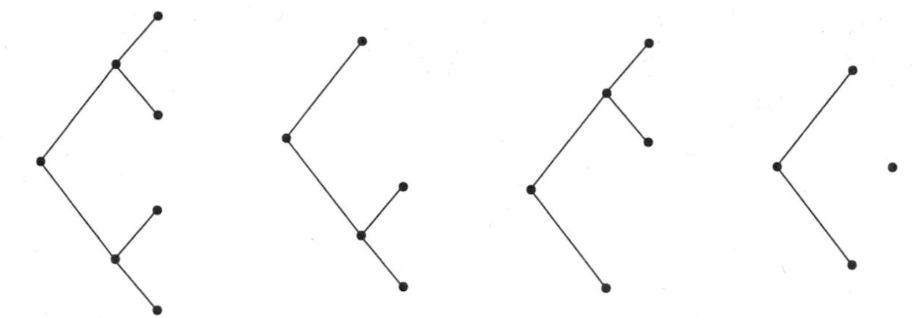

Figure 3.11 All possible combinations of tree-structured filter banks of depth 2. Symbolically, a fork stands for a two-channel filter bank with the lowpass on the bottom. From left to right is the full tree (STFT like), the octave-band tree (wavelet), the tree where only the highpass is split further, the two-band tree and finally the nil-tree tree (no split at all). Note that all smaller trees are pruned versions of the full tree.

grown tree of depth J. Consider, for example, Figure 3.11. It shows all possible tree structures of depth less or equal to two.

Note in particular the full tree, which yields a linear division of the spectrum similar to the short-time Fourier transform, and the octave-band tree, which performs a two-step discrete-time wavelet series expansion. Such arbitrary tree structures were recently introduced as a family of orthonormal bases for discrete-time signals, and are known under the name of *wavelet packets* [63]. The potential of wavelet packets lies in the capacity to offer a rich menu of orthonormal bases, from which the "best" one can be chosen ("best" according to a particular criterion). This will be discussed in more detail in Chapter 7 when applications in compression are considered. What we will do here, is define the basis functions and write down the appropriate orthogonality relations; however, since the octave-band case was discussed in detail, the proofs will be omitted (for a proof, see [274]).

Denote the equivalent filters by $g_i^{(j)}[n]$, $i = 0, \ldots, 2^j - 1$. In other words, $g_i^{(j)}$ is the ith equivalent filter going through one of the possible paths of length j. The ordering is somewhat arbitrary, and we will choose the one corresponding to a full tree with a lowpass in the lower branch of each fork, and start numbering from the bottom.

Example 3.10

Let us find all equivalent filters in Figure 3.11, or the filters corresponding to depth-1 and depth-2 trees. Since we will be interested in the basis functions, we consider the synthesis filter banks. For simplicity, we do it in z-domain.

$$\begin{aligned} G_0^{(1)}(z) &= G_0(z), & G_1^{(1)}(z) &= G_1(z), \\ G_0^{(2)}(z) &= G_0(z)\, G_0(z^2), & G_1^{(2)}(z) &= G_0(z)\, G_1(z^2), \end{aligned} \qquad (3.3.23)$$

$$G_2^{(2)}(z) = G_1(z)\, G_0(z^2), \qquad G_3^{(2)}(z) = G_1(z)\, G_1(z^2). \tag{3.3.24}$$

Note that with the ordering chosen in (3.3.23–3.3.24), increasing index does not always correspond to increasing frequency. It can be verified that for ideal filters, $G_2^{(2)}(e^{j\omega})$ chooses the range $[3\pi/4, \pi]$, while $G_3^{(2)}(e^{j\omega})$ covers the range $[\pi/2, 3\pi/4]$ (see Problem 3.16). Beside the identity basis, which corresponds to the no-split situation, we have four possible orthonormal bases, corresponding to the four trees in Figure 3.11. Thus, we have a family $W = \{W_0, W_1, W_2, W_3, W_4\}$, where W_4 is simply $\{\delta[n-k]\}_{k\in\mathcal{Z}}$.

$$W_0 = \{g_0^{(2)}[n-2^2k], g_1^{(2)}[n-2^2k], g_2^{(2)}[n-2^2k], g_3^{(2)}[n-2^2k]\}_{k\in\mathcal{Z}},$$

corresponds to the full tree.

$$W_1 = \{g_1^{(1)}[n-2k], g_0^{(2)}[n-2^2k], g_1^{(2)}[n-2^2k]\}_{k\in\mathcal{Z}},$$

corresponds to the octave-band tree.

$$W_2 = \{g_0^{(1)}[n-2k], g_2^{(2)}[n-2^2k], g_3^{(2)}[n-2^2k]\}_{k\in\mathcal{Z}},$$

corresponds to the tree with the highband split twice, and

$$W_3 = \{g_0^{(0)}[n-2k], g_1^{(1)}[n-2k]\}_{k\in\mathcal{Z}},$$

is simply the usual two-channel filter bank basis.

This small example should have given the intuition behind orthonormal bases generated from tree-structured filter banks. In the general case, with filter banks of depth J, it can be shown that, counting the no-split tree, the number of orthonormal bases satisfies

$$M_J = M_{J-1}^2 + 1. \tag{3.3.25}$$

Among this myriad of bases, there are the STFT-like basis, given by

$$W_0 = \{g_0^{(J)}[n-2^Jk], \ldots, g_{2^J-1}^{(J)}[n-2^Jk]\}_{k\in\mathcal{Z}}, \tag{3.3.26}$$

and the wavelet-like basis,

$$W_1 = \{g_1^{(1)}[n-2k], g_1^{(2)}[n-2^2k], \ldots, g_1^{(J)}[n-2^Jk], g_0^{(J)}[n-2^Jk]\}_{k\in\mathcal{Z}}. \tag{3.3.27}$$

It can be shown that the sets of basis functions in (3.3.26) and (3.3.27), as well as in all other bases generated by the filter bank tree, are orthonormal (for example, along the lines of the proof in the discrete-time wavelet series case). However, this would be quite cumbersome. A more immediate proof is sketched here. Note that we have a perfect reconstruction system by construction, and that the synthesis and the analysis filters are related by time reversal. That is, the inverse operator of the analysis filter bank (whatever its particular structure) is its transpose, or equivalently, the overall filter bank is orthonormal. Therefore, the impulse responses of all equivalent filters and their appropriate shifts form an orthonormal basis for $l_2(\mathcal{Z})$.

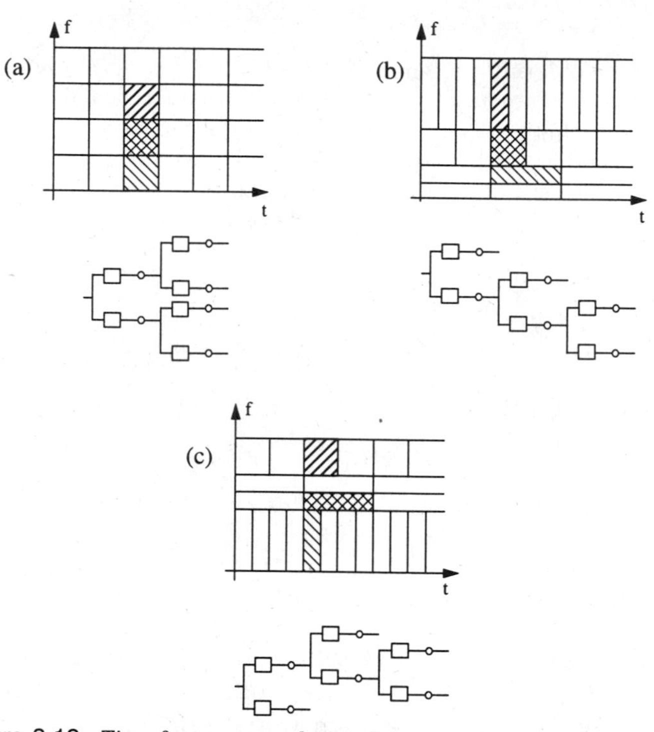

Figure 3.12 Time-frequency analysis achieved by different binary subband trees. The trees are on bottom, the time-frequency tilings on top. (a) Full tree or STFT. (b) Octave-band tree or wavelet series. (c) Arbitrary tree or one possible wavelet packet.

It is interesting to consider the time-frequency analysis performed by various filter banks. This is shown schematically in Figure 3.12 for three particular cases of binary trees. Note the different trade-offs in time and frequency resolutions.

Figure 3.13 shows a dynamic time-frequency analysis, where the time and frequency resolutions are modified as time evolves. This is achieved by modifying the frequency split on the fly [132], and can be used for signal compression as discussed in Section 7.3.4.

3.4 Multichannel Filter Banks

In the previous section, we have seen how one can obtain multichannel filter banks by cascading two-channel ones. Although this is a very easy way of achieving the goal, one might be interested in designing multichannel filter banks directly. Therefore, in this section we will present a brief analysis of N-channel filter banks, as given in Figure 3.14. We start the section by

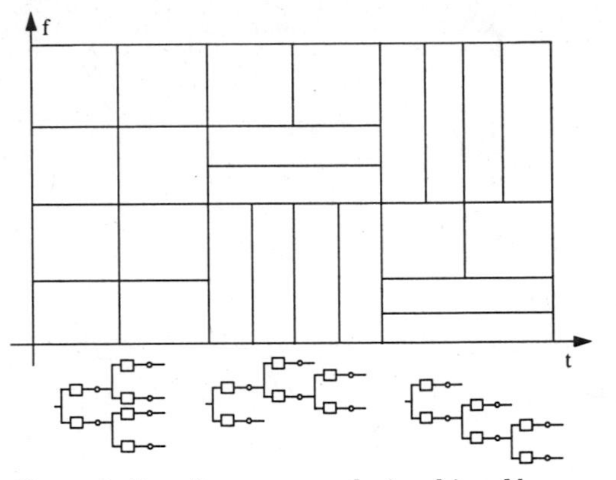

Figure 3.13 Dynamic time-frequency analysis achieved by concatenating the analyses from Figure 3.12. The tiling and the evolving tree are shown.

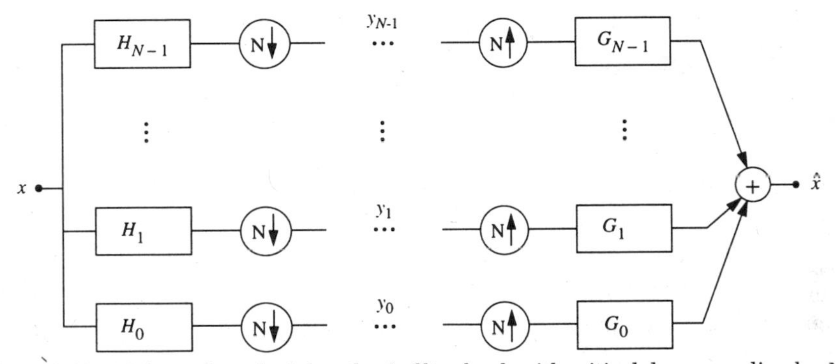

Figure 3.14 N-channel analysis/synthesis filter bank with critical downsampling by N.

discussing two special cases which are of interest in applications: the first, block transforms, and the second, lapped orthogonal transforms. Then, we will formalize our treatment of N-channel filter banks (time-, modulation- and polyphase-domain analyses). Finally, a particular class of multichannel filter banks, where all filters are obtained by modulating a single, prototype filter — called modulated filter banks — is presented.

3.4.1 Block and Lapped Orthogonal Transforms

Block Transforms Block transforms, which are used quite frequently in signal compression (for example, the discrete cosine transform), are a special case of filter banks with N channels, filters of length N, and downsampling by N. Moreover, when such transforms are unitary or orthogonal, they are the simplest examples of orthogonal (also called paraunitary or lossless) N-channel filter banks. Let us analyze such filter banks in a manner similar to Section 3.2.

Therefore, the channel signals, after filtering and sampling can be expressed as

$$\begin{pmatrix} \vdots \\ y_0[0] \\ \vdots \\ y_{N-1}[0] \\ y_0[1] \\ \vdots \\ y_{N-1}[1] \\ \vdots \end{pmatrix} = \begin{pmatrix} & \vdots & \vdots & \\ \cdots & \boldsymbol{A}_0 & \boldsymbol{0} & \cdots \\ \cdots & \boldsymbol{0} & \boldsymbol{A}_0 & \cdots \\ & \vdots & \vdots & \end{pmatrix} \begin{pmatrix} \vdots \\ x[0] \\ x[1] \\ \vdots \end{pmatrix}, \tag{3.4.1}$$

where the block $\boldsymbol{A}_0$ is equal to (similarly to (3.2.3))

$$\boldsymbol{A}_0 = \begin{pmatrix} h_0[N-1] & \cdots & h_0[0] \\ \vdots & & \vdots \\ h_{N-1}[N-1] & \cdots & h_{N-1}[0] \end{pmatrix} = \begin{pmatrix} g_0[0] & \cdots & g_0[N-1] \\ \vdots & & \vdots \\ g_{N-1}[0] & \cdots & g_{N-1}[N-1] \end{pmatrix}. \tag{3.4.2}$$

The second equality follows since the transform is unitary, that is,

$$\boldsymbol{A}_0\, \boldsymbol{A}_0^T = \boldsymbol{A}_0^T\, \boldsymbol{A}_0 = \boldsymbol{I}. \tag{3.4.3}$$

We can see that (3.4.2–3.4.3) imply that

$$\langle h_i[kN-n], h_j[lN-n]\rangle = \langle g_i[n-kN], g_j[n-lN]\rangle = \delta[i-j]\,\delta[k-l],$$

that is, we obtained the orthonormality relations for this case. Denoting by $\varphi_{kN+i}[n] = g_i[n-kN]$, we have that the set of basis functions $\{\varphi_{kN+i}[n]\} = \{g_0[n-kN], g_1[n-kN], \ldots, g_{N-1}[n-kN]\}$, with $i = 0, \ldots, N-1$, and $k \in \mathcal{Z}$, is an orthonormal basis for $l_2(\mathcal{Z})$.

Lapped Orthogonal Transforms Lapped orthogonal transforms (LOT's), introduced by Cassereau [43] and Malvar [189, 188] are a class of N-channel unitary filter banks where some additional constraints are imposed. In particular, the length of the filters is restricted to $L = 2N$, or twice the number of channels (or downsampling rate), and thus, it is easy to interpret LOT's as an extension of block transforms where neighboring filters overlap. Usually, the number of channels is even and sometimes they are all obtained from a single prototype window by modulation. In this case, fast algorithms taking advantage of the modulation relation between the filters reduce the order N^2 operations per N outputs of the filter bank to $cN\log_2 N$ (see also Chapter 6). This computational efficiency, as well as the simplicity and close relationship to block transforms, has made LOT's quite popular. A related class of filter banks, called time-domain aliasing cancellation filter banks, studied by Princen and Bradley [229] can be seen as another interpretation of LOT's. For an excellent treatment of LOT's, see the book by Malvar [188], to which we refer for more details.

Let us examine the lapped orthogonal transform. First, the fact that the filter length is $2N$, means that the time-domain matrix analogous to the one in (3.4.1), has the following form:

$$\boldsymbol{T}_a = \begin{pmatrix} & \vdots & \vdots & \vdots & \vdots & \\ \cdots & \boldsymbol{A}_0 & \boldsymbol{A}_1 & 0 & 0 & \cdots \\ \cdots & 0 & \boldsymbol{A}_0 & \boldsymbol{A}_1 & 0 & \cdots \\ & \vdots & \vdots & \vdots & \vdots & \end{pmatrix}, \qquad (3.4.4)$$

that is, it has a double block diagonal. The fact that $\boldsymbol{T}_a$ is orthogonal, or $\boldsymbol{T}_a\boldsymbol{T}_a^T = \boldsymbol{T}_a^T\boldsymbol{T}_a = \boldsymbol{I}$, yields

$$\boldsymbol{A}_0^T\boldsymbol{A}_0 + \boldsymbol{A}_1^T\boldsymbol{A}_1 = \boldsymbol{A}_0\boldsymbol{A}_0^T + \boldsymbol{A}_1\boldsymbol{A}_1^T = \boldsymbol{I}, \qquad (3.4.5)$$

as well as

$$\boldsymbol{A}_0^T\boldsymbol{A}_1 = \boldsymbol{A}_1^T\boldsymbol{A}_0 = 0, \quad \boldsymbol{A}_0\boldsymbol{A}_1^T = \boldsymbol{A}_1\boldsymbol{A}_0^T = 0. \qquad (3.4.6)$$

The property (3.4.6) is called *orthogonality of tails* since overlapping tails of the basis functions are orthogonal to each other. Note that these conditions characterize nothing but an N-channel orthogonal filter bank, with filters of length $2N$ and downsampling by N. To obtain certain classes of LOT's, one imposes additional constraints. For example, in Section 3.4.3, we will consider a cosine modulated filter bank.

Generalizations What we have seen in these two simple cases, is how to obtain N-channel filter banks with filters of length N (block transforms) and filters of length $2N$ (lapped orthogonal transforms). It is obvious that by allowing longer filters, or more blocks $\boldsymbol{A}_i$ in (3.4.4), we can obtain general N-channel filter banks.

3.4.2 Analysis of Multichannel Filter Banks

The analysis of N-channel filter banks is in many ways analogous to that of two-channel filter banks; therefore, the treatment here will be fairly brisk, with references to Section 3.2.

Time-Domain Analysis We can proceed here exactly as in Section 3.2.1. Thus, we can say that the channel outputs (or transform coefficients) in Figure 3.14 can be expressed as in (3.2.1)

$$\boldsymbol{y} = \boldsymbol{X} = \boldsymbol{T}_a\,\boldsymbol{x},$$

where the vector of transform coefficients is $\boldsymbol{X}$, with $X[Nk+i] = y_i[k]$. The analysis matrix $\boldsymbol{T}_a$ is given as in (3.2.2) with blocks $\boldsymbol{A}_i$ of the form

$$\boldsymbol{A}_i = \begin{pmatrix} h_0[Nk-1-Ni] & \cdots & h_0[Nk-N-Ni] \\ \vdots & & \vdots \\ h_{N-1}[Nk-1-Ni] & \cdots & h_{N-1}[Nk-N-Ni] \end{pmatrix}.$$

When the filters are of length $L = KN$, there are K blocks $\boldsymbol{A}_i$ of size $N \times N$ each. Similarly to (3.2.4–3.2.5), we see that the basis functions of the first basis corresponding to the analysis are

$$\varphi_{Nk+i}[n] = h_i[Nk-n].$$

Defining the synthesis matrix as in (3.2.7), we obtain the basis functions of the dual basis

$$\tilde{\varphi}_{Nk+i}[n] = g_i[n-Nk],$$

and they satisfy the following biorthogonality relations:

$$\langle \varphi_k[n], \tilde{\varphi}_l[n] \rangle = \delta[k-l],$$

which can be expressed in terms of analysis/synthesis matrices as

$$\boldsymbol{T}_s \boldsymbol{T}_a = \boldsymbol{I}.$$

As was done in Section 3.2, we can define single operators for each branch. If the operator $\boldsymbol{H}_i$ represents filtering by h_i followed by downsampling by N, its matrix representation is

$$\boldsymbol{H}_i = \begin{pmatrix} & \vdots & & \vdots & \vdots & \\ \cdots & h_i[L-1] & \cdots & h_i[L-N] & h_i[L-N-1] & \cdots \\ \cdots & 0 & \cdots & 0 & h_i[L-1] & \cdots \\ & \vdots & & \vdots & \vdots & \end{pmatrix}.$$

Defining $\boldsymbol{G}_i$ similarly to $\boldsymbol{H}_i$ (except that there is no time reversal), the output of the system can then be written as

$$\hat{\boldsymbol{x}} = \left(\sum_{i=0}^{N-1} \boldsymbol{G}_i^T \boldsymbol{H}_i \right) \boldsymbol{x}.$$

Then, the condition for perfect reconstruction is

$$\sum_{i=0}^{N-1} \boldsymbol{G}_i^T \boldsymbol{H}_i = \boldsymbol{I}.$$

We leave the details and proofs of the above relationships as an exercise (Problem 3.21), since they are simple extensions of the two-channel case seen in Section 3.2.

Modulation-Domain Analysis Let us turn our attention to filter banks represented in the modulation domain. We write directly the expressions we need in the z-domain. One can verify that downsampling a signal $x[n]$ by N followed by upsampling by N (that is, replacing $x[n], n \bmod N \neq 0$ by 0) produces a signal $y[n]$ with z-transform $Y(z)$ equal to

$$Y(z) = \frac{1}{N} \sum_{i=0}^{N-1} X(W_N^i z), \quad W_N = e^{-j2\pi/N}, \quad j = \sqrt{-1}$$

because of the orthogonality of the roots of unity. Then, the output of the system in Figure 3.14 becomes, in a similar fashion to (3.2.14)

$$\hat{X}(z) = \frac{1}{N} \boldsymbol{g}^T(z) \, \boldsymbol{H}_m(z) \, \boldsymbol{x}_m(z),$$

where $\boldsymbol{g}^T(z) = (\, G_0(z) \quad \ldots \quad G_{N-1}(z) \,)$ is the vector containing synthesis filters, $\boldsymbol{x}_m(z) = (\, X(z) \quad \ldots \quad X(W_N^{N-1}z) \,)^T$ and the ith line of $\boldsymbol{H}_m(z)$ is equal to $(\, H_i(z) \quad \ldots \quad H_i(W_N^{N-1}z) \,)$, $i = 0, \ldots, N-1$. Then, similarly to the two-channel case, to cancel aliasing, $\boldsymbol{g}^T \boldsymbol{H}_m$ has to have all elements equal to zero, except for the first one. To obtain perfect reconstruction, this only nonzero element has to be equal to a scaled pure delay.

As in the two-channel case, it can be shown that the perfect reconstruction condition is equivalent to the system being biorthogonal, as given earlier. The proof is left as an exercise for the reader (Problem 3.21). For completeness, let us define $\boldsymbol{G}_m(z)$ as the matrix with the ith row equal to

$$(\, G_0(W_N^i z) \quad G_1(W_N^i z) \quad \ldots \quad G_{N-1}(W_N^i z) \,).$$

Polyphase-Domain Analysis The gist of the polyphase analysis of two-channel filter banks downsampled by 2 was to expand signals and filter impulse responses into even- and odd-indexed components (together with some adequate phase terms). Quite naturally, in the N-channel case with downsampling by N, there will be N polyphase components. We follow the same definitions as in Section 3.2.1 (the choice of the phase in the polyphase component is arbitrary, but consistent).

Thus, the input signal can be decomposed into its polyphase components as

$$X(z) = \sum_{j=0}^{N-1} z^{-j} X_j(z^N),$$

where

$$X_j(z) = \sum_{n=-\infty}^{\infty} x[nN + j] \, z^{-n}.$$

Define the polyphase vector as

$$\boldsymbol{x}_p(z) = (X_0(z)\ X_1(z) \ldots X_{N-1}(z))^T.$$

The polyphase components of the synthesis filter g_i are defined similarly, that is

$$G_i(z) = \sum_{j=0}^{N-1} z^{-j} G_{ij}(z^N),$$

where

$$G_{ij}(z) = \sum_{n=-\infty}^{\infty} g_i[nN+j]\ z^{-n}.$$

The polyphase matrix of the synthesis filter bank is given by

$$[\boldsymbol{G}_p(z)]_{ji} = G_{ij}(z),$$

where the implicit transposition should be noticed. Up to a phase factor and a transpose, the analysis filter bank is decomposed similarly. The filter is written as

$$H_i(z) = \sum_{j=0}^{N-1} z^j H_{ij}(z^N), \tag{3.4.7}$$

where

$$H_{ij}(z) = \sum_{n=-\infty}^{\infty} h_i[nN-j]\ z^{-n}. \tag{3.4.8}$$

The analysis polyphase matrix is then defined as follows:

$$[\boldsymbol{H}_p(z)]_{ij} = H_{ij}(z).$$

For example, the vector of channel signals,

$$\boldsymbol{y}(z) = (y_0(z)\quad y_1(z) \ldots y_{N-1}(z))^T,$$

can be compactly written as

$$\boldsymbol{y}(z) = \boldsymbol{H}_p(z)\, \boldsymbol{x}_p(z).$$

Putting it all together, the output of the analysis/synthesis filter bank in Figure 3.14 can be written as

$$\hat{X}(z) = (1 \quad z^{-1} \quad z^{-1} \quad \ldots \quad z^{-N+1}) \cdot \boldsymbol{G}_p(z^N) \cdot \boldsymbol{H}_p(z^N) \cdot \boldsymbol{x}_p(z^N).$$

Similarly to the two-channel case, we can define the transfer function matrix $\boldsymbol{T}_p(z) = \boldsymbol{G}_p(z)\boldsymbol{H}_p(z)$. Then, the same results hold as in the two-channel case. Here, we just state them (the proofs are N-channel counterparts of the two-channel ones).

THEOREM 3.16 *Multichannel Filter Banks*

(a) Aliasing in a one-dimensional system is cancelled *if and only if* the transfer function matrix is pseudo-circulant [311].

(b) Given an analysis filter bank downsampled by N with polyphase matrix $\boldsymbol{H}_p(z)$, alias-free reconstruction is possible *if and only if* the normal rank of $\boldsymbol{H}_p(z)$ is equal to N.

(c) Given a critically sampled FIR analysis filter bank, perfect reconstruction with FIR filters is possible *if and only if* $\det(\boldsymbol{H}_p(z))$ is a pure delay.

Note that the modulation and polyphase representations are related via the Fourier matrix. For example, one can verify that

$$\boldsymbol{x}_p(z^N) = \frac{1}{N} \begin{pmatrix} 1 & & & \\ & z & & \\ & & \ddots & \\ & & & z^{N-1} \end{pmatrix} \boldsymbol{F}\boldsymbol{x}_m(z), \tag{3.4.9}$$

where $\boldsymbol{F}_{kl} = W_N^{kl} = e^{-j(2\pi/N)kl}$. Similar relationships hold between $\boldsymbol{H}_m(z)$, $\boldsymbol{G}_m(z)$ and $\boldsymbol{H}_p(z)$, $\boldsymbol{G}_p(z)$, respectively (see Problem 3.22). The important point to note is that modulation and polyphase matrices are related by unitary operations (such as $\boldsymbol{F}$ and delays as in (3.4.9)).

Orthogonal Multichannel FIR Filter Banks Let us now consider the particular but important case when the filter bank is unitary or orthogonal. This is an extension of the discussion in Section 3.2.3 to the N-channel case. The idea is to implement an orthogonal transform using an N-channel filter bank, or in other words, we want the following set:

$$\{g_0[n - NK], \ldots, g_{N-1}[n - NK]\}, \quad n \in \mathcal{Z}$$

to be an orthonormal basis for $l_2(\mathcal{Z})$. Then

$$\langle g_i[n - Nk], g_j[n - Nl] \rangle = \delta[i-j]\, \delta[l-k]. \tag{3.4.10}$$

Since in the orthogonal case analysis and synthesis filters are identical up to a time reversal, (3.4.10) holds for $h_i[Nk - l]$ as well. By using (2.5.19), (3.4.10) can be expressed in z-domain as

$$\sum_{k=0}^{N-1} G_i(W_N^k z)\, G_j(W_N^{-k} z^{-1}) = N\delta[i-j], \tag{3.4.11}$$

or

$$\boldsymbol{G}_{m*}^T(z^{-1})\, \boldsymbol{G}_m(z) = N\boldsymbol{I},$$

where the subscript $_*$ stands for conjugation of the coefficients but not of z (this is necessary since $\boldsymbol{G}_m(z)$ has complex coefficients). Thus, as in the two-channel case, having an orthogonal transform is equivalent to having a paraunitary modulation matrix. Unlike the two-channel case, however, not all of the filters are obtained from a single prototype filter.

Since modulation and polyphase matrices are related, it is easy to check that having a paraunitary modulation matrix is equivalent to having a paraunitary polyphase matrix, that is

$$\boldsymbol{G}_{m*}^T(z^{-1})\,\boldsymbol{G}_m(z) \;=\; N\,\boldsymbol{I} \iff \boldsymbol{G}_p^T(z^{-1})\,\boldsymbol{G}_p(z) \;=\; \boldsymbol{I}. \tag{3.4.12}$$

Finally, in time domain

$$\boldsymbol{G}_i\,\boldsymbol{G}_j^T \;=\; \delta[i-j]\,\boldsymbol{I}, \qquad i,j=0,1,$$

or

$$\boldsymbol{T}_a^T\,\boldsymbol{T}_a \;=\; \boldsymbol{I}.$$

The above relations lead to a direct extension of Theorem 3.8, where the particular case $N=2$ was considered.

Thus, according to (3.4.12), designing an orthogonal filter bank with N channels reduces to finding $N\times N$ paraunitary matrices. Just as in the two-channel case, where we saw a lattice realization of orthogonal filter banks (see (3.2.60)), $N\times N$ paraunitary matrices can be parametrized in terms of cascades of elementary matrices (2×2 rotations and delays). Such parametrizations have been investigated by Vaidyanathan, and we refer to his book [308] for a thorough treatment. An overview can be found in Appendix 3.A.2. As an example, we will see how to construct three-channel paraunitary filter banks.

Example 3.11

We use the factorization given in Appendix 3.A.2, (3.A.8). Thus, we can express the 3×3 polyphase matrix as

$$\boldsymbol{G}_p(z) \;=\; \boldsymbol{U}_0\left[\prod_{i=1}^{K-1}\begin{pmatrix} z^{-1} & & \\ & 1 & \\ & & 1\end{pmatrix}\boldsymbol{U}_i\right],$$

where

$$\boldsymbol{U}_0 \;=\; \begin{pmatrix} 1 & 0 & 0 \\ 0 & \cos\alpha_{00} & -\sin\alpha_{00} \\ 0 & \sin\alpha_{00} & \cos\alpha_{00}\end{pmatrix}\begin{pmatrix} \cos\alpha_{01} & 0 & -\sin\alpha_{01} \\ 0 & 1 & 0 \\ \sin\alpha_{01} & 0 & \cos\alpha_{01}\end{pmatrix}$$
$$\times\begin{pmatrix} \cos\alpha_{02} & -\sin\alpha_{02} & 0 \\ \sin\alpha_{02} & \cos\alpha_{02} & 0 \\ 0 & 0 & 1\end{pmatrix},$$

and U_i are given by

$$U_i = \begin{pmatrix} \cos\alpha_{i0} & -\sin\alpha_{i0} & 0 \\ \sin\alpha_{i0} & \cos\alpha_{i0} & 0 \\ 0 & 0 & 1 \end{pmatrix} \begin{pmatrix} 1 & 0 & 0 \\ 0 & \cos\alpha_{i1} & -\sin\alpha_{i1} \\ 0 & \sin\alpha_{i1} & \cos\alpha_{i1} \end{pmatrix}.$$

The degrees of freedom are given by the angles α_{ij}. To obtain the three analysis filters, we upsample the polyphase matrix, and thus

$$[G_0(z)\; G_1(z)\; G_2(z)] = [1\; z^{-1}\; z^{-2}]\; G_p(z^3).$$

To design actual filters, one could minimize an objective function as the one given in [306], where the sum of all the stopbands was minimized.

It is worthwhile mentioning that N-channel orthogonal filter banks with more than two channels have greater design freedom. It is possible to obtain orthogonal linear phase FIR solutions [275, 321], a solution which was impossible for two channels (see Appendix 3.A.2).

3.4.3 Modulated Filter Banks

We will now examine a particular class of N channel filter banks — modulated filter banks. The name stems from the fact that all the filters in the analysis bank are obtained by modulating a single prototype filter. If we impose orthogonality as well, the synthesis filters will obviously be modulated as well. The first class we consider imitates the short-time Fourier transform (STFT), but in the discrete-time domain. The second one — cosine modulated filter banks, is an interesting counterpart to the STFT, and when the length of the filters is restricted to $2N$, it is an example of a modulated LOT.

Short-Time Fourier Transform in the Discrete-Time Domain The short-time Fourier or Gabor transform [204, 226] is a very popular tool for nonstationary signal analysis (see Section 2.6.3). It has an immediate filter bank interpretation. Assume a window function $h_{pr}[n]$ with a corresponding z-transform $H_{pr}(z)$. This window function is a prototype lowpass filter with a bandwidth of $2\pi/N$, which is then modulated evenly over the frequency spectrum using consecutive powers of the Nth root of unity

$$H_i(z) = H_{pr}(W_N^i z), \qquad i = 0, \ldots, N-1, \quad W_N = e^{-j2\pi/N}, \tag{3.4.13}$$

or

$$h_i[n] = W_N^{-in}\, h_{pr}[n]. \tag{3.4.14}$$

That is, if $H_{pr}(e^{j\omega})$ is a lowpass filter centered around $\omega = 0$, then $H_i(e^{j\omega})$ is a bandpass filter centered around $\omega = (i2\pi)/N$. Note that the prototype window is usually real, but the bandpass filters are complex.

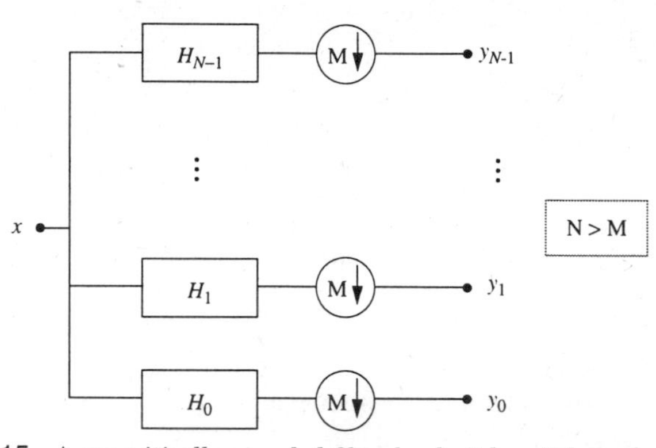

Figure 3.15 A noncritically sampled filter bank; it has N branches followed by sampling by M ($N > M$). When the filters are modulated versions (by the Nth root of unity), then this implements a discrete-time version of the short-time Fourier transform.

In the short-time Fourier transform, the window is advanced by M samples at a time, which corresponds to a downsampling by M of the corresponding filter bank. This filter bank interpretation of the short-time Fourier transform analysis is depicted in Figure 3.15. The short-time Fourier transform synthesis is achieved similarly with a modulated synthesis filter bank. Usually, M is chosen smaller than N (for example, $N/2$), and then, it is obviously an oversampled scheme or a noncritically sampled filter bank. Let us now consider what happens if we critically sample such a filter bank, that is, downsample by N. Compute a critically sampled discrete short-time Fourier (or Gabor) transform, where the window function is given by the prototype filter. It is easy to verify the following negative result [315] (which is a discrete-time equivalent of the Balian-Low theorem, given in Section 5.3.3):

THEOREM 3.17

There are no finite-support bases with filters as in (3.4.13) (except trivial ones with only N nonzero coefficients).

PROOF

The proof consists in analyzing the polyphase matrix $\boldsymbol{H}_p(z)$. Write the prototype filter $H_{pr}(z)$ in terms of its polyphase components (see (3.4.7–3.4.8))

$$H_{pr}(z) = \sum_{j=0}^{N-1} z^j H_{pr_j}(z^N),$$

where $H_{pr_j}(z)$ is the jth polyphase component of $H_{pr}(z)$.

Obviously, following (3.4.7) and (3.4.13),

$$H_i(z) = \sum W_N^{ij} \, z^j \, H_{pr_j}(z^N).$$

Therefore, the polyphase matrix $\boldsymbol{H}_p(z)$ has entries

$$[\boldsymbol{H}_p(z)]_{ij} \;=\; W_N^{ij}\, H_{pr_j}(z).$$

Then, $\boldsymbol{H}_p(z)$ can be factored as

$$\boldsymbol{H}_p(z) \;=\; \boldsymbol{F} \begin{pmatrix} H_{pr_0}(z) & & & \\ & H_{pr_1}(z) & & \\ & & \ddots & \\ & & & H_{pr_{N-1}}(z) \end{pmatrix}, \tag{3.4.15}$$

where $F_{kl} = W_N^{kl} = e^{-j(2\pi/N)kl}$. For FIR perfect reconstruction, the determinant of $H_p(z)$ has to be a delay (by Theorem 3.16). Now,

$$\det(\boldsymbol{H}_p(z)) \;=\; c \prod_{j=0}^{N-1} H_{pr_j}(z),$$

where c is a complex number equal to $\det(\boldsymbol{F})$. Therefore, for perfect FIR reconstruction, $H_{pr_j}(z)$ has to be of the form $\alpha_i \cdot z^{-m}$, that is, the prototype filter has exactly N nonzero coefficients. For an orthogonal solution, the α_i's have to be unit-norm constants.

What happens if we relax the FIR requirement? For example, one can choose the following prototype:

$$H_{pr}(z) \;=\; \sum_{i=0}^{N-1} P_i(z^N)\, z^i, \tag{3.4.16}$$

where $P_i(z)$ are allpass filters. The factorization (3.4.15) still holds, with $H_{pr_i}(z) = P_i(z)$, and since $P_i(z^{-1}) \cdot P_i(z) = 1$, $\boldsymbol{H}_p(z)$ is paraunitary. While this gives an orthogonal modulated filter bank, it is IIR (either analysis or synthesis will be noncausal), and the quality of the filter in (3.4.16) can be poor.

Cosine Modulated Filter Banks The problems linked to complex modulated filter banks can be solved by using appropriate cosine modulation. Such cosine-modulated filter banks are very important in practice, for example in audio compression (see Section 7.2.2). Since they are often of length $L = 2N$ (where N is the downsampling rate), they are sometimes referred to as modulated LOT's, or MLT's. A popular version was proposed in [229] and thus called the Princen-Bradley filter bank. We will study one class of cosine modulated filter banks in some depth, and refer to [188, 308] for a more general and detailed treatment. The cosine modulated filter banks we consider here are a particular case of *pseudoquadrature mirror filter banks* (PQMF) when the filter length is restricted to twice the number of channels $L = 2N$. Pseudo QMF filters have been proposed as an extension to N channels of the classical two-channel QMF filters. Pseudo QMF analysis/synthesis systems achieve in general only cancellation of the main aliasing term (aliasing from neighboring channels). However, when the filter length is restricted to $L = 2N$, they can achieve perfect reconstruction. Due to the modulated structure and just as in the STFT case, there are fast computational algorithms, making such filter banks attractive for implementations.

A family of PQMF filter banks that achieves cancellation of the main aliasing term is of the form [188, 321][11]

$$h_k[n] = \frac{1}{\sqrt{N}} h_{pr}[n] \cos\left(\frac{\pi(2k+1)}{2N}\left(n - \left(\frac{L-1}{2}\right)\right) + \phi_k\right), \qquad (3.4.17)$$

for the analysis filters ($h_{pr}[n]$ is the impulse response of the window). The modulating frequencies of the cosines are at $\pi/2N, 3\pi/2N, \ldots, (2N-1)\pi/2N$, and the prototype window is a lowpass filter with support $[-\pi/2N, \pi/2N]$. Then, the kth filter is a bandpass filter with support from $k\pi/N$ to $(k+1)\pi/N$ (and a mirror image from $-k\pi/N$ to $-(k+1)\pi/N$), thus covering the range from 0 to π evenly. Note that for $k = 0$ and $N-1$, the two lobes merge into a single lowpass and highpass filter respectively. In the general case, the main aliasing term is canceled for the following possible value of the phase:

$$\phi_k = \frac{\pi}{4} + k\frac{\pi}{2}.$$

For this value of phase, and in the special case $L = 2N$, exact reconstruction is achieved. This yields filters of the form

$$h_k[n] = \frac{1}{\sqrt{N}} h_{pr}[n] \cos\left(\frac{2k+1}{4N}(2n - N + 1)\pi\right), \qquad (3.4.18)$$

for $k = 0, \ldots, N-1$, $n = 0, \ldots, 2N-1$. Since the filter length is $2N$, we have an LOT, and we can use the formalism in (3.4.4). It can be shown that, due to the particular structure of the filters, if $h_{pr}[n] = 1$, $n = 0, \ldots, 2N-1$, (3.4.5–3.4.6) hold. The idea of the proof is the following (we assume N to be even): Being of length $2N$, each filter has a left and a right tail of length N. It can be verified that with the above choice of phase, all the filters have symmetric left tails ($h_k[N/2-1-l] = h_k[N/2+l]$, for $l = 0, \ldots, N/2-1$) and antisymmetric right tails ($h_k[3N/2-1-l] = h_k[3N/2+l]$, for $l = 0, \ldots, N/2-1$). Then, orthogonality of the tails (see (3.4.6)) follows because the product of the left and right tail is an odd function, and therefore, sums to zero. Additionally, each filter is orthogonal to its modulated versions and has norm 1, and thus, we have an orthonormal LOT. The details are left as an exercise (see Problem 3.24).

Suppose now that we use a symmetric window $h_{pr}[n]$. We want to find conditions under which (3.4.5–3.4.6) still hold. Call $\boldsymbol{B}_i$ the blocks in (3.4.5–

[11] The derivation of this type of filter bank is somewhat technical and thus less explicit at times than other filter banks seen so far.

3.4.6) when no windowing is used, or $h_{pr}[n] = 1$, $n = 0, \ldots, 2N-1$, and $\boldsymbol{A}_i$ the blocks, with a general symmetric window $h_{pr}[n]$. Then, we can express $\boldsymbol{A}_0$ in terms of $\boldsymbol{B}_0$ as

$$\boldsymbol{A}_0 = \begin{pmatrix} h_0[2N-1] & \cdots & h_0[N] \\ \vdots & & \vdots \\ h_{N-1}[2N-1] & \cdots & h_{N-1}[N] \end{pmatrix} \tag{3.4.19}$$

$$= \boldsymbol{B}_0 \cdot \begin{pmatrix} h_{pr}[2N-1] & & \\ & \ddots & \\ & & h_{pr}[N] \end{pmatrix} \tag{3.4.20}$$

$$= \boldsymbol{B}_0 \cdot \underbrace{\begin{pmatrix} h_{pr}[0] & & \\ & \ddots & \\ & & h_{pr}[N-1] \end{pmatrix}}_{\boldsymbol{W}} \tag{3.4.21}$$

since h_{pr} is symmetric, that is $h_{pr}[n] = h_{pr}[2N-1-n]$, and $\boldsymbol{W}$ denotes the window matrix. Using the antidiagonal matrix $\boldsymbol{J}$,

$$\boldsymbol{J} = \begin{pmatrix} & & 1 \\ & \cdots & \\ 1 & & \end{pmatrix},$$

it is easy to verify that $\boldsymbol{A}_1$ is related to $\boldsymbol{B}_1$, in a similar fashion, up to a reversal of the entries of the window function, or

$$\boldsymbol{A}_1 = \boldsymbol{B}_1 \boldsymbol{J} \boldsymbol{W} \boldsymbol{J}. \tag{3.4.22}$$

Note also that due to the particular structure of the cosines involved, the following are true as well:

$$\boldsymbol{B}_0 \boldsymbol{B}_0^T = \boldsymbol{B}_0^T \boldsymbol{B}_0 = \frac{1}{2}(\boldsymbol{I} - \boldsymbol{J}), \qquad \boldsymbol{B}_1 \boldsymbol{B}_1^T = \boldsymbol{B}_1^T \boldsymbol{B}_1 = \frac{1}{2}(\boldsymbol{I} + \boldsymbol{J}). \tag{3.4.23}$$

The proof of the above fact is left as an exercise to the reader (see Problem 3.24). Therefore, take (3.4.5) and substitute the expressions for $\boldsymbol{A}_0$ and $\boldsymbol{A}_1$ given in (3.4.19) and (3.4.22)

$$\boldsymbol{A}_0^T \boldsymbol{A}_0 + \boldsymbol{A}_1^T \boldsymbol{A}_1 = \boldsymbol{W} \boldsymbol{B}_0^T \boldsymbol{B}_0 \boldsymbol{W} + \boldsymbol{J} \boldsymbol{W} \boldsymbol{J} \boldsymbol{B}_1^T \boldsymbol{B}_1 \boldsymbol{J} \boldsymbol{W} \boldsymbol{J} = \boldsymbol{I}.$$

Using now (3.4.23), this becomes

$$\frac{1}{2} \boldsymbol{W}^2 + \frac{1}{2} \boldsymbol{J} \boldsymbol{W}^2 \boldsymbol{J} = \boldsymbol{I},$$

where we used the fact that $\boldsymbol{J}^2 = \boldsymbol{I}$. In other words, for perfect reconstruction, the following has to hold:

$$h_{pr}^2[i] + h_{pr}^2[N-1-i] = 2, \tag{3.4.24}$$

Table 3.4 Values of a power complementary window used for generating cosine modulated filter banks (the window satisfies (3.4.24)). It is symmetric ($h_{pr}[16-k-1] = h_{pr}[k]$).

$h_{pr}[0]$	0.125533	$h_{pr}[4]$	1.111680
$h_{pr}[1]$	0.334662	$h_{pr}[5]$	1.280927
$h_{pr}[2]$	0.599355	$h_{pr}[6]$	1.374046
$h_{pr}[3]$	0.874167	$h_{pr}[7]$	1.408631

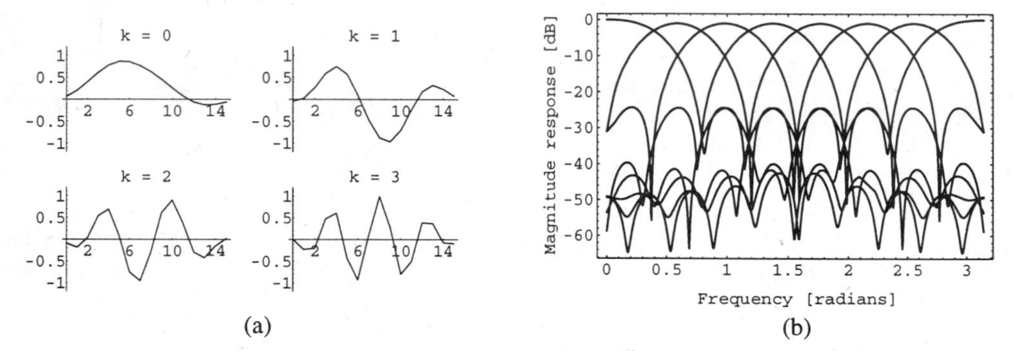

Figure 3.16 An example of a cosine modulated filter bank with $N = 8$. (a) Impulse responses for the first four filters. (b) The magnitude responses of all the filters are given. The symmetric prototype window is of length 16 with the first 8 coefficients given in Table 3.4.

that is, a power complementary property. Using the expressions for $\boldsymbol{A}_0$ and $\boldsymbol{A}_1$, one can easily prove that (3.4.6) holds as well.

Condition (3.4.24) also regulates the shape of the window. For example, if instead of length $2N$, one uses shorter window of length $2N-2M$, then the outer M coefficients of each "tail" (the symmetric nonconstant half of the window) are set to zero, and the inner M ones are set to $\sqrt{2}$ according to (3.4.24).

Example 3.12

Consider the case $N = 8$. The center frequency of the modulated filter $h_k[n]$ is $(2k+1)2\pi/32$, and since this is a cosine modulation and the filters are real, there is a mirror lobe at $(32-2k-1)2\pi/32$. For the filters $h_0[n]$ and $h_7[n]$, these two lobes overlap to form a single lowpass and highpass, respectively, while $h_1[n], \ldots, h_6[n]$ are bandpass filters. A possible symmetric window of length 16 and satisfying (3.4.24) is given in Table 3.4, while the impulse responses of the first four filters as well as the magnitude responses of all the modulated filters are given in Figure 3.16.

Note that cosine modulated filter banks which are orthogonal have been recently generalized to lengths $L = KN$ where K can be larger than 2. For more details, refer to [159, 188, 235, 308].

3.5 Pyramids and Overcomplete Expansions

In this section, we will consider expansions that are overcomplete, that is, the set of functions used in the expansion is larger than actually needed. In other words, even if the functions play the role of a set of "basis functions", they are actually linearly dependent. Of course, we are again interested in structured overcomplete expansions and will consider the ones implementable with filter banks. In filter bank terminology, overcomplete means we have a noncritically sampled filter bank, as the one given in Figure 3.15.

In compression applications, such redundant representations tend to be avoided, even if an early example of a multiresolution overcomplete decomposition (the pyramid scheme to be discussed below) has been used for compression. Such schemes are also often called *hierarchical transforms* in the compression literature.

In some other applications, overcomplete expansions might be more appropriate than bases. One of the advantages of such expansions is that, due to oversampling, the constraints on the filters used are relaxed. This can result in filters of a superior quality than those in critically sampled systems. Another advantage is that time variance can be reduced, or in the extreme case of no downsampling, avoided. One such example is the oversampled discrete-time wavelet series which is also explained in what follows.

3.5.1 Oversampled Filter Banks

The simplest way to obtain a noncritically sampled filter bank is not to sample at all, producing an overcomplete expansion. Thus, let us consider a two-channel filter bank with no downsampling. In the scheme given in Figure 3.15 this means that $N = 2$ and $M = 1$. Then, the output is (see also Example 5.2)

$$\hat{X}(z) = [G_0(z)\, H_0(z) + G_1(z)\, H_1(z)]\, X(z), \tag{3.5.1}$$

and perfect reconstruction is easily achievable. For example, in the FIR case if $H_0(z)$ and $H_1(z)$ have no zeros in common (that is, the polynomials in z^{-1} are coprime), then one can use Euclid's algorithm [32] to find $G_0(z)$ and $G_1(z)$ such that

$$G_0(z)\, H_0(z) + G_1(z)\, H_1(z) = 1$$

is satisfied leading to $\hat{X}(z) = X(z)$ in (3.5.1). Note how coprimeness of $H_0(z)$ and $H_1(z)$, used in Euclid's algorithm, is also a very natural requirement in terms of signal processing. A common zero would prohibit FIR reconstruction, or even IIR reconstruction (if the common zero is on the unit circle). Another case appears when we have two filters $G_0(z)$ and $G_1(z)$ which have unit norm and satisfy

$$G_0(z)\, G_0(z^{-1}) + G_1(z)\, G_1(z^{-1}) \;=\; 2, \tag{3.5.2}$$

since then with $H_0(z) = G_0(z^{-1})$ and $H_1(z) = G_1(z^{-1})$ one obtains

$$\hat{X}(z) \;=\; [G_0(z)\, G_0(z^{-1}) + G_1(z)\, G_1(z^{-1})]\; X(z) \;=\; 2X(z).$$

Writing this in time domain (see Example 5.2), we realize that the set $\{g_i[n-k]\}$, $i = 0, 1$, and $k \in \mathcal{Z}$, forms a tight frame for $l_2(\mathcal{Z})$ with a redundancy factor $R = 2$.

The fact that $\{g_i[n-k]\}$ form a tight frame simply means that they can uniquely represent any sequence from $l_2(\mathcal{Z})$ (see also Section 5.3). However, the basis vectors are not linearly independent and thus they do not form an orthonormal basis. The redundancy factor indicates the oversampling rate; we can indeed check that it is two in this case, that is, there are twice as many basis functions than actually needed to represent sequences from $l_2(\mathcal{Z})$. This is easily seen if we remember that until now we needed only the even shifts of $g_i[n]$ as basis functions, while now we use the odd shifts as well. Also, the expansion formula in a tight frame is similar to that in the orthogonal case, except for the redundancy (which means the functions in the expansion are not linearly independent). There is an energy conservation relation, or Parseval's formula, which says that the energy of the expansion coefficients equals R times the energy of the original. In our case, calling $y_i[n]$ the output of the filter $h_i[n]$, we can verify (Problem 3.26) that

$$\|x\|^2 \;=\; 2(\|y_0\|^2 + \|y_1\|^2). \tag{3.5.3}$$

To design such a tight frame for $l_2(\mathcal{Z})$ based on filter banks, that is, to find solutions to (3.5.2), one can find a unit norm[12] filter $G_0(z)$ which satisfies

$$0 \;\leq\; |G_0(e^{j\omega})|^2 \;\leq\; 2,$$

and then take the spectral factorization of the difference $2 - G_0(z)G_0(z^{-1}) = G_1(z)G_1(z^{-1})$ to find $G_1(z)$. Alternatively, note that (3.5.2) means the 2×1 vector $(\, G_0(z) \quad G_1(z)\,)^T$ is lossless, and one can use a lattice structure for its factorization, just as in the 2×2 lossless case [308]. On the unit circle, (3.5.2) becomes

$$|G_0(e^{j\omega})|^2 + |G_1(e^{j\omega})|^2 \;=\; 2,$$

that is, $G_0(z)$ and $G_1(z)$ are power complementary. Note that (3.5.2) is less restrictive than the usual orthogonal solutions we have seen in Section 3.2.3. For example, odd-length filters are possible.

[12]Note that the unit norm requirement is not necessary for constructing a tight frame.

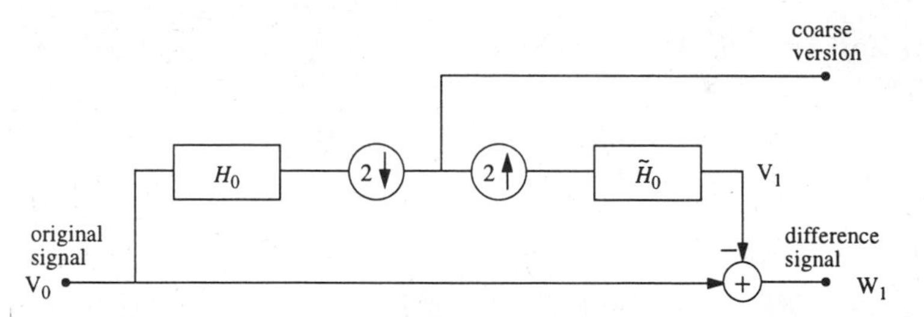

Figure 3.17 Pyramid scheme involving a coarse lowpass approximation and a difference between the coarse approximation and the original. We show the case where an orthogonal filter is used and therefore, the coarse version (after interpolation) is a projection onto V_1, while the difference is a projection onto W_1. This indicates the multiresolution behavior of the pyramid.

Of course, one can iterate such nondownsampled two-channel filter banks, and get more general solutions. In particular, by adding two-channel non-downsampled filter banks with filters $\{H_0(z^2), H_1(z^2)\}$ to the lowpass analysis channel and iterating (raising z to the appropriate power) one can devise a discrete-time wavelet series. This is a very redundant expansion, since there is no downsampling. However, unlike the critically sampled wavelet series, this expansion is shift-invariant and is useful in applications where shift invariance is a requirement (for example, object recognition).

More general cases of noncritically sampled filter banks, that is, N-channel filter banks with downsampling by M where $M < N$, have not been much studied (except for the Fourier case discussed below). While some design methods are possible (for example, embedding into larger lossless systems), there are still open questions.

3.5.2 Pyramid Scheme

In computer vision and image coding, a successive approximation or multiresolution technique called an image pyramid is frequently used. This scheme was introduced by Burt and Adelson [41] and was recognized by the wavelet community to have a strong connection to multiresolution analysis as well as orthonormal bases of wavelets. It consists of deriving a low-resolution version of the original, then predicting the original based on the coarse version, and finally taking the difference between the original and the prediction (see Figure 3.17). At the reconstruction, the prediction is added back to the difference, guaranteeing perfect reconstruction. A shortcoming of this scheme is the oversampling, since we end up with a low-resolution version and a full-resolution difference signal (at the initial rate). Obviously, the scheme can be iterated, decomposing the coarse version repeatedly, to obtain a coarse version at level J plus J detailed versions. From the above description, it is obvious that the scheme is inherently multiresolution. Consider, for example, the coarse and detailed versions at the first level (one stage). The coarse version is now at twice the scale (downsampling has contracted it by 2) and half the resolution (information loss has occurred), while the detailed version is also of half resolution but of the same scale as the original. Also, a successive approximation flavor is easily seen: One could start with the coarse version at level J, and by adding difference signals, obtain versions at levels $J-1, \ldots, 1, 0$, (that is, the original).

An advantage of the pyramid scheme in image coding is that nonlinear interpolation and decimation operators can be used. A disadvantage, however, as we have already mentioned, is that the scheme is oversampled, although the overhead in number of samples decreases as the dimensionality increases. In n dimensions, oversampling s as a function of the number of levels L in the pyramid is given by

$$s = \sum_{i=0}^{L-1} \left(\frac{1}{2^n}\right)^i < \frac{2^n}{2^n - 1}, \tag{3.5.4}$$

which is an overhead of 50–100% in one dimension. It goes down to 25–33% in two dimensions, and further down to 12.5–14% in three dimensions. However, we will show below [240, 319] that if the system is linear and the lowpass filter is orthogonal to its even translates, then one can actually downsample the difference signal after filtering it. In that case, the pyramid reduces exactly to a critically downsampled orthogonal subband coding scheme.

First, the prediction of the original, based on the coarse version, is simply the projection onto the space spanned by $\{h_0[2k-n], k \in \mathcal{Z}\}$. That is, calling the prediction $\bar{\boldsymbol{x}}$

$$\bar{\boldsymbol{x}} = \boldsymbol{H}_0^T \boldsymbol{H}_0 \, \boldsymbol{x}.$$

The difference signal is thus

$$\boldsymbol{d} = (\boldsymbol{I} - \boldsymbol{H}_0^T \boldsymbol{H}_0) \, \boldsymbol{x}.$$

But, because it is a perfect reconstruction system

$$\boldsymbol{I} - \boldsymbol{H}_0^T \boldsymbol{H}_0 = \boldsymbol{H}_1^T \boldsymbol{H}_1,$$

that is, $\boldsymbol{d}$ is the projection onto the space spanned by $\{h_1[2k-n], k \in \mathcal{Z}\}$. Therefore, we can filter and downsample $\boldsymbol{d}$ by 2, since

$$\boldsymbol{H}_1 \boldsymbol{H}_1^T \boldsymbol{H}_1 = \boldsymbol{H}_1.$$

In that case, the redundancy of $\boldsymbol{d}$ is removed ($\boldsymbol{d}$ is now critically sampled) and the pyramid is equivalent to an orthogonal subband coding system.

The signal $\boldsymbol{d}$ can be reconstructed by upsampling by 2 and filtering with $h_1[n]$. Then we have

$$\boldsymbol{H}_1^T (\boldsymbol{H}_1 \boldsymbol{H}_1^T \boldsymbol{H}_1) \, \boldsymbol{x} = \boldsymbol{H}_1^T \boldsymbol{H}_1 \, \boldsymbol{x} = \boldsymbol{d}$$

and this, added to $\bar{\boldsymbol{x}} = \boldsymbol{H}_0^T \boldsymbol{H}_0 \boldsymbol{x}$, is indeed equal to $\boldsymbol{x}$. In the notation of the multiresolution scheme the prediction $\bar{\boldsymbol{x}}$ is the projection onto the space V_1 and $\boldsymbol{d}$ is the projection onto W_1. This is indicated in Figure 3.17. We have thus shown that pyramidal schemes can be critically sampled as well, that is, in Figure 3.17 the difference signal can be followed by a filter $h_1[n]$ and a downsampler by 2 without any loss of information.

Note that we assumed an orthogonal filter and no quantization of the coarse version. The benefit of the oversampled pyramid comes from the fact that arbitrary filters (including nonlinear ones) can be used, and that quantization of the coarse version does not influence perfect reconstruction (see Section 7.3.2).

This scheme is very popular in computer vision, not so much because perfect reconstruction is desired but because it is a computationally efficient way to obtain multiple resolution of an image. As a lowpass filter, an approximation to a Gaussian, bell-shaped filter is often used and because the difference signal resembles the original filtered by the Laplace operator, such a scheme is usually called a Laplacian pyramid.

3.5.3 Overlap-Save/Add Convolution and Filter Bank Implementations

Filter banks can be used to implement algorithms for the computation of convolutions (see also Section 6.5.1). Two classic examples are block processing schemes — the overlap-save and overlap-add algorithms for computing a running convolution [211]. Essentially, a block of input is processed at a time (typically with frequency-domain circular convolution) and the output is merged so as to achieve true linear running convolution. Since the processing advances by steps (which corresponds to downsampling the input by the step size), these two schemes are multirate in nature and have an immediate filter bank interpretation [317].

Overlap-Add Scheme This scheme performs the following task: Assuming a filter of length L, the overlap-add algorithm takes a block of input samples of length $M = N - L + 1$, and feeds it into a size-N FFT ($N > L$). This results in a linear convolution of the signal with the filter. Since the size of the FFT is N, there will be $L - 1$ samples overlapping with adjacent blocks of size M, which are then added together (thus the name overlap-add). One can see that such a scheme can be implemented with an N-channel analysis filter bank downsampled by M, followed by multiplication (convolution in Fourier domain), upsampling by M and an N-channel synthesis filter bank, as shown in Figure 3.18.

For the details on computational complexity of the filter bank, refer to Sections 6.2.3 and 6.5.1. Also, note, that the filters used are based on the

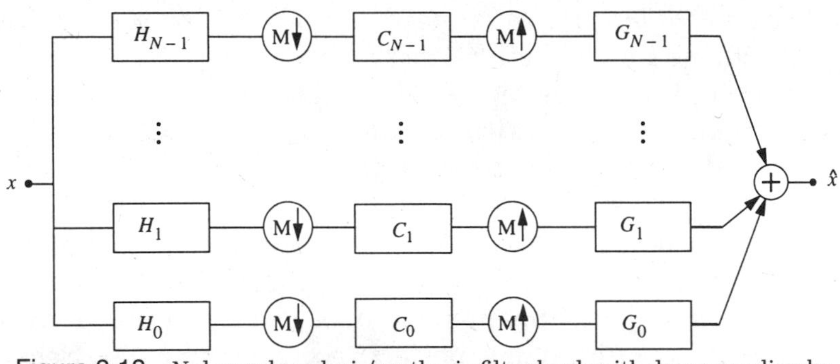

Figure 3.18 N-channel analysis/synthesis filter bank with downsampling by M and filtering of the channel signals. The downsampling by M is equivalent to moving the input by M samples between successive computations of the output. With filters based on the Fourier transform, and filtering of the channels chosen to perform frequency-domain convolution, such a filter bank implements overlap-save/add running convolution.

short-time Fourier transform.

Overlap-Save Scheme Given a length-L filter, the overlap-save algorithm performs the following: It takes N input samples, computes a circular convolution of which $N - L + 1$ samples are valid linear convolution outputs and $L - 1$ samples are wrap-around effects. These last $L - 1$ samples are discarded. The $N - L + 1$ valid ones are kept and the algorithm moves up by $N - L + 1$ samples. The filter bank implementation is similar to the overlap-add scheme, except that analysis and synthesis filters are interchanged [317].

Generalizations The above two schemes are examples from a general class of oversampled filter banks which compute running convolution. For example, the pointwise multiplication in the above schemes can be replaced by a true convolution and will result in a longer overall convolution if adequately chosen. Another possibility is to use analysis and synthesis filters based on fast convolution algorithms other than Fourier ones. For more details, see [276, 317] and Section 6.5.1.

3.6 Multidimensional Filter Banks

It seems natural to ask if the results we have seen so far on expansion of one-dimensional discrete-time signals can be generalized to multiple dimensions. This is both of theoretical interest as well as relevant in practice, since popular applications such as image compression often rely on signal decompositions. One easy solution to the multidimensional problem is to apply all known one-dimensional techniques separately along one dimension at a time. Although a very simple solution, it suffers from some drawbacks: First, only separable (for example, two-dimensional) filters are obtained in this way, leading to fairly constrained designs (nonseparable filters of size $N_1 \times N_2$ would offer $N_1 \cdot N_2$ free design variables versus $N_1 + N_2$ in the separable case). Then, only rectangular divisions of the spectrum are possible, though one might need divisions that would better capture the signal's energy concentration (for example, close to circular).

Choosing nonseparable solutions, while solving some of these problems, comes at a price: the design is more difficult, and the complexity is substantially higher.

The first step toward using multidimensional techniques on multidimensional signals is to use the same kind of sampling as before (that is, in the case of an image, sample first along the horizontal and then along the vertical dimension), but use nonseparable filters. A second step consists in using nonseparable sampling as well as nonseparable filters. This calls for the development of a new theory that starts by pointing out the major difference between one- and multidimensional cases — sampling. Sampling in multiple dimensions is represented by lattices. An excellent presentation of lattice sampling can be found in the tutorial by Dubois [86] (Appendix 3.B gives a brief overview). Filter banks using nonseparable downsampling were studied in [11, 314]. The generalization of one-dimensional analysis methods to multidimensional filter banks using lattice downsampling was done in [155, 325]. The topic has been quite active recently (see [19, 47, 48, 161, 257, 264, 288]).

In this section, we will give an overview of the field of multidimensional filter banks. We will concentrate mostly on two cases: the separable case with downsampling by 2 in two dimensions, and the quincunx case, that is, the simplest multidimensional nonseparable case with overall sampling density of 2. Both of these cases are of considerable practical interest, since these are the ones mostly used in image processing applications.

3.6.1 Analysis of Multidimensional Filter Banks

In Appendix 3.B, a brief account of multidimensional sampling is given. Using the expressions given for sampling rate changes, analysis of multidimensional systems can be performed in a similar fashion to their one-dimensional counterparts. Let us start with the simplest case, where both the filters and the sampling rate change are separable.

Example 3.13 *Separable Case with Sampling by 2 in Two Dimensions*

If one uses the schemeas in Figure 3.19 then all one-dimensional results are trivially extended to two dimensions. However, all limitations appearing in one dimension, will appear in two dimensions as well. For example, we know that there are no real two-

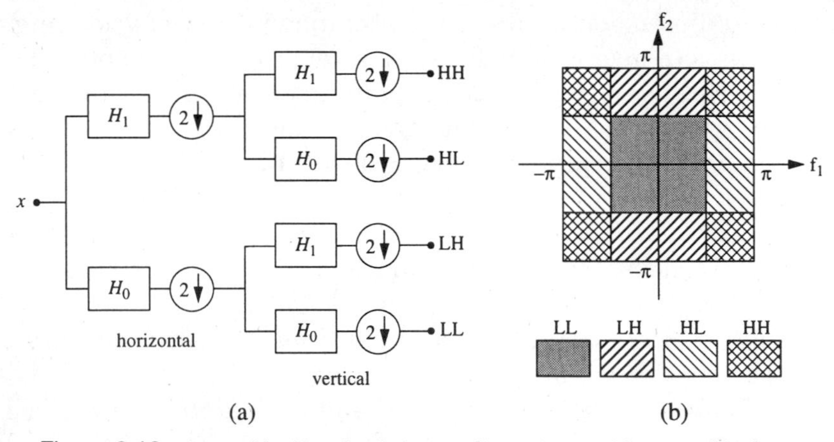

Figure 3.19 Separable filter bank in two dimensions, with separable downsampling by 2. (a) Cascade of horizontal and vertical decompositions. (b) Division of the frequency spectrum.

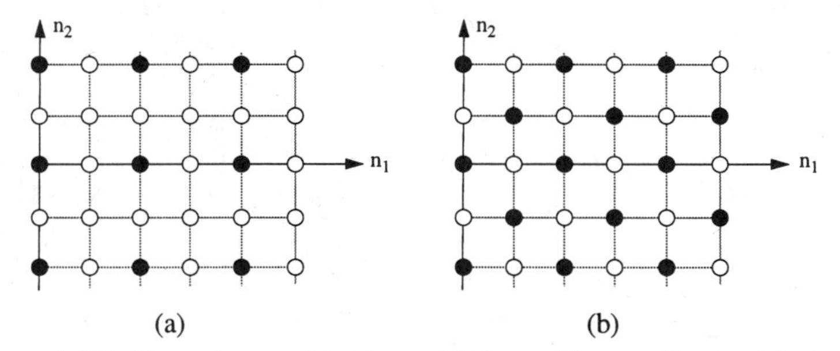

Figure 3.20 Two often used lattices. (a) Separable sampling by 2 in two dimensions. (b) Quincunx sampling.

channel perfect reconstruction filter banks, being orthogonal and linear phase at the same time. This implies that the same will hold in two dimensions if separable filters are used.

Alternatively, one could still sample separately (see Figure 3.20(a)) and yet use nonseparable filters. In other words, one could have a direct four-channel implementation of Figure 3.19 where the four filters could be H_0, H_1, H_2, H_3. While before, $H_i(z_1, z_2) = H_{i_1}(z_1)H_{i_2}(z_2)$ where $H_i(z)$ is a one-dimensional filter, $H_i(z_1, z_2)$ is now a true two-dimensional filter. This solution, while more general, is more complex to design and implement. It is possible to obtain an orthogonal linear phase FIR solution [155, 156], which cannot be achieved using separable filters (see Example 3.15 below).

Similarly to the one-dimensional case, one can define polyphase decompositions of signals and filters. Recall that in one dimension, the polyphase decomposition of the signal with respect to N was simply the subsignals which have the same indexes modulo N. The generalization in multiple dimensions are cosets with respect to a downsampling lattice. There is no natural ordering such as in one dimension but as long as all N cosets are included, the decomposition is valid. In separable downsampling by 2 in two dimensions, we can take as coset representatives the points $\{(0,0),(1,0),(0,1),(1,1)\}$. Then the signal $X(z_1, z_2)$ can be written as

$$X(z_1,z_2) = X_{00}(z_1^2,z_2^2) + z_1^{-1}X_{10}(z_1^2,z_2^2) + z_2^{-1}X_{01}(z_1^2,z_2^2) + z_1^{-1}z_2^{-1}X_{11}(z_1^2,z_2^2), \tag{3.6.1}$$

where

$$X_{ij}(z_1,z_2) = \sum_m \sum_n z_1^{-m} z_2^{-n} x[2m+i, 2n+j].$$

Thus, the polyphase component with indexes i, j corresponds to a square lattice downsampled by 2, and with the origin shifted to (i, j). The recombination of $X(z_1, z_2)$ from its polyphase components as given in (3.6.1) corresponds to an inverse polyphase transform and its dual is therefore the forward polyphase transform. The polyphase decomposition of analysis and synthesis filter banks follow similarly.

The synthesis filters are decomposed just as the signal (see (3.6.1)), while the analysis filters have reverse phase. We shall not dwell longer on these decompositions since they follow easily from their one-dimensional counterparts but tend to involve a bit of algebra. The result, as to be expected, is that the output of an analysis/synthesis filter bank can be written in terms of the input polyphase components times the product of the polyphase matrices.

The output of the system could also be written in terms of modulated versions of the signal and filters. For example, downsampling by 2 in two dimensions, and then upsampling by 2 again (zeroing out all samples except the ones where both indexes are even) can be written in z-domain as

$$\frac{1}{4}(X(z_1,z_2) + X(-z_1,z_2) + X(z_1,-z_2) + X(-z_1,-z_2)).$$

Therefore, it is easy to verify that the output of a four-channel filter bank with separable downsampling by 2 has an output that can be written as

$$Y(z_1,z_2) = \frac{1}{4}\, \boldsymbol{g}^T(z_1,z_2)\, \boldsymbol{H}_m(z_1,z_2)\, \boldsymbol{x}_m(z_1,z_2),$$

where

$$\boldsymbol{g}^T(z_1,z_2) = \begin{pmatrix} G_0(z_1,z_2) & G_1(z_1,z_2) & G_2(z_1,z_2) & G_3(z_1,z_2) \end{pmatrix}, \tag{3.6.2}$$

$$\boldsymbol{H}_m(z_1,z_2) =$$

$$\begin{pmatrix} H_0(z_1,z_2) & H_0(-z_1,z_2) & H_0(z_1,-z_2) & H_0(-z_1,-z_2) \\ H_1(z_1,z_2) & H_1(-z_1,z_2) & H_1(z_1,-z_2) & H_1(-z_1,-z_2) \\ H_2(z_1,z_2) & H_2(-z_1,z_2) & H_2(z_1,-z_2) & H_2(-z_1,-z_2) \\ H_3(z_1,z_2) & H_3(-z_1,z_2) & H_3(z_1,-z_2) & H_3(-z_1,-z_2) \end{pmatrix}, \quad (3.6.3)$$

$$\boldsymbol{x}_m(z_1,z_2) = \\ \left(X(z_1,z_2) \quad X(-z_1,z_2) \quad X(z_1,-z_2) \quad X(-z_1,-z_2) \right).$$

Let us now consider an example involving nonseparable downsampling. We examine quincunx sampling (see Figure 3.20(b)) because it is the simplest multidimensional nonseparable lattice. Moreover, it samples by 2, that is, it is the counterpart of the one-dimensional two-channel case we discussed in Section 3.2.

Example 3.14 *Quincunx Case*

It is easy to verify that, given $X(z_1,z_2)$, quincunx downsampling followed by quincunx upsampling (that is, replacing the locations with empty circles in Figure 3.20(b) by 0) results in a z-transform equal to $1/2(X(z_1,z_2)+X(-z_1,-z_2))$. From this, it follows that a two-channel analysis/synthesis filter bank using quincunx sampling has an input/output relationship given by

$$Y(z_1,z_2) = \frac{1}{2}\left(G_0(z_1,z_2)\; G_1(z_1,z_2)\right) \begin{pmatrix} H_0(z_1,z_2) & H_0(-z_1,-z_2) \\ H_1(z_1,z_2) & H_1(-z_1,-z_2) \end{pmatrix} \\ \begin{pmatrix} X(z_1,z_2) \\ X(-z_1,-z_2) \end{pmatrix}.$$

Similarly to the one-dimensional case, it can be verified that the orthogonality of the system is achieved when the lowpass filter satisfies

$$H_0(z_1,z_2)H_0(z_1^{-1},z_2^{-1}) + H_0(-z_1,-z_2)H_0(-z_1^{-1},-z_2^{-1}) = 2, \quad (3.6.4)$$

that is, the lowpass filter is orthogonal to its shifts on the quincunx lattice. Then, a possible highpass filter is given by

$$H_1(z_1,z_2) = -z_1^{-1}H_0(-z_1^{-1},-z_2^{-1}). \quad (3.6.5)$$

The synthesis filters are the same (within shift reversal, or $G_i(z_1,z_2) = H_i(z_1^{-1},z_2^{-1})$). In polyphase domain, define the two polyphase components of the filters as

$$H_{i0}(z_1,z_2) = \sum_{(n_1,n_2)\in\mathcal{Z}^2} h_i[n_1+n_2, n_1-n_2] z_1^{-n_1} z_2^{-n_2},$$

$$H_{i1}(z_1,z_2) = \sum_{(n_1,n_2)\in\mathcal{Z}^2} h_i[n_1+n_2+1, n_1-n_2] z_1^{-n_1} z_2^{-n_2},$$

with

$$H_i(z_1,z_2) = H_{i0}(z_1z_2, z_1z_2^{-1}) + z_1^{-1}H_{i1}(z_1z_2, z_1z_2^{-1}).$$

The results on alias cancellation and perfect reconstruction are very similar to their one-dimensional counterparts. For example, perfect reconstruction with FIR filters is achieved *if and only if* the determinant of the analysis polyphase matrix is a monomial, that is,

$$\boldsymbol{H}_p(z_1, \ldots, z_n) \;=\; c \cdot z_1^{-K_1} \cdot \ldots \cdot z_n^{-K_n}.$$

Since the results are straightforward extensions of one-dimensional results, we rather discuss two cases of interest in more detail, while the reader is referred to [48, 164, 308, 325] for a more in-depth discussion of multidimensional results.

3.6.2 Synthesis of Multidimensional Filter Banks

The design of nonseparable systems is more challenging than the one-dimensional cases. Designs based on cascade structures as well as one- to multidimensional transformations are discussed next.

Cascade Structures When synthesizing filter banks, one of the most obvious approaches is to try to find cascade structures that would generate filters of the desired form. This is because cascade structures (a) usually have low complexity, (b) higher-order filters are easily derived from lower-order ones, and (c) the coefficients can be quantized without affecting the desired form. However, unlike in one dimension, there are very few results on completeness of cascade structures in multiple dimensions.

While cascades of orthogonal building blocks (that is, orthogonal matrices and diagonal delay matrices) obviously will yield orthogonal filter banks, producing linear phase solutions needs more care. For example, one can make use of the linear phase testing condition given in [155] or [164] to obtain possible cascades. As one of the possible approaches consider the generalization of the linear phase cascade structure proposed in [155, 156, 321]. Suppose that a linear phase system has been already designed and a higher-order one is needed. Choosing

$$\boldsymbol{H}_p''(\boldsymbol{z}) \;=\; \boldsymbol{R}\,\boldsymbol{D}(\boldsymbol{z})\,\boldsymbol{H}_p'(\boldsymbol{z}),$$

where $\boldsymbol{D}(\boldsymbol{z}) = \boldsymbol{z}^{-\boldsymbol{k}}\boldsymbol{J}\boldsymbol{D}(\boldsymbol{z}^{-1})\boldsymbol{J}$ and $\boldsymbol{R}$ is persymmetric ($\boldsymbol{R} = \boldsymbol{J}\boldsymbol{R}\boldsymbol{J}$), another linear phase system is obtained, where the filters have the same symmetry as in $\boldsymbol{H}_p'$. Although this cascade is by no means complete, it can produce very useful filters. Let us also point out that when building cascades in the polyphase domain, one must bear in mind that using different sampling matrices for the same lattice will greatly affect the geometry of the filters obtained.

Example 3.15 *Separable Case*

Let us first present a cascade structure, that will generate four linear phase/ orthogonal filters of the same size, where two of them are symmetric and the other two antisymmetric [156]

$$\boldsymbol{H}_p(z_1, z_2) = \left[\prod_{i=K-1}^{1} \boldsymbol{R}_i \, \boldsymbol{D}(z_1, z_2) \right] \boldsymbol{S}_0.$$

In the above, $\boldsymbol{D}$ is the matrix of delays containing $(1 \;\; z_1^{-1} \;\; z_2^{-1} \;\; (z_1 z_2)^{-1})$ along the diagonal, and $\boldsymbol{R}_i$ and $\boldsymbol{S}_0$ are scalar persymmetric matrices, that is, they satisfy

$$\boldsymbol{R}_i = \boldsymbol{J} \boldsymbol{R}_i \boldsymbol{J}. \tag{3.6.6}$$

Equation (3.6.6) along with the requirement that the $\boldsymbol{R}_i$ be unitary, allows one to design filters being both linear phase and orthogonal. Recall that in the two-channel one-dimensional case these two requirements are mutually exclusive, thus one cannot design separable filters satisfying both properties in this four-channel two-dimensional case. This shows how using a true multidimensional solution offers greater freedom in design. To obtain both linear phase and orthogonality, one has to make sure that, on top of being persymmetric, matrices $\boldsymbol{R}_i$ have to be unitary as well. These two requirements lead to

$$\boldsymbol{R}_i = \frac{1}{2} \begin{pmatrix} \boldsymbol{I} & \\ & \boldsymbol{J} \end{pmatrix} \begin{pmatrix} \boldsymbol{I} & \boldsymbol{I} \\ \boldsymbol{I} & -\boldsymbol{I} \end{pmatrix} \begin{pmatrix} \boldsymbol{R}_{2i} & \\ & \boldsymbol{R}_{2i+1} \end{pmatrix} \begin{pmatrix} \boldsymbol{I} & \boldsymbol{I} \\ \boldsymbol{I} & -\boldsymbol{I} \end{pmatrix} \begin{pmatrix} \boldsymbol{I} & \\ & \boldsymbol{J} \end{pmatrix},$$

where $\boldsymbol{R}_{2i}, \; \boldsymbol{R}_{2i+1}$ are 2×2 rotation matrices, and

$$\boldsymbol{S}_0 = \begin{pmatrix} \boldsymbol{R}_0 & \\ & \boldsymbol{R}_1 \end{pmatrix} \begin{pmatrix} \boldsymbol{I} & \boldsymbol{I} \\ \boldsymbol{I} & -\boldsymbol{I} \end{pmatrix} \begin{pmatrix} \boldsymbol{I} & \\ & \boldsymbol{J} \end{pmatrix}.$$

This cascade is a two-dimensional counterpart of the one given in [275, 321], and will be shown to be useful in producing regular wavelets being both linear phase and orthonormal [166] (see Chapter 4).

Example 3.16 *Quincunx Cascades*

Let us first present a cascade structure that can generate filters being either orthogonal or linear phase. It is obtained by the following:

$$\boldsymbol{H}_p(z_1, z_2) = \left[\prod_{i=K-1}^{1} \boldsymbol{R}_{2_i} \begin{pmatrix} 1 & 0 \\ 0 & z_2^{-1} \end{pmatrix} \boldsymbol{R}_{1_i} \begin{pmatrix} 1 & 0 \\ 0 & z_1^{-1} \end{pmatrix} \right] \boldsymbol{R}_0.$$

For the filters to be orthogonal the matrices $\boldsymbol{R}_{j_i}$ have to be unitary. To be linear, phase matrices have to be symmetric. In the latter case the filters obtained will have opposite symmetry. Consider, for example, the orthogonal case. The smallest lowpass filter obtained from the above cascade would be

$$h_0[n_1, n_2] = \begin{pmatrix} & -a_1 & -a_0 a_1 & \\ -a_2 & -a_0 a_2 & -a_0 & 1 \\ & a_0 a_1 a_2 & -a_1 a_2 & \end{pmatrix}, \tag{3.6.7}$$

where a_i are free variables, and $h_0[n_1, n_2]$ is denormalized for simplicity. The highpass filter is obtained by modulation and time reversal (see (3.6.5)). This filter, with some additional constraints, will be shown to be the smallest regular two-dimensional filter (the counterpart of the Daubechies' D_2 filter [71]). Note that this cascade has its generalization in more than two dimensions (its one-dimensional counterpart is the lattice structure given in (3.2.60)).

One to Multidimensional Transformations Because of the difficulty of designing good filters in multiple dimensions, transformations to map one-dimensional designs into multidimensional ones have been used for some time, the most popular being the McClellan transformation [88, 191].

For purely discrete-time purposes, the only requirement that we impose is that perfect reconstruction be preserved when transforming a one-dimensional filter bank into a multidimensional one. We will see later, that in the context of building continuous-time wavelet bases, one needs to preserve the order of zeros at aliasing frequencies. Two methods are presented: the first is based on separable polyphase components and the second on the McClellan transformation.

Separable Polyphase Components A first possible transform is obtained by designing a multidimensional filter having separable polyphase components, given as products of the polyphase components of a one-dimensional filter [11, 47]. To be specific, consider the quincunx downsampling case. Start with a one-dimensional filter having polyphase components $H_0(z)$ and $H_1(z)$, that is, a filter with a z-transform $H(z) = H_0(z^2) + z^{-1}H_1(z^2)$. Derive separable polyphase components

$$H_i(z_1, z_2) \;=\; H_i(z_1)\, H_i(z_2), \qquad i = 0, 1.$$

Then, the two-dimensional filter with respect to the quincunx lattice is given as (by upsampling the polyphase components with respect to the quincunx lattice)

$$H(z_1, z_2) \;=\; H_0(z_1 z_2)\, H_0(z_1 z_2^{-1}) \;+\; z_1^{-1} H_1(z_1 z_2)\, H_1(z_1 z_2^{-1}).$$

It can be verified that an Nth-order zero at π in $H(e^{j\omega})$, maps into an Nth-order zero at (π, π) in $H(e^{j\omega_1}, e^{j\omega_2})$ (we will come back to this property in Chapter 4). However, an orthogonal filter bank is mapped into an orthogonal two-dimensional bank, *if and only if* the polyphase components of the one-dimensional filter are allpass functions (that is, $H_i(e^{j\omega})H_i(e^{-j\omega}) = c$). Perfect reconstruction is thus not conserved in general. Note that the separable polyphase components lead to efficient implementations, reducing the number of operations from $O[L^2]$ to $O[L]$ per output, where L is the filter size.

McClellan Transformation [191] The second transformation is the well-known McClellan transformation, which has recently become a popular way to design linear phase multidimensional filter banks (see [47, 164, 257, 288] among others). The Fourier transform of a zero-phase symmetric filter ($h[n] = h[-n]$), can be written as a function of $\cos(n\omega)$ [211]

$$H(\omega) = \sum_{n=-L}^{L} a[n] \cos(n\omega),$$

where $a[0] = h[0]$ and $a[n] = 2h[n], \ n \neq 0$. Using Tchebycheff polynomials, one can replace $\cos(n\omega)$ by $T_n[\cos(\omega)]$, where $T_n[.]$ is the nth Tchebycheff polynomial, and thus $H(\omega)$ can be written as a polynomial of $\cos(\omega)$

$$H(\omega) \ = \ \sum_{n=-L}^{L} a[n] \ T_n[\cos(\omega)].$$

The idea of the McClellan transformation is to replace $\cos(\omega)$ by a zero-phase two-dimensional filter $F(\omega_1, \omega_2)$. This results in an overall zero-phase two-dimensional filter [88, 191]

$$H(\omega_1, \omega_2) \ = \ \sum_{n=-L}^{L} a[n] \ T_n[F(\omega_1, \omega_2)].$$

In the context of filter banks, this transformation can only be applied to the biorthogonal case (because of the zero-phase requirement). Typically, in the case of quincunx downsampling, $F(\omega_1, \omega_2)$ is chosen as [57]

$$F(\omega_1, \omega_2) \ = \ \frac{1}{2}(\cos(\omega_1) + \cos(\omega_2)). \tag{3.6.8}$$

That the perfect reconstruction is preserved, can be checked by considering the determinant of the polyphase matrix. This is a monomial in the one-dimensional case since one starts with a perfect reconstruction filter bank. The transformation in (3.6.8) leads to a determinant which is also a monomial, and thus, perfect reconstruction is conserved.

In addition to this, it is easy to see that pairs of zeroes at π (factors of the form $1 + \cos(w)$) map into zeroes of order two at (π, π) in the transformed domain (or factors of the form $1 + \cos(\omega_1)/2 + \cos(\omega_2)/2$).

Therefore, the McClellan transformation is a powerful method to map one-dimensional biorthogonal solutions to multidimensional biorthogonal solutions, and this while conserving zeroes at aliasing frequencies. We will show how important this is in trying to build continuous-time wavelet bases.

Remarks We have given a rapid overview of multidimensional filter bank results and relied on simple examples in order to give the intuition rather than developing the full algebraic framework. We refer the interested reader to [47, 48, 161, 164, 308], among others, for more details.

3.7 TRANSMULTIPLEXERS AND ADAPTIVE FILTERING IN SUBBANDS

3.7.1 Synthesis of Signals and Transmultiplexers

So far, we have been mostly interested in decomposing a given signal into components, from which the signal can be recovered. This is essentially an *analysis* problem.

The dual problem is to start from some components and to synthesize a signal from which the components can be recovered. This has some important applications, in particular in telecommunications. For example, several users share a common channel to transmit information. Two obvious ways to solve the problem are to either *multiplex in time* (each user receives a time slot out of a period) or *multiplex in frequency* (each user gets a subchannel). In general, the problem can be seen as one of designing (orthogonal) functions that are assigned to the different users within a time window so that each user can use "his" function for signaling (for example, by having it on or off). Since the users share the channel, the functions are added together, but because of orthogonality,[13] each user can monitor "his" function at the receiving end. The next time period looks exactly the same. Therefore, the problem is to design an orthogonal set of functions over a window, possibly meeting some boundary constraints as well. Obviously, time- and frequency-division multiplexing are just two particular cases.

Because of the fact that the system is invariant to shifts by a multiple of the time window, it is also clear that, in discrete time, this is a multirate filter bank problem. Below, we describe briefly the analysis of such systems, which is very similar to its dual problem, as well as some applications.

Analysis of Transmultiplexers A device synthesizing a single signal from several signals, followed by the inverse operation of recovering the initial signals, is usually called a *transmultiplexer*. This is because a main application is in telecommunications for going from time-division multiplexing (TDM) to frequency-division multiplexing (FDM) [25]. Such a device is shown in Figure 3.21.

It is clear that since this scheme involves multirate analysis and synthesis filter banks, all the algebraic tools developed for analysis/synthesis systems can be used here as well. We will not go through the details, since they are very similar to the familiar case, but will simply discuss a few key results [316].

It is easiest to look at the polyphase decomposition of the two filter banks, shown in Figure 3.21(b). The definitions of $\boldsymbol{H}_p(z)$ and $\boldsymbol{G}_p(z)$ are as given in Section 3.2. Note that they are of sizes $N \times M$ and $M \times N$, respectively. It

[13] Orthogonality is not necessary, but makes the system simpler.

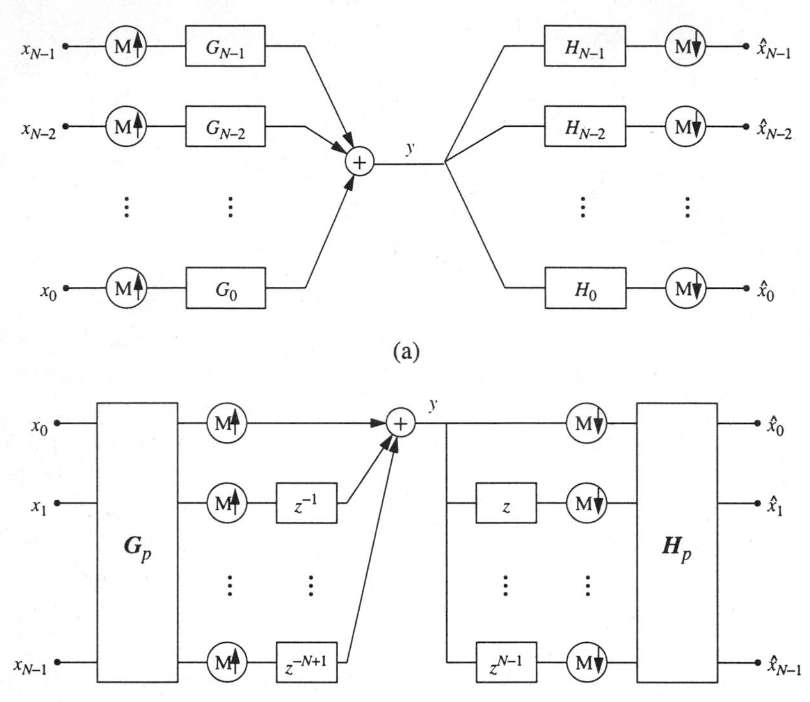

Figure 3.21 Transmultiplexer. (a) General scheme. (b) Polyphase-domain implementation.

is clear that the two polyphase transforms in the middle of the system cancel each other, and therefore, defining the input vector as

$$\boldsymbol{x}(z) \;=\; (X_0(z)\; X_1(z) \ldots X_{N-1}(z))^T,$$

and similarly the output vector as

$$\tilde{\boldsymbol{x}}(z) \;=\; \left(\tilde{X}_0(z)\; \tilde{X}_1(z) \ldots \tilde{X}_{N-1}(z)\right)^T,$$

we have the following input/output relationship:

$$\tilde{\boldsymbol{x}}(z) \;=\; \boldsymbol{H}_p(z)\, \boldsymbol{G}_p(z)\, \boldsymbol{x}(z). \tag{3.7.1}$$

We thus immediately get the following result:

PROPOSITION 3.18

In a transmultiplexer with polyphase matrices $\boldsymbol{H}_p(z)$ and $\boldsymbol{G}_p(z)$, the following holds:

(a) Perfect reconstruction is achieved *if and only if* $\boldsymbol{H}_p(z)\boldsymbol{G}_p(z) \;=\; \boldsymbol{I}$.

(b) There is no crosstalk between channels *if and only if* $\boldsymbol{H}_p(z)\boldsymbol{G}_p(z)$ is diagonal.

The above result holds for any M and N. One can show that $M \geq N$ is a necessary condition for crosstalk cancellation and perfect reconstruction. In the critical sampling case, or $M = N$, there is a simple duality result between transmultiplexers and analysis/synthesis systems seen earlier.

PROPOSITION 3.19

In the critically sampled case (number of channels equal to sampling rate change), a perfect reconstruction subband coding system is equivalent to a perfect reconstruction transmultiplexer.

PROOF

Since $\boldsymbol{G}_p(z)\boldsymbol{H}_p(z) = \boldsymbol{I}$ and they are square, it follows that $\boldsymbol{H}_p(z)\boldsymbol{G}_p(z) = \boldsymbol{I}$ as well.

Therefore, the design of perfect subband coding systems and of perfect transmultiplexers is equivalent, at least in theory. A problem in the transmultiplexer case is that the channel over which y is transmitted can be far from ideal. In order to highlight the potential problem, consider the following simple case: Multiplex two signals $X_0(z)$ and $X_1(z)$ by upsampling by 2, delaying the second one by 2 and adding them. This gives a channel signal

$$Y(z) = X_0(z^2) + z^{-1}X_1(z^2).$$

Obviously, $X_0(z)$ and $X_1(z)$ can be recovered by a polyphase transform (downsampling $Y(z)$ by 2 yields $X_0(z)$, downsampling $zY(z)$ by 2 yields $X_1(z)$). However, if $Y(z)$ has been delayed by z^{-1}, then the two signals will be interchanged at the output of the transmultiplexer. A solution to this problem is obtained if the signals $X_0(z^2)$ and $X_1(z^2)$ are filtered by perfect lowpass and highpass filters, respectively, and similarly at the reconstruction. Therefore, transmultiplexers usually use very good bandpass filters. In practice, critical sampling is not attempted. Instead, N signals are upsampled by $M > N$ and filtered by good bandpass filters. This higher upsampling rate allows guard bands to be placed between successive bands carrying the useful signals and suppresses crosstalk between channels even without using ideal filters. Note that all filter banks used in transmultiplexers are based on modulation of a prototype window to an evenly spaced set of bandpass filters, and can thus be very efficiently implemented using FFT's [25] (see also Section 6.2.3).

3.7.2 Adaptive Filtering in Subbands

A possible application of multirate filter banks is in equalization problems. The purpose is to estimate and apply an inverse filter (typically, a nonideal channel has to be compensated). The reason to use a multirate implementation rather than a direct time-domain version is related to computational complexity and convergence behavior. Since a filter bank computes a form of frequency analysis, subband adaptive filtering is a version of frequency-domain adaptive filtering. See [263] for an excellent overview on the topic.

We will briefly discuss a simple example. Assume that a filter with z-transform $F(z)$ is to be implemented in the subbands of a two-channel perfect reconstruction filter bank with critical sampling. Then, it can be shown that the channel transfer function between the analysis and synthesis filter banks, $C(z)$, is not diagonal in general [112]. That is, one has to estimate four components, two direct channel components, and two crossterms. These components can be relatively short (especially the crossterms) and run at half the sampling rate, and thus, the scheme can be computationally attractive. Yet, the crossterms turn out to be difficult to estimate accurately (they correspond to aliasing terms). Therefore, it is more interesting to implement an oversampled system, that is, decompose into N channels and downsample by $M < N$. Then, the matrix $C(z)$ can be well approximated by a diagonal matrix, making the estimation of the components easier. We refer to [112, 263], and to references therein for more details and discussions of applications such as acoustic echo cancellation.

APPENDIX 3.A LOSSLESS SYSTEMS

We have seen in (3.2.60) a very simple, yet powerful factorization yielding orthogonal solutions and pointed to the relation to lossless systems. Here, the aim is to give a brief review of lossless systems and two-channel as well as N-channel factorizations. Lossless systems have been thoroughly studied in classical circuit theory. Many results, including factorizations of lossless matrices, can be found in the circuit theory literature, for example in the text by Belevitch [23]. For a description of this topic in the context of filter banks and detailed derivations of factorizations, we refer to [308].

The general definition of a *paraunitary* matrix is [309]

$$\tilde{\boldsymbol{H}}(z)\,\boldsymbol{H}(z) \;=\; c\boldsymbol{I}, \qquad c \neq 0,$$

where $\tilde{\boldsymbol{H}}(z) = \boldsymbol{H}_*^T(z^{-1})$ and subscript $_*$ means conjugation[14] of the coefficients (but not of z). If all entries are stable, such a matrix is called *lossless*. The interpretation of losslessness, a concept very familiar in classical circuit theory [23], is that the energy of the signals is conserved through the system given by

[14]Here we give the general definition, which includes complex-valued filter coefficients, whereas we considered mostly the real case in the main text.

$\boldsymbol{H}(z)$. Note that the losslessness of $\boldsymbol{H}(z)$ implies that $\boldsymbol{H}(e^{j\omega})$ is unitary

$$\boldsymbol{H}^*(e^{j\omega})\,\boldsymbol{H}(e^{j\omega}) \;=\; c\boldsymbol{I},$$

where the superscript $*$ stands for hermitian conjugation (note that $\boldsymbol{H}^*(e^{j\omega}) = \boldsymbol{H}_*^T(e^{-j\omega})$). For the scalar case (single input/single output), lossless transfer functions are allpass filters given by [211]

$$F(z) \;=\; \frac{a(z)}{z^{-k}a_*(z^{-1})}, \tag{3.A.1}$$

where $k = \deg(a(z))$ (possibly, there is a multiplicative delay and scaling factor equal to cz^{-k}). Thus, to any zero at $z = a$ corresponds a pole at $z = 1/a^*$, that is, at a mirror location with respect to the unit circle. This guarantees a perfect transmission at all frequencies (in amplitude) and only phase distortion. It is easy to verify that (3.A.1) is lossless (assuming all poles inside the unit circle) since

$$F_*(z^{-1})\,F(z) \;=\; \frac{a_*(z^{-1})}{z^k a(z)}\,\frac{a(z)}{z^{-k}a_*(z^{-1})} \;=\; 1.$$

Obviously, nontrivial scalar allpass functions are IIR, and are thus not linear phase. Interestingly, matrix allpass functions exist that are FIR, and linear phase behavior is possible. Trivial examples of matrix allpass functions are unitary matrices, as well as diagonal matrices of delays.

3.A.1 Two-Channel Factorizations

We will first give an expression for the most general form of a 2×2 causal FIR lossless system of an arbitrary degree. Then, based on this, we will derive a factorization of a lossless system (already given in (3.2.60)).

PROPOSITION 3.20

The most general causal, FIR, 2×2 lossless system of arbitrary degree and real coefficients, can be written in the form [309]

$$L(z) \;=\; \begin{pmatrix} L_0(z) & L_2(z) \\ L_1(z) & L_3(z) \end{pmatrix} \;=\; \begin{pmatrix} L_0(z) & cz^{-K}\tilde{L}_1(z) \\ L_1(z) & -cz^{-K}\tilde{L}_0(z) \end{pmatrix}, \tag{3.A.2}$$

where $L_0(z)$ and $L_1(z)$ satisfy the power complementary property, c is a real scalar constant with $|c| = 1$, and K is a large enough positive integer so as to make the entries of the right column in (3.A.2) causal.

PROOF

Let us first demonstrate the following fact: If the polyphase matrix is orthogonal, then L_0 and L_1 are relatively prime. Similarly, L_2 and L_3 are relatively prime. Let us prove the first statement (the second one follows similarly). Expand $\tilde{L}(z)L(z)$ as follows:

$$\begin{aligned}
\tilde{L}_0(z)L_0(z) + \tilde{L}_1(z)L_1(z) &= 1, && (3.A.3)\\
\tilde{L}_0(z)L_2(z) + \tilde{L}_1(z)L_3(z) &= 0, && (3.A.4)\\
\tilde{L}_2(z)L_0(z) + \tilde{L}_3(z)L_1(z) &= 0, && (3.A.5)\\
\tilde{L}_2(z)L_2(z) + \tilde{L}_3(z)L_3(z) &= 1. && (3.A.6)
\end{aligned}$$

Suppose now that L_0 and L_1 are not coprime, and call their common factor $P(z)$, that is, $L_0(z) = P(z)L_0'(z)$, $L_1(z) = P(z)L_1'(z)$. Substituting this into (3.A.3)

$$P(z)\tilde{P}(z) \cdot (\tilde{L}_0'(z)L_0'(z) + \tilde{L}_1'(z)L_1'(z)) = 1,$$

which for all zeros of $P(z)$ goes to 0, contradicting the fact the right side is identically 1.

Consider (3.A.4). Since L_0 and L_1, as well as L_2 and L_3 are coprime, we have that $L_3(z) = C_1 z^{-K}\tilde{L}_0(z)$ and $L_2(z) = C_2 z^{-K'}\tilde{L}_1(z)$ where K and K' are large enough integers to make L_3 and L_2 causal. Take now (3.A.5). This implies that $K = K'$ and $C_1 = -C_2$. Finally, (3.A.3) or (3.A.6) imply that $C_1 = \pm 1$.

To obtain a cascade-form realization of (3.A.2), we find such a realization for the left column of (3.A.2) and then use it derive a cascade form of the whole matrix. To that end, a result from [309] will be used. It states that for two, real-coefficient polynomials P_{K-1} and Q_{K-1} of degree $(K-1)$, with $p_{K-1}(0)\ p_{K-1}(K-1) \neq 0$ (and P_{K-1}, Q_{K-1} are power complementary), there exists another pair P_{K-2}, Q_{K-2} such that

$$\begin{pmatrix} P_{K-1}(z) \\ Q_{K-1}(z) \end{pmatrix} = \begin{pmatrix} \cos\alpha & -\sin\alpha \\ \sin\alpha & \cos\alpha \end{pmatrix} \begin{pmatrix} P_{K-2}(z) \\ z^{-1}Q_{K-2}(z) \end{pmatrix}. \tag{3.A.7}$$

Repeatedly applying the above result to (3.A.2) one obtains the lattice factorization given in (3.2.60), that is,

$$\begin{aligned}
\begin{pmatrix} L_0(z) & L_2(z) \\ L_1(z) & L_3(z) \end{pmatrix} &= \begin{pmatrix} \cos\alpha_0 & -\sin\alpha_0 \\ \sin\alpha_0 & \cos\alpha_0 \end{pmatrix} \\
&\times \left[\prod_{i=1}^{K-1} \begin{pmatrix} 1 & \\ & z^{-1} \end{pmatrix} \begin{pmatrix} \cos\alpha_i & -\sin\alpha_i \\ \sin\alpha_i & \cos\alpha_i \end{pmatrix} \right].
\end{aligned}$$

A very important point is that the above structure is complete, that is, all orthogonal systems with filters of length $2K$ can be generated in this fashion. The lattice factorization was given in Figure 3.6.

3.A.2 Multichannel Factorizations

Here, we will present a number of ways in which one can design N-channel orthogonal systems. Some of the results are based on lossless factorizations (for factorizations of unitary matrices, see Appendix 2.B in Chapter 2).

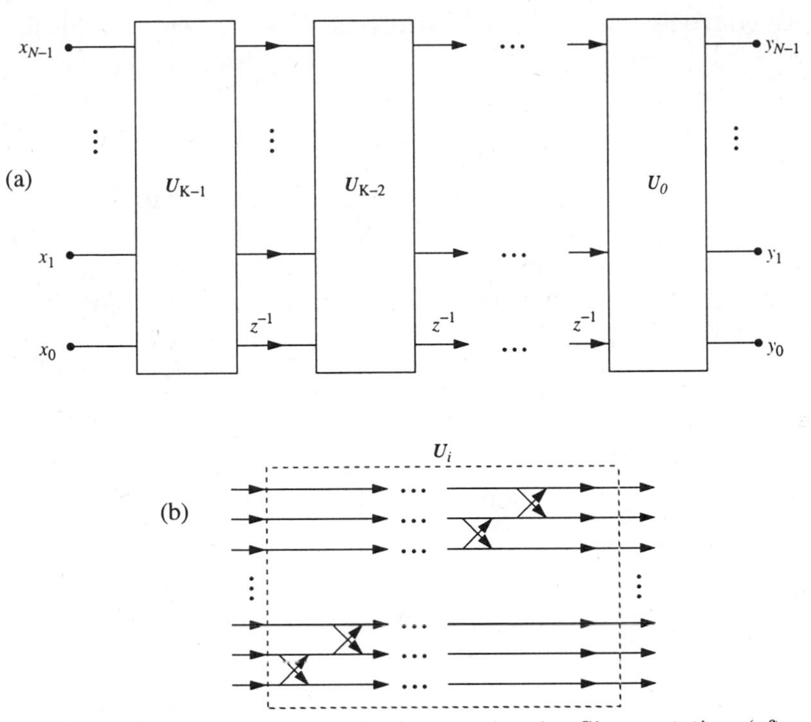

Figure 3.22 Factorization of a lossless matrix using Givens rotations (after [306]). (a) General lossless transfer matrix $H(z)$ of size $N \times N$. (b) Constrained orthogonal matrix for $U_1, \ldots, U_{K-1}$, where each cross represents a rotation as in (3.A.7).

Givens Factorization We have seen in Appendix 3.A.1 a lattice factorization for the two-channel case. Besides delays, the key building blocks were 2×2 rotation matrices, also called Givens rotations. An extension of that construction, holds in the N-channel case as well. More precisely, a real lossless FIR matrix $L(z)$ of size $N \times N$ can be written as [306]

$$L(z) \;=\; \boldsymbol{U}_0 \left[\prod_{i=1}^{K-1} \boldsymbol{D}_i(z) \boldsymbol{U}_i \right], \tag{3.A.8}$$

where $\boldsymbol{U}_1 \ldots \boldsymbol{U}_{K-1}$ are special orthogonal matrices as given in Figure 3.22(b) (each cross is a rotation as in (3.A.7)). $\boldsymbol{U}_0$ is a general orthogonal matrix as given in Figure 2.13 with $n = N$, and $\boldsymbol{D}(z)$ are delay matrices of the form

$$\boldsymbol{D}(z) \;=\; \text{diag}(z^{-1}\, 1\, 1 \ldots 1).$$

Such a general, real, lossless, FIR, N-input N-output system, is shown in Figure 3.22(a). Figure 3.22(b) indicates the form of the matrices $\boldsymbol{U}_1 \ldots \boldsymbol{U}_{K-1}$. Note that $\boldsymbol{U}_0$ is characterized by $\binom{N}{2}$ rotations [202] while the other orthogonal matrices are characterized by $N-1$ rotations. Thus, a real FIR lossless system of degree $K-1$ has the following number of free parameters:

$$p = (K-1)(N-1) + \binom{N}{2}.$$

It is clear that these structures are lossless, and the completeness is demonstrated in [85]. In order to obtain good filters, one can optimize the various angles in the rotation matrices, derive the filters corresponding to the resulting polyphase matrix, and evaluate an objective cost function measuring the quality of the filters (such as the stopband energy).

Householder Factorization An alternative representation of FIR lossless systems based on products of Householder matrices, which turns out to be more convenient for optimization, was presented in [312]. There it is shown that an $N \times N$ causal FIR system of degree $K-1$ is lossless *if and only if* it can be written in the form

$$L_{N-1}(z) = \boldsymbol{V}_{K-1}(z) \cdot \boldsymbol{V}_{K-2}(z) \cdots \boldsymbol{V}_1(z) L_0,$$

where L_0 is a general $N \times N$ unitary matrix (see Appendix 2.B) and

$$\boldsymbol{V}_k(z) = (\boldsymbol{I} - (1-z^{-1})\boldsymbol{v}_k\boldsymbol{v}_k^*),$$

with $\boldsymbol{v}_k$ a size-N vector of unit norm (recall that superscript * denotes hermitian conjugation). It is easy to verify that $\boldsymbol{V}_k(z)$ is lossless, since

$$\begin{aligned}
\boldsymbol{V}_{k*}^T(z^{-1})\boldsymbol{V}_k(z) &= (\boldsymbol{I} - (1-z)\boldsymbol{v}_k\boldsymbol{v}_k^*) \cdot (\boldsymbol{I} - (1-z^{-1})\boldsymbol{v}_k\boldsymbol{v}_k^*) \\
&= \boldsymbol{I} + \boldsymbol{v}_k\boldsymbol{v}_k^*((z-1) + (z^{-1}-1) + (1-z)(1-z^{-1})) \\
&= \boldsymbol{I},
\end{aligned}$$

where we used $\boldsymbol{v}_k\boldsymbol{v}_k^*\boldsymbol{v}_k\boldsymbol{v}_k^* = \boldsymbol{v}_k\boldsymbol{v}_k^*$, and for the completeness issues, we refer to [312]. Note that these structures can be extended to the IIR case as well, simply by replacing the delay element z^{-1} with a first-order scalar allpass section $(1-az^{-1})/(z^{-1}-a^*)$. Again, it is easy to verify that such structures are lossless (assuming $|a| > 1$) and completeness can be demonstrated similarly to the FIR case.

Orthogonal and Linear Phase Factorizations Recently, a factorization for a large class of paraunitary, linear phase systems has been developed [275]. It is a complete factorization for linear phase paraunitary filter banks with an even number of channels N $(N > 2)$ where the polyphase matrix is described by the following [321] (see also (3.2.69))

$$\boldsymbol{H}_p(z) = z^{-L}\boldsymbol{a}\, \boldsymbol{H}_p(z^{-1})\, \boldsymbol{J}, \tag{3.A.9}$$

where $\boldsymbol{a}$ is the diagonal matrix of symmetries (+1 for a symmetric filter and −1 for an antisymmetric filter), L is the filter length and $\boldsymbol{J}$ is an antidiagonal matrix. Note that there exist linear phase systems which cannot be described by (3.A.9) but many useful solutions do satisfy it. The cascade is given by

$$\begin{aligned} \boldsymbol{H}_p(z) &= \boldsymbol{S}\left[\boldsymbol{P}\prod_{i=K-1}^{1}\boldsymbol{W}\begin{pmatrix}\boldsymbol{U}_{2i} & \\ & \boldsymbol{U}_{2i+1}\end{pmatrix}\boldsymbol{W}\boldsymbol{D}(z)\right] \\ &\quad \times \boldsymbol{W}\begin{pmatrix}\boldsymbol{U}_0 & \\ & \boldsymbol{U}_1\end{pmatrix}\boldsymbol{W}\boldsymbol{P}, \end{aligned}$$

where

$$\boldsymbol{S} = \frac{1}{\sqrt{2}}\begin{pmatrix}\boldsymbol{S}_0 & \\ & \boldsymbol{S}_1\end{pmatrix}\begin{pmatrix}\boldsymbol{I} & \boldsymbol{J} \\ \boldsymbol{I} & -\boldsymbol{J}\end{pmatrix},$$

is a unitary matrix. $\boldsymbol{S}_0$, $\boldsymbol{S}_1$ are unitary matrices of size $N/2$,

$$\boldsymbol{P} = \begin{pmatrix}\boldsymbol{I} & \\ & \boldsymbol{J}\end{pmatrix}, \quad \boldsymbol{W} = \frac{1}{\sqrt{2}}\begin{pmatrix}\boldsymbol{I} & \boldsymbol{I} \\ \boldsymbol{I} & -\boldsymbol{I}\end{pmatrix}, \quad \boldsymbol{D}(z) = \begin{pmatrix}\boldsymbol{I} & \\ & z^{-1}\boldsymbol{I}\end{pmatrix},$$

and $\boldsymbol{U}_i$ are all size-$(N/2)$ unitary matrices. Note that all subblocks in the above matrices are of size $N/2$. In the same paper [275], the authors develop a cascade structure for filter banks with an odd number of channels as well.

State-Space Description It is interesting to consider the lossless property in state-space description. If we call $\boldsymbol{v}[n]$ the state vector, then a state space description is given by [150]

$$\begin{aligned} \boldsymbol{v}[n+1] &= \boldsymbol{A}\boldsymbol{v}[n] + \boldsymbol{B}\boldsymbol{x}[n], \\ \boldsymbol{y}[n] &= \boldsymbol{C}\boldsymbol{v}[n] + \boldsymbol{D}\boldsymbol{x}[n], \end{aligned}$$

where $\boldsymbol{A}$ is of size $d \times d$ ($d \geq K-1$, the degree of the system), $\boldsymbol{D}$ of size $M \times N$, $\boldsymbol{C}$ of size $M \times d$ and $\boldsymbol{B}$ of size $d \times N$. A minimal realization satisfies $d = K-1$. The transfer function matrix is equal to

$$\boldsymbol{H}(z) = \boldsymbol{D} + \boldsymbol{C}(z\boldsymbol{I} - \boldsymbol{A})^{-1}\boldsymbol{B},$$

and the impulse response is given by

$$[\boldsymbol{D}, \boldsymbol{C}\boldsymbol{B}, \boldsymbol{C}\boldsymbol{A}\boldsymbol{B}, \boldsymbol{C}\boldsymbol{A}^2\boldsymbol{B}, \ldots].$$

The fundamental nature of the losslessness property appears in the following result [304, 309]: A stable transfer matrix $\boldsymbol{H}(z)$ is lossless *if and only if* there exists a minimal realization such that

$$\boldsymbol{R} = \begin{pmatrix}\boldsymbol{A} & \boldsymbol{B} \\ \boldsymbol{C} & \boldsymbol{D}\end{pmatrix},$$

is unitary. This gives another way to parametrize lossless transfer function matrices. In particular, $\boldsymbol{H}(z)$ will be FIR if $\boldsymbol{A}$ is lower triangular with a zero diagonal, and thus, it is sufficient to find orthogonal matrices with an upper right triangular corner of size $K-1$ with only zeros to find all lossless transfer matrices of a given size and degree [85].

APPENDIX 3.B Sampling in Multiple Dimensions and Multirate Operations

Sampling in multiple dimensions is represented by a lattice, defined as the set of all linear combinations of n basis vectors $\boldsymbol{a}_1, \boldsymbol{a}_2, \ldots, \boldsymbol{a}_n$, with integer coefficients [42, 86], that is, a lattice is the set of all vectors generated by $\boldsymbol{D}\boldsymbol{k}$, $\boldsymbol{k} \in \mathcal{Z}^n$, where $\boldsymbol{D}$ is the matrix characterizing the sampling process. Note that $\boldsymbol{D}$ is not unique for a given sampling pattern and that two matrices representing the same sampling process are related by a linear transformation represented by a unimodular matrix [42]. We will call input and output lattice the set of points reached by $\boldsymbol{k}$ and $\boldsymbol{D}\boldsymbol{k}$, respectively. The input lattice is often $\mathcal{Z}^n$ (like above) but need not be.

A *separable* lattice is a lattice that can be represented by a diagonal matrix and it will appear when one-dimensional systems are used in a separable fashion along each dimension. The *unit cell* is a set of points such that the union of copies of the output lattice shifted to all points in the cell yields the input lattice. The number of input lattice points contained in the unit cell represents the reciprocal of the sampling density and is given by $N = \det(\boldsymbol{D})$. An important unit cell is the *fundamental parallelepiped* $\mathcal{U}_c$ (the parallelepiped formed by n basis vectors). In what follows $\mathcal{U}_c^T$ will denote the fundamental parallelepiped of the transposed lattice. Shifting the origin of the output lattice to any of the points of the input lattice yields a *coset*. Clearly there are exactly N distinct cosets obtained by shifting the origin of the output lattice to all of the points of the parallelepiped. The union of all cosets for a given lattice yields the input lattice.

Another important notion is that of the *reciprocal lattice* [42, 86]. This lattice is actually the Fourier transform of the original lattice, and its points represent the points of replicated spectrums in the frequency domain. If the matrix corresponding to the reciprocal lattice is denoted by $\boldsymbol{D}_r$, then $\boldsymbol{D}_r^T\boldsymbol{D} = \boldsymbol{I}$. Observe that the determinant of the matrix $\boldsymbol{D}$ represents the hypervolume of any unit cell of the corresponding lattice, as well as the reciprocal of the sampling density. One of the possible unit cells is the *Voronoi cell* which is actually the set of points closer to the origin than to any other lattice point. The meaning of the unit cell in the frequency domain is extremely important since if the signal to be sampled is bandlimited to that cell, no overlapping of spectrums will occur and the signal can be reconstructed from its samples.

Let us now examine multidimensional counterparts of some operations involving sampling that are going to be used later. First, *downsampling* will mean that the points on the sampling lattice are kept while all the others are discarded. The time- and Fourier-domain expressions for the output of a downsampler are given by [86, 325]

$$
\begin{aligned}
y[\boldsymbol{n}] &= x[\boldsymbol{D}\boldsymbol{n}], \\
Y(\boldsymbol{\omega}) &= \frac{1}{N} \sum_{\boldsymbol{k}\in\mathcal{U}_c^t} X((\boldsymbol{D}^t)^{-i}(\boldsymbol{\omega} - 2\pi\boldsymbol{k})),
\end{aligned}
$$

where $N = \det(\boldsymbol{D})$, $\boldsymbol{\omega}$ is an n-dimensional real vector, and $\boldsymbol{n}, \boldsymbol{k}$ are n-dimensional integer vectors.

Next consider *upsampling*, that is, the process that maps a signal on the input lattice to another one that is nonzero only at the points of the sampling lattice

$$
\begin{aligned}
y[\boldsymbol{n}] &= \begin{cases} x[\boldsymbol{D}^{-1}\boldsymbol{n}] & \text{if } \boldsymbol{n} = \boldsymbol{D}\boldsymbol{k}, \\ 0 & \text{otherwise,} \end{cases} \\
Y(\boldsymbol{\omega}) &= X(\boldsymbol{D}^t\boldsymbol{\omega}).
\end{aligned}
$$

Let us finish this discussion with examples often encountered in practice.

Example 3.17 *Separable Case: Sampling by 2 in Two Dimensions*

Let us start with the separable case with sampling by 2 in each of the two dimensions. The sampling process is then represented by the following matrix:

$$
\boldsymbol{D}_S = \begin{pmatrix} 2 & 0 \\ 0 & 2 \end{pmatrix} = 2\boldsymbol{I}. \tag{3.B.1}
$$

The unit cell consists of the following points:

$$
(n_1, n_2) \in \{(0,0), (1,0), (0,1), (1,1)\}.
$$

In z-domain, these correspond to

$$
\{1, z_1^{-1}, z_2^{-1}, (z_1 z_2)^{-1}\}.
$$

Its Voronoi cell is a square and the corresponding critically sampled filter bank will have $N = \det(\boldsymbol{D}) = 4$ channels. This is the case most often used in practice in image coding, since it represents separable one-dimensional treatment of an image. Looking at it this way (in terms of lattices), however, will give us the additional freedom to design nonseparable filters even if sampling is separable. The expression for upsampling in this case is

$$
Y(\omega_1, \omega_2) = X(2\omega_1, 2\omega_2),
$$

while downsampling followed by upsampling gives

$$
Y(\omega_1, \omega_2) = \frac{1}{4}(X(\omega_1, \omega_2) + X(\omega_1 + \pi, \omega_2) + X(\omega_1, \omega_2 + \pi) + X(\omega_1 + \pi, \omega_2 + \pi)),
$$

that is, samples where both n, and n_2 are even are kept, while all others are put to zero.

Example 3.18 *Quincunx Sampling*

Consider next the quincunx case, that is, the simplest multidimensional sampling structure that is nonseparable. It is generated using, for example,

$$\boldsymbol{D}_Q = \begin{pmatrix} 1 & 1 \\ 1 & -1 \end{pmatrix}. \tag{3.B.2}$$

Since its determinant equals 2, the corresponding critically sampled filter bank will have two channels. The Voronoi cell for this lattice is a diamond (tilted square). Since the reciprocal lattice for this case is again quincunx, its Voronoi cell will have the same diamond shape. This fact has been used in some image and video coding schemes [12, 320] since, if restricted to this region, (a) the spectrums of the signal and its repeated occurrences that appear due to sampling will not overlap and (b) due to the fact that the human eye is less sensitive to resolution along diagonals, it is more appropriate for the lowpass filter to have diagonal cutoff. Note that the two vectors belonging to the unit cell are

$$\boldsymbol{n}_0 = \begin{pmatrix} 0 \\ 0 \end{pmatrix}, \quad \boldsymbol{n}_1 = \begin{pmatrix} 1 \\ 0 \end{pmatrix},$$

while their z-domain counterparts are 1 and z_1^{-1} and are the same for the unit cell of the transposed lattice. Shifting the origin of the quincunx lattice to points determined by the unit cell vectors yields the two cosets for this lattice. Obviously, their union gives back the original lattice. Write now the expression for the output of an upsampler in Fourier domain

$$Y(\omega_1, \omega_2) = X(\omega_1 + \omega_2, \omega_1 - \omega_2).$$

Similarly, the output of a downsampler followed by an upsampler can be expressed as

$$Y(\omega_1, \omega_2) = \frac{1}{2}(X(\omega_1, \omega_2) + X(\omega_1 + \pi, \omega_2 + \pi)).$$

It is easy to see that all the samples at locations where $(n_1 + n_2)$ is even are kept, while where $(n_1 + n_2)$ is odd, they are put to zero.

PROBLEMS

3.1 *Orthogonality and completeness of the sinc basis (Section 3.1.3):*

(a) Prove the orthogonality relations (3.1.27) and (3.1.28).

(b) Prove that the set $\{\varphi_k\}$ given in (3.1.24) is complete in $l_2(\mathcal{Z})$. *Hint:* Use the same argument as in Proposition 3.1. Take first the even terms and find the Fourier transform of $\langle \varphi_{2k}[n], x[n] \rangle = 0$. Do the same for the odd terms. Combining the two, you should get $\|x\| = 0$ violating the assumption and proving completeness.

3.2 Show that $g_0[n] = 1/\sqrt{2}\ \sin((\pi/2)n)/((\pi/2)n)$ and $g_1[n] = (-1)^n g_0[-n]$ and their even translates do not form an orthogonal basis for $l_2(\mathcal{Z})$, that is, the shift by 1 in (3.1.24) is necessary for completeness. *Hint:* Show incompleteness by finding a counterexample based on $\sin((\pi/2)n)$ with proper normalization.

3.3 Show that Proposition 3.3 does not hold in the nonorthogonal case, that is, there exist nonorthogonal time-invariant expansions with frequency selectivity.

3.4 Prove the equivalences of (a)–(e) in Theorem 3.7.

3.5 Based on the fact that in an orthogonal FIR filter bank, the autocorrelation of the lowpass filter satisfies $P(z) + P(-z) = 2$, show that the length of the filter has to be even.

3.6 For $A(z) = (1+z)^3(1+z^{-1})^3$, verify that $B(z) = 1/256(3z^2 - 18z + 38 - 18z^{-1} + 3z^{-2})$ is the solution such that $P(z) = A(z)\ B(z)$ is valid. If you have access to adequate software (for example, Matlab), do the spectral factorization (obviously, only $B(z)$ needs to be factored). Give the filters of this orthogonal filter bank.

3.7 Prove the equivalences (a)–(e) in Theorem 3.8.

3.8 Prove the three statements on the structure of linear phase solutions given in Proposition 3.11. *Hint:* Use $P(z) = H_0(z)\ G_0(z) = z^{-k} H_0(z)\ H_1(-z)$, and determine when it is valid.

3.9 Show that, when the filters $H_0(z)$ and $H_1(z)$ are of the same length and linear phase, the linear phase testing condition given by (3.2.69), holds. *Hint:* Find out the form of the polyphase components of each linear phase filter.

3.10 In Proposition 3.12, it was shown that there are no real symmetric/antisymmetric orthogonal FIR filter banks.

(a) Show that if the filters can be complex valued, then solutions exist.

(b) For length-6 filters, find the solution with a maximum numbers of zeros at $\omega = \pi$. *Hint:* Refactor the $P(z)$ that leads to the D_3 filter into complex-valued symmetric/antisymmetric filters.

3.11 *Spectral factorization method for two-channel filter banks:* Consider the factorization of $P(z)$ in order to obtain orthogonal or biorthogonal filter banks.

(a) Take

$$P(z) = -1/4z^3 + 1/2z + 1 + 1/2z^{-1} - 1/4z^{-3}.$$

Build an orthogonal filter bank based on this $P(z)$. If the function is not positive on the unit circle, apply an adequate correction (see Smith-Barnwell method in Section 3.2.3).

(b) Alternatively, compute a linear phase factorization of $P(z)$. In particular, choose $H_0(z) = z + 1 + z^{-1}$. Give the other filters in this biorthogonal filter bank.

(c) Assume now that a particular $P(z)$ was designed using the Parks-McClellan algorithm (which leads to equiripple pass and stopbands). Show that if $P(z)$ is not positive on the unit circle, then the correction to make it greater or equal to zero places all stopband zeros on the unit circle.

3.12 Using Proposition 3.13, prove that the filter $H_0(z) = (1 + z^{-1})^N$ has always a complementary filter.

3.13 Prove that in the orthogonal lattice structure, the sum of angles has to be equal to $\pi/4$ or $5\pi/4$ in order to have one zero at $\omega = \pi$ in $H_0(e^{j\omega})$. *Hint:* There are several ways to prove this, but an intuitive one is to consider the sequence $x[n] = (-1)^n$ at the input, or, to consider z-transforms at $z = e^{j\omega} = -1$. See also Example 3.3.

3.14 *Interpolation followed by decimation:* Given an input $x[n]$, consider upsampling by 2, followed by interpolation with a filter having z-transform $H(z)$ for magnification of the signal. Then, to recover the original signal size, apply filtering by a decimation filter $G(z)$ followed by downsampling by 2, in order to obtain a reconstruction $\hat{x}[n]$.

(a) What does the product filter $P(z) = H(z) \cdot G(z)$ have to satisfy in order for $\tilde{x}[n]$ to be a perfect replica of $x[n]$ (possibly with a shift).

(b) Given an interpolation filter $H(z)$, what condition does it have to satisfy so that one can find a decimation filter $G(z)$ in order to achieve perfect reconstruction. *Hint:* This is similar to the complementary filter problem in Section 3.2.3.

(c) For the following two filters,

$$H'(z) \;=\; 1 + z^{-1} + z^{-2} + z^{-3}, \quad H''(z) \;=\; 1 + z^{-1} + z^{-2} + z^{-3} + z^{-4},$$

give filters $G'(z)$ and $G''(z)$ so that perfect reconstruction is achieved (if possible, give shortest such filter, if not, say why).

3.15 Prove the orthogonality relations (3.3.16) and (3.3.17) for an octave-band filter bank, using similar arguments as in the proof of (3.3.15).

3.16 Consider tree-structured orthogonal filter banks as discussed in Example 3.10, and in particular the full tree of depth 2.

(a) Assume ideal sinc filters, and give the frequency response magnitude of $G_{i0}^{(2)}(e^{j\omega}), i = 0, \ldots, 3$. Note that this is not the natural ordering one would expect.

(b) Now take the Haar filters, and give $g_i^{(2)}[n], i = 0, \ldots, 3$. These are the discrete-time Walsh-Hadamard functions of length 4.

(c) Given that $\{g_0[n], g_1[n]\}$ is an orthogonal pair, prove orthogonality for any of the equivalent filters with respect to shifts by 4.

3.17 In the general case of a full-grown binary tree of depth J, define the equivalent filters such that their indexes increase as the center frequency increases. In Example 3.10, it would mean interchanging $G_3^{(2)}$ with $G_2^{(2)}$ (see (3.3.23)).

3.18 Show that in a filter bank with linear phase filters, the iterated filters are also linear phase. In particular, consider the case where $h_0[n]$ and $h_1[n]$ are of even length, symmetric and antisymmetric respectively. Consider a four-channel bank, with $H_a(z) = H_0(z)H_0(z^2)$, $H_b(z) = H_0(z)H_1(z^2)$, $H_c(z) = H_1(z)H_0(z^2)$, and $H_d(z) = H_1(z)H_1(z^2)$. What are the lengths and symmetries of these four filters?

3.19 Consider a general perfect reconstruction filter bank (not necessary orthogonal). Build a tree-structured filter bank. Give and prove the biorthogonality relations for the equivalent impulse responses of the analysis and synthesis filters. For simplicity, consider a full tree of depth 2 rather than an arbitrary tree. *Hint:* The method is similar to the orthogonal case, except that now analysis and synthesis filters are involved.

3.20 Prove that the number of wavelet packet bases generated from a depth-J binary tree is equal to (3.3.25).

3.21 Prove that the perfect reconstruction condition given in terms of the modulation matrix for the N-channel case, is equivalent to the system being biorthogonal. *Hint:* Mimic the proof for the two-channel case given in Section 3.2.1.

3.22 Give the relationship between $G_p(z)$ and $G_m(z)$, which is similar to (3.4.9), as well as between $H_p(z)$ and $H_m(z)$ and this in the general N-channel case.

3.23 Consider a modulated filter bank with filters $H_0(z) = H(z)$, $H_1(z) = H(W_3 z)$, and $H_2(z) = H(W_3^2 z)$. The modulation matrix $H_m(z)$ is circulant. (Note that $W_3 = e^{-j2\pi/3}$).

(a) Show how to diagonalize $H_m(z)$.

(b) Give the form of the determinant $\det(H_m(z))$.

(c) Relate the above to the special form of $H_p(z)$.

3.24 *Cosine modulated filter banks:*

(a) Prove that (3.4.5–3.4.6) hold for the cosine modulated filter bank with filters given in (3.4.18) and $h_{pr}[n] = 1, n = 0, \ldots, 2N-1$.

(b) Prove that in this case (3.4.23) holds as well.

Hint: Show that left and right tails are symmetric/antisymmetric, and thus the tails are orthogonal.

3.25 *Orthogonal pyramid:* Consider a pyramid decomposition as discussed in Section 3.5.2 and shown in Figure 3.17. Now assume that $h[n]$ is an "orthogonal" filter, that is, $\langle h[n], h[n-2l]\rangle = \delta_l$. Perfect reconstruction is achieved by upsampling the coarse version, filtering it by $\tilde{h}$, and adding it to the difference signal.

(a) Analyze the above system in time domain and in z-transform domain, and show perfect reconstruction.

(b) Take $h[n] = (1/\sqrt{2})[1,\ 1]$. Show that $y_1[n]$ can be filtered by $(1/\sqrt{2})[1,\ 1]$ and downsampled by 2 while still allowing perfect reconstruction.

(c) Show that (b) is equivalent to a two-channel perfect reconstruction filter bank with filters $h_0[n] = (1/\sqrt{2})[1,\ 1]$ and $h_1[n] = (1/\sqrt{2})[1,\ -1]$.

(d) Show that (b) and (c) are true for general orthogonal lowpass filters, that is, $y_1[n]$ can be filtered by $g[n] = (-1)^n h[-n+L-1]$ and downsampled by 2, and reconstruction is still perfect using an appropriate filter bank.

3.26 Verify Parseval's formula (3.5.3) in the tight frame case given in Section 3.5.1.

3.27 Consider a two-dimensional two-channel filter bank with quincunx downsampling. Assume that $H_0(z_1, z_2)$ and $H_1(z_1, z_2)$ satisfy (3.6.4–3.6.5). Show that their impulse responses with shifts on a quincunx lattice form an orthonormal basis for $l_2(\mathcal{Z}^2)$.

3.28 *Linear phase diamond-shaped quincunx filters:* We want to construct a perfect reconstruction linear phase filter bank for quincunx sampling and the matrix

$$D = \begin{pmatrix} 1 & 1 \\ 1 & -1 \end{pmatrix}.$$

To that end, we start with the following filters $h_0[n_1, n_2]$ and $h_1[n_1, n_2]$:

$$h_0[n_1, n_2] = \begin{pmatrix} & b & \\ 1 & a & 1 \\ & b & \end{pmatrix},$$

$$h_1[n_1, n_2] = \begin{pmatrix} & & 1 & & \\ & b+\frac{c}{a} & a & b+\frac{c}{a} & \\ \frac{bc}{a} & c & d & c & \frac{bc}{a} \\ & b+\frac{c}{a} & a & b+\frac{c}{a} & \\ & & 1 & & \end{pmatrix},$$

where the origin is where the leftmost coefficient is.

(a) Using the sampling matrix above, identify the polyphase components and verify that perfect FIR reconstruction is possible (the determinant of the polyphase matrix has to be a monomial).

(b) Instead of only having top-bottom, left-right symmetry, impose circular symmetry on the filters. What are b, c? If $a = -4, d = -28$, what type of filters do we obtain (lowpass/highpass)?

4

Series Expansions Using Wavelets and Modulated Bases

"All this time, the guard was looking at her, first through a telescope, then through a microscope, and then through an opera glass"
— Lewis Carroll, *Through the Looking Glass*

Series expansions of continuous-time signals of functions go back at least to Fourier's original expansion of periodic functions. The idea of representing a signal as a sum of elementary basis functions or equivalently, to find orthonormal bases for certain function spaces, is very powerful. However, classic approaches have limitations, in particular, there are no "good" local Fourier series that have both good time and frequency localization.

An alternative is the Haar basis where, in addition to time shifting, one uses scaling instead of modulation in order to obtain an orthonormal basis for $L_2(R)$ [126]. This interesting construction was somewhat of a curiosity (together with a few other special constructions) until wavelet bases were found in the 1980's [71, 180, 194, 21, 22, 175, 283]. Not only are there "good" orthonormal bases, but there also exist efficient algorithms to compute the wavelet coefficients. This is due to a fundamental relation between the continuous-time wavelet series and a set of (discrete-time) sequences. These correspond to a discrete-time filter bank which can be used, under certain conditions, to compute the wavelet series expansion. These relations follow from multiresolution analysis; a framework for analyzing wavelet bases [180, 194]. The emphasis of this chapter is on the construction of wavelet series. We also discuss local Fourier series and the construction of local cosine bases, which are "good" modulated bases [61]. Note that in this chapter we construct bases for $L_2(\mathcal{R})$; however, these bases have much stronger characteristics as they are actually unconditional bases for L_p spaces, $1 < p < \infty$ [73].

The development of wavelet orthonormal bases has been quite explosive in the last decade. While the initial work focused on the continuous wavelet transform (see Chapter 5), the discovery of orthonormal bases by Daubechies [71], Meyer [194], Battle [21, 22], Lemarié [175], Stromberg [283], and others, lead to a wealth of subsequent work.

Compactly supported wavelets, following Daubechies' construction, are based on discrete-time filter banks, and thus many filter banks studied in Chapter 3 can lead to wavelets. We list below, without attempting to be exhaustive, a few such constructions. Cohen, Daubechies and Feaveau [58] and Vetterli and Herley [318, 319] considered biorthogonal wavelet bases. Bases with more than one wavelet were studied by Zou and Tewfik [343, 344], Steffen, Heller, Gopinath and Burrus [277], and Soman, Vaidyanathan and Nguyen [275], among others. Multidimensional, nonseparable wavelets following from filter banks were constructed by Cohen and Daubechies [57] and Kovačević and Vetterli [164]. Recursive filter banks leading to wavelets with exponential decay were derived by Herley and Vetterli [133, 130]. Rioul studied regularity of iterated filter banks [239], complexity of wavelet decomposition algorithms [245], and design of "good" wavelet filters [246]. More constructions relating filter banks and wavelets can be found, for example, in the work of Akansu and Haddad [3, 4], Blu [33], Cohen [55], Evangelista [96, 95], Gopinath [115], Herley [130], Lawton [170, 171], Rioul [240, 242, 243, 244] and Soman and Vaidyanathan [274].

The study of the regularity of the iterated filter that leads to wavelets was done by Daubechies and Lagarias [74, 75], Cohen [55], and Rioul [239] and is related to work on recursive subdivision schemes which was done independently of wavelets (see [45, 80, 87, 92]). The regularity condition and approximation property occurring in wavelets are related to the Strang-Fix condition first derived in the context of finite-element methods [282].

Direct wavelet constructions followed the work of Meyer [194], Battle [21, 22] and Lemarié [175]. They rely on the multiresolution framework established by Mallat [181, 179, 180] and Meyer [194]. In particular, the case of wavelets related to splines was studied by Chui [52, 49, 50] and by Aldroubi and Unser [7, 296, 297]. The extension of the wavelet construction for rational rather than integer dilation factors was done by Auscher [16] and Blu [33]. Approximation properties of wavelet expansions have been studied by Donoho [83], and DeVore and Lucier [82]. These results have interesting consequences for compression.

The computation of the wavelet series coefficients using filter banks was studied by Mallat [181, 179] and Shensa [261], among others. Wavelet sampling theorems are given by Aldroubi and Unser [6], Walter [328] and Xia and Zhang [340]. Local cosine bases were derived by Coifman and Meyer [61] (see also

[17]). The wavelet framework has also proven useful in the context of analysis and synthesis of stochastic processes, see for example [20, 178, 338, 339].

The material in this chapter is covered in more depth in Daubechies' book [73] to which we refer for more details. Our presentation is less formal and based mostly on signal processing concepts.

The outline of the chapter is as follows: First, we discuss series expansions in general and the need for structured series expansion with good time and frequency localization. In particular, the local Fourier series is contrasted with the Haar expansion and a proof that the Haar system is an orthonormal basis for $L_2(R)$ is given. In Section 4.2, we introduce multiresolution analysis and show how a wavelet basis can be constructed. As an example, the sinc (or Littlewood-Paley) wavelet is derived. Section 4.3 gives wavelet bases constructions in the Fourier domain, using the Meyer and Battle-Lemarié wavelets as important examples. Section 4.4 gives the construction of wavelets based on iterated filter banks. The regularity (conditions under which filter banks generate wavelet bases) of the discrete-time filters is studied. In particular, the Daubechies' family of compactly supported wavelets is given. Section 4.5 discusses some of the properties of orthonormal wavelet series expansions as well as the computation of the expansion coefficients. Variations on the theme of wavelets from filter banks are explored in Section 4.6, where biorthogonal bases, wavelets based on IIR filter banks and wavelets with integer dilation factors greater than 2 are given. Section 4.7 discusses multidimensional wavelets obtained from multidimensional filter banks. Finally, Section 4.8 gives an interesting alternative to local Fourier series in the form of local cosine bases which have better time-frequency behavior than their Fourier counterparts.

4.1 Definition of the Problem

4.1.1 Series Expansions of Continuous-Time Signals

In the last chapter orthonormal bases were built for discrete-time sequences, that is, sets of orthogonal sequences $\{\varphi_k[n]\}_{k\in\mathcal{Z}}$ were found such that any signal $x[n] \in l_2(\mathcal{Z})$ could be written as

$$x[n] \;=\; \sum_{k=-\infty}^{\infty} \langle \varphi_k[m], x[m]\rangle \; \varphi_k[n],$$

where

$$\langle \varphi_k[m], x[m]\rangle \;=\; \sum_{m=-\infty}^{\infty} \varphi_k^*[m] \; x[m].$$

In this chapter the aim is to represent continuous-time functions in terms of a *series expansion*. We intend to find sets of orthonormal continuous-time functions $\{\varphi_k(t)\}$ such that signals $f(t)$ belonging to a certain class (for example, $L_2(\mathcal{R})$) can be expressed as

$$f(t) = \sum_{k=-\infty}^{\infty} \langle \varphi_k(u), f(u) \rangle \, \varphi_k(t),$$

where

$$\langle \varphi_k(u), f(u) \rangle = \int_{-\infty}^{\infty} \varphi_k^*(u) \, f(u) \, du.$$

In other words, $f(t)$ can be written as the sum of its orthogonal projections onto the basis vectors $\varphi_k(t)$. Beside having to meet orthonormality constraints, or

$$\langle \varphi_k(u), \varphi_l(u) \rangle = \delta[k-l],$$

the set $\{\varphi_k(t)\}$ has also to be complete. Its span has to cover the space of functions to be represented.

We start by briefly reviewing two standard series expansions that were studied in Section 2.4. The better-known series expansion is certainly the Fourier series. A periodic function, $f(t+nT) = f(t)$, can be written as a linear combination of sines and cosines or complex exponentials, as

$$f(t) = \sum_{k=-\infty}^{\infty} F[k] \, e^{j(2\pi kt)/T}, \tag{4.1.1}$$

where the $F[k]$'s are the Fourier coefficients obtained as

$$F[k] = \frac{1}{T} \int_{-T/2}^{T/2} e^{-j(2\pi kt)/T} f(t) \, dt, \tag{4.1.2}$$

that is, the Fourier transform of one period evaluated at integer multiples of $\omega_0 = 2\pi/T$. It is easy to see that the set of functions $\{e^{j(2\pi kt)/T}, k \in \mathcal{Z}, t \in [-T/2, T/2]\}$ is an orthogonal set, that is,

$$\langle e^{j(2\pi kt)/T}, e^{j(2\pi lt)/T} \rangle_{[-T/2,T/2]} = T\delta[k-l].$$

Since the set is also complete, it is an orthonormal basis for functions belonging to $L_2([-T/2, T/2])$ (up to a scale factor of $1/\sqrt{T}$).

The other standard series expansion is that of bandlimited signals (see also Section 2.4.5). Provided that $|X(\omega)| = 0$ for $|\omega| \geq \omega_s/2 = \pi/T$, then sampling $x(t)$ by multiplying with Dirac impulses at integer multiples of T leads to the function $x_s(t)$ given by

$$x_s(t) = \sum_{n=-\infty}^{\infty} x(nT) \, \delta(t - nT).$$

The Fourier transform of $x_s(t)$ is periodic with period ω_s and is given by (see Section 2.4.5)

$$X_s(\omega) = \frac{1}{T}\sum_{k=-\infty}^{\infty} X(\omega - k\omega_s). \tag{4.1.3}$$

From (4.1.3) it follows that the Fourier transforms of $x(t)$ and $x_s(t)$ coincide over the interval $(-\omega_s/2, \omega_s/2)$ (up to a scale factor), that is, $X(\omega) = TX_s(\omega), \ |\omega| < \omega_s/2$. Thus, to reconstruct the original signal $X(\omega)$, we have to window the sampled signal spectrum $X_s(\omega)$, or $X(\omega) = G(\omega)X_s(\omega)$, where $G(\omega)$ is the window function

$$G(\omega) = \begin{cases} T & |\omega| < \omega_s/2, \\ 0 & \text{otherwise.} \end{cases}$$

Its inverse Fourier transform,

$$g(t) = \text{sinc}_T(t) = \frac{\sin(\pi t/T)}{\pi t/T}, \tag{4.1.4}$$

is called the sinc function.[1] In time domain, we convolve the sampled function $x_s(t)$ with the window function $g(t)$ to recover $x(t)$:

$$x(t) = x_s(t) * g(t) = \sum_{n=-\infty}^{\infty} x(nT)\ \text{sinc}_T(t - nT). \tag{4.1.5}$$

This is usually referred to as the sampling theorem (see Section 2.4.5). Note that the interpolation functions $\{\text{sinc}_T(t-nT)\}_{n\in\mathcal{Z}}$, form an orthogonal set, that is

$$\langle \text{sinc}_T(t - mT), \text{sinc}_T(t - nT)\rangle = T\ \delta[m - n].$$

Then, since $x(t)$ is bandlimited, the process of sampling at times nT can be written as

$$x(nT) = \frac{1}{T}\langle \text{sinc}_T(u - nT), x(u)\rangle,$$

or convolving $x(t)$ with $\text{sinc}_T(-t)$ and sampling the resulting function at times nT. Thus, (4.1.5) is an expansion of a signal into an orthogonal basis

$$x(t) = \frac{1}{T}\sum_{n=-\infty}^{\infty} \langle \text{sinc}_T(u - nT), x(u)\rangle\ \text{sinc}_T(t - nT). \tag{4.1.6}$$

Moreover, if a signal is not bandlimited, then (4.1.6) performs an orthogonal projection onto the space of signals bandlimited to $(-\omega_s/2, \omega_s/2)$ (see Section 2.4.5).

[1]The standard definition from the digital signal processing literature is used here, even if it would make sense to divide the sinc by $1/\sqrt{T}$ to make it of unit norm.

4.1.2 Time and Frequency Resolution of Expansions

Having seen two possible series expansions (Fourier series and sinc expansion), let us discuss some of their properties. First, both cases deal with a limited signal space — periodic or bandlimited. In what follows, we will be interested in representing more general signals. Then, the basis functions, while having closed-form expressions, have poor decay in time (no decay in the Fourier series case, $1/t$ decay in the sinc case). Local effects spread over large regions of the transform domain. This is often undesirable if one wants to detect some local disturbance in a signal which is a classic task in nonstationary signal analysis.

In this chapter, we construct alternative series expansions, mainly based on wavelets. But first, let us list a few desirable features of basis functions [238]:

(a) Simple characterization.

(b) Desirable localization properties in both time and frequency, that is, appropriate decay in both domains.

(c) Invariance under certain elementary operations (for example, shifts in time).

(d) Smoothness properties (continuity, differentiability).

(e) Moment properties (zero moments, see Section 4.5).

However, some of the above requirements conflict with each other and ultimately, the application at hand will greatly influence the choice of the basis.

In addition, it is often desirable to look at a signal at different *resolutions*, that is, both globally and locally. This feature is missing in classical Fourier analysis. Such a *multiresolution* approach is not only important in many applications (ranging from signal compression to image understanding), but is also a powerful theoretical framework for the construction and analysis of wavelet bases as alternatives to Fourier bases.

In order to satisfy some of the above requirements, let us first review how one can modify Fourier analysis so that local signal behavior in time can be seen even in the transform domain. We thus reconsider the short-time Fourier (STFT) or Gabor transform introduced in Section 2.6. The idea is to window the signal (that is, multiply the signal by an appropriate windowing function centered around the point of interest), and then take its Fourier transform. To analyze the complete signal, one simply shifts the window over the whole time range in sufficiently small steps so as to have substantial overlap between adjacent windows. This is a very redundant representation (the signal has been mapped into an infinite set of Fourier transforms) and thus it can be sampled. This scheme will be further analyzed in Section 5.3.

As an alternative, consider a "local Fourier series" obtained as follows: Starting with an infinite and arbitrary signal, divide it into pieces of length T and expand each piece in terms of a Fourier series. Note that at the boundary between two intervals the expansion will in general be incorrect because the periodization creates a discontinuity. However, this error has zero energy, and therefore this simple scheme is a possible orthogonal expansion which has both a frequency index (corresponding to multiples of $\omega_0 = 2\pi/T$) and a time index (corresponding to the interval number, or the multiple of the interval length T). That is, we can expand $x(t)$ as (following (4.1.1), (4.1.2))

$$\hat{x}(t) = \sum_{m=-\infty}^{\infty} \sum_{n=-\infty}^{\infty} \langle \varphi_{m,n}(u), x(u) \rangle \, \varphi_{m,n}(t), \tag{4.1.7}$$

where

$$\varphi_{m,n}(u) = \begin{cases} 1/\sqrt{T} e^{j2\pi n(u-mT)/T} & u \in [mT - T/2, mT + T/2), \\ 0 & \text{otherwise.} \end{cases}$$

The $1/\sqrt{T}$ factor makes the basis functions of unit norm. The expansion $\hat{x}(t)$ is equal to $x(t)$ almost everywhere (except at $t = (m + 1/2)T$) and thus, the L_2 norm of the difference $x(t) - \hat{x}(t)$ is equal to zero. We call this transform a *piecewise Fourier series.*

Consider what has been achieved. The expansion in (4.1.7) is valid for arbitrary functions. Then, instead of an integral expansion as in the Fourier transform, we have a double-sum expansion, and the set of basis functions is orthonormal and complete. Time locality is now achieved and there is some frequency localization (not very good, however, because the basis functions are rectangular windowed sinusoids and therefore discontinuous; their Fourier transforms decay only as $1/\omega$). In terms of time-frequency resolution, we have the rectangular tiling of the time-frequency plane that is typical of the short-time Fourier transform (as was shown in Figure 2.12(b)).

However, there is a price to be paid. The size of the interval T (that is, the location of the boundaries) is arbitrary and leads to problems. The reconstruction $\hat{x}(t)$ has singular points even if $x(t)$ is continuous and the transform of $x(t)$ can have infinitely many "high frequency" components even if $x(t)$ is a simple sinusoid (for example, if its period T_s is such that T_s/T is irrational). Therefore, the expansion will converge slowly to the function. In other words, if one wants to approximate the signal with a truncated series, the quality of the approximation will depend on the choice of T. In particular, the convergence at points of discontinuity (created by periodization) is poor due to the Gibbs phenomenon [218]. Finally, a shift of the signal can lead to completely different transform coefficients and the transform is thus time-variant.

In short, we have gained the flexibility of a double-indexed transform indicating time and frequency, but we have lost time invariance and convergence is sometimes poor. Note that some of these problems are inherent to local Fourier bases and can be solved with local cosine bases discussed in Section 4.8.

4.1.3 Haar Expansion

We explore the Haar expansion because it is the simplest example of a wavelet expansion, yet it contains all the ingredients of such constructions. It also addresses some of the problems we mentioned for the local Fourier series. The arbitrariness of a single window of fixed length T, as discussed, is avoided by having a variable size window. Time invariance is not obtained (actually, requiring locality in time implies time variance). The Haar wavelet, or prototype basis function, has finite support in time and $1/\omega$ decay in frequency. Note that it has its dual in the so-called sinc wavelet (discussed in Section 4.2) which has finite support in frequency and $1/t$ decay in time. We will see that the Haar and sinc wavelets are two extreme examples and that all the other examples of interest will have a behavior that lies in between.

The Haar wavelet is defined as

$$\psi(t) = \begin{cases} 1 & 0 \leq t < \frac{1}{2}, \\ -1 & \frac{1}{2} \leq t < 1, \\ 0 & \text{otherwise}, \end{cases} \tag{4.1.8}$$

and the whole set of basis functions is obtained by dilation and translation as

$$\psi_{m,n}(t) = 2^{-m/2}\psi(2^{-m}t - n), \qquad m, n \in \mathcal{Z}. \tag{4.1.9}$$

We call m the scale factor, since $\psi_{m,n}(t)$ is of length 2^m, while n is called the shift factor, and the shift is scale dependent ($\psi_{m,n}(t)$ is shifted by $2^m n$). The normalization factor $2^{-m/2}$ makes $\psi_{m,n}(t)$ of unit norm. The Haar wavelet is shown in Figure 4.1(c) (part (a) shows the scaling function which will be introduced shortly). A few of the basis functions are shown in Figure 4.2(a). It is easy to see that the set is orthonormal. At a given scale, $\psi_{m,n}(t)$ and $\psi_{m,n'}(t)$ have no common support. Across scales, even if there is common support, the larger basis function is constant over the support of the shorter one. Therefore, the inner product amounts to the average of the shorter one which is zero (see Figure 4.2(b)). Therefore,

$$\langle \psi_{m,n}(t), \psi_{m',n'}(t) \rangle = \delta[m - m']\,\delta[n - n'].$$

The advantage of these basis functions is that they are well localized in time (the support is finite). Actually, as $m \to -\infty$, they are arbitrarily sharp in time,

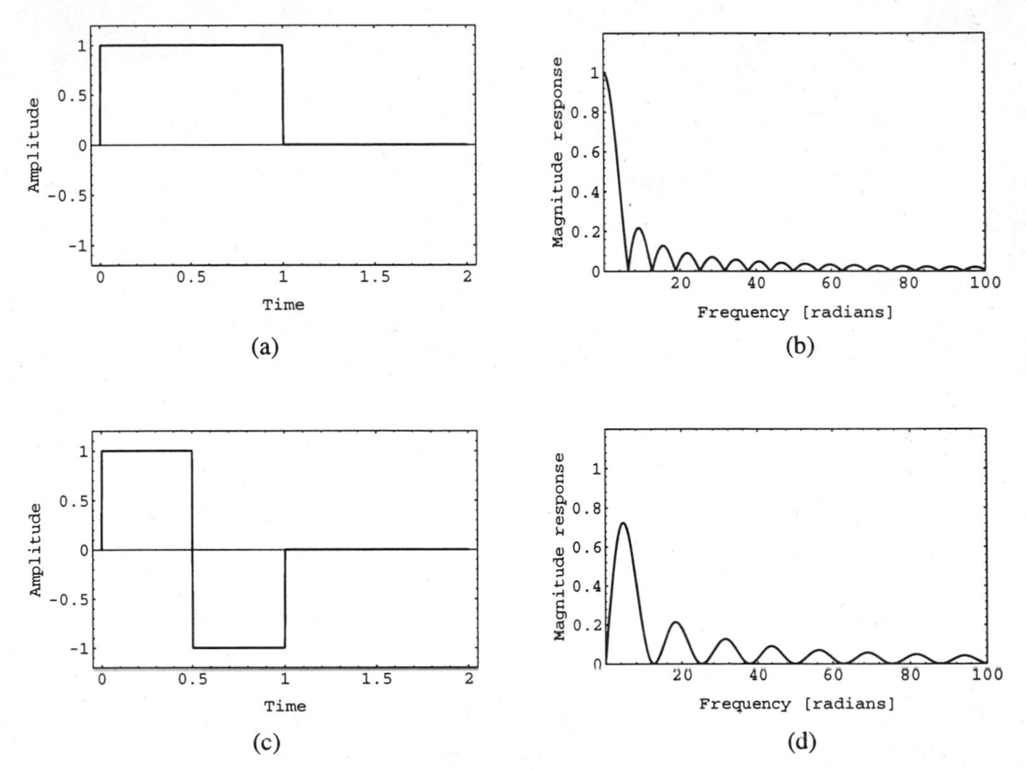

Figure 4.1 The Haar scaling function and wavelet, given in Table 4.1. (a) The scaling function $\varphi(t)$. (b) Fourier transform magnitude $|\Phi(\omega)|$. (c) Wavelet $\psi(t)$. (d) Fourier transform magnitude $|\Psi(\omega)|$.

since the length goes to zero. That is, a discontinuity (for example, a step in a function) will be localized with arbitrary precision. However, the frequency localization is not very good since the Fourier transform of (4.1.8) decays only as $1/\omega$ when $\omega \to \infty$. The basis functions are not smooth, since they are not even continuous.

One of the fundamental characteristics of the wavelet type expansions which we will discuss in more detail later is that they are series expansions with a double sum. One is for shifts, the other is for scales and there is a trade-off between time and frequency resolutions. This resolution is what differentiates this double-sum expansion from the one given in (4.1.7). Now, long basis functions (for m large and positive) are sharp in frequency (with corresponding loss of time resolution), while short basis functions (for negative m with large absolute value) are sharp in time. Conceptually, we obtain a tiling of the time-frequency plane as was shown in Figure 2.12(d), that is, a dyadic tiling rather than the rectangular tiling of the short-time Fourier transform

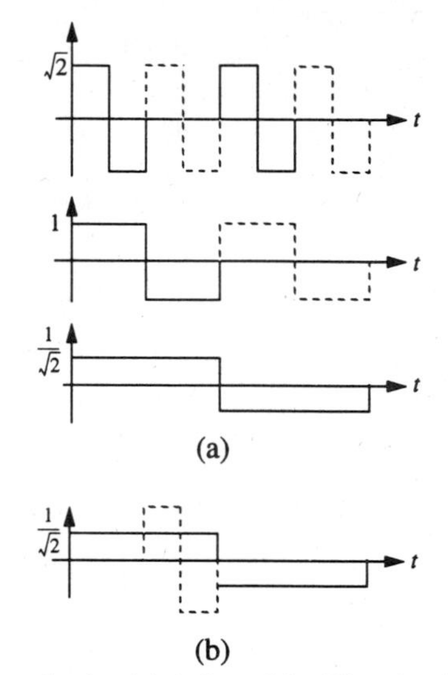

Figure 4.2 The Haar basis. (a) A few of the Haar basis functions. (b) Haar wavelets are orthogonal across scales since the inner product is equal to the average of the shorter one.

shown in Figure 2.12(b).

In what follows, the proof that the Haar system is a basis for $L_2(\mathcal{R})$ is given using a multiresolution flavor [73]. Thus, it has more than just technical value; the intuition gained and concepts introduced will be used again in later wavelet constructions.

THEOREM 4.1

The set of functions $\{\psi_{m,n}(t)\}_{m,n\in\mathcal{Z}}$, with $\psi(t)$ and $\psi_{m,n}(t)$ as in (4.1.8–4.1.9), is an orthonormal basis for $L_2(\mathcal{R})$.

PROOF

The idea is to consider functions which are constant on intervals $[n2^{-m_0},(n+1)2^{-m_0})$ and which have finite support on $[-2^{m_1},2^{m_1})$, as shown in Figure 4.3(a). By choosing m_0 and m_1 large enough, one can approximate any $L_2(\mathcal{R})$ function arbitrarily well. Call such a piecewise constant function $f^{(-m_0)}(t)$. Introduce a unit norm indicator function for the interval $[n2^{-m_0},(n+1)2^{-m_0})$

$$\varphi_{-m_0,n}(t) \;=\; \begin{cases} 2^{\frac{m_0}{2}} & n2^{-m_0} \le t < (n+1)2^{-m_0}, \\ 0 & \text{otherwise.} \end{cases} \tag{4.1.10}$$

This is called the *scaling function* in the Haar case. Obviously, $f^{(-m_0)}(t)$ can be written as a linear combination of indicator functions from (4.1.10)

$$f^{(-m_0)}(t) \;=\; \sum_{n=-N}^{N-1} f_n^{(-m_0)}\varphi_{-m_0,n}(t), \tag{4.1.11}$$

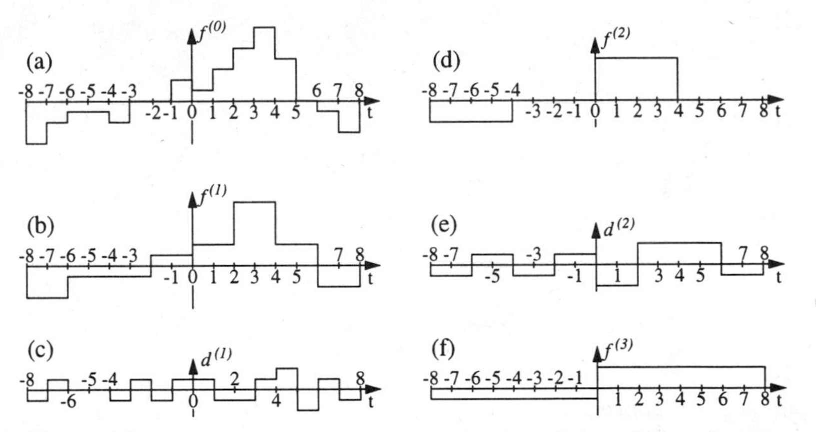

Figure 4.3 Haar wavelet decomposition of a piecewise continuous function. Here, $m_0 = 0$ and $m_1 = 3$. (a) Original function $f^{(0)}$. (b) Average function $f^{(1)}$. (c) Difference $d^{(1)}$ between (a) and (b). (d) Average function $f^{(2)}$. (e) Difference $d^{(2)}$. (f) Average function $f^{(3)}$.

where $N = 2^{m_0+m_1}$, and $f_n^{(-m_0)} = 2^{-m_0/2} f^{(-m_0)}(n \cdot 2^{-m_0})$. Now comes the key step: Examine two intervals $[2n \cdot 2^{-m_0}, (2n+1)2^{-m_0})$ and $[(2n+1) \cdot 2^{-m_0}, (2n+2)2^{-m_0})$. The function over these two intervals is from (4.1.11)

$$f_{2n}^{(-m_0)} \varphi_{-m_0,2n}(t) + f_{2n+1}^{(-m_0)} \varphi_{-m_0,2n+1}(t). \tag{4.1.12}$$

However, the same function can be expressed as the average over the two intervals plus the difference needed to obtain (4.1.12). The average is given by

$$\frac{f_{2n}^{(-m_0)} + f_{2n+1}^{(-m_0)}}{2} \cdot \sqrt{2} \cdot \varphi_{-m_0+1,n}(t),$$

while the difference can be expressed with the Haar wavelet as

$$\frac{f_{2n}^{(-m_0)} - f_{2n+1}^{(-m_0)}}{2} \cdot \sqrt{2} \cdot \psi_{-m_0+1,n}(t).$$

Note that here we have used the wavelet and the scaling function of twice the length. Their support is from $[n \cdot 2^{-m_0+1}, (n+1)2^{-m_0+1}) = [2n \cdot 2^{-m_0}, (2n+2)2^{-m_0})$. Also note that the factor $\sqrt{2}$ is due to $\psi_{-m_0+1,n}(t)$ and $\varphi_{-m_0+1,n}(t)$ having height $2^{(m_0-1)/2} = 2^{m_0/2}/\sqrt{2}$, instead of $2^{m_0/2}$ with which we started. Calling now

$$f_n^{(-m_0+1)} = \frac{1}{\sqrt{2}}(f_{2n}^{(-m_0)} + f_{2n+1}^{(-m_0)}),$$

and

$$d_n^{(-m_0+1)} = \frac{1}{\sqrt{2}}(f_{2n}^{(-m_0)} - f_{2n+1}^{(-m_0)}),$$

we can rewrite (4.1.12) as

$$f_n^{(-m_0+1)} \varphi_{-m_0+1,n}(t) + d_n^{(-m_0+1)} \psi_{-m_0+1,n}(t).$$

Applying the above to the pairs of intervals of the whole function, we finally obtain

$$f^{(-m_0)}(t) = f^{(-m_0+1)}(t) + d^{(-m_0+1)}(t)$$

$$= \sum_{n=-\frac{N}{2}}^{\frac{N}{2}-1} f_n^{(-m_0+1)} \varphi_{-m_0+1,n}(t) + \sum_{n=-\frac{N}{2}}^{\frac{N}{2}-1} d_n^{(-m_0+1)} \psi_{-m_0+1,n}(t).$$

This decomposition in local "average" and "difference" is shown in Figures 4.3(b) and (c) respectively. In order to obtain $f^{(-m_0+2)}(t)$ plus some linear combination of $\psi_{-m_0+2,n}(t)$, one can iterate the averaging process on the function $f^{(-m_0+1)}(t)$ exactly as above (see Figures 4.3(d),(e)). Repeating the process until the average is over intervals of length 2^{m_1} leads to

$$f^{(-m_0)}(t) = f^{(m_1)}(t) + \sum_{m=-m_0+1}^{m_1} \sum_{n=-2^{m_1-m}}^{2^{m_1-m}-1} d_n^{(m)} \psi_{m,n}(t). \tag{4.1.13}$$

The function $f^{(m_1)}(t)$ is equal to the average of $f^{(-m_0)}(t)$ over the intervals $[-2^{m_1}, 0)$ and $[0, 2^{m_1})$, respectively (see Figure 4.3(f)). Consider the right half, which equals $f_0^{(m_1)}$ from 0 to 2^{m_1}. It has L_2 norm equal to $|f_0^{(m_1)}|2^{m_1/2}$. This function can further be decomposed as the average over the interval $[0, 2^{m_1+1})$ plus a Haar function. The new average function has norm $(|f_0^{(m_1)}|2^{m_1/2}/\sqrt{2} = |f_0^{(m_1)}|2^{(m_1-1)/2}$ (since there is no contribution from $[2^{m_1}, 2^{m_1+1})$). Iterating this M times shows that the norm of the average function decreases as $(|f_0^{(m_1)}|2^{m_1/2})/2^{M/2} = |f_0^{(m_1)}|2^{(m_1-M)/2}$. The same argument holds for the left side as well and therefore, $f^{(-m_0)}(t)$ can be approximated from (4.1.13), as

$$f^{(-m_0)}(t) = \sum_{m=-m_0+1}^{m_1+M} \sum_{n=-2^{m_1-m}}^{2^{m_1-m}-1} d_n^{(m)} \psi_{m,n}(t) + \varepsilon_M,$$

where $\|\varepsilon_M\| = (|f_{-1}^{(m_1)}| + |f_0^{(m_1)}|) \cdot 2^{(m_1-M)/2}$. The approximation error ϵ_M can thus be made arbitrarily small since $|f_n^{(m_1)}|$, $n = -1, 0$, are bounded and M can be made arbitrarily large. This, together with the fact that m_0 and m_1 can be arbitrarily large completes the proof that any $L_2(\mathcal{R})$ function can be represented as a linear combination of Haar wavelets.

The key in the above proof was the decomposition into a coarse approximation (the *average*) and a detail (the *difference*). Since the norm of the coarse version goes to zero as the scale goes to infinity, any $L_2(\mathcal{R})$ function can be represented as a succession of multiresolution details. This is the crux of the multiresolution analysis presented in Section 4.2 and will prove to be a general framework, of which the Haar case is a simple but enlightening example.

Let us point out a few features of the Haar case above. First, we can define spaces V_m of piecewise constant functions over intervals of length 2^m. Obviously, V_m is included in V_{m-1}, and an orthogonal basis for V_m is given by φ_m and its shifts by multiples of 2^m. Now, call W_m the orthogonal complement of V_m in V_{m-1}. An orthogonal basis for W_m is given by ψ_m and its shifts by multiples of 2^m. The proof above relied on decomposing V_{-m_0} into V_{-m_0+1} and W_{-m_0+1}, and then iterating the decomposition again on V_{-m_0+1} and so on. It is important to note that once we had a signal in V_{-m_0}, the rest of the decomposition involved only discrete-time computations (average and difference operations on previous

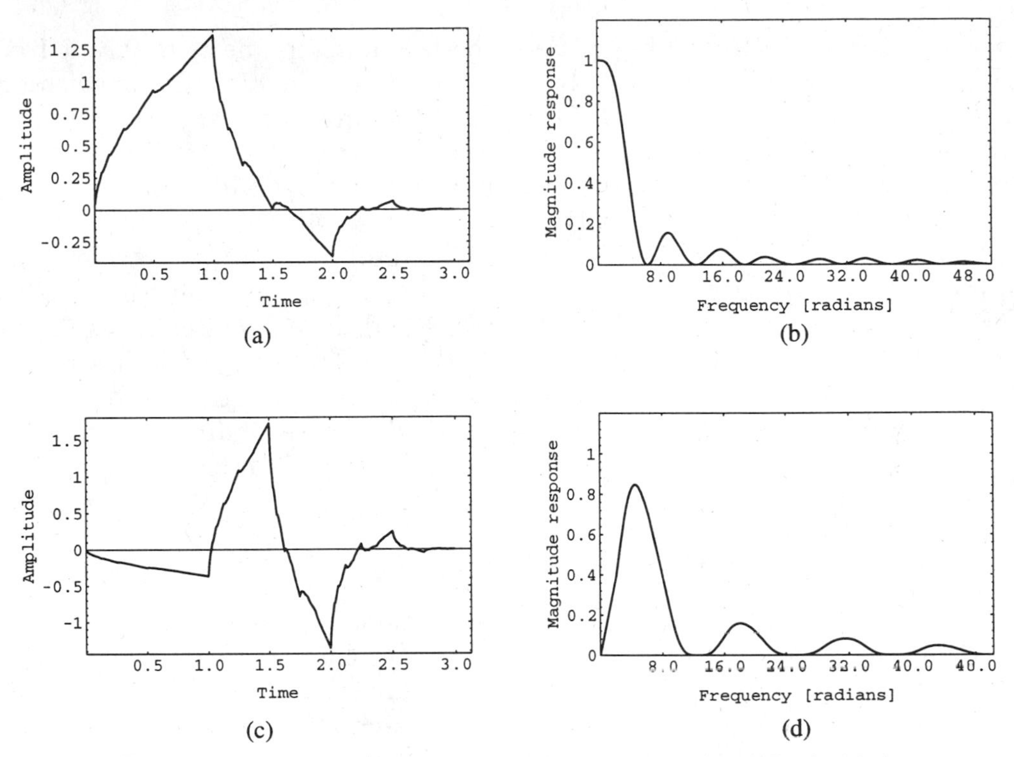

(a) (b) (c) (d)

Figure 4.4 Scaling function and wavelet obtained from iterating Daubechies' 4-tap filter. (a) Scaling function $\varphi(t)$. (b) Fourier transform magnitude $|\Phi(\omega)|$. (c) Wavelet $\psi(t)$. (d) Fourier transform magnitude $|\Psi(\omega)|$.

coefficients). This is a fundamental and attractive feature of wavelet series expansions which holds in general, as we shall see.

4.1.4 Discussion

As previously mentioned, the Haar case (seen above) and the sinc case (in Section 4.2.3) are two extreme cases, and the purpose of this chapter is to construct "intermediate" solutions with additional desirable properties. For example, Figure 4.4 shows a wavelet constructed first by Daubechies [71] which has finite (compact) support (its length is $L = 3$, that is, less local than the Haar wavelet which has length 1) but is continuous and has better frequency resolution than the Haar wavelet. While not achieving a frequency resolution comparable to the sinc wavelet, its time resolution is much improved since it has finite length. This is only one of many possible wavelet constructions, some of which will be shown in more detail later.

We have shown that it is possible to construct series expansions of general functions. The resulting tiling of the time-frequency plane is different from that of a local Fourier series. It has the property that high frequencies are analyzed with short basis functions, while low frequencies correspond to long basis functions. While this trade-off is intuitive for many "natural" functions or signals, it is not the only one; therefore, alternative tilings will also be explored. One elegant property of wavelet type bases is the self-similarity of the basis functions, which are all obtained from a single prototype "mother" wavelet using scaling and translation. This is unlike local Fourier analysis, where modulation is used instead of scaling. The basis functions and the associated tiling for the local Fourier analysis (short-time Fourier transform) were seen in Figures 2.12 (a) and (b). Compare these to the wavelet-type tiling and the corresponding basis functions given in Figures 2.12(c) and (d) where scaling has replaced modulation. One can see that a dyadic tiling has been obtained.

4.2 Multiresolution Concept and Analysis

In this section, we analyze signal decompositions which rely on successive approximation (the Haar case is a particular example). A given signal will be represented by a coarse approximation plus added details. We show that the coarse and detail subspaces are orthogonal to each other. In other words, the detail signal is the difference between the fine and the coarse version of the signal. By applying the successive approximation recursively, we will see that the space of input signals $L_2(\mathcal{R})$ can be spanned by spaces of successive details at all resolutions. This follows because, as the detail resolution goes to infinity, the approximation error goes to zero.

Note that this multiresolution approach, pioneered by Mallat [180] and Meyer [194], is not only a set of tools for deriving wavelet bases, but also a mathematical framework which is very useful in conceptualizing problems linked to wavelet and subband decompositions of signals. We will also see that multiresolution analysis leads to particular orthonormal bases, with basis functions being self-similar at different scales. We will also show that a multiresolution analysis leads to the two-scale equation property and that some special discrete-time sequences play a special role in that they are equivalent to the filters in an orthogonal filter bank.

4.2.1 Axiomatic Definition of Multiresolution Analysis

Let us formally define multiresolution analysis. We will adhere to the choice of axioms as well as the ordering of spaces adopted by Daubechies in [73].

DEFINITION 4.2

A multiresolution analysis consists of a sequence of embedded closed subspaces

$$\ldots V_2 \subset V_1 \subset V_0 \subset V_{-1} \subset V_{-2} \ldots \tag{4.2.1}$$

such that

(a) *Upward Completeness*

$$\overline{\bigcup_{m \in \mathcal{Z}} V_m} = L_2(\mathcal{R}). \tag{4.2.2}$$

(b) *Downward Completeness*

$$\bigcap_{m \in \mathcal{Z}} V_m = \{0\}. \tag{4.2.3}$$

(c) *Scale Invariance*

$$f(t) \in V_m \Longleftrightarrow f(2^m t) \in V_0. \tag{4.2.4}$$

(d) *Shift Invariance*

$$f(t) \in V_0 \Longrightarrow f(t-n) \in V_0, \text{ for all } n \in \mathcal{Z}. \tag{4.2.5}$$

(e) *Existence of a Basis* There exists $\varphi \in V_0$, such that

$$\{\varphi(t-n) \mid n \in \mathcal{Z}\} \tag{4.2.6}$$

is an orthonormal basis for V_0.

Remarks

(a) If we denote by $\text{Proj}_{V_m}[f(t)]$, the orthogonal projection of $f(t)$ onto V_m, then (4.2.2) states that $\lim_{m \to -\infty} \text{Proj}_{V_m}[f(t)] = f(t)$.

(b) The multiresolution notion comes into play only with (4.2.4), since all the spaces are just scaled versions of the central space V_0 [73].

(c) As seen earlier for the Haar case, the function $\varphi(t)$ in (4.2.6) is called the *scaling function*.

(d) Using the Poisson formula, the orthonormality of the family $\{\varphi(t-n)\}_{n \in \mathcal{Z}}$ as given in (4.2.6) is equivalent to the following in the Fourier domain (see (2.4.31)):

$$\sum_{k=-\infty}^{\infty} |\Phi(\omega + 2k\pi)|^2 = 1. \tag{4.2.7}$$

(e) Using (4.2.4–4.2.6), one obtains that $\{2^{m/2}\varphi(2^m t - n) \mid n \in \mathcal{Z}\}$ is a basis for V_{-m}.

(f) The orthogonality of $\varphi(t)$ is not necessary, since a nonorthogonal basis (with the shift property) can always be orthogonalized [180] (see also Section 4.3.2).

As an example, define V_m as the space of functions which are piecewise constant over intervals of length 2^m and define $\varphi(t)$ as the indicator function of the unit interval. Then, it is easy to verify that the Haar example in the previous section satisfies the axioms of multiresolution analysis (see Example 4.1 below).

Because of the embedding of spaces (4.2.1) and the scaling property (4.2.4), we can verify that the scaling function $\varphi(t)$ satisfies a two-scale equation. Since V_0 is included in V_{-1}, $\varphi(t)$, which belongs to V_0, belongs to V_{-1} as well. As such, it can be written as a linear combination of basis functions from V_{-1}. However, we know that $\{\sqrt{2}\varphi(2t - n) \mid n \in \mathcal{Z}\}$ is an orthonormal basis for V_{-1}; thus, $\varphi(t)$ can be expressed as

$$\varphi(t) = \sqrt{2} \sum_{n=-\infty}^{\infty} g_0[n] \, \varphi(2t - n). \tag{4.2.8}$$

Note that with the above normalization, $\|g_0[n]\| = 1$ and $g_0[n] = \sqrt{2} \cdot \langle \varphi(2t - n), \varphi(t) \rangle$ (see Problem 4.2). Taking the Fourier transform of both sides, we obtain

$$\begin{aligned}
\Phi(\omega) &= \int \varphi(t) e^{-j\omega t} dt = \sqrt{2} \int \sum_{n=-\infty}^{\infty} g_0[n] \, \varphi(2t - n) e^{-j\omega t} \, dt \\
&= \sqrt{2} \sum_{n=-\infty}^{\infty} g_0[n] \, \frac{1}{2} \int \varphi(t) e^{-j\omega t/2} e^{-j\omega n/2} \, dt \\
&= \frac{1}{\sqrt{2}} \sum_{n=-\infty}^{\infty} g_0[n] \, e^{-j(\omega/2)n} \int \varphi(t) e^{-j(\omega/2)t} \, dt \\
&= \frac{1}{\sqrt{2}} G_0(e^{j\omega/2}) \, \Phi(\omega/2),
\end{aligned} \tag{4.2.9}$$

where

$$G_0(e^{j\omega}) = \sum_{n \in \mathcal{Z}} g_0[n] \, e^{-j\omega n}.$$

It will be shown that this function characterizes a multiresolution analysis. It is obviously 2π-periodic and can be viewed as a discrete-time Fourier transform of a discrete-time filter $g_0[n]$. This last observation links discrete and continuous time, and allows one to construct continuous-time wavelet bases starting from discrete iterated filters. It also allows one to compute continuous-time wavelet expansions using discrete-time algorithms.

An important property of $G_0(e^{j\omega})$ is the following:

$$|G_0(e^{j\omega})|^2 + |G_0(e^{j(\omega+\pi)})|^2 = 2. \tag{4.2.10}$$

Note that (4.2.10) was already given in (3.2.54) (again a hint that there is a strong connection between discrete and continuous time). Equation (4.2.10) can be proven by using (4.2.7) for 2ω:

$$\sum_{k=-\infty}^{\infty} |\Phi(2\omega + 2k\pi)|^2 = 1. \tag{4.2.11}$$

Substituting (4.2.9) into (4.2.11)

$$\begin{aligned}
1 &= \frac{1}{2}\sum_k |G_0(e^{j(\omega+k\pi)})|^2 |\Phi(\omega + k\pi)|^2 \\
&= \frac{1}{2}\sum_k |G_0(e^{j(\omega+2k\pi)})|^2 |\Phi(\omega + 2k\pi)|^2 \\
&\quad + \frac{1}{2}\sum_k |G_0(e^{j(\omega+(2k+1)\pi)})|^2 |\Phi(\omega + (2k+1)\pi)|^2 \\
&= \frac{1}{2}|G_0(e^{j\omega})|^2 \sum_k |\Phi(\omega + 2k\pi)|^2 + \frac{1}{2}|G_0(e^{j(\omega+\pi)})|^2 \sum_k |\Phi(\omega + (2k+1)\pi)|^2 \\
&= \frac{1}{2}(|G_0(e^{j\omega})|^2 + |G_0(e^{j(\omega+\pi)})|^2),
\end{aligned}$$

which completes the proof of (4.2.10). With a few restrictions on the Fourier transform $\Phi(\omega)$ (bounded, continuous in $\omega = 0$, and $\Phi(0) \neq 0$), it can be shown that $G_0(e^{j\omega})$ satisfies

$$|G_0(1)| = \sqrt{2}$$
$$G_0(-1) = 0$$

(see Problem 4.3). Note that the above restrictions on $\Phi(\omega)$ are always satisfied in practice.

4.2.2 Construction of the Wavelet

We have shown that a multiresolution analysis is characterized by a 2π-periodic function $G_0(e^{j\omega})$ with some additional properties. The axioms (4.2.1–4.2.6) guarantee the existence of bases for approximation spaces V_m. The importance of multiresolution analysis is highlighted by the following theorem. We outline the proof and show how it leads to the construction of wavelets.

THEOREM 4.3

Whenever the sequence of spaces satisfy (4.2.1–4.2.6), there exists an orthonormal basis for $L_2(\mathcal{R})$:

$$\psi_{m,n}(t) = 2^{-m/2}\psi(2^{-m}t - n), \qquad m, n \in \mathcal{Z},$$

such that $\{\psi_{m,n}\}$, $n \in \mathcal{Z}$ is an orthonormal basis for W_m, where W_m is the orthogonal complement of V_m in V_{m-1}.

PROOF

To prove the theorem, let us first establish a couple of important facts. First, we defined W_m as the orthogonal complement of V_m in V_{m-1}. In other words

$$V_{m-1} = V_m \oplus W_m.$$

By repeating the process and using (4.2.2) we obtain that

$$L_2(\mathcal{R}) = \bigoplus_{m \in \mathcal{Z}} W_m. \tag{4.2.12}$$

Also, due to the scaling property of the V_m spaces (4.2.4), there exists a scaling property for the W_m spaces as well:

$$f(t) \in W_m \Longleftrightarrow f(2^m t) \in W_0. \tag{4.2.13}$$

Our aim here is to explicitly construct[2] a wavelet $\psi(t) \in W_0$, such that $\psi(t-n), n \in \mathcal{Z}$ is an orthonormal basis for W_0. If we have such a wavelet $\psi(t)$, then by the scaling property (4.2.13), $\psi_{m,n}(t), n \in \mathcal{Z}$ will be an orthonormal basis for W_m. On the other hand, (4.2.12) together with upward/downward completeness properties (4.2.2–4.2.3), imply that $\{\psi_{m,n}\}$, $m, n \in \mathcal{Z}$ is an orthonormal basis for $L_2(\mathcal{R})$, proving the theorem. Thus, we start by constructing the wavelet $\psi(t)$, such that $\psi \in W_0 \subset V_{-1}$. Since $\psi \in V_{-1}$

$$\psi(t) = \sqrt{2}\sum_{n \in \mathcal{Z}} g_1[n]\varphi(2t - n). \tag{4.2.14}$$

Taking the Fourier transform one obtains

$$\Psi(\omega) = \frac{1}{\sqrt{2}} G_1(e^{j\omega/2}) \cdot \Phi\left(\frac{\omega}{2}\right), \tag{4.2.15}$$

where $G_1(e^{j\omega})$ is a 2π-periodic function from $L_2([0, 2\pi])$. The fact that $\psi(t)$ belongs to W_0, which is orthogonal to V_0, implies that

$$\langle \varphi(t-k), \psi(t) \rangle = 0, \text{ for all } k.$$

This can also be expressed as (in the Fourier domain)

$$\int \Psi(\omega)\, \Phi^*(\omega)\, e^{j\omega k} = 0,$$

[2]Note that the wavelet we construct is not unique.

or equivalently,

$$\int_0^{2\pi} e^{j\omega k} d\omega \sum_l \Psi(\omega + 2\pi l)\ \Phi^*(\omega + 2\pi l) \ = \ 0.$$

This further implies that

$$\sum_l \Psi(\omega + 2\pi l)\Phi^*(\omega + 2\pi l) \ = \ 0. \tag{4.2.16}$$

Now substitute (4.2.9) and (4.2.15) into (4.2.16) and split the sum over l into two sums over even and odd l's

$$\begin{aligned}
&\frac{1}{2}\sum_l G_1(e^{j(\omega/2+2l\pi)})\ \Phi(\omega/2 + 2l\pi)\ G_0^*(e^{j(\omega/2+2l\pi)})\ \Phi^*(\omega/2+2l\pi) \\
&\quad + \frac{1}{2}\sum_l G_1(e^{j(\omega/2+(2l+1)\pi)})\ \Phi(\omega/2 + (2l+1)\pi)\ G_0^*(e^{j(\omega/2+(2l+1)\pi)})\ \Phi^*(\omega/2+(2l+1)\pi) \\
&= 0.
\end{aligned}$$

However, since G_0 and G_1 are both 2π-periodic, substituting Ω for $\omega/2$ gives

$$G_1(e^{j\Omega})\ G_0^*(e^{j\Omega}) \sum_l |\Phi(\Omega + 2l\pi)|^2 + G_1(e^{j(\Omega+\pi)})\ G_0^*(e^{j(\Omega+\pi)}) \sum_l |\Phi(\Omega + (2l+1)\pi)|^2 \ = \ 0.$$

Using now (4.2.7), the sums involving $\Phi(\omega)$ become equal to 1, and thus

$$G_1(e^{j\Omega})\ G_0^*(e^{j\Omega}) + G_1(e^{j(\Omega+\pi)})\ G_0^*(e^{j(\Omega+\pi)}) \ = \ 0. \tag{4.2.17}$$

Note how (4.2.17) is the same as (3.2.48) in Chapter 3 (on the unit circle). Again, this displays the connection between discrete and continuous time. Since $G_0^*(e^{j\omega})$ and $G_0^*(e^{j(\omega+\pi)})$ cannot go to zero at the same time (see (4.2.10)), it means that

$$G_1(e^{j\omega}) \ = \ \lambda(e^{j\omega})\ G_0^*(e^{j(\omega+\pi)}),$$

where $\lambda(e^{j\omega})$ is 2π-periodic and

$$\lambda(e^{j\omega}) + \lambda(e^{j(\omega+\pi)}) \ = \ 0.$$

We can choose $\lambda(e^{j\omega}) = -e^{-j\omega}$ to obtain

$$G_1(e^{j\omega}) \ = \ -e^{-j\omega} G_0^*(e^{j(\omega+\pi)}), \tag{4.2.18}$$

or, in time domain

$$g_1[n] \ = \ (-1)^n\ g_0[-n+1].$$

Finally, the wavelet is obtained as (see (4.2.15))

$$\begin{aligned}
\Psi(\omega) &= -\frac{1}{\sqrt{2}} e^{-j\omega/2}\ G_0^*(e^{j(\omega/2+\pi)})\ \Phi(\omega/2), \qquad (4.2.19) \\
\psi(t) &= \sqrt{2} \sum_{n \in \mathcal{Z}} (-1)^n\ g_0[-n+1]\ \varphi(2t - n).
\end{aligned}$$

To prove that this wavelet, together with its integer shifts, indeed generates an orthonormal basis for W_0, one would have to prove the orthogonality of basis functions $\psi_{0,n}(t)$ as well as completeness; that is, that any $f(t) \in W_0$ can be written as $f(t) = \sum_n \alpha_n \psi_{0,n}$. This part is omitted here and can be found in [73], pp. 134-135.

4.2.3 Examples of Multiresolution Analyses

In this section we will discuss two examples: Haar, which we encountered in Section 4.1, and sinc, as a dual of the Haar case. The aim is to indicate the embedded spaces in these two example cases, as well as to show how to construct the wavelets in these cases.

Example 4.1 *Haar Case*

Let us go back to Section 4.1.3. Call V_m the space of functions which are constant over intervals $[n2^m, (n+1)2^m)$. Using (4.1.10), one has

$$f^{(m)} \in V_m \Leftrightarrow f^{(m)} = \sum_{n=-\infty}^{\infty} f_n^{(m)} \varphi_{m,n}(t).$$

The process of taking the average over two successive intervals creates a function $f^{(m+1)} \in V_{m+1}$ (since it is a function which is constant over intervals $[n2^{m+1}, (n+1)2^{m+1})$). Also, it is clear that

$$V_{m+1} \subset V_m.$$

The averaging operation is actually an orthogonal projection of $f^{(m)} \in V_m$ onto V_{m+1}, since the difference $d^{(m+1)} = f^{(m)} - f^{(m+1)}$ is orthogonal to V_{m+1} (the inner product of $d^{(m+1)}$ with any function from V_{m+1} is equal to zero). In other words, $d^{(m+1)}$ belongs to a space W_{m+1} which is orthogonal to V_{m+1}. The space W_{m+1} is spanned by translates of $\psi_{m+1,n}(t)$

$$d^{(m+1)} \in W_{m+1} \Leftrightarrow d^{(m+1)} = \sum_{n=-\infty}^{\infty} d_n^{(m+1)} \psi_{m+1,n}(t).$$

This difference function is again the orthogonal projection of $f^{(m)}$ onto W_{m+1}. We have seen that any function $f^{(m)}$ can be written as an "average" plus a "difference" function

$$f^{(m)}(t) = f^{(m+1)}(t) + d^{(m+1)}(t). \tag{4.2.20}$$

Thus, W_{m+1} is the orthogonal complement of V_{m+1} in V_m. Therefore,

$$V_m = V_{m+1} \oplus W_{m+1}$$

and (4.2.20) can be written as

$$f^{(m)}(t) = \text{Proj}_{V_{m+1}}[f^{(m)}(t)] + \text{Proj}_{W_{m+1}}[f^{(m)}(t)].$$

Repeating the process (decomposing V_{m+1} into $V_{m+2} \oplus W_{m+2}$ and so on), the following is obtained:

$$V_m = W_{m+1} \oplus W_{m+2} \oplus W_{m+3} \oplus \cdots$$

Since piecewise constant functions are dense in $L_2(\mathcal{R})$, as the step size goes to zero (4.2.2) is satisfied as well as (4.2.12), and thus the Haar wavelets form a basis for $L_2(\mathcal{R})$. Now, let us see how we can construct the Haar wavelet using the technique from the previous section. As we said before, the basis for V_0 is $\{\varphi(t-n)\}_{n\in\mathcal{Z}}$ with

$$\varphi(t) = \begin{cases} 1 & 0 \le t < 1, \\ 0 & \text{otherwise.} \end{cases}$$

To find $G_0(e^{j\omega})$, write

$$\varphi(t) = \varphi(2t) + \varphi(2t-1),$$

hence

$$\Phi(\omega) = \frac{1}{\sqrt{2}} \frac{1+e^{-j\omega/2}}{\sqrt{2}} \Phi\left(\frac{\omega}{2}\right),$$

from which

$$G_0(e^{j\omega}) = \frac{1}{\sqrt{2}}(1+e^{-j\omega}).$$

Then by using

$$G_1(e^{j\omega}) = -e^{-j\omega}\, G_0(e^{j(\omega+\pi)}) = -e^{-j\omega}\, \frac{1+e^{j(\omega+\pi)}}{\sqrt{2}} = \frac{1-e^{-j\omega}}{\sqrt{2}},$$

one obtains

$$\Psi(\omega) = \frac{1}{\sqrt{2}}\, G_1(e^{j\omega/2})\, \Phi\left(\frac{\omega}{2}\right).$$

Finally

$$\psi(t) = \varphi(2t) - \varphi(2t-1),$$

or

$$\psi(t) = \begin{cases} 1 & 0 \leq t < \frac{1}{2}, \\ -1 & \frac{1}{2} \leq t < 1, \\ 0 & \text{otherwise.} \end{cases}$$

The Haar wavelet and scaling function, as well as their Fourier transforms, were given in Figure 4.1.

Example 4.2 *Sinc Case*

In order to derive the sinc wavelet,[3] we will start with the sequence of embedded spaces. Instead of piecewise constant functions, we will consider bandlimited functions. Call V_0 the space of functions bandlimited to $[-\pi, \pi]$ (to be precise, V_0 includes $\cos(\pi t)$ but not $\sin(\pi t)$). Thus, V_{-1} is the space of functions bandlimited to $[-2\pi, 2\pi]$. Then, call W_0 the space of functions bandlimited to $[-2\pi, -\pi] \cup [\pi, 2\pi]$ (again, to be precise, W_0 includes $\sin(\pi t)$ but not $\cos(\pi t)$). Therefore

$$V_{-1} = V_0 \oplus W_0,$$

since V_0 is orthogonal to W_0 and together they span the same space as V_{-1} (see Figure 4.5). Obviously, a projection of a function $f^{(-1)}$ from V_{-1} onto V_0 will be a lowpass approximation $f^{(0)}$, while the difference $d^{(0)} = f^{(-1)} - f^{(0)}$ will exist in W_0. Repeating the above decomposition leads to

$$V_{-1} = \bigoplus_{m=0}^{\infty} W_m,$$

as shown in Figure 4.5. This is an *octave-band decomposition* of V_{-1}. It is also called a constant-Q filtering, since each band has a *constant relative bandwidth*. It is clear that an orthogonal basis for V_0 is given by $\{\text{sinc}_1(t-n)\}$ (see (4.1.4), or

$$\varphi(t) = \frac{\sin \pi t}{\pi t},$$

[3]In the mathematical literature, this is often referred to as the Littlewood-Paley wavelet [73].

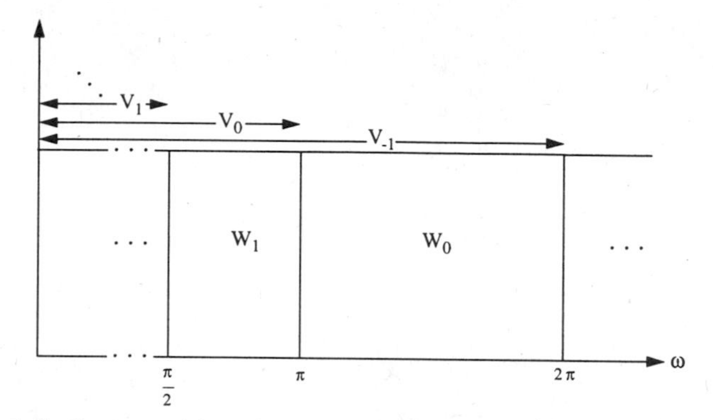

Figure 4.5 Decomposition of V_0 into successive octave bands. Actually, there is a scaling factor for V_j and W_j by $2^{j/2}$ to make the subspaces of unit norm.

which is thus the *scaling function* for the sinc case and the space V_0 of functions bandlimited to $[-\pi, \pi]$. Using (4.2.9) one gets that

$$g_0[n] = \frac{1}{\sqrt{2}} \frac{\sin(\pi n/2)}{\pi n/2}, \qquad (4.2.21)$$

that is,

$$G_0(e^{j\omega}) = \begin{cases} \sqrt{2} & -\frac{\pi}{2} \leq \omega \leq \frac{\pi}{2}, \\ 0 & \text{otherwise,} \end{cases}$$

or, $G_0(e^{j\omega})$ is an ideal lowpass filter. Then $G_1(e^{j\omega})$ becomes (use (4.2.18))

$$G_1(e^{j\omega}) = \begin{cases} -\sqrt{2}e^{-j\omega} & \omega \in [-\pi, -\frac{\pi}{2}] \cup [\frac{\pi}{2}, \pi], \\ 0 & \text{otherwise,} \end{cases}$$

which is an ideal highpass filter with a phase shift. The sequence $g_1[n]$ is then

$$g_1[n] = (-1)^n g_0[-n+1], \qquad (4.2.22)$$

whereupon

$$\psi(t) = \sqrt{2} \sum_n (-1)^{-n+1} g_0[n] \, \varphi(2t+n-1).$$

Alternatively, we can construct the wavelet directly by taking the inverse Fourier transform of the indicator function of the intervals $[-2\pi, -\pi] \cup [\pi, 2\pi]$:

$$\psi(t) = \frac{1}{2\pi} \int_{-2\pi}^{-\pi} e^{j\omega t} d\omega + \frac{1}{2\pi} \int_{\pi}^{2\pi} e^{j\omega t} d\omega = 2\frac{\sin(2\pi t)}{2\pi t} - \frac{\sin(\pi t)}{\pi t} = \frac{\sin(\pi t/2)}{\pi t/2} \cos(3\pi t/2). \qquad (4.2.23)$$

This function is orthogonal to its translates by integers, or $\langle \psi(t), \psi(t-n) \rangle = \delta[n]$, as can be verified using Parseval's formula (2.4.11). To be coherent with our definition of W_0 (which excludes $\cos(\pi t)$), we need to shift $\psi(t)$ by 1/2, and thus $\{\psi(t-n-1/2)\}$, $n \in \mathcal{Z}$, is an orthogonal basis for W_0. The wavelet basis is now given by

$$\psi_{m,n}(t) = \left\{2^{-m/2}\psi(2^{-m}t - n - 1/2)\right\}, \quad m, n \in \mathcal{Z},$$

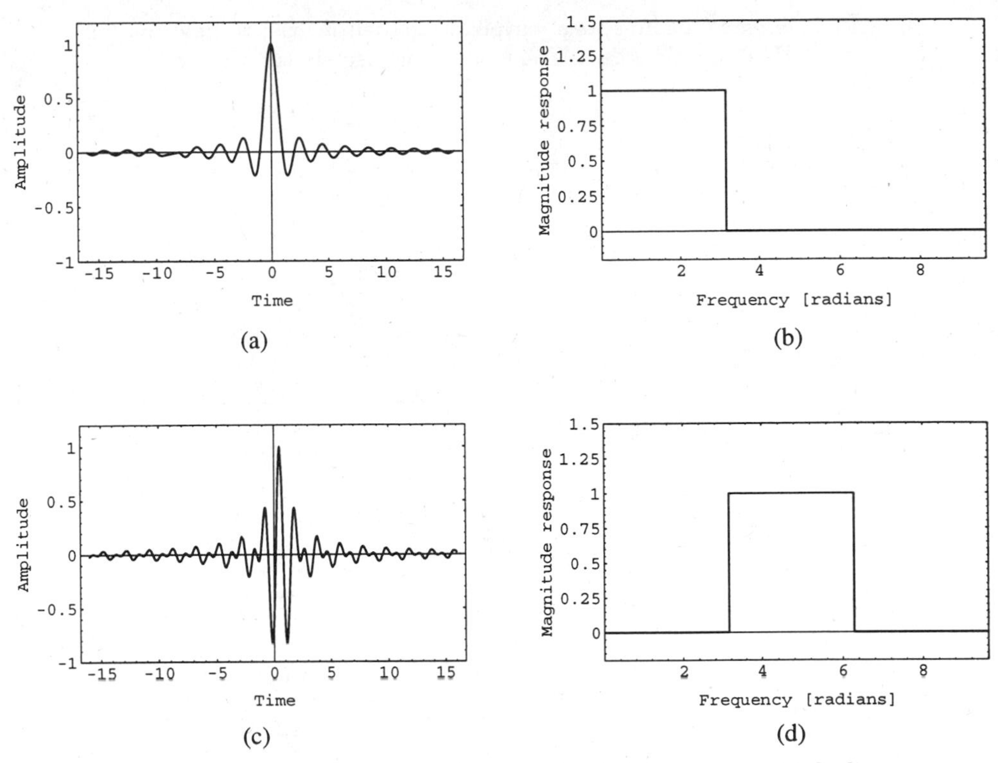

Figure 4.6 Scaling function and the wavelet in the sinc case. (a) Scaling function $\varphi(t)$. (b) Fourier transform magnitude $|\Phi(\omega)|$. (c) Wavelet $\psi(t)$. (d) Fourier transform magnitude $|\Psi(\omega)|$.

where $\psi_{m,n}(t), n \in \mathcal{Z}$, is a basis for functions supported on

$$[-2^{-m+1}\pi, -2^{-m}\pi] \cup [2^{-m}\pi, 2^{-m+1}\pi].$$

Since m can be arbitrarily large (positive or negative), it is clear that we have a basis for $L_2(\mathcal{R})$ functions. The wavelet, scaling function, and their Fourier transforms are shown in Figure 4.6. The slow decay of the time-domain function ($1/t$ as $t \to \infty$) can be seen in the figure, while the frequency resolution is obviously ideal.

To conclude this section, we summarize the expressions for the scaling function and the wavelet as well as their Fourier transforms in Haar and sinc cases in Table 4.1. The underlying discrete-time filters were given in Table 3.1.

4.3 Construction of Wavelets Using Fourier Techniques

What we have seen until now, is the conceptual framework for building orthonormal bases with the specific structure of multiresolution analysis, as well as two particular cases of such bases: Haar and sinc. We will now concentrate on ways of building such bases in the Fourier domain. Two constructions are

Table 4.1 Scaling functions, wavelets and their Fourier transforms in the Haar and sinc cases. The underlying discrete-time filters are given in Table 3.1.

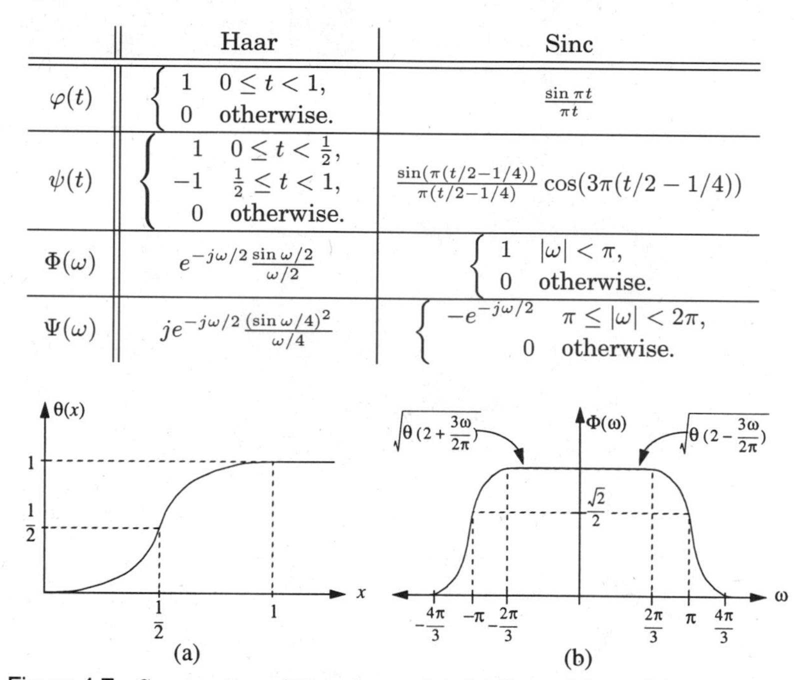

	Haar	Sinc
$\varphi(t)$	$\begin{cases} 1 & 0 \le t < 1, \\ 0 & \text{otherwise.} \end{cases}$	$\frac{\sin \pi t}{\pi t}$
$\psi(t)$	$\begin{cases} 1 & 0 \le t < \frac{1}{2}, \\ -1 & \frac{1}{2} \le t < 1, \\ 0 & \text{otherwise.} \end{cases}$	$\frac{\sin(\pi(t/2-1/4))}{\pi(t/2-1/4)} \cos(3\pi(t/2-1/4))$
$\Phi(\omega)$	$e^{-j\omega/2} \frac{\sin \omega/2}{\omega/2}$	$\begin{cases} 1 & \lvert\omega\rvert < \pi, \\ 0 & \text{otherwise.} \end{cases}$
$\Psi(\omega)$	$je^{-j\omega/2} \frac{(\sin \omega/4)^2}{\omega/4}$	$\begin{cases} -e^{-j\omega/2} & \pi \le \lvert\omega\rvert < 2\pi, \\ 0 & \text{otherwise.} \end{cases}$

Figure 4.7 Construction of Meyer's wavelet. (a) General form of the function $\theta(x)$. (b) $|\Phi(\omega)|$ in Meyer's construction.

indicated, both of which rely on the multiresolution framework derived in the previous section. First, Meyer's wavelet is derived, showing step by step how it verifies the multiresolution axioms. Then, wavelets for spline spaces are constructed. In this case, one starts with the well-known spaces of piecewise polynomials and shows how to construct an orthonormal wavelet basis.

4.3.1 Meyer's Wavelet

The idea behind Meyer's wavelet is to soften the ideal — sinc case. Recall that the sinc scaling function and the wavelet are as given in Figure 4.6. The idea of the proof is to construct a scaling function $\varphi(t)$ that satisfies the orthogonality and scaling requirements of the multiresolution analysis and then construct the wavelet using the standard method. In order to soften the sinc scaling function, we find a smooth function (in frequency) that satisfies (4.2.7).

We are going to show the construction step by step, leading first to the scaling function and then to the associated wavelet.

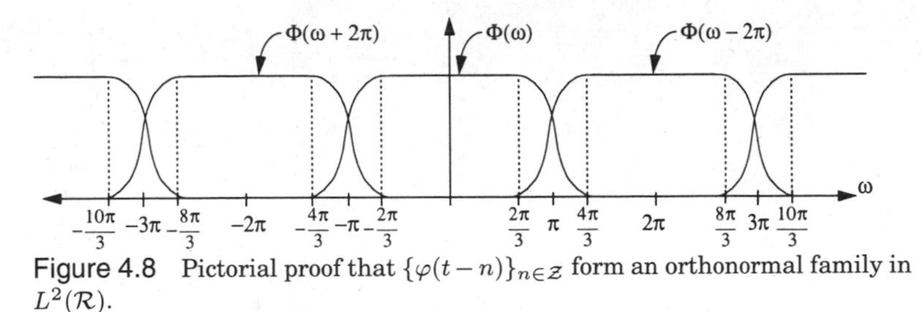

Figure 4.8 Pictorial proof that $\{\varphi(t-n)\}_{n\in\mathcal{Z}}$ form an orthonormal family in $L^2(\mathcal{R})$.

(a) Start with a nonnegative function $\theta(x)$ that is differentiable (maybe several times) and such that (see Figure 4.7(a))

$$\theta(x) = \begin{cases} 0 & x \le 0, \\ 1 & 1 \le x. \end{cases} \tag{4.3.1}$$

and satisfying $\theta(x) + \theta(1-x) = 1$ for $0 \le x \le 1$. There exist various choices for $\theta(x)$, one of them being

$$\theta(x) = \begin{cases} 0 & x \le 0, \\ 3x^2 - 2x^3 & 0 \le x \le 1, \\ 1 & 1 \le x. \end{cases} \tag{4.3.2}$$

(b) Construct the scaling function $\Phi(\omega)$ such that (see Figure 4.7(b))

$$\Phi(\omega) = \begin{cases} \sqrt{\theta(2 + \frac{3\omega}{2\pi})} & \omega \le 0, \\ \sqrt{\theta(2 - \frac{3\omega}{2\pi})} & 0 \le \omega. \end{cases}$$

To show that $\Phi(\omega)$ is indeed a scaling function with a corresponding multiresolution analysis, one has to show that (4.2.1–4.2.6) hold. As a preliminary step, let us first demonstrate the following:

(c) $\{\varphi(t-n)\}_{n\in\mathcal{Z}}$ is an orthonormal family from $L_2(\mathcal{R})$. To that end, we use the Poisson formula and instead show that (see (4.2.7))

$$\sum_{k\in\mathcal{Z}} |\Phi(\omega + 2k\pi)|^2 \ = \ 1. \tag{4.3.3}$$

From Figure 4.8 it is clear that for $\omega \in [-(2\pi/3) - 2n\pi, (2\pi)/3 - 2n\pi]$

$$\sum_k |\Phi(\omega + 2k\pi)|^2 = |\Phi(\omega + 2n\pi)|^2 \ = \ 1.$$

The only thing left is to show (4.3.3) holds in overlapping regions. Thus, take for example, $\omega \in [(2\pi)/3, (4\pi)/3]$:

$$\Phi(\omega)^2 + \Phi(\omega - 2\pi)^2 \ = \ \theta\left(2 - \frac{3\omega}{2\pi}\right) + \theta\left(2 + \frac{3(\omega - 2\pi)}{2\pi}\right)$$

$$\begin{aligned} &= \theta\left(2-\frac{3\omega}{2\pi}\right)+\theta\left(-1+\frac{3\omega}{2\pi}\right) \\ &= \theta\left(2-\frac{3\omega}{2\pi}\right)+\theta\left(1-\left(2-\frac{3\omega}{2\pi}\right)\right) \\ &= 1. \end{aligned}$$

The last equation follows from the definition of θ (see (4.3.2)).

(d) Define as V_0 the subspace of $L_2(\mathcal{R})$ generated by $\varphi(t-n)$ and define as V_m's those satisfying (4.2.4).

Now we are ready to show that the V_m's form a multiresolution analysis. Until now, by definition, we have taken care of (4.2.4–4.2.6), those left to be shown are (4.2.1–4.2.3).

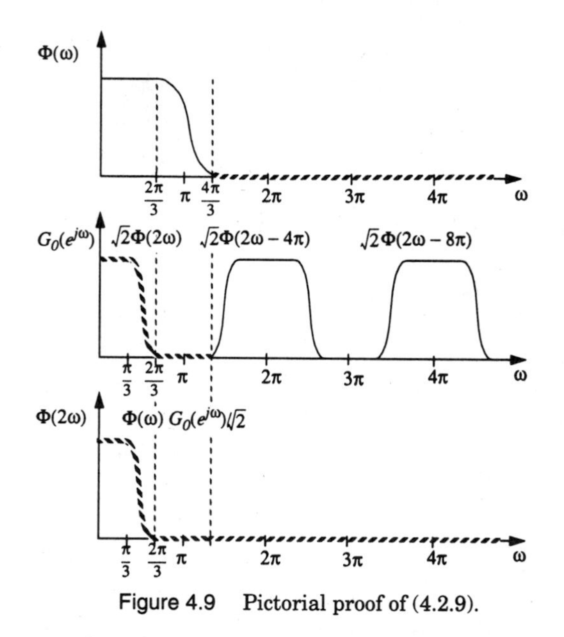

Figure 4.9 Pictorial proof of (4.2.9).

(e) Prove (4.2.1): It is enough to show that $V_1 \subset V_0$, or $\varphi\,(t/2) = \sum_n c_n \varphi(t-n)$. This is equivalent to saying that there exists a periodic function $G_0(e^{j\omega}) \in L_2([0,2\pi])$ such that $\Phi(2\omega) = (1/\sqrt{2})G_0(e^{j\omega})\Phi(\omega)$ (see 4.2.9). Then choose

$$G_0(e^{j\omega}) = \sqrt{2}\sum_{k\in\mathcal{Z}} \Phi(2\omega + 4k\pi). \qquad (4.3.4)$$

A pictorial proof is given in Figure 4.9.

(f) Show (4.2.2): In this case, it is enough to show that if

$$\langle f, \varphi_{m,n} \rangle = 0, \quad m, n \in \mathcal{Z} \Rightarrow f = 0,$$

then

$$\langle f, \varphi_{m,n} \rangle = 0 \iff \sum_{k \in \mathcal{Z}} F(2^m(\omega + 2k\pi))\ \Phi^*(\omega + 2k\pi) = 0.$$

Take for example, $\omega \in [-(2\pi)/3, (2\pi)/3]$. Then for any k

$$F(2^m(\omega + 2k\pi))\ \Phi(\omega + 2k\pi) = 0,$$

and for $k = 0$

$$F(2^m\omega)\ \overline{\Phi(\omega)} = 0.$$

For any m

$$F(2^m\omega) = 0, \quad \omega \in [-\tfrac{2\pi}{3}, \tfrac{2\pi}{3}]\ ,$$

and thus

$$F(\omega) = 0,\ \omega \in \mathcal{R},$$

or

$$f - 0.$$

(g) Show (4.2.3): If $f \in \bigcap_{m \in \mathcal{Z}} V_m$ then $F \in \bigcap_{m \in \mathcal{Z}} F\{V_m\}$ where $F\{V_m\}$ is the Fourier transform of V_m with the basis $2^{m/2}e^{-jk\omega 2^{-m}}\Phi(2^{-m}\omega)$. Since $\Phi(2^{-m}\omega)$ has its support in the interval

$$I = \left[-\frac{4\pi}{3}2^m, \frac{4\pi}{3}2^m\right],$$

it follows that $I \to \{0\}$ as $m \to -\infty$.

In other words,

$$F(\omega) \in \bigcap_{m \in \mathcal{Z}} F\{V_m\} = 0,$$

or

$$f(t) = 0.$$

(h) Finally, one just has to find the corresponding wavelet using (4.2.19): $\Psi(\omega) = -(1/\sqrt{2})\ e^{-j\omega/2}\ G_0^*(e^{j(\omega/2+\pi)})\ \Phi(\omega/2)$.

Thus using (4.3.4) one gets

$$\Psi(\omega) \;=\; -\frac{1}{\sqrt{2}}e^{-j\omega/2}\sum_{k \in \mathcal{Z}} \Phi(\omega + (4k+1)\pi)\ \Phi\left(\frac{\omega}{2}\right).$$

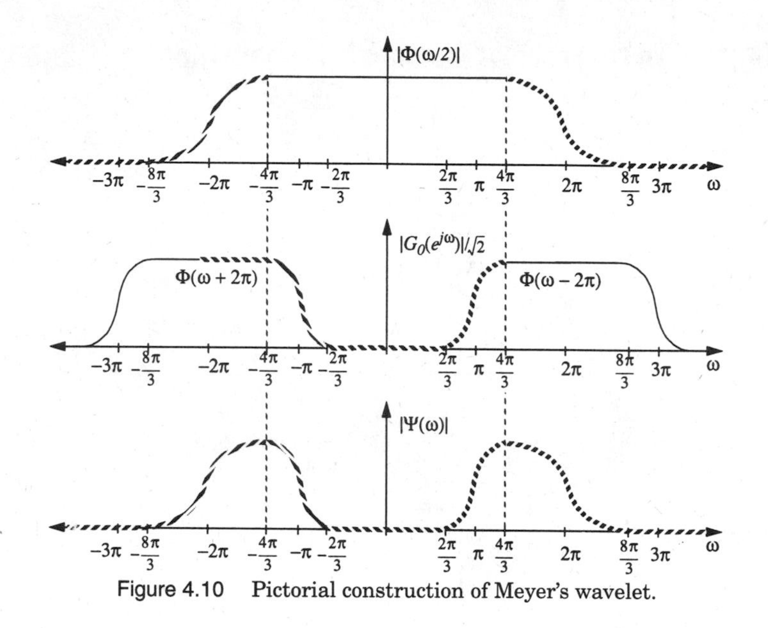

Figure 4.10 Pictorial construction of Meyer's wavelet.

Hence $\Psi(\omega)$ is defined as follows (see Figure 4.10):

$$\Psi(\omega) = \begin{cases} 0 & 0 \le \omega \le \frac{2\pi}{3}, \\ \frac{-1}{\sqrt{2}} e^{-j\omega/2}\Phi(\omega - 2\pi) & \frac{2\pi}{3} \le \omega \le \frac{4\pi}{3}, \\ \frac{-1}{\sqrt{2}} e^{-j\omega/2}\Phi(\frac{\omega}{2}) & \frac{4\pi}{3} \le \omega \le \frac{8\pi}{3}, \\ 0 & \frac{8\pi}{3} \le \omega, \end{cases} \tag{4.3.5}$$

and $\Psi(\omega)$ is an even function of ω (except for a phase factor $e^{-j\omega/2}$). Note that (see Problem 4.4)

$$\sum_{k \in \mathcal{Z}} |\Psi(2^k \omega)|^2 = 1. \tag{4.3.6}$$

An example of Meyer's scaling function and wavelet is shown in Figure 4.11. A few remarks can be made on Meyer's wavelet. The time-domain function, while of infinite support, can have very fast decay. The discrete-time filter $G_0(e^{j\omega})$ which is involved in the two-scale equation, corresponds (by inverse Fourier transform) to a sequence $g_0[n]$ which has similarly fast decay. However, $G_0(e^{j\omega})$ is not a rational function of $e^{j\omega}$ and thus, the filter $g_0[n]$ cannot be efficiently implemented. Thus, Meyer's wavelet is more of theoretical interest.

4.3.2 Wavelet Bases for Piecewise Polynomial Spaces

Spline or Piecewise Polynomial Spaces Spaces which are both interesting and easy to characterize are the spaces of piecewise polynomial functions. To be

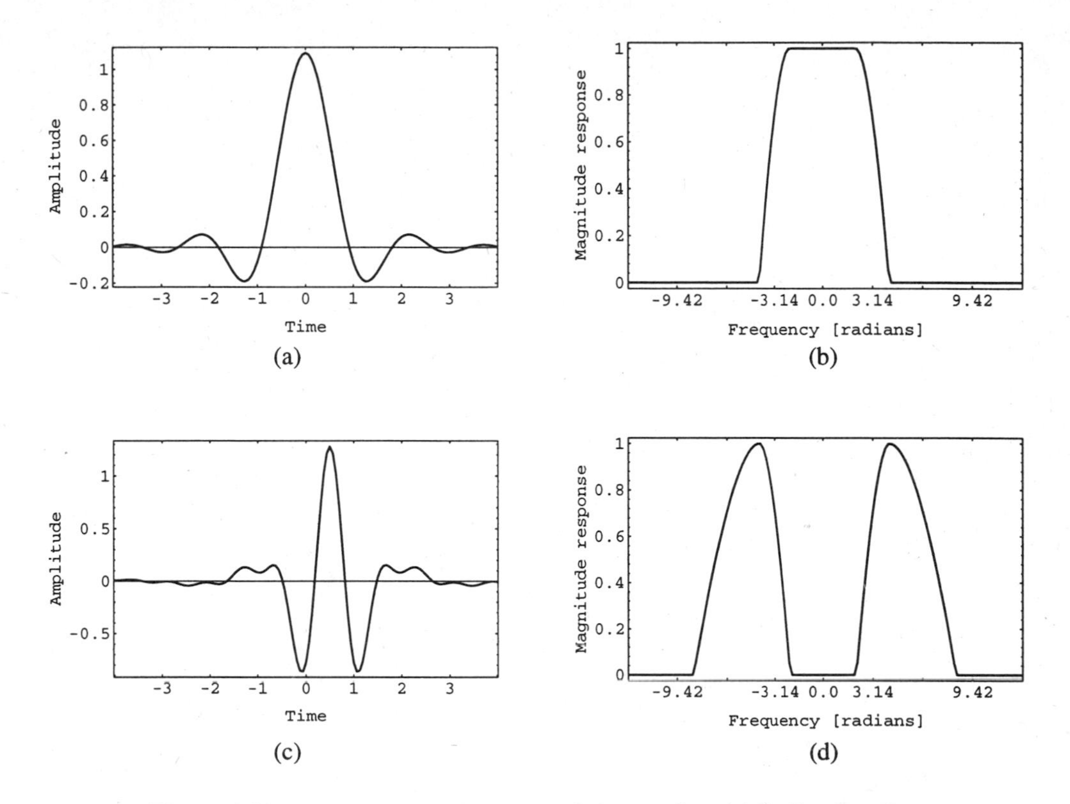

Figure 4.11 Meyer's scaling function and the wavelet. (a) Scaling function $\varphi(t)$. (b) Fourier transform magnitude $|\Phi(\omega)|$. (c) Wavelet $\psi(t)$. (d) Fourier transform magnitude $|\Psi(\omega)|$.

more precise, they are polynomials of degree l over fixed length intervals and at the knots (the boundary between intervals) they have continuous derivatives up to order $l-1$. Two characteristics of such spaces make them well suited for the development of wavelet bases. First, there is a ladder of spaces as required for a multiresolution construction of wavelets. Functions which are piecewise polynomial of degree l over intervals $[k2^i, (k+1)2^i)$ are obviously also piecewise polynomial over subintervals $[k2^j, (k+1)2^j]$, $j < i$. Second, there exist simple bases for such spaces, namely the B-splines. Call:

$$V_i^{(l)} = \left\{ \begin{array}{l} \text{functions which are piecewise polynomial of degree } l \\ \text{over intervals } [k2^i, (k+1)2^i) \text{ and having } l-1 \\ \text{continuous derivatives at } k2^i, \quad k \in \mathcal{Z} \end{array} \right\}.$$

For example, $V_{-1}^{(1)}$ is the space of all functions which are linear over half-integer intervals and continuous at the interval boundaries. Consider first, the spaces

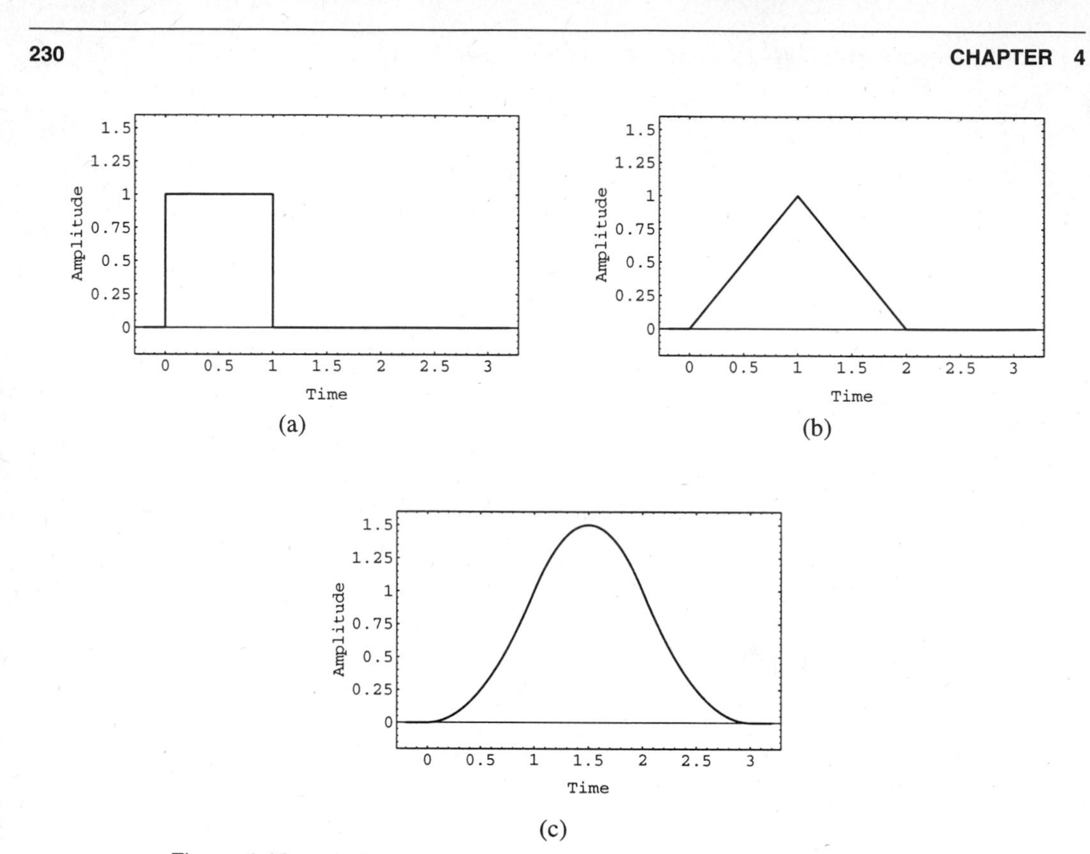

Figure 4.12 B-splines, for $N = 0, 1, 2$. (a) Constant spline. (b) Linear spline. (c) Quadratic spline.

with unit intervals, that is, $V_0^{(l)}$. Then, bases for these spaces are given by the B-splines [76, 255]. These are obtained by convolution of box functions (indicator functions of the unit interval) with themselves. For example, the hat function, which is a box function convolved with itself, is a (nonorthogonal) basis for piecewise linear functions over unit intervals, that is $V_0^{(1)}$.

The idea of the wavelet construction is to start with these nonorthogonal bases for the $V_0^{(l)}$'s and apply a suitable orthogonalization procedure in order to get an orthogonal scaling function. Then, the wavelet follows from the usual construction. Below, we follow the approach and notation of Unser and Aldroubi [6, 298, 299, 296]. Note that the relation between splines and digital filters has also been exploited in [118].

Call $I(t)$ the indicator function of the interval $[-1/2, 1/2]$ and $I^{(k)}(t)$ the k-time convolution of $I(t)$ with itself, that is, $I^{(k)}(t) = I(t) * I^{(k-1)}(t)$, $I^{(0)}(t) = I(t)$. Denote by $\beta^{(N)}(t)$ the B-spline of order N where

(a) for N odd:

$$\beta^{(N)}(t) = I^{(N)}(t), \tag{4.3.7}$$

$$\mathcal{B}^{(N)}(\omega) = \left(\frac{\sin(\omega/2)}{\omega/2}\right)^{N+1}, \tag{4.3.8}$$

(b) and for N even:

$$\beta^{(N)}(t) = I^{(N)}\left(t-\frac{1}{2}\right), \tag{4.3.9}$$

$$\mathcal{B}^{(N)}(\omega) = e^{-j\omega/2}\left(\frac{\sin(\omega/2)}{\omega/2}\right)^{N+1}. \tag{4.3.10}$$

The shift by $1/2$ in (4.3.9) is necessary so that the nodes of the spline are at integer intervals. The first few examples, namely $N = 0$ (constant spline), $N = 1$ (linear spline), and $N = 2$ (quadratic spline) are shown in Figure 4.12.

Orthogonalization Procedure While the B-spline $\beta^{(N)}(t)$ and its integer translates form a basis for $V_0^{(N)}$, it is not an orthogonal basis (except for $N = 0$). Therefore, we have to apply an orthogonalization procedure. Recall that a function $f(t)$ that is orthogonal to its integer translates satisfies (see (4.2.7))

$$\langle f(t), f(t-n)\rangle_{n\in\mathcal{Z}} = \delta[n] \iff \sum_{k\in\mathcal{Z}} |F(\omega+2k\pi)|^2 = 1.$$

Starting with a nonorthogonal $\beta^{(N)}(t)$, we can evaluate the following 2π-periodic function:

$$B^{(2N+1)}(\omega) = \sum_{k\in\mathcal{Z}} |\mathcal{B}^{(N)}(\omega+2k\pi)|^2. \tag{4.3.11}$$

In this case[4] $B^{(2N+1)}(\omega)$ is the discrete-time Fourier transform of the discrete-time B-spline $b^{(2N+1)}[n]$, which is the sampled version of the continuous-time B-spline [299],

$$b^{(2N+1)}[n] = \beta^{(2N+1)}(t)\Big|_{t=n}. \tag{4.3.12}$$

Because $\{\beta^{(N)}(t-n)\}$ is a basis for $V_0^{(N)}$, one can show that there exist two positive constants A and C such that [71]

$$0 < A \leq B^{(2N+1)}(\omega) \leq C < \infty. \tag{4.3.13}$$

[4]Note that $\beta^{(N)}(t)$ has a Fourier transform $\mathcal{B}^{(N)}(\omega)$. On the other hand, $b^{(2N+1)}[n]$ has a discrete-time Fourier transform $B^{(2N+1)}(\omega)$. $\mathcal{B}^{(N)}(\omega)$ and $B^{(2N+1)}(\omega)$ should not be confused. Also, $B^{(2N+1)}(\omega)$ is a function of $e^{j\omega}$.

One possible choice for a scaling function is

$$\Phi(\omega) = \frac{\mathcal{B}^{(N)}(\omega)}{\sqrt{B^{(2N+1)}(\omega)}}. \tag{4.3.14}$$

Because of (4.3.13), $\Phi(\omega)$ is well defined. Obviously

$$\sum_k |\Phi(\omega + 2k\pi)|^2 = \frac{1}{B^{(2N+1)}(\omega)} \sum_k |\mathcal{B}^{(N)}(\omega + 2k\pi)|^2 = 1,$$

and thus, the set $\{\varphi(t-n)\}$ is orthogonal. That it is a basis for $V_0^{(N)}$ follows from the fact that (from (4.3.14)) $\beta^{(N)}(t)$ can be written as a linear combination of $\varphi(t-n)$ and therefore, since any $f(t) \in V_0^{(N)}$ can be written in terms of $\beta^{(N)}(t-n)$, it can be expressed in terms of $\varphi(t-n)$ as well.

Now, both $\beta^{(N)}(t)$ and $\varphi(t)$ satisfy a two-scale equation because they belong to $V_0^{(N)}$ and thus $V_{-1}^{(N)}$; therefore, they can be expressed in terms of $\beta^{(N)}(2t-n)$ and $\varphi(2t-n)$, respectively. In Fourier domain we have

$$\mathcal{B}^{(N)}(\omega) = M\left(\frac{\omega}{2}\right) \mathcal{B}^{(N)}\left(\frac{\omega}{2}\right), \tag{4.3.15}$$

$$\Phi(\omega) = \frac{1}{\sqrt{2}} G_0(e^{j\omega/2})\, \Phi\left(\frac{\omega}{2}\right), \tag{4.3.16}$$

where we used (4.2.9) for $\Phi(\omega)$. Using (4.3.14) and (4.3.15), we find that

$$\begin{aligned} \Phi(\omega) &= \frac{\mathcal{B}^{(N)}(\omega)}{\sqrt{B^{(2N+1)}(\omega)}} = \frac{M(\omega/2)\sqrt{B^{(2N+1)}(\omega/2)}\Phi(\omega/2)}{\sqrt{B^{(2N+1)}(\omega)}} \\ &= \frac{1}{\sqrt{2}}\, G_0(e^{j\omega/2})\, \Phi\left(\frac{\omega}{2}\right), \end{aligned} \tag{4.3.17}$$

that is,

$$G_0(e^{j\omega}) = \sqrt{2}\, \frac{M(\omega)\, \sqrt{B^{(2N+1)}(\omega)}}{\sqrt{B^{(2N+1)}(2\omega)}}. \tag{4.3.18}$$

Then, following (4.2.19), we have the following expression for the wavelet:

$$\Psi(\omega) = -\frac{1}{\sqrt{2}}\, e^{-j\omega/2}\, G_0^*\left(e^{j(\omega/2+\pi)}\right)\, \Phi\left(\frac{\omega}{2}\right). \tag{4.3.19}$$

Note that the orthogonalization method just described is quite general and can be applied whenever we have a multiresolution analysis with nested spaces and a basis for V_0. In particular, it indicates that in Definition 4.2, $\varphi(t)$ in (4.2.6) need not be from an orthogonal basis since it can be orthogonalized using the above method. That is, given $g(t)$ which forms a (nonorthogonal) basis for V_0 and satisfies a two-scale equation, compute a 2π-periodic function $D(\omega)$

$$D(\omega) = \sum_{k \in \mathcal{Z}} |G(\omega + 2k\pi)|^2, \tag{4.3.20}$$

where $G(\omega)$ is the Fourier transform $g(t)$. Then

$$\Phi(\omega) = \frac{G(\omega)}{\sqrt{D(\omega)}}$$

corresponds to an orthogonal scaling function for V_0 and the rest of the procedure follows as above.

Orthonormal Wavelets for Spline Spaces We will apply the method just described to construct wavelets for spaces of piecewise polynomial functions introduced at the beginning of this section. This construction was done by Battle [21, 22] and Lemarié [175], and the resulting wavelets are often called Battle-Lemarié wavelets. Earlier work by Stromberg [283, 284] also derived orthogonal wavelets for piecewise polynomial spaces. We will start with a simple example of the linear spline, given by

$$\beta^{(1)}(t) = \begin{cases} 1 - |t| & |t| \leq 1, \\ 0 & \text{otherwise.} \end{cases}$$

It satisfies the following two-scale equation:

$$\beta^{(1)}(t) = \frac{1}{2}\,\beta^{(1)}(2t+1) + \beta^{(1)}(2t) + \frac{1}{2}\,\beta^{(1)}(2t-1). \tag{4.3.21}$$

The Fourier transform, from (4.3.7), is

$$\mathcal{B}^{(1)}(\omega) = \left(\frac{\sin(\omega/2)}{\omega/2}\right)^2. \tag{4.3.22}$$

In order to find $B^{(2N+1)}(\omega)$ (see (4.3.11)), we note that its inverse Fourier transform is equal to

$$\begin{aligned} b^{(2N+1)} &= \frac{1}{2\pi}\int_0^{2\pi} e^{jn\omega} \sum_{k \in \mathcal{Z}} |\mathcal{B}^{(N)}(\omega + 2\pi k)|^2 \, d\omega \\ &= \frac{1}{2\pi}\int_{-\infty}^{\infty} e^{jn\omega} |\mathcal{B}^{(N)}(\omega)|^2 \, d\omega \\ &= \int_{-\infty}^{\infty} \beta^{(N)}(t)\, \beta^{(N)}(t-n)\, dt, \end{aligned} \tag{4.3.23}$$

by Parseval's formula (2.4.11). In the linear spline case, we find $b^{(3)}[0] = 2/3$ and $b^{(3)}[1] = b^{(3)}[-1] = 1/6$, or

$$B^{(3)}(\omega) = \frac{2}{3} + \frac{1}{6}e^{j\omega} + \frac{1}{6}e^{-j\omega} = \frac{2}{3} + \frac{1}{3}\cos(\omega) = 1 - \frac{2}{3}\sin^2\left(\frac{\omega}{2}\right),$$

which is the discrete-time cubic spline [299]. From (4.3.14) and (4.3.22), one gets

$$\Phi(\omega) = \frac{\sin^2(\omega/2)}{(\omega/2)^2(1-(2/3)\sin^2(\omega/2))^{1/2}},$$

which is an orthonormal scaling function for the linear spline space $V_0^{(1)}$. Observation of the inverse Fourier transform of the 2π-periodic function $(1 - (2/3)\sin^2(\omega/2))^{1/2}$, which corresponds to a sequence $\{\alpha_n\}$, indicates that $\varphi(t)$ can be written as a linear combination of $\{\beta^{(1)}(t-n)\}$:

$$\varphi(t) = \sum_{n \in \mathcal{Z}} \alpha_n \beta^{(1)}(t-n).$$

This function is thus piecewise linear as can be verified in Figure 4.13(a). Taking the Fourier transform of the two-scale equation (4.3.21) leads to

$$\mathcal{B}^{(1)}(\omega) = \left(\frac{1}{4}e^{-j\frac{\omega}{2}} + \frac{1}{2} + \frac{1}{4}e^{j\omega/2}\right)\mathcal{B}^{(1)}\left(\frac{\omega}{2}\right) = \frac{1}{2}\left(1 + \cos\left(\frac{\omega}{2}\right)\right)\mathcal{B}^{(1)}\left(\frac{\omega}{2}\right),$$

and following the definition of $M(\omega)$ in (4.3.15), we get

$$M(\omega) = \frac{1}{2}\left(1 + \cos\left(\frac{\omega}{2}\right)\right) = \cos^2\left(\frac{\omega}{2}\right).$$

Therefore, $G_0(e^{j\omega})$ is equal to (following (4.3.18)),

$$G_0(e^{j\omega}) = \sqrt{2}\,\frac{\cos^2(\omega/2)(1-(2/3)\sin^2(\omega/2))^{1/2}}{(1-(2/3)\sin^2(\omega))^{1/2}},$$

and the wavelet follows from (4.3.19) as

$$\Psi(\omega) = -e^{-j\omega/2}\,\frac{\sin^2(\omega/4)(1-(2/3)\cos^2(\omega/4))^{1/2}}{(1-(2/3)\sin^2(\omega/2))^{1/2}} \cdot \Phi(\omega/2),$$

or

$$\Psi(\omega) = -e^{-j\omega/2}\,\frac{\sin^4(\omega/4)}{(\omega/4)^2}\left(\frac{1-(2/3)\cos^2(\omega/4)}{(1-(2/3)\sin^2(\frac{\omega}{2}))(1-(2/3)\sin^2(\frac{\omega}{4}))}\right)^{1/2}. \quad (4.3.24)$$

Rewrite the above as

$$\Psi(\omega) = \frac{\sin^2(\omega/4)}{(\omega/4)^2}Q(\omega) \quad (4.3.25)$$

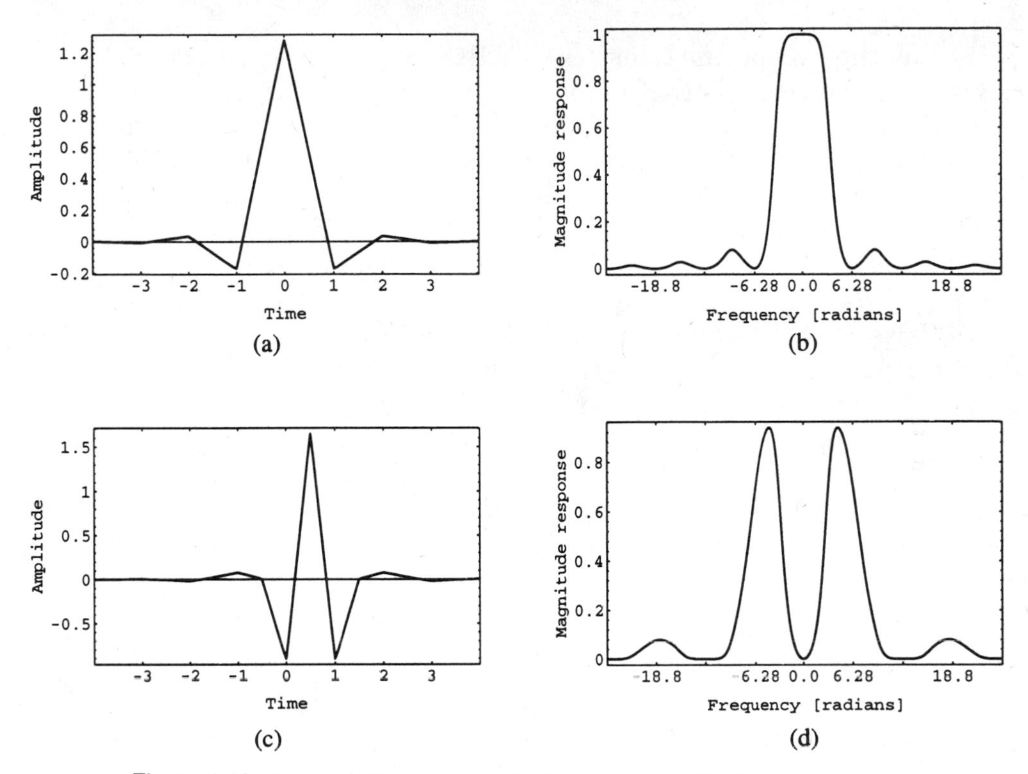

Figure 4.13 Linear spline basis. (a) Scaling function $\varphi(t)$. (b) Fourier transform magnitude $|\Phi(\omega)|$. (c) Wavelet $\psi(t)$. (d) Fourier transform magnitude $|\Psi(\omega)|$.

where the definition of $Q(\omega)$, which is 4π-periodic, follows from (4.3.24). Taking the inverse Fourier transform of (4.3.25) leads to

$$\psi(t) = \sum_{n \in \mathcal{Z}} q[n]\, \beta^{(1)}(2t - n),$$

with the sequence $\{q[n]\}$ being the inverse Fourier transform of $Q(\omega)$. Therefore, $\psi(t)$ is piecewise linear over half-integer intervals, as can be seen in Figure 4.13(b).

In this simple example, the multiresolution approximation is particularly clear. As said at the outset, $V_0^{(1)}$ is the space of functions piecewise linear over integer intervals, and likewise, $V_{-1}^{(1)}$ has the same property but over half-integer intervals. Therefore, $W_0^{(1)}$ (which is the orthogonal complement to $V_0^{(1)}$ in $V_{-1}^{(1)}$) contains the difference between a function in $V_{-1}^{(1)}$ and its approximation in $V_0^{(1)}$. Such a difference is obviously piecewise linear over half-integer intervals.

With the above construction, we have obtained orthonormal bases for $V_0^{(1)}$

and $W_0^{(1)}$ as the sets of functions $\{\varphi(t-n)\}$ and $\{\psi(t-n)\}$ respectively. What was given up, however, is the compact support that $\beta^{(N)}(t)$ has. But it can be shown that the scaling function and the wavelet have exponential decay. The argument begins with the fact that $\varphi(t)$ is a linear combination of functions $\beta^{(N)}(t-n)$. Because $\beta^{(N)}(t)$ has compact support, a finite number of functions from the set $\{\beta^{(N)}(t-n)\}_{n\in\mathcal{Z}}$ contribute to $\varphi(t)$ for a given t (for example, two in the linear spline case). That is, $|\varphi(t)|$ is of the same order as $|\sum_{l=0}^{L-1}\alpha_{k+l}|$ where $k = \lfloor t \rfloor$. Now, $\{\alpha_k\}$ is the impulse response of a stable filter (noncausal in general) because it has no poles on the unit circle (this follows from (4.3.13)). Therefore, the sequence α_k decays exponentially and so does $\varphi(t)$. The same argument holds for $\psi(t)$ as well. For a formal proof of this result, see [73]. While the compact support of $\beta^{(N)}(t)$ has been lost, the fast decay indicates that $\varphi(t)$ and $\psi(t)$ are concentrated around the origin, as is clear from Figures 4.13(a) and (c). The above discussion on orthogonalization was limited to the very simple linear spline case. However, it is clear that it works for the general B-spline case since it is based on the orthogonalization (4.3.14). For example, the quadratic spline, given by

$$\mathcal{B}^{(2)}(\omega) = e^{-j\omega/2}\left(\frac{\sin(\omega/2)}{(\omega/2)}\right)^3, \tag{4.3.26}$$

leads to a function $B^{(5)}(\omega)$ (see 4.3.11) equal to

$$B^{(5)}(\omega) = 66 + 26(e^{j\omega} + e^{-j\omega}) + e^{j2\omega} + e^{-j2\omega}, \tag{4.3.27}$$

which can be used to orthogonalize $\mathcal{B}^{(2)}(\omega)$ (see Problem 4.7).

Note that instead of taking a square root of $B^{(2N+1)}(\omega)$ in the orthogonalization of $\mathcal{B}^{(N)}(\omega)$ (see (4.3.14)), one can use spectral factorization which leads to wavelets based on IIR filters [133, 296] (see also Section 4.6.2 and Problem 4.8). Alternatively, it is possible to give up intrascale orthogonality (but keep interscale orthogonality). See [299] for such a construction where a possible scaling function is a B-spline. One advantage of keeping a scaling function that is a spline is that, as the order increases, its localization in time and frequency rapidly approaches the optimum since it tends to a Gaussian [297].

An interesting limiting result occurs in the case of orthogonal wavelets for B-spline space. As the order of splines goes to infinity, the scaling function tends to the ideal lowpass or sinc function [7, 175]. In our B-spline construction with $N = 0$ and $N \to \infty$, we thus recover the Haar and sinc cases discussed in Section 4.2.3 as extreme cases of a multiresolution analysis.

4.4 WAVELETS DERIVED FROM ITERATED FILTER BANKS AND REGULARITY

In the previous section, we constructed orthonormal families of functions where each function was related to a single prototype wavelet through shifting and scaling. The construction was a direct continuous-time approach based on the axioms of multiresolution analysis. In this section, we will take a different, indirect approach that also leads to orthonormal families derived from a prototype wavelet. Instead of a direct continuous-time construction, we will start with discrete-time filters. They can be iterated and under certain conditions will lead to continuous-time wavelets. This important construction, pioneered by Daubechies [71], produces very practical wavelet decomposition schemes, since they are implementable with finite-length discrete-time filters.

In this section, we will first review the Haar and sinc wavelets as limits of discrete-time filters. Then we extend this construction to general orthogonal filters, showing how to obtain a scaling function φ and a wavelet ψ as limits of an appropriate graphical function. This will lead to a discussion of basic properties of φ and ψ, namely orthogonality and two-scale equations. It will be indicated that the function system $\{2^{-m/2}\varphi(2^m t - n)\}$, $m, n \in \mathcal{Z}$, forms an orthonormal basis for $L_2(\mathcal{R})$.

A key property that the discrete-time filter has to satisfy is the *regularity* condition, which we explore first by way of examples. A discrete-time filter will be called *regular* if it converges (through the iteration scheme we will discuss) to a scaling function and wavelet with some degree of regularity (for example, piecewise smooth, continuous, or derivable). We show conditions that have to be met by the filter and describe regularity testing methods. Then, Daubechies' family of maximally regular filters will be derived.

4.4.1 Haar and Sinc Cases Revisited

As seen earlier, the Haar and sinc cases are two particular examples which are duals of each other, or two extreme cases. Both are useful to explain the iterated filter bank construction. The Haar case is most obvious in time domain, while the sinc case is immediate in frequency domain.

Haar Case Consider the discrete-time Haar filters (see also Section 4.1.3). The lowpass is the average of two neighboring samples, while the highpass is their difference. The corresponding orthogonal filter bank has filters $g_0[n] = [1/\sqrt{2}, 1/\sqrt{2}]$ and $g_1[n] = [1/\sqrt{2}, -1/\sqrt{2}]$ which are the basis functions of the discrete-time Haar expansion. Now consider what happens if we iterate the filter bank on the lowpass channel, as shown in Figure 4.14. In order to derive an equivalent filter bank, we recall the following result from multirate signal processing (Section 2.5.3): Filtering by $g_0[n]$ followed by upsampling by two is

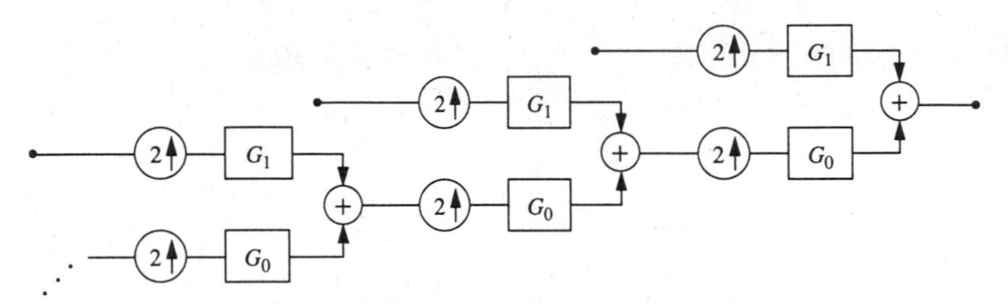

Figure 4.14 Filter bank iterated on the lowpass channel: connection between discrete- and continuous-time cases.

equivalent to upsampling by two, followed by filtering by $g_0'[n]$, where $g_0'[n]$ is the upsampled version of $g_0[n]$.

Using this equivalence, we can transform the filter-bank tree into one equivalent to the one depicted in Figure 3.8 where we assumed three stages and Haar filters. It is easy to verify that this corresponds to an orthogonal filter bank (it is the cascade of orthogonal filter banks). This is a size-8 discrete Haar transform on successive blocks of 8 samples. Iterating the lowpass channel in Figure 4.14 i times, will lead to the equivalent last two filters

$$g_0^{(i)}[n] = \begin{cases} 2^{-i/2} & n = 0, \ldots, 2^i - 1, \\ 0 & \text{otherwise}, \end{cases}$$

$$g_1^{(i)}[n] = \begin{cases} 2^{-i/2} & n = 0, \ldots, 2^{i-1} - 1, \\ -2^{-i/2} & n = 2^{i-1}, \ldots, 2^i - 1, \\ 0 & \text{otherwise}, \end{cases}$$

where $g_0^{(i)}[n]$ is a lowpass filter and $g_1^{(i)}[n]$ a bandpass filter. Note also that $g_0^{(1)}[n] = g_0[n]$ and $g_1^{(1)}[n] = g_1[n]$. As we can see, as i becomes large the length grows exponentially and the coefficients go to zero.

Let us now define a continuous-time function associated with $g_0^{(i)}[n]$ and $g_1^{(i)}[n]$ in the following way:

$$\varphi^{(i)}(t) = 2^{i/2} g_0^{(i)}[n] \quad \tfrac{n}{2^i} \leq t < \tfrac{n+1}{2^i}, \tag{4.4.1}$$

$$\psi^{(i)}(t) = 2^{i/2} g_1^{(i)}[n] \quad \tfrac{n}{2^i} \leq t < \tfrac{n+1}{2^i}.$$

These functions are piecewise constant and because the interval diminishes at the same speed as the length of $g_0^{(i)}[n]$ and $g_1^{(i)}[n]$ increases, their lengths remain bounded.

For example, $\varphi^{(3)}(t)$ and $\psi^{(3)}(t)$ (the functions associated with the two bottom filters of Figure 3.8) are simply the indicator functions of the $[0,1]$ interval and the difference between the indicator functions of $[0, \frac{1}{2}]$ and $[\frac{1}{2}, 1]$, respectively. Of course, in this particular example, it is clear that $\varphi^{(i)}(t)$ and

$\psi^{(i)}(t)$ are all identical, regardless of i. What is also worth noting is that $\varphi^{(i)}(t)$ and $\psi^{(i)}(t)$ are orthogonal to each other and to their translates. Note that

$$\varphi^{(i)}(t) = 2^{1/2}(g_0[0]\, \varphi^{(i-1)}(2t) + g_0[1]\, \varphi^{(i-1)}(2t-1))$$

or, because $\varphi^{(i)}(t) = \varphi^{(i-1)}(t)$ in this particular example,

$$\varphi(t) = 2^{1/2}(g_0[0]\, \varphi(2t) + g_0[1]\, \varphi(2t-1)).$$

Thus, the scaling function $\varphi(t)$ satisfies a two-scale equation.

Sinc Case Recall the sinc case (see Example 4.2). Take an orthogonal filter bank where the lowpass and highpass filters are ideal half-band filters. The impulse response of the lowpass filter is

$$g_0[n] = \frac{1}{\sqrt{2}} \frac{\sin(\pi/2n)}{\pi/2n}, \tag{4.4.2}$$

(see also (4.2.21)) which is orthogonal to its even translates and of norm 1. Its 2π-periodic Fourier transform is equal to $\sqrt{2}$ for $|\omega| \leq \pi/2$, and 0 for $\pi/2 < |\omega| < \pi$. A perfect half-band highpass can be obtained by modulating $g_0[n]$ with $(-1)^n$, since this shifts the passband by π. For completeness, a shift by one is required as well. Thus (see (4.2.22))

$$g_1[n] = (-1)^{(n)} g_0[-n+1].$$

Its 2π-periodic Fourier transform is

$$G_1(e^{j\omega}) = \begin{cases} -\sqrt{2}e^{-j\omega} & \pi/2 \leq |\omega| \leq \pi, \\ 0 & 0 \leq |\omega| < \pi/2. \end{cases} \tag{4.4.3}$$

Now consider the iterated filter bank as in Figure 4.14 with ideal filters. Upsampling the filter impulse response by 2 (to pass it across the upsampler) leads to a filter $g_0'[n]$ with discrete-time Fourier transform (see Section 2.5.3)

$$G_0'(e^{j\omega}) = G_0(e^{j2\omega}),$$

which is π-periodic. It is easy to check that $G_0'(e^{j\omega})G_0(e^{j\omega})$ is a quarter-band filter. Similarly, with $G_1'(e^{j\omega}) = G_1(e^{j2\omega})$, it is clear that $G_1'(e^{j\omega})G_0(e^{j\omega})$ is a bandpass filter with a passband from $\pi/4$ to $\pi/2$. Figure 4.15 shows the amplitude frequency responses of the equivalent filters for a three-step division.

Let us emulate the Haar construction with $g_0^{(i)}[n]$ and $g_1^{(i)}[n]$ which are the lowpass and bandpass equivalent filters for the cascade of i-banks. In Figures 4.15(c) and (d), we have thus the frequency responses of $g_1^{(3)}[n]$ and $g_0^{(3)}[n]$, respectively. Then, we define $\varphi^{(i)}(t)$ as in (4.4.1). The procedure for obtaining $\varphi^{(i)}(t)$ can be described by the following two steps:

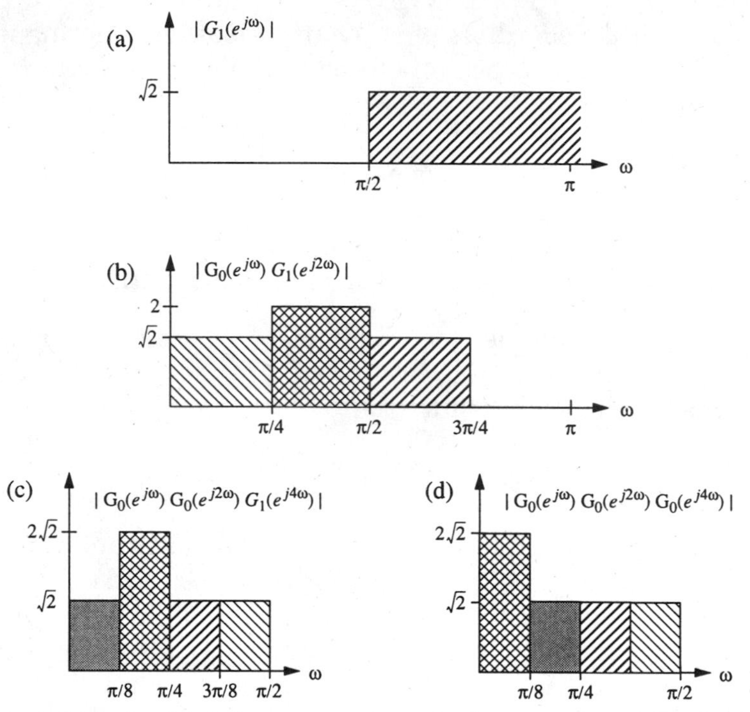

Figure 4.15 Amplitude frequency response of a three-step iterated filter bank with ideal half-band low and highpass filters. (a) $|G_1(e^{j\omega})|$, (b) $|G_0(e^{j\omega})\, G_1(e^{j2\omega})|$, (c) $|G_0(e^{j\omega})\, G_0(e^{j2\omega})\, G_1(e^{j4\omega})|$, (d) $|G_0(e^{j\omega})\, G_0(e^{j2\omega})\, G_0(e^{j4\omega})|$.

(a) Associate with $g_0^{(i)}[n]$ a sequence of weighted Dirac pulses spaced 2^{-i} apart. This sequence has a $2^i 2\pi$-periodic Fourier transform.

(b) Convolve this pulse sequence with an indicator function for the interval $[0, 2^{-i}]$ of height $2^{i/2}$ (so it is of norm 1).

Therefore the Fourier transform of $\varphi^{(i)}(t)$ is

$$\Phi^{(i)}(\omega) \;=\; 2^{-i/2}\, G_0^{(i)}(e^{j\omega/2^i})\, e^{-j\omega/2^{i+1}}\, \frac{\sin(\omega/2^{i+1})}{\omega/2^{i+1}}.$$

Now,

$$G_0^{(i)}(e^{j\omega}) \;=\; G_0(e^{j\omega})\, G_0(e^{j2\omega}) \cdots G_0(e^{j2^{i-1}\omega}). \tag{4.4.4}$$

We introduce the shorthand

$$M_0(\omega) \;=\; \frac{1}{\sqrt{2}} G_0(e^{j\omega}). \tag{4.4.5}$$

Note that $M_0(0) = 1$. We can rewrite $\Phi^{(i)}(\omega)$ as

$$\Phi^{(i)}(\omega) = \left[\prod_{k=1}^{i} M_0\left(\frac{\omega}{2^k}\right)\right] \cdot e^{-j\omega/2^{i+1}} \frac{\sin(\omega/2^{i+1})}{\omega/2^{i+1}}. \tag{4.4.6}$$

The important part in (4.4.6) is the product inside the square brackets (the rest is just a phase factor and the interpolation function). In particular, as i becomes large, the second part tends toward 1 for any finite ω. Thus, let us consider the product involving $M_0(\omega)$ in (4.4.6). Because of the definitions of $M_0(\omega)$ in (4.4.5) and of $G_0(e^{j\omega})$ following (4.4.2), we get

$$M_0\left(\frac{\omega}{2^k}\right) = \begin{cases} 1 & (2l - \frac{1}{2})2^k\pi \le \omega \le (2l + \frac{1}{2})2^k\pi, l \in \mathcal{Z} \\ 0 & \text{otherwise.} \end{cases}$$

The product

$$M_0\left(\frac{\omega}{2}\right) M_0\left(\frac{\omega}{4}\right) \cdots M_0\left(\frac{\omega}{2^i}\right),$$

is $2^i 2\pi$ periodic and equal to 1 for ω between $-\pi$ and π, 0 elsewhere. Therefore, as i goes to infinity, we are left with a perfect lowpass from $-\pi$ to π, that is

$$\lim_{i\to\infty} \varphi^{(i)}(t) = \frac{\sin(\pi t)}{\pi t},$$

or a sinc scaling function. What happens to the function $\psi^{(i)}(t)$? The iterated filter becomes

$$G_1^{(i)}(e^{j\omega}) = G_0(e^{j\omega}) \cdots G_0(e^{j2^{i-2}\omega})\, G_1(e^{j2^{i-1}\omega}),$$

where $G_1(e^{j\omega})$ is given by (4.4.3). The Fourier transform of the wavelet is thus

$$\Psi^{(i)}(\omega) = M_1\left(\frac{\omega}{2}\right) \left[\prod_{k=2}^{i} M_0\left(\frac{\omega}{2^k}\right)\right] e^{-j\omega/2^{i+1}} \frac{\sin(\omega/2^{i+1})}{\omega/2^{i+1}}, \tag{4.4.7}$$

where, similarly to (4.4.5),

$$M_1(\omega) = \frac{1}{\sqrt{2}} G_1(e^{j\omega}). \tag{4.4.8}$$

Suppose that we have $i = 3$. Note that $M_1(\omega/2)$ produces, following (4.4.3), a phase shift of $e^{-j\omega/2}$ or a time-domain delay of $1/2$. It is clear that as i goes to infinity, $\Psi^{(i)}(\omega)$ converges to the indicator function of the interval $[\pi, 2\pi]$ (with a phase shift of $e^{-j\omega/2}$). Thus

$$\lim_{i\to\infty} \psi^{(i)}(t) = 2\frac{\sin(2\pi(t - \frac{1}{2}))}{2\pi(t - \frac{1}{2})} - \frac{\sin(\pi(t - \frac{1}{2}))}{\pi(t - \frac{1}{2})}.$$

This is of course the sinc wavelet we had introduced in Section 4.2 (see (4.2.23)). What we have just seen seems a cumbersome way to rederive a known result.

However, it is an instance of a general construction and some properties can be readily seen. For example, assuming that the infinite product converges, the scaling function satisfies (from (4.4.6))

$$\Phi(\omega) = \lim_{i\to\infty} \Phi^{(i)}(\omega) = \prod_{k=1}^{\infty} M_0\left(\frac{\omega}{2^k}\right) = M_0\left(\frac{\omega}{2}\right)\Phi\left(\frac{\omega}{2}\right),$$

or, in time domain

$$\varphi(t) = \sqrt{2}\sum_{n=-\infty}^{\infty} g_0[n]\,\varphi(2t-n),$$

and similarly, the wavelet satisfies

$$\psi(t) = \sqrt{2}\sum_{n=-\infty}^{\infty} g_1[n]\,\varphi(2t-n).$$

That is, the two-scale equation property is implicit in the construction of the iterated function. The key in this construction is the behavior of the infinite product of the $M_0(\omega/2^k)$'s. This leads to the fundamental *regularity property* of the discrete-time filters involved, which will be studied below. But first, we formalize the iterated filter bank construction.

4.4.2 Iterated Filter Banks

We will now show that the above derivation of the Haar and sinc wavelets using iterated filter banks can be used in general to obtain wavelet bases, assuming that the filters satisfy some regularity constraints. In our discussion, we will concentrate mainly on the well-behaved cases, namely when the graphical function $\varphi^{(i)}(t)$ (associated with the iterated impulse response $g_0^{(i)}[n]$) converges in $L_2(\mathcal{R})$ to a piecewise smooth[5] function $\varphi(t)$ (possibly with more regularity, such as continuity). In this case, the Fourier transform $\Phi^{(i)}(\omega)$ converges in $L_2(\mathcal{R})$ to $\Phi(\omega)$ (the Fourier transform of $\varphi(t)$). That is, one can study the behavior of the iteration either in time or in frequency domain. A counter-example to this "nice" behavior is discussed in Example 4.3 below.

To demonstrate the construction, we start with a two-channel orthogonal filter bank as given in Section 3.2. Let $g_0[n]$ and $g_1[n]$ denote lowpass and highpass filters, respectively. Similarly to the Haar and sinc cases, the filter bank is iterated on the branch with the lowpass filter (see Figure 4.14) and the process is iterated to infinity. The constructions in the previous section indicate how to proceed. First, express the two equivalent filters after i steps

[5]This is more restrictive than necessary, but makes the treatment easier.

of iteration as (use the fact that filtering with $G_i(z)$ followed by upsampling by 2 is equivalent to upsampling by 2 followed by filtering with $G_i(z^2)$)

$$G_0^{(i)}(z) = \prod_{k=0}^{i-1} G_0\left(z^{2^k}\right), \tag{4.4.9}$$

$$G_1^{(i)}(z) = G_1(z^{2^{i-1}}) \prod_{k=0}^{i-2} G_0\left(z^{2^k}\right), \qquad i = 1, 2, \ldots$$

These filters are preceded by upsampling by 2^i (note that $G_0^{(0)}(z) = G_1^{(0)}(z) = 1$). Then, associate the discrete-time iterated filters $g_0^{(i)}[n], g_1^{(i)}[n]$ with the continuous-time functions $\varphi^{(i)}(t), \psi^{(i)}(t)$ as follows:

$$\varphi^{(i)}(t) = 2^{i/2} g_0^{(i)}[n], \quad n/2^i \leq t < \tfrac{n+1}{2^i}, \tag{4.4.10}$$

$$\psi^{(i)}(t) = 2^{i/2} g_1^{(i)}[n], \quad n/2^i \leq t < \tfrac{n+1}{2^i}. \tag{4.4.11}$$

Note that the elementary interval is divided by $1/2^i$. This rescaling is necessary because if the length of the filter $g_0[n]$ is L then the length of the iterated filter $g_0^{(i)}[n]$ is

$$L^{(i)} = (2^i - 1)(L - 1) + 1$$

which will become infinite as $i \to \infty$. Thus, the normalization ensures that the associated continuous-time function $\varphi^{(i)}(t)$ stays compactly supported (as $i \to \infty$, $\varphi^{(i)}(t)$ will remain within the interval $[0, L-1]$). The factor $2^{i/2}$ which multiplies $g_0^{(i)}[n]$ and $g_1^{(i)}[n]$ is necessary to preserve the L_2 norm between discrete and continuous-time cases. If $\|g_0^{(i)}[n]\| = 1$, then $\|\varphi^{(i)}(t)\| = 1$ as well, since each piecewise constant block has norm $|g_0^{(i)}[n]|$.

In Figure 4.16 we show the graphical function for the first four iterations of a length-4 filter. This indicates the piecewise constant approximation and the halving of the interval.

In Fourier domain, using $M_0(\omega) = G_0(e^{j\omega})/\sqrt{2}$ and $M_1(\omega) = G_1(e^{j\omega})/\sqrt{2}$, we can write (4.4.10) and (4.4.11) as (from (4.4.6))

$$\Phi^{(i)}(\omega) = \left[\prod_{k=1}^{i} M_0\left(\frac{\omega}{2^k}\right)\right] \Theta^{(i)}(\omega),$$

where

$$\Theta^{(i)}(\omega) = e^{-j\omega/2^{i+1}} \frac{\sin(\omega/2^{i+1})}{\omega/2^{i+1}},$$

as well as (from (4.4.7)

$$\Psi^{(i)}(\omega) = M_1\left(\frac{\omega}{2}\right) \left[\prod_{k=2}^{i} M_0\left(\frac{\omega}{2^k}\right)\right] \Theta^{(i)}(\omega).$$

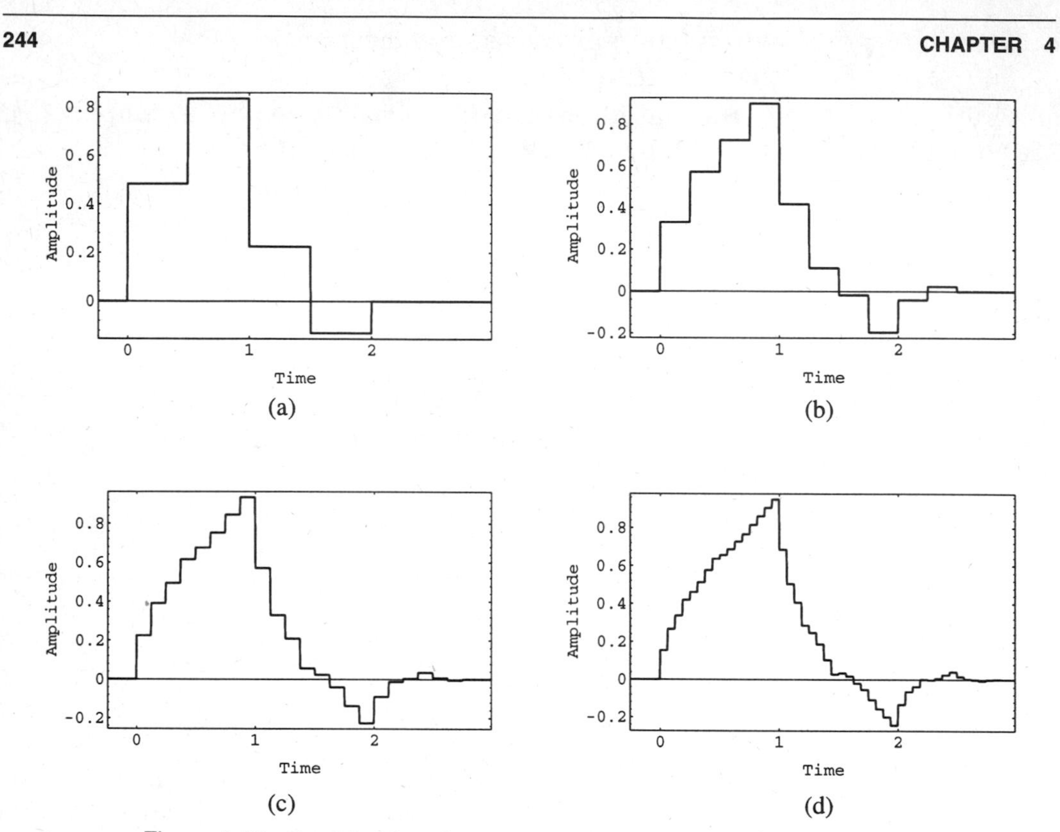

Figure 4.16 Graphical functions corresponding to the first four iterations of an orthonormal 4-tap filter with two zeros at $\omega = \pi$. The filter is given in the first column of Table 4.3. (a) $\varphi^{(1)}(t)$. (b) $\varphi^{(2)}(t)$. (c) $\varphi^{(3)}(t)$. (d) $\varphi^{(4)}(t)$.

A fundamental question is: To what, if anything, do the functions $\varphi^{(i)}(t)$ and $\psi^{(i)}(t)$ converge as $i \to \infty$? We will proceed by assuming convergence to piecewise smooth functions in $L^2(\mathcal{R})$:

$$\varphi(t) = \lim_{i\to\infty} \varphi^{(i)}(t), \tag{4.4.12}$$

$$\psi(t) = \lim_{i\to\infty} \psi^{(i)}(t). \tag{4.4.13}$$

In Fourier domain, the above equations become

$$\Phi(\omega) = \lim_{i\to\infty} \Phi^{(i)}(\omega) = \prod_{k=1}^{\infty} M_0\left(\frac{\omega}{2^k}\right), \tag{4.4.14}$$

$$\Psi(\omega) = \lim_{i\to\infty} \Psi^{(i)}(\omega) = M_1\left(\frac{\omega}{2}\right) \prod_{k=2}^{\infty} M_0\left(\frac{\omega}{2^k}\right), \tag{4.4.15}$$

since $\Theta^{(i)}(\omega)$ becomes 1 for any finite ω as $i \to \infty$. Next, we demonstrate that the functions $\varphi(t)$ and $\psi(t)$, obtained as limits of discrete-time iterated filters, are actually a scaling function and a wavelet, and that they carry along an underlying multiresolution analysis.

Two-Scale Equation Property Let us show that the scaling function $\varphi(t)$ satisfies a two-scale equation, as required by (4.2.8). Following (4.4.9), one can write the equivalent filter after i steps in terms of the equivalent filter after $(i-1)$ steps as

$$g_0^{(i)}[n] = \sum_k g_0[k]\, g_0^{(i-1)}[n - 2^{i-1}k]. \tag{4.4.16}$$

Using (4.4.10), express the previous equation in terms of iterated functions:

$$g_0^{(i)}[n] = 2^{-\frac{i}{2}}\, \varphi^{(i)}(t), \tag{4.4.17}$$

$$g_0^{(i-1)}[n - 2^{i-1}k] = 2^{-(i-1)/2}\, \varphi^{(i-1)}(2t-k), \tag{4.4.18}$$

both for $n/2^i \le t < (n+1)/2^i$. Substituting (4.4.17) and (4.4.18) into (4.4.16) yields

$$\varphi^{(i)}(t) = \sqrt{2} \sum_k g_0[k]\, \varphi^{(i-1)}(2t-k). \tag{4.4.19}$$

By assumption, the iterated function $\varphi^{(i)}(t)$ converges to the scaling function $\varphi(t)$. Hence, take limits on both sides of (4.4.19) to obtain

$$\varphi(t) = \sqrt{2} \sum_k g_0[k]\, \varphi(2t-k), \tag{4.4.20}$$

that is, the limit of the discrete-time iterated filter (4.4.12) satisfies a two-scale equation. Similarly,

$$\psi(t) = \sqrt{2} \sum_k g_1[k]\, \varphi(2t-k).$$

These relations also follow directly from the Fourier-domain expressions $\Phi(\omega)$ and $\Psi(\omega)$, since, for example, from (4.4.14) we get

$$\begin{aligned}\Phi(\omega) &= \prod_{k=1}^{\infty} M_0\left(\frac{\omega}{2^k}\right) = M_0\left(\frac{\omega}{2}\right) \prod_{k=2}^{\infty} M_0\left(\frac{\omega}{2^k}\right) \\ &= M_0\left(\frac{\omega}{2}\right) \Phi\left(\frac{\omega}{2}\right) = \frac{1}{\sqrt{2}}\, G_0\left(e^{j\omega/2}\right) \Phi\left(\frac{\omega}{2}\right).\end{aligned}$$

Orthogonality and Completeness of the Wavelet Basis We want to show that the wavelets constitute a basis for $L_2(\mathcal{R})$. To that end, we will have to prove the orthogonality as well as the completeness of the basis functions. First, however, let us recall a few facts that are going to be used in our discussion. We will assume that we have an orthonormal filter bank as seen in Section 3.2.3. We will also assume the following:

(a) $\langle g_0[k], g_1[k+2n]\rangle = 0$, $\langle g_0[k], g_0[k+2n]\rangle = \langle g_1[k], g_1[k+2n]\rangle = \delta[n]$, that is, filters g_0 and g_1 are orthogonal to each other and their even translates as given in Section 3.2.3.

(b) $G_0(z)|_{z=1} = \sqrt{2}$, $G_0(z)|_{z=-1} = 0$, that is, the lowpass filter has a zero at the aliasing frequency π (see also the next section).

(c) The filters are FIR.

(d) $g_1[n] = (-1)^n g_0[-n+1]$, as given in Section 3.2.3.

(e) The scaling function and the wavelet are given by (4.4.12) and (4.4.13).

In the Haar case, it was shown that the scaling function and the wavelet were orthogonal to each other. Using appropriate shifts and scales, it was shown that the wavelets formed an orthonormal set. Here, we demonstrate these relations in the general case, starting from discrete-time iterated filters. The proof is given only for the first fact, the others would follow similarly.

PROPOSITION 4.4 *Orthogonality Relations for the Scaling Function and Wavelet*

(a) The scaling function is orthogonal to its appropriate translates at a given scale

$$\langle \varphi(2^m t - n), \varphi(2^m t - n')\rangle = 2^{-m}\delta[n-n'].$$

(b) The wavelet is orthogonal to its appropriate translates at all scales

$$\langle \psi(2^m t - n), \psi(2^m t - n')\rangle = 2^{-m}\delta[n-n'].$$

(c) The scaling function is orthogonal to the wavelet and its integer shifts

$$\langle \varphi(t), \psi(t-n)\rangle = 0.$$

(d) Wavelets are orthogonal across scales and with respect to shifts

$$\langle \psi(2^m t - n), \psi(2^{m'} t - n')\rangle = 2^{-m-m'}\delta[m-m']\,\delta[n-n'].$$

PROOF

To prove the first fact, we use induction on $\varphi^{(i)}$ and then take the limit (which exists by assumption). For clarity, this fact will be proven only for scale 0 (scale m would follow similarly). The first step $\langle \varphi^{(0)}(t), \varphi^{(0)}(t-l)\rangle = \delta[n]$ is obvious since, by definition, $\varphi^{(0)}(t)$ is just the indicator function of the interval $[0,1)$. For the inductive step, write

$$\langle \varphi^{(i+1)}(t), \varphi^{(i+1)}(t-l)\rangle = \langle \sqrt{2}\sum_k g_0[k]\,\varphi^{(i)}(2t-k), \sqrt{2}\sum_m g_0[m]\,\varphi^{(i)}(2t-2l-m)\rangle$$

$$= 2\sum_k \sum_m g_0[k]\, g_0[m]\, \langle \varphi^{(i)}(2t-k), \varphi^{(i)}(2t-2l-m)\rangle$$

$$= \sum_m g_0[m]\, g_0[2l+m] \;=\; \langle g_0[m], g_0[2l+m]\rangle \;=\; \delta[l],$$

where the orthogonality relations between discrete-time filters, given at the beginning of this subsection, were used. Taking the limits of both sides of the previous equation, the first fact is obtained. The proofs of the other facts follow similarly.

We have thus verified that

$$S \;=\; \{2^{-\frac{m}{2}}\psi(2^{-m}t-n) \mid m,n \in \mathcal{Z},\; t \in \mathcal{R}\},$$

is an orthonormal set. The only remaining task is to show that the members of the set S constitute an orthonormal basis for $L_2(\mathcal{R})$, as stated in the following theorem.

THEOREM 4.5 *[71]*

The orthonormal set of functions $S = \{\psi_{m,n} \mid m,n \in \mathcal{Z}, t \in \mathcal{R}\}$ where $\psi_{m,n}(t) = 2^{-\frac{m}{2}}\psi(2^{-m}t-n)$ is a basis for $L_2(\mathcal{R})$, that is, for every $f \in L_2(\mathcal{R})$

$$\sum_{m,n\in\mathcal{Z}} |\langle \psi_{m,n}, f\rangle|^2 \;=\; \|f\|^2.$$

Since the proof is rather technical and does not have an immediate intuitive interpretation, an outline is given in Appendix 4.A. For more details, the reader is referred to [71, 73]. Note that the statement of the theorem is nothing else but the Parseval's equality as given by (d) in Theorem 2.4.

4.4.3 Regularity

We have seen that the conditions under which (4.4.12–4.4.13) exist are critical. We will loosely say that they exist and lead to piecewise smooth functions if the filter $g_0[n]$ is *regular*. In other words, a regular filter leads, through iteration, to a scaling function with some degree of smoothness or regularity.

Given a filter $G_0(z)$ and an iterated filter bank scheme, the limit function $\varphi(t)$ depends on the behavior of the product

$$\prod_{k=1}^{i} M_0\left(\frac{\omega}{2^k}\right), \tag{4.4.21}$$

for large i, where $M_0(\omega) = G_0(e^{j\omega})/G_0(1)$ so that $M_0(0) = 1$. This normalization is necessary since otherwise either the product blows up at $\omega = 0$ (if $M_0(0) > 1$) or goes to zero (if $M_0(0) < 1$) which would mean that $\varphi(t)$ is not a lowpass function.

Key questions are: Does the product converge (and in what sense)? If it converges, what are the properties of the limit function (continuity, differentiability, etc.)? It can be shown that if $|M_0(\omega)| \leq 1$ and $M_0(0) = 1$, then we have pointwise convergence of the infinite product to a limit function $\Phi(\omega)$ (see Problem 4.12). In particular, if $M_0(\omega)$ corresponds to the normalized lowpass filter in an orthonormal filter bank, then this condition is automatically satisfied. However, pointwise convergence is not sufficient. To build orthonormal bases we need L_2 convergence. This can be obtained by imposing some additional constraints on $M_0(\omega)$. Finally, beyond mere L_2 convergence, we would like to have a limit $\Phi(\omega)$ corresponding to a smooth function $\varphi(t)$. This can be achieved with further constraints of $M_0(\omega)$. Note that we will concentrate on the regularity of the lowpass filter, which leads to the scaling function $\varphi(t)$ in iterated filter bank schemes. The regularity of the wavelet $\psi(t)$ is equal to that of the scaling function when the filters are of finite length since $\psi(t)$ is a finite linear combination of $\varphi(2t-n)$.

First, it is instructive to reconsider a few examples. In the case of the perfect half-band lowpass filter, the limit function associated with the iterated filter converged to $\sin(\pi t)/\pi t$ in time. Note that this limit function is infinitely differentiable. In the Haar case, the lowpass filter, after normalization, gives

$$M_0(\omega) = \frac{1+e^{-j\omega}}{2},$$

which converged to the box function, that is, it converged to a function with two discontinuous points. In other words, the product in (4.4.21) converges to

$$\prod_{k=1}^{\infty} M_0\left(\frac{\omega}{2^k}\right) = \prod_{k=1}^{\infty}\left(\frac{1+e^{-j\omega/2^k}}{2}\right) = e^{-j\omega/2}\,\frac{\sin(\omega/2)}{\omega/2}. \tag{4.4.22}$$

For an alternative proof of this formula, see Problem 4.11. Now consider a filter with impulse response $[\frac{1}{2}, 1, \frac{1}{2}]$, that is, the Haar lowpass filter convolved with itself. The corresponding $M_0(\omega)$ is

$$M_0(\omega) = \frac{1+2e^{-j\omega}+e^{-j2\omega}}{4} = \left(\frac{1+e^{-j\omega}}{2}\right)^2. \tag{4.4.23}$$

The product (4.4.21) can thus be split into two parts; each of which converges to the Fourier transform of the box function. Therefore, the limit function $\varphi(t)$ is the convolution of two boxes, or, the hat function. This is a continuous function and is differentiable except at the points $t = 0, 1$ and 2. It is easy to see that if we have the Nth power instead of the square in (4.4.23), the limit function will be the $(N-1)$-time convolution of the box with itself. This function is $(N-1)$-times differentiable (except at integers where it is once less differentiable).

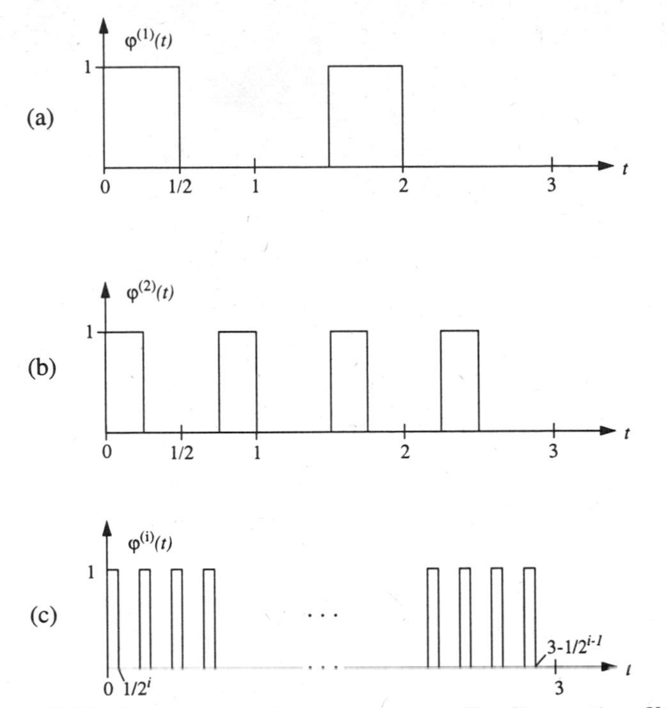

Figure 4.17 Counter-example to convergence. The discrete-time filter has impulse response $[1/\sqrt{2}, 0, 0, 1/\sqrt{2}]$. (a) $\varphi^{(1)}(t)$. (b) $\varphi^{(2)}(t)$. (c) $\varphi^{(i)}(t)$.

These are the well-known B-spline functions [76, 255] (see also Section 4.3.2). An important fact to note is that each additional factor $(1 + e^{j\omega})/2$ leads to one more degree of regularity. That is, zeros at $\omega = \pi$ in the discrete-time filter play an important role. However, zeros at $\omega = \pi$ are not sufficient to insure regularity. We can see this in the following counter-example [71]:

Example 4.3 *Convergence Problems*

Consider the orthonormal filter $g_0[n] = [1/\sqrt{2}, 0, 0, 1/\sqrt{2}]$ or $M_0(\omega) = (1 + e^{-j3\omega})/2$. The infinite product in frequency becomes, following (4.4.22),

$$\Phi(\omega) = \prod_{k=1}^{\infty} M_0\left(\frac{\omega}{2^k}\right) = e^{-j3\omega/2}\,\frac{\sin(3\omega/2)}{3\omega/2}, \qquad (4.4.24)$$

which is the Fourier transform of $1/3$ times the indicator function of the interval $[0, 3]$. This function is clearly not orthogonal to its integer translates, even though every finite iteration of the graphical function is. That is, (4.2.21) is not satisfied by the limit. Also, while every finite iteration is of norm 1, the limit is not. Therefore, we have failure of L_2 convergence of the infinite product.

Looking at the time-domain graphical function (see Figure 4.17), it is easy to check that $\varphi^{(i)}(t)$ takes only the values 0 or 1, and therefore, there is no pointwise convergence

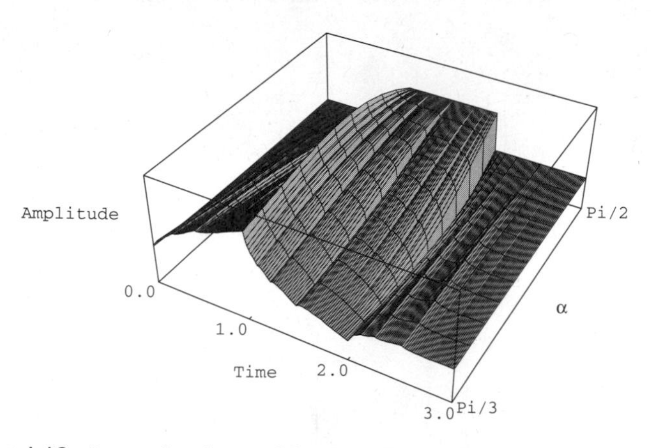

Figure 4.18 Iterated orthogonal lowpass filters of length 4 with one zero at $\omega = \pi$ (or $\alpha_1 = \pi/4 - \alpha_0$). For $\alpha_0 = \pi/3$, there are two zeros at π and this leads to a regular iterated filter of length 4. This corresponds to the Daubechies' scaling function. The sixth iteration is shown.

on the interval $[0, 3]$. Note that $\varphi^{(i)}(t)$ is not of bounded variation as $i \to \infty$. Thus, even though $\varphi^{(i)}(t)$ and $\Phi^{(i)}(\omega)$ are valid Fourier transform pairs for any finite i, their limits are not; since $\varphi(t)$ does not exist while $\Phi(\omega)$ is given by (4.4.24). This simple example indicates that the convergence problem is nontrivial.

A main point of the previous example is that failure of L_2 convergence indicates a breakdown of the orthonormal basis construction that is based on iterated filter banks. Several sufficient conditions for L_2 convergence have been given. Mallat shows in [180] that a sufficient condition is

$$|M_0(\omega)| > 0, \qquad |\omega| < \frac{\pi}{2}.$$

It is easy to verify that the above example does not meet it since $M_0(\pi/3) = 0$. Another sufficient condition by Daubechies also allows one to impose regularity. This will be discussed in Proposition 4.7. Necessary and sufficient conditions for L_2 convergence are more involved, and were derived by Cohen [55] and Lawton [169, 170] (see [73] for a discussion of these conditions).

The next example considers the orthogonal filter family that was derived in Section 3.2.3. It shows that very different behavior can be obtained within a family.

Example 4.4 *Iteration of Length-4 Orthogonal Family*

Consider a 4-tap orthogonal filter bank. From the cascade structure discussed in Section 3.2.3 (Example 3.3) the 4-tap lowpass filter has the impulse response

$$g_0[n] = [\cos\alpha_0 \cos\alpha_1, \cos\alpha_1 \sin\alpha_0, -\sin\alpha_0 \sin\alpha_1, \cos\alpha_0 \sin\alpha_1]. \tag{4.4.25}$$

In order to force this filter to have a zero at $\omega = \pi$, it is necessary that $\alpha_0 + \alpha_1 = \pi/4$.

Choosing $\alpha_0 = \pi/3$ and $\alpha_1 = -\pi/12$ leads to a double zero at $\omega = \pi$ and corresponds to a Daubechies' filter of length 4. In Figure 4.18, we show iterates of the orthogonal filter in (4.4.25) from $\alpha_0 = \pi/3$ (the Daubechies' filter) to $\alpha_0 = \pi/2$ (the Haar filter), with α_1 being equal to $\pi/4 - \alpha_0$. As can be seen, iterated filters around the Daubechies' filter look regular as well. The continuity of the Daubechies' scaling function will be shown below.

The above example should give an intuition for the notion of regularity. The Haar filter, leading to a discontinuous function, is less regular than the Daubechies filter. In the literature, regularity is somewhat loosely defined (continuity in [194], continuity and differentiability in [181]). As hinted in the spline example, zeros at the aliasing frequency $\omega = \pi$ (or $z = -1$) play a key role for the regularity of the filter. First, let us show that a zero at $\omega = \pi$ is necessary for the limit function to exist. There are several proofs of this result (for example in [92]) and we follow Rioul's derivation [239].

Given a lowpass filter $G_0(z)$ and its iteration $G_0^{(i)}(z)$ (see (4.4.9)), consider the associated graphical function $\varphi^{(i)}(t)$ (see (4.4.10)).

PROPOSITION 4.6 *Necessity of a Zero at Aliasing Frequency*

For the limit $\varphi(t) = \lim_{i\to\infty} \varphi^{(i)}(t)$ to exist, it is necessary that $G_0(-1) = 0$.

PROOF

For the limit of $\varphi^{(i)}(t)$ to exist it is necessary that, as i increases, the even and odd samples of $g_0^{(i)}[n]$ tend to the same limit sequence. This limit sequence has an associated limit function $\varphi(2t)$. Use the fact that (see 4.4.4)

$$G_0^{(i)}(z) = G_0(z)\, G_0^{(i-1)}(z^2) = (G_e(z^2) + z^{-1}G_o(z^2))\, G_0^{(i-1)}(z^2),$$

where the subscripts e and o stand for even and odd-indexed samples of $g_0[n]$, respectively. We can write the even and odd-indexed samples of $g_0^{(i)}[n]$ in z-transform domain as

$$\begin{aligned} G_e^{(i)}(z) &= G_e(z)\, G_0^{(i-1)}(z), \\ G_o^{(i)}(z) &= G_o(z)\, G_0^{(i-1)}(z), \end{aligned}$$

or, in time domain

$$g_0^{(i)}[2n] = \sum_k g_0[2k]\, g_0^{(i-1)}[n-k], \tag{4.4.26}$$

$$g_0^{(i)}[2n+1] = \sum_k g_0[2k+1]\, g_0^{(i-1)}[n-k]. \tag{4.4.27}$$

When considering the associated continuous function $\varphi^{(i)}(t)$ and its limit as i goes to infinity, the left side of the above two equations tends to $\varphi(2t)$. For the right side, note that k is bounded while n is not. Because the intervals for the interpolation diminish as $1/2^i$, the shift by k vanishes as i goes to infinity and $g_0^{(i-1)}[n-k]$ leads also to $\varphi(2t)$. That is,

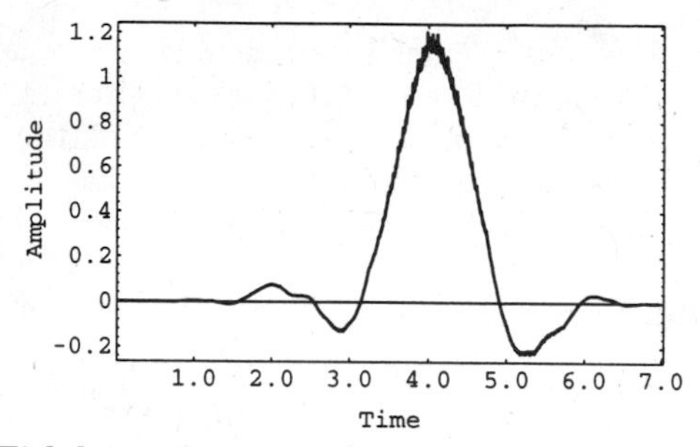

Figure 4.19 Eighth iteration of the filter which fails to converge because of the absence of an exact zero at $\omega = \pi$. The filter is a Smith and Barnwell filter of length 8 [271] (see Table 3.2).

(4.4.26) and (4.4.27) become equal and

$$\left[\sum_k g_0[2k]\right] \varphi(2t) \;=\; \left[\sum_k g_0[2k+1]\right] \varphi(2t),$$

which, assuming that $\varphi(2t)$ is not zero for some t, leads to

$$\sum_k g_0[2k] \;=\; \sum_k g_0[2k+1].$$

Since $G_0(e^{j\omega})|_{\omega=\pi} \;=\; \sum_k g_0[2k] - \sum_k g_0[2k+1]$, we have verified that if $\varphi(t)$ is to exist, the filter has necessarily a zero at $\omega \;=\; \pi$.

Note that a zero at $\omega \;=\; \pi$ is not sufficient, as demonstrated by the filter with impulse response $[1/\sqrt{2}, 0, 0, 1/\sqrt{2}]$ (see Example 4.3). Another interpretation of the above result can be made in Fourier domain, when considering the product in (4.4.21). This product is $2\pi \cdot 2^i$-periodic. Consider its value at $\omega \;=\; \pi 2^i$:

$$\prod_{k=1}^{i} M_0(\pi 2^{(i-k)}) \;=\; M_0(\pi) \prod_{k=1}^{i-1} M_0(2\pi 2^{(i-k-1)}) \;=\; M_0(\pi),$$

since $M_0(\omega)$ is 2π-periodic and $M_0(0) \;=\; 1$. That is, unless $M_0(\pi)$ is exactly zero, there is a nonzero Fourier component at an arbitrary high frequency. This indicates that $g_0^{(i)}[n]$ and $g_0^{(i)}[2n+1]$ will never be the same. This results in highest frequency "wiggles" in the iterated impulse response. As an example, we show, in Figure 4.19, the iteration of a filter which is popular in subband coding [271], but which does not have an exact zero at $\omega \;=\; \pi$. The resulting iterated function has small wiggles and will not converge. Note that most filters designed for subband coding have high (but maybe not infinite) attenuation at $\omega = \pi$, thus the problem is usually minor.

A Sufficient Condition for Regularity In [71], Daubechies studies the regularity of iterated filters in detail and gives a very useful sufficient condition for an iterated filter and its associated graphical function to converge to a continuous function. Factor $M_0(\omega)$ as

$$M_0(\omega) = \left(\frac{1+e^{j\omega}}{2}\right)^N R(\omega).$$

Because of the above necessary condition, we know that N has to be at least equal to 1. Define B as

$$B = \sup_{\omega\in[0,2\pi]} |R(\omega)|.$$

Then the following result due to Daubechies holds [71]:

PROPOSITION 4.7

If

$$B < 2^{N-1}, \tag{4.4.28}$$

then the limit function $\varphi^{(i)}(t)$ as $i \to \infty$ converges pointwise to a continuous function $\varphi(t)$ with the Fourier transform

$$\Phi(\omega) = \prod_{k=1}^{\infty} M_0\left(\frac{\omega}{2^k}\right). \tag{4.4.29}$$

PROOF

It is sufficient to show that for a large enough ω, the decay of $\Phi(\omega)$ is faster than $C(1+|\omega|)^{-1}$. This indicates that $\varphi(t)$ will be continuous. Rewrite (4.4.29) as follows:

$$\prod_{k=1}^{\infty} M_0\left(\frac{\omega}{2^k}\right) = \prod_{k=1}^{\infty}\left(\frac{1+e^{j\omega/2^k}}{2}\right)^N \prod_{k=1}^{\infty} R\left(\frac{\omega}{2^k}\right).$$

In the above, the first product on the right side is a smoothing part and equals

$$\frac{\sin(\omega/2)}{\omega/2}^N,$$

which leads to a decay of the order of $C'(1+|\omega|)^{-N}$. But then, there is the effect of the remainder $R(\omega)$. Recall that $|R(0)| = 1$. Now, $|R(\omega)|$ can be bounded above by $1 + c|\omega|$, for some c, and thus $|R(\omega)| \le e^{c|\omega|}$. Consider now $\prod_{k=1}^{\infty} R\left(\omega/2^k\right)$, for $|\omega| < 1$. In particular,

$$\sup_{|\omega|<1} \prod_{k=1}^{\infty}\left|R\left(\frac{\omega}{2^k}\right)\right| \le \prod_{k=1}^{\infty} e^{c(\omega/2^k)} = e^{c\omega(1/2+1/4+\ldots)} \le e^c.$$

Thus, for $|\omega| < 1$, we have an upper bound. For any ω, $|\omega| > 1$, there exists $J \ge 1$ such that $2^{J-1} \le |\omega| 2^J$. Therefore, split the infinite product into two parts:

$$\prod_{k=1}^{\infty}\left|R\left(\frac{\omega}{2^k}\right)\right| = \prod_{k=1}^{J}\left|R\left(\frac{\omega}{2^k}\right)\right| \prod_{k=1}^{\infty}\left|R\left(\frac{\omega}{2^J 2^k}\right)\right|.$$

Since $|\omega| < 2^J$, we can bound the second product by e^c. The first product is smaller than, or equal to B^J. Thus

$$\prod_{k=1}^{\infty} \left| R\left(\frac{\omega}{2^k}\right) \right| \leq B^J e^c.$$

Now, $B < 2^{N-1}$ thus,

$$B^J e^c \; < \; c' 2^{J(N-1-\epsilon)} \; < \; c''(1+|\omega|)^{N-1-\epsilon}.$$

Putting all this together, we finally get

$$\prod_{k=1}^{\infty} M_0\left(\frac{\omega}{2^k}\right) < (1+|\omega|)^{-1-\epsilon}.$$

Let us check the Haar filter, or $M_0(\omega) = \frac{1}{2}(1+e^{j\omega})^2 \times 1$. Here, $N = 1$ and the supremum of the remainder is one. Therefore, the inequality (4.4.28) is not satisfied. Since the bound in (4.4.28) is sufficient but not necessary, we cannot infer discontinuity of the limit. However, we know that the Haar function is discontinuous at two points. On the other hand, the length-4 Daubechies' filter (see Example 4.4) yields

$$M_0(\omega) = \frac{1}{4}(1+e^{-jw})^2 \frac{1}{2}(1+\sqrt{3}+(1-\sqrt{3})e^{-jw})$$

and the maximum of $|R(\omega)|$, attained at $\omega = \pi$, is $B = \sqrt{3}$. Since $N = 2$, the bound (4.4.28) is met and continuity of the limit function $\varphi(t)$ is proven.

A few remarks are in place. First, there are variations in using the above sufficient condition. For example, one can test the cascade of l filters with respect to upsampling by 2^l. Calling B_l the following supremum,

$$B_l \; = \; \sup_{\omega \in [0,2\pi]} \prod_{k=0}^{l-1} R(2^k \omega),$$

the bound (4.4.28) becomes

$$B_l \; < \; 2^{l(N-1)}.$$

Obviously, as l becomes large, we get a better approximation since the cascade resembles the iterated function. Another variation consists in leaving some of the zeros at $\omega = \pi$ in the remainder, so as to attenuate the supremum B. If there is a factorization that meets the bound, continuity is shown.

Then, additional zeros at $\omega = \pi$ (beyond the one to ensure that the limit exists) will ensure continuity, differentiability and so on. More precisely if, instead of (4.4.28), we have

$$B \; < \; 2^{N-1-l}, \; l = 1, 2, \ldots$$

then $\varphi(t)$ is l-times continuously differentiable.

Other Methods for Investigating Regularity Daubechies' sufficient condition might fail and the filter might still be regular. Another criterion will give a lower bound on regularity. It is the Cohen's fixed-point method [55] which we describe briefly with an example.

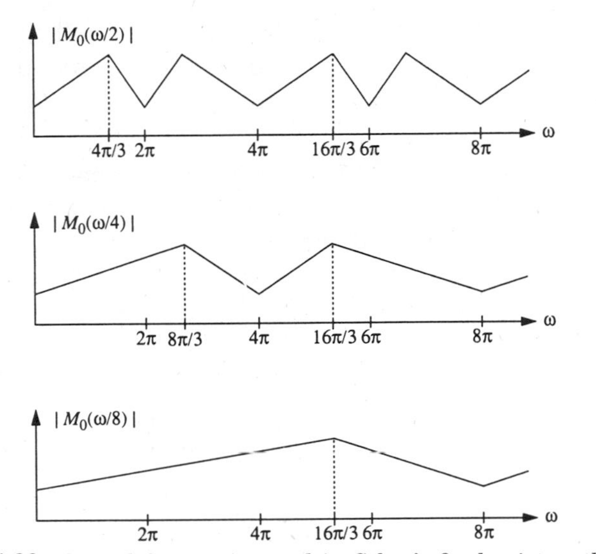

Figure 4.20 Critical frequencies used in Cohen's fixed point method (the shape of the Fourier transform is only for the sake of example).

When evaluating the product (4.4.21), certain critical frequencies will align. These are fixed points of the mapping $\omega \mapsto 2\omega$ modulo 2π. For example, $\omega = \pm 2\pi/3$ is a critical frequency. This can be seen in Figure 4.20 where we show $M_0(\omega/2)$, $M_0(\omega/4)$ and $M_0(\omega/8)$. It is clear from this figure that the absolute value of the product of $M_0(\omega/2)$, $M_0(\omega/4)$ and $M_0(\omega/8)$ evaluated at $\omega = 16\pi/3$ is equal to $|M_0(2\pi/3)|^3$. In general

$$\prod_{k=1}^{i} \left| M_0\left(\frac{\omega}{2^k}\right)\right| \Bigg|_{\omega=2^i\pi/3} = \left| M_0\left(\frac{2\pi}{3}\right)\right|^i .$$

From this, it is clear that if $|M_0\,(2\pi/3)|$ is larger than $1/2$, the decay of the Fourier transform will not be of the order of $1/\omega$ and continuity would be disproved. Because it involves only certain values of the Fourier transform, the fixed-point method can be used to test large filters quite easily. For a thorough discussion of the fixed-point method, we refer to [55, 57].

Another possible method for studying regularity uses $L \times L$ matrices corresponding to a length-L filter downsampled by 2 (that is, the rows contain the filter coefficients but are shifted by 2). By considering a subset of eigenvalues of these matrices, it is possible to estimate the regularity of the scaling function using Littlewood-Paley theory (which divides the Fourier domain into dyadic blocks and uses norms on these dyadic blocks to characterize, for example, continuity). These methods are quite sophisticated and we refer to [57, 73] for details.

Finally, Rioul [239, 242] derived direct regularity estimates on the iterated filters which not only give sharp estimates but are quite intuitive. The idea is to consider iterated filters $g_0^{(i)}[n]$ and the maximum difference between successive coefficients. For continuity, it is clear that this difference has to go to zero. The normalization is now different because we consider the discrete-time sequences directly. Normalizing $G_0(z)$ such that $G_0(1) = 2$ and requiring again the necessary condition $G_0(-1) = 0$, we have

$$\lim_{i\to\infty} \max_n \left| g_0^{(i)}[n+1] - g_0^{(i)}[n] \right| \; = \; 0,$$

where $g_0^{(i)}[n]$ is the usual iterated sequence. For the limit function $\varphi(t)$ to be continuous, Rioul shows that the convergence has to be uniform in n and that the following bound has to be satisfied for a positive α:

$$\max \left| g_0^{(i)}[n+1] - g_0^{(i)}[n] \right| \; \leq \; C 2^{-i\alpha}.$$

Taking higher-order differences leads to testing differentiability as well [239, 242]. The elegance of this method is that it deals directly with the iterated sequences, and associates discrete successive differences with continuous-time derivatives in an intuitive manner. Because it is computationally oriented, it can be run easily on large filters.

4.4.4 Daubechies' Family of Regular Filters and Wavelets

To conclude the discussion of iterated filters and regularity, we give the explicit construction of Daubechies' family of orthonormal wavelets. For more details, the reader is referred to [71, 73]. Note that this is another derivation of the maximally flat orthogonal filters studied in Chapter 3. Recall that perfect reconstruction together with orthogonality can be expressed as (see Section 3.2.3)

$$|M_0(e^{j\omega})|^2 + |M_0(e^{j(\omega+\pi)})|^2 \; = \; 1, \tag{4.4.30}$$

where $M_0(e^{j\omega}) = G_0(e^{j\omega})/\sqrt{2}$ is normalized so that $M_0(1) = 1$ and we assume $M_0(\pi) = 0$. For regularity, the following is imposed on $M_0(e^{j\omega})$:

$$M_0(e^{j\omega}) \; = \; \left[\frac{1}{2}(1+e^{j\omega})\right]^N R(e^{j\omega})$$

where $N \geq 1$. Note that $R(1) = 1$ and that $|M_0(e^{j\omega})|^2$ can be written as

$$|M_0(e^{j\omega})|^2 = \left[\cos^2 \frac{\omega}{2}\right]^N |R(e^{j\omega})|^2. \tag{4.4.31}$$

Since $|R(e^{j\omega})|^2 = R(e^{j\omega}) \cdot R^*(e^{j\omega}) = R(e^{j\omega})R(e^{-j\omega})$, it can be expressed as a polynomial in $\cos\omega$ or of $\sin^2 \omega/2 = (1 - \cos\omega)/2$. Using the shorthands $y = \cos^2(\omega/2)$ and $P(1-y) = |R(e^{j\omega})|^2$, we can write (4.4.30) using (4.4.31) as

$$y^N P(1-y) + (1-y)^N P(y) = 1, \tag{4.4.32}$$

where

$$P(y) \geq 0 \text{ for } y \in [0,1]. \tag{4.4.33}$$

Suppose that we have a polynomial $P(y)$ satisfying (4.4.32) and (4.4.33) and moreover

$$\sup_\omega |R(e^{j\omega})| = \sup_{y\in[0,1]} |P(y)|^{\frac{1}{2}} < 2^{N-1}.$$

Then, there exists an orthonormal basis associated with $G_0(e^{j\omega})$, since the iterated filter will converge to a continuous scaling function (following Proposition 4.7) from which a wavelet basis can be obtained (Theorem 4.5).

Thus, the problem becomes to find $P(y)$ satisfying (4.4.32) and (4.4.33) followed by extracting $R(e^{j\omega})$ as the "root" of P. Daubechies shows [71, 73] that any polynomial P solving (4.4.32) is of the form

$$P(y) = \sum_{j=0}^{N-1} \binom{N-1+j}{j} y^j + y^N Q(y), \tag{4.4.34}$$

where Q is an antisymmetric polynomial. For the specific family in question, Daubechies constructs filters of minimum order, that is, with $Q \equiv 0$ (see also Problem 4.13). Note that such maximally flat filters (they have a maximum number of zeros at $\omega = \pi$) have been derived long before filter banks and wavelets by Herrmann [134], in the context of FIR filter design.

With such a P, the remaining task is to determine R. Using spectral factorization, one can construct such a R from a given P as explained in Section 2.5.2. Systematically choosing zeros inside the unit circle for $R(e^{j\omega})$ one obtains the minimum phase solution for $G_0(e^{j\omega})$. Choosing zeros inside and outside the unit circle leads to mixed phase filters. There is no linear phase solution (except the Haar case when $N = 1$ and $R(e^{j\omega}) = 1$).

Example 4.5

Let us illustrate the construction for the case $N = 2$. Using (4.4.34) with $N = 2, Q = 0$,

$$P(y) = 1 + 2y.$$

Table 4.2 Minimum degree remainder polynomials $R(z)$ such that $P(z) = 2^{-2N+1}$ $(1+z)^N$ $(1+z^{-1})^N$ $R(z)$ is valid.

N	Coefficients of $R(z)$
2	$2^{-1}[-1, 4, -1]$
3	$2^{-3}[3, -18, 38, -18, 3]$
4	$2^{-4}[-5, 40, -131, 208, -131, 40, -5]$
5	$2^{-7}[35, -350, 1520, -3650, 5018, -3650, 1520, -350, 35]$
6	$2^{-8}[-63, 756, -4067, 12768, -25374, 32216, -25374, 12768, -4067, 756, -63]$

From (4.4.32),

$$|R(e^{j\omega})|^2 = P(1-y) = 3 - 2\cos^2(\omega/2) = 2 - \cos\omega = 2 - \frac{1}{2}e^{j\omega} - \frac{1}{2}e^{-j\omega},$$

where we used $y = \cos^2\omega/2 = 1/2(1+\cos\omega)$. Now take the spectral factorization of $|R(e^{j\omega})|^2$. The roots are $r_1 = 2+\sqrt{3}$ and $r_2 = 2-\sqrt{3} = 1/r_1$. Thus

$$|R(e^{j\omega})|^2 = \frac{1}{4-2\sqrt{3}}[e^{j\omega} - (2-\sqrt{3})][e^{-j\omega} - (2-\sqrt{3})].$$

A possible $R(e^{j\omega})$ is therefore,

$$R(e^{j\omega}) = \frac{1}{\sqrt{3}-1}[e^{j\omega} - (2-\sqrt{3})] = \frac{1}{2}[(1+\sqrt{3})e^{j\omega} + 1 - \sqrt{3}]$$

and the resulting $M_0(e^{j\omega})$ is

$$\begin{aligned} M_0(e^{j\omega}) &= \left[\frac{1}{2}(1+e^{j\omega})\right]^2 \frac{1}{2}[(1+\sqrt{3})e^{j\omega} + 1 - \sqrt{3}] \\ &= \frac{1}{8}[(1+\sqrt{3})e^{j3\omega} + (3+\sqrt{3})e^{j2\omega} + (3-\sqrt{3})e^{j\omega} + 1 - \sqrt{3}]. \end{aligned}$$

This filter is the 4-tap Daubechies' filter (within a phase shift to make it causal and a scale factor of $1/\sqrt{2}$). That is, by computing the iterated filters and the associated continuous-time functions (see (4.4.12)- (4.4.13)), one obtains the D_2 wavelet and scaling function as shown in Figure 4.4. The regularity (continuity) of this filter was discussed after Proposition 4.7.

Figure 4.21 gives the iterated graphical functions for $N = 3, \ldots, 6$ (the eighth iteration is plotted and they converge to their corresponding scaling functions). Recall that the case $N = 2$ is given in Figure 4.4. Table 4.2 gives the $R(z)$ functions for $N = 2, \ldots, 6$, which can be factored into maximally regular filters. The lowpass filters obtained by a minimum phase factorization are given in Table 4.3. Table 4.4 gives the regularity of the first few Daubechies' filters.

This concludes our discussion of iterated filter bank constructions leading to wavelet bases. Other variations are possible by looking at other filter banks

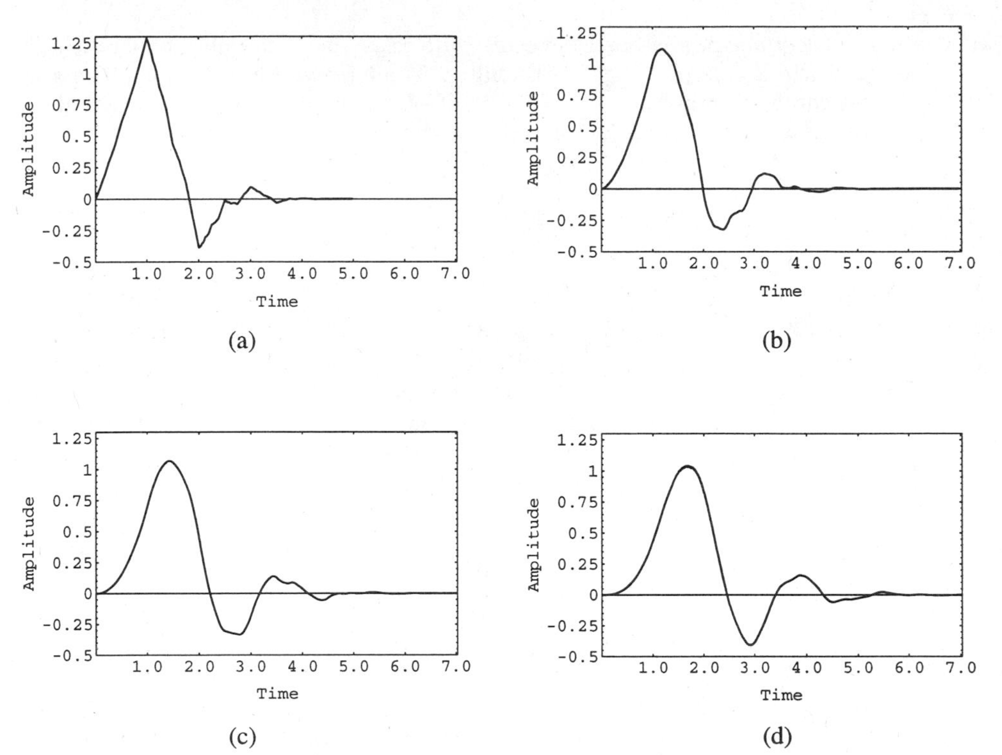

Figure 4.21 Daubechies' iterated graphical functions for $N = 3, \ldots, 6$ (eighth iteration is plotted and they converge to their corresponding scaling functions). Their regions of support are from 0 to $2N - 1$ and thus only for $N = 3, 4$, they are plotted in their entirety. For $N = 5, 6$, after $t = 7.0$, their amplitude is negligible. Recall that the case $N = 2$ is given in Figure 4.4. (a) $N = 3$. (b) $N = 4$. (c) $N = 5$. (d) $N = 6$.

such as biorthogonal filter banks or IIR filter banks. Assuming regularity, they lead to biorthogonal wavelet bases with compact support and wavelets with exponential decay (see Section 4.6 for more details).

4.5 Wavelet Series and Its Properties

Until now, we have seen ways of building orthonormal bases with structure. It was shown how such bases arise naturally from the multiresolution framework. We also discussed ways of constructing these bases; both directly, in the Fourier domain, and starting from discrete-time bases — filter banks.

The aim in this section is to define the wavelet series expansion together with its properties, enumerate some general properties of the basis functions, and demonstrate how one computes wavelet series expansion of a function.

Table 4.3 First few maximally flat Daubechies' filters. N is the number of zeros at $\omega = \pi$ and equals $L/2$ where L is the length of the filter. The lowpass filter $g_0[n]$ is given and the highpass filter can be obtained as $g_1[n] = (-1)^n g_0[-n + 2N - 1]$. These are obtained from a minimum phase factorization of $P(z)$ corresponding to Table 4.2.

$g_0[n]$	N = 2	N = 3	N = 4	N = 5	N = 6
$g_0[0]$	0.48296291	0.33267	0.230377813309	0.16010239	0.111540743350
$g_0[1]$	0.8365163	0.806891	0.714846570553	0.60382926	0.494623890398
$g_0[2]$	0.22414386	0.459877	0.630880767930	0.724308528	0.751133908021
$g_0[3]$	-0.129409522	-0.135011	-0.027983769417	0.13842814	0.315250351709
$g_0[4]$		-0.08544	-0.187034811719	-0.24229488	-0.226264693965
$g_0[5]$		0.03522	0.030841381836	-0.03224486	-0.129766867567
$g_0[6]$			0.032883011667	0.07757149	0.097501605587
$g_0[7]$			-0.010597401785	-0.00624149	0.027522865530
$g_0[8]$				-0.01258075	-0.031582039318
$g_0[9]$				0.003335725	0.000553842201
$g_0[10]$					0.004777257511
$g_0[11]$					-0.001077301085

Table 4.4 Hölder regularity estimates for the first few Daubechies' filters (from [73]). The estimates given below are lower bounds. For example, for $N = 3$, finer estimates show that the function is actually differentiable [73].

N	$\alpha(N)$
2	0.500
3	0.915
4	1.275
5	1.596
6	1.888

4.5.1 Definition and Properties

DEFINITION 4.8

Assuming a multiresolution analysis defined by Axioms (4.2.1–4.2.6) and the mother wavelet $\psi(t)$ given in (4.2.14), any function $f \in L_2(\mathcal{R})$ can be expressed as

$$f(t) = \sum_{m,n \in \mathcal{Z}} F[m,n]\, \psi_{m,n}(t), \tag{4.5.1}$$

where

$$F[m,n] = \langle \psi_{m,n}(t), f(t) \rangle = \int_{-\infty}^{\infty} \psi_{m,n}(t)\, f(t) dt. \tag{4.5.2}$$

We have assumed a real wavelet (otherwise, a conjugate is necessary). Equation (4.5.2) is the analysis and (4.5.1) the synthesis formula. We will list several important properties of the wavelet series expansion.

Linearity Suppose that the operator T is defined as

$$T[f(t)] = F[m,n] = \langle \psi_{m,n}(t), f(t) \rangle.$$

Then for any $a, b \in \mathcal{R}$

$$T[a\, f(t) + b\, g(t)] = a\, T[f(t)] + b\, T[g(t)],$$

that is, the wavelet series operator is linear. Its proof follows from the linearity of the inner product.

Shift Recall that the Fourier transform has the following shift property: If a signal and its Fourier transform pair are denoted by $f(t)$ and $F(\omega)$ respectively, then the signal $f(t-\tau)$ will have $e^{-j\omega\tau}F(\omega)$ as its Fourier transform (see Section 2.4.2).

Consider now what happens in the wavelet series case. Suppose that the function and its transform coefficient are denoted by $f(t)$ and $F[m,n]$ respectively. If we shift the signal by τ, that is, $f(t-\tau)$,

$$\begin{aligned} F'[m,n] &= \int_{-\infty}^{\infty} \psi_{m,n}(t)\, f(t-\tau) dt \\ &= \int_{-\infty}^{\infty} 2^{-m/2}\psi(2^{-m}t - n + 2^{-m}\tau)\, f(t) dt. \end{aligned}$$

For the above to be a coefficient from the original transform $F[m,n]$, one must have that

$$2^{-m}\tau \in \mathcal{Z},$$

or $\tau = 2^m k, k \in \mathcal{Z}$. Therefore, the wavelet series expansion possesses the following shift property: If a signal and its transform coefficient are denoted by $f(t)$ and $F[m,n]$, then the signal $f(t-\tau)$, $\tau = 2^m k$, $k \in \mathcal{Z}$, will have $F[m', n - 2^{-m'}\tau]$, $m' \leq m$ as its transform coefficient, that is,

$$f(t - 2^m k) \longleftrightarrow F[m', n - 2^{m-m'}k], \qquad k \in \mathcal{Z}, \quad m' \leq m.$$

Thus, if a signal has a scale-limited expansion

$$f(t) = \sum_{n \in \mathcal{Z}} \sum_{m=-\infty}^{M_2} F[m,n]\, \psi_{m,n}(t),$$

then this signal will possess the weak shift property with respect to the shifts by $2^{M_2}k$, that is

$$f(t - 2^{M_2}k) \longleftrightarrow F[m, n - 2^{M_2 - m}k], \; -\infty \leq m \leq M_2.$$

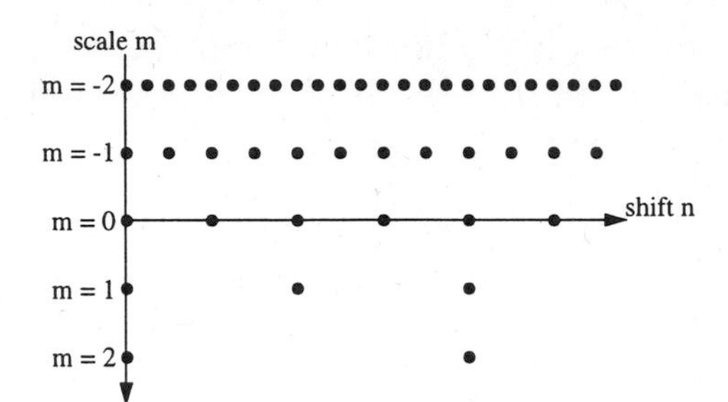

Figure 4.22 Dyadic sampling of the time-frequency plane in the wavelet series expansion. The dots indicate the center of the wavelets $\psi_{m,n}(t)$.

Scaling Recall the scaling property of the Fourier transform: If a signal and its Fourier transform pair are denoted by $f(t)$ and $F(\omega)$, then the scaled version of the signal $f(at)$ will have $(1/|a|) \cdot F(\omega/a)$ as its transform (see Section 2.4.2).

The wavelet series expansion $F'[m,n]$ of $f'(t) = f(at),\ a > 0$, is

$$\begin{aligned} F'[m,n] &= \int_{-\infty}^{\infty} \psi_{m,n}(t)\ f(at)\ dt \\ &= \frac{1}{a}\int_{-\infty}^{\infty} 2^{-m/2}\psi\left(\frac{2^{-m}t}{a} - n\right) f(t)\ dt. \end{aligned}$$

Thus, when $2^{-m}/a = 2^{-p},\ p \in \mathcal{Z}$, or $a = 2^{-k},\ k \in \mathcal{Z}$, then $F'[m,n]$ can be obtained from $F[m,n]$, the wavelet transform of $f(t)$:

$$f(2^{-k}t) \longleftrightarrow 2^{k/2}F[m-k,n], \qquad k \in \mathcal{Z}.$$

Scaling by factors which are not powers of two require reinterpolation. That is, either one reinterpolates the signal and then takes the wavelet expansion, or some interpolation of the wavelet series coefficients is made. The former method is more immediate.

Parseval's Identity The Parseval's identity, as seen for the Fourier-type expansions (see Section 2.4), holds for the wavelet series as well. That is, the orthonormal family $\{\psi_{m,n}\}$ satisfies (see Theorem 4.5)

$$\sum_{m,n\in\mathcal{Z}} |\langle\psi_{m,n}, f\rangle|^2 = \|f\|^2, \qquad f \in L_2(\mathcal{R}).$$

Dyadic Sampling and Time-Frequency Tiling When considering a series expansion, it is important to locate the basis functions in the time-frequency plane. The sampling in time, at scale m, is done with a period of 2^m, since $\psi_{m,n}(t) = \psi_{m,0}(t - 2^m n)$. In scale, powers of two are considered. Since frequency is the inverse of scale, we find that if the wavelet is centered around ω_0, then $\Psi_{m,n}(\omega)$ is

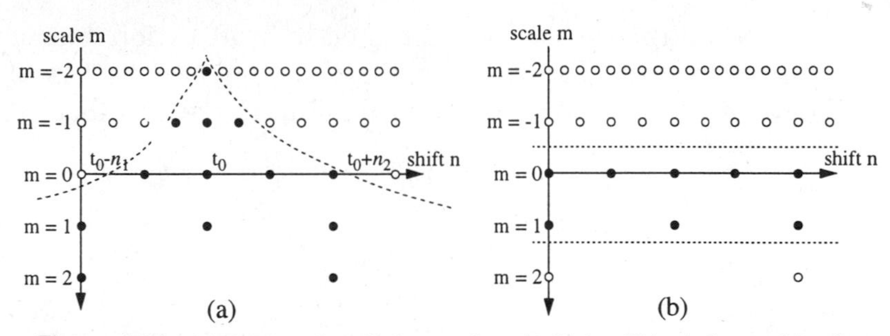

Figure 4.23 (a) Region of coefficients $F[m, n]$ which will be influenced by the value of the function at t_0. (b) Region of influence of the Fourier component $F(\omega_0)$.

centered around $\omega_0/2^m$. This leads to a *dyadic sampling* of the time-frequency plane, as shown in Figure 4.22. Note that the scale axis (or inverse frequency) is logarithmic. On a linear scale, we have the equivalent time-frequency tiling as was shown in Figure 2.12(d).

Localization One of the reasons why wavelets are so popular is due to their ability to have good time and frequency localization. We will discuss this next.

Time Localization Suppose that one is interested in the signal around $t = t_0$. Then a valid question is: Which values $F[m, n]$ will carry some information about the signal $f(t)$ at t_0, that is, which region of the (m, n) grid will give information about $f(t_0)$?

Suppose a wavelet $\psi(t)$ is compactly supported on the interval $[-n_1, n_2]$. Thus, $\psi_{m,0}(t)$ is supported on $[-n_1 2^m, n_2 2^m]$ and $\psi_{m,n}(t)$ is supported on $[(-n_1 + n)2^m, (n_2 + n)2^m]$. Therefore, at scale m, wavelet coefficients with index n satisfying

$$(-n_1 + n)2^m \ \le \ t_0 \ \le \ (n_2 + n)2^m,$$

will be influenced. This can be rewritten as

$$2^{-m}t_0 - n_2 \ \le \ n \ \le \ 2^{-m}t_0 + n_1.$$

Figure 4.23(a) shows this region on the (m, n) grid.

The converse question is: Given a point $F[m_0, n_0]$ in the wavelet series expansion, which region of the signal contributed to it? From the support of $\psi_{m,n}(t)$, it follows that $f(t)$ for t satisfying

$$(-n_1 + n_0)2^{m_0} \ \le \ t \ \le \ (n_2 + n_0)2^{m_0}$$

influences $F[m_0, n_0]$.

Frequency Localization Suppose we are interested in localization, but now in the frequency domain. Since the Fourier transform of $\psi_{m,n}(t) = 2^{-m/2}\psi(2^{-m}t - n)$ is $2^{m/2} \cdot \Psi(2^m\omega) \cdot e^{-j2^m n\omega}$, we can write $F[m,n]$ using Parseval's formula as

$$\begin{aligned} F[m,n] &= \int_{-\infty}^{\infty} \psi_{m,n}(t)\, f(t)\, dt \\ &= \frac{1}{2\pi} 2^{m/2} \int_{-\infty}^{\infty} F(\omega)\, \Psi^*(2^m\omega)\, e^{j2^m n\omega}\, d\omega. \end{aligned}$$

Now, suppose that a wavelet $\psi(t)$ vanishes in the Fourier domain outside the region $[\omega_{min}, \omega_{max}]$.[6] At scale m, the support of $\Psi_{m,n}(\omega)$ will be $[\omega_{min}/2^m, \omega_{max}/2^m]$. Therefore, a frequency component at ω_0 influences the wavelet series at scale m if

$$\frac{\omega_{\min}}{2^m} \le \omega_0 \le \frac{\omega_{\max}}{2^m}$$

is satisfied or if the following range of scales is influenced:

$$\log_2\left(\frac{\omega_{\min}}{\omega_0}\right) \le m \le \log_2\left(\frac{\omega_{\max}}{\omega_0}\right).$$

This is shown in Figure 4.23(b). Conversely, given a scale m_0, all frequencies of the signal between $\omega_{\min}/2^{m_0}$ and $\omega_{\max}/2^{m_0}$ will influence the expansion at that scale.

Existence of Scale-Limited Signals Because of the importance of bandlimited signals in signal processing, a natural question is: Are there any scale-limited signals? An easy way to construct such a signal would be to add, for example, Haar wavelets from a range of scales $m_0 \le m \le m_1$. Thus, the wavelet series expansion will posses a limited number of scales; or transform coefficients $F[m,n]$ will exist only for $m_0 \le m \le m_1$.

However, note what happens with the signal $f(t - \varepsilon)$, for ε not a multiple of 2^{m_1}. The scale-limitedness property is lost, and the expansion can have an infinite number of coefficients. For more details, see [116] and Problem 4.1. Note that the sinc wavelet expansion does not have this problem, since it is intrinsically band/scale limited.

Characterization of Singularities The Fourier transform and Fourier series can be used to characterize the regularity of a signal by looking at the decay of the transform or series coefficients (see Appendix 2.C.2). One can use the wavelet transform and the wavelet series behavior in a similar way. There is one notable advantage over the Fourier case, however, in that one can characterize *local*

[6]Therefore, the wavelet cannot be compactly supported. However, the discussion holds approximately for wavelets which have most of their energy in the band $[\omega_{min}, \omega_{max}]$.

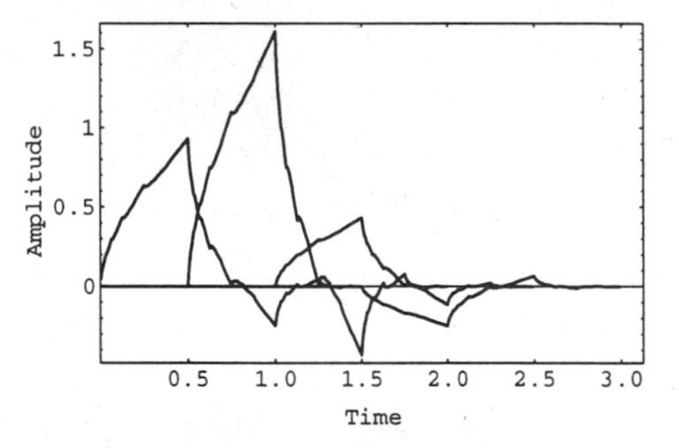

Figure 4.24 Two-scale equation for the D_2 scaling function given in Figure 4.4(a).

regularity. Remember that the Fourier transform gives a global characterization only. The wavelet transform and the wavelet series, because of the fact that high frequency basis functions become arbitrarily sharp in time, allow one to look at the regularity at a particular location independent of the regularity elsewhere. This property will be discussed in more detail for the continuous-time wavelet transform in Chapter 5. The basic properties of regularity characterization carry over to the wavelet series case since it is a sampled version of the continuous wavelet transform, and since the sampling grid becomes arbitrarily dense at high frequencies (we consider "well-behaved" functions only, that is, of bounded variation).

In a dual manner, we can make statements about the decay of the wavelet series coefficients depending on the regularity of the analyzed signal. This gives a way to quantify the approximation property of the wavelet series expansion for a signal of a given regularity. Again, the approximation property is local (since regularity is local).

Note that in all these discussions, one assumes that the wavelet is more regular than the signal (otherwise, the wavelet's regularity interferes). Also, because of the sampling involved in the wavelet series, one might have to go to very fine scales in order to get good estimates. Therefore, it is easier to use the continuous wavelet transform or a highly oversampled discrete-time wavelet transform (see Chapter 5 and [73]).

4.5.2 Properties of Basis Functions

Let us summarize some of the important properties of the wavelet series basis functions. While some of them (such as the two-scale equation property) have been seen earlier, we will summarize them here for completeness.

Two-Scale Equation Property The scaling function can be built from itself (see Figure 4.24). Recall the definition of a multiresolution analysis. The scaling function $\varphi(t)$ belongs to V_0. However, since $V_0 \subset V_{-1}$, $\varphi(t)$ belongs to V_{-1} as well. We know that $\varphi(t-n)$ is an orthonormal basis for V_0 and thus, $\sqrt{2}\varphi(2t-n)$ is an orthonormal basis for V_{-1}. This means that any function from V_0, including $\varphi(t)$, can be expressed as a linear combination of the basis functions from V_{-1}, that is, $\varphi(2t-n)$. This leads to the following two-scale equation

$$\varphi(t) = \sqrt{2}\sum_n g_0[n]\, \varphi(2t-n). \tag{4.5.3}$$

On the other hand, using the same argument for the wavelet $\psi(t) \in W_0 \subset V_{-1}$, one can see that

$$\psi(t) = \sqrt{2}\sum_n g_1[n]\, \varphi(2t-n). \tag{4.5.4}$$

These two relations can be expressed in the Fourier domain as

$$\Phi(\omega) = \frac{1}{\sqrt{2}}\sum_n g_0[n]e^{-jn(\omega/2)}\,\Phi\left(\frac{\omega}{2}\right) = M_0\left(\frac{\omega}{2}\right)\Phi\left(\frac{\omega}{2}\right), \tag{4.5.5}$$

$$\Psi(\omega) = \frac{1}{\sqrt{2}}\sum_n g_1[n]e^{-jn(\omega/2)}\,\Phi\left(\frac{\omega}{2}\right) = M_1\left(\frac{\omega}{2}\right)\Phi\left(\frac{\omega}{2}\right). \tag{4.5.6}$$

As an illustration, consider the two-scale equation in the case of the Daubechies' scaling function. Figure 4.24 shows how the D_2 scaling function is built using four scaled and shifted versions of itself.

The functions $M_0(\omega)$ and $M_1(\omega)$ in (4.5.5) and (4.5.6) are 2π-periodic functions and correspond to scaled versions of filters $g_0[n]$, $g_1[n]$ (see (4.4.5) and (4.4.8)) which can be used to build filter banks (see Section 4.5.2 below).

The two-scale equation can also be used as a starting point in constructing a multiresolution analysis. In other words, instead of starting from an axiomatic definition of a multiresolution analysis, choose $\varphi(t)$ such that (4.5.3) holds, with $\sum_n |g_0[n]|^2 < \infty$ and $0 < A \le \sum_n |\Phi(\omega + 2\pi n)|^2 \le B < \infty$. Then define V_m to be the closed subspace spanned by $2^{-m/2}\varphi(2^{-m}t - n)$. All the other axioms follow (an orthogonalization step is involved if $\varphi(t)$ is not orthogonal to its integer translates). For more details, refer to [73].

Moment Properties of Wavelets Recall that the lowpass filter $g_0[n]$, in an iterated filter bank scheme, has at least one zero at $\omega = \pi$ and thus, $g_1[n]$ has at least one zero at $\omega = 0$. Since $\Phi(0) = 1$ (from the normalization of $M_0(\omega)$) it follows that $\Psi(\omega)$ has at least one zero at $\omega = 0$. Therefore,

$$\int_{-\infty}^{\infty} \psi(t)\, dt = \Psi(0) = 0,$$

which is to be expected since $\psi(t)$ is a bandpass function. In general, if $G_0(e^{j\omega})$ has an Nth-order zero at $\omega = \pi$, the wavelet $\Psi(\omega)$ has an Nth-order zero at

$\omega = 0$. Using the moment theorem of the Fourier transform (see Section 2.4.2), it follows that

$$\int_{-\infty}^{\infty} t^n \psi(t)\, dt \;=\; 0, \qquad n = 0, \ldots, N-1,$$

that is, the first N moments of the wavelet are zero. Besides wavelets constructed from iterated filter banks, we have seen Meyer's and Battle-Lemarié wavelets. Meyer's wavelet, which is not based on the iteration of a rational function, has by construction an infinite "zero" at the origin, that is, an infinite number of zero moments. The Battle-Lemarié wavelet, on the other hand, is based on the Nth-order B-spline function. The orthogonal filter $G_0(e^{j\omega})$ has an $(N+1)$th-order zero at π (see (4.3.18)) and the wavelet thus has $N+1$ zero moments.

The importance of zero moments comes from the following fact. Assume a length L wavelet with N zero moments. Assume that the function $f(t)$ to be represented by the wavelet series expansion is polynomial of order $N-1$ in an interval $[t_0, t_1]$. Then, for sufficiently small scales (such that $2^m L < (t_1 - t_0)/2$) the wavelet expansion coefficients will automatically vanish in the region corresponding to $[t_0, t_1]$ since the inner product with each term of the polynomial will be zero. Another view is to consider the Taylor expansion of a function around a point t_0,

$$f(t_0 + \epsilon) \;=\; f(t_0) + \frac{f'(t_0)}{1!}\epsilon + \frac{f''(t_0)}{2!}\epsilon^2 + \ldots.$$

The wavelet expansion around t_0 now depends only on the terms of degree N and higher of the Taylor expansion since the terms 0 through $N-1$ are zeroed out because of the N zero moments of the wavelet. If the function is smooth, the high-order terms of the Taylor expansion are very small. Because the wavelet series coefficients now depend only on Taylor coefficients of order N and larger, they will be very small as well.

These approximation features of wavelets with zero moments are important in approximation of smooth functions and operators and also in signal compression (see Chapter 7).

Smoothness and Decay Properties of Wavelets In discussing the iterated filters leading to the Daubechies' wavelets, we pointed out that besides convergence, continuity or even differentiability of the wavelet was often desirable. While this regularity of the wavelet is linked to the number of zeros at $\omega = \pi$ of the lowpass filter $G_0(e^{j\omega})$, the link is not as direct as in the case of the zero-moment property seen above. In particular, there is no direct relation between these two properties.

The regularity of all the wavelets discussed so far is indicated in Table 4.5. Regularity r means that the rth derivative exists almost everywhere. The

Table 4.5 Zero moments, regularity, and decay of various wavelets. $\alpha(N)$ is a linearly increasing function of N which approaches $0.2075 \cdot N$ for large N. The Battle-Lemarié wavelet of order N is based on a B-spline of order $N-1$. The Daubechies' wavelet of order N corresponds to a length-$2N$ maximally flat orthogonal filter.

Wavelet	# of zero moments	regularity r	decay or support in time	decay or support in frequency
Haar	1	0	[0,1]	$1/\omega$
Sinc	∞	∞	$1/t$	$[\pi, 2\pi]$
Meyer	∞	∞	1/poly.	$[2\pi/3, 8\pi/3]$
Battle-Lemarié	N	N	exponential	$1/\omega^N$
Daubechies N	N	$\lfloor \alpha(N) \rfloor$	$[0, 2N-1]$	$1/\omega^{\alpha(N)}$

localization or decay in time and frequency of all these wavelets is also indicated in the table.

Filter Banks Obtained from Wavelets Consider again (4.5.3) and (4.5.4). An interesting fact is that using the coefficients $g_0[n]$ and $g_1[n]$ for the synthesis lowpass and highpass filters respectively, one obtains a perfect reconstruction orthonormal filter bank (as defined in Section 3.2.3). To check the orthonormality conditions for these filters use the orthonormality conditions of the scaling function and the wavelet. Thus, start from

$$\langle \varphi(t+l), \varphi(t+k) \rangle = \delta[k-l],$$

or

$$\begin{aligned}\langle \varphi(t+l), \varphi(t+k) \rangle &= \left\langle \sum_n g_0[n]\, \varphi(2t+2l-n), \sum_m g_0[m]\, \varphi(2t+2k-m) \right\rangle \\ &= \left\langle \sum_{n'} g_0[n'+2l]\, \varphi(2t-n'), \sum_{m'} g_0[m'+2k]\, \varphi(2t-m') \right\rangle \\ &= \frac{1}{2} \sum_{n'} g_0[n'+2l]\, g_0[n'+2k] = \delta[l-k],\end{aligned}$$

that is, the lowpass is orthogonal to its even translates. In a similar fashion, one can show that the lowpass filter is orthogonal to the highpass and its even translates. The highpass filter is orthogonal to its even translates as well. That is, $\{g_i[n-2k]\}$, $i = 0, 1$, is an orthonormal set, and it can be used to build an orthogonal filter bank (see Section 3.2.3).

4.5.3 Computation of the Wavelet Series and Mallat's Algorithm

An attractive feature of the wavelet series expansion is that the underlying multiresolution structure leads to an efficient discrete-time algorithm based on a filter bank implementation. This connection was pointed out by Mallat [181]. The computational procedure is therefore referred to as *Mallat's algorithm*.

Assume we start with a function $f(t) \in V_0$ and we are given the sequence $f^{(0)}[n] = \langle \varphi(t-n), f(t) \rangle, n \in \mathcal{Z}$. That is

$$f(t) = \sum_{n=-\infty}^{\infty} f^{(0)}[n]\, \varphi(t-n). \tag{4.5.7}$$

We also assume that the axioms of multiresolution analysis hold. In searching for projections of $f(t)$ onto V_1 and W_1, we use the fact that $\varphi(t)$ and $\psi(t)$ satisfy two-scale equations. Consider first the projection onto V_1, that is

$$f^{(1)}[n] = \left\langle \frac{1}{\sqrt{2}}\, \varphi\left(\frac{t}{2} - n\right),\, f(t) \right\rangle. \tag{4.5.8}$$

Because $\varphi(t) = \sqrt{2} \sum_k g_0[k]\, \varphi(2t-k)$,

$$\frac{1}{\sqrt{2}}\, \varphi\left(\frac{t}{2} - n\right) = \sum_k g_0[k]\, \varphi(t - 2n - k). \tag{4.5.9}$$

Thus, (4.5.8) becomes

$$f^{(1)}[n] = \sum_k g_0[k]\, \langle \varphi(t - 2n - k), f(t) \rangle,$$

and using (4.5.7),

$$f^{(1)}[n] = \sum_k \sum_l g_0[k]\, f^{(0)}[l]\, \langle \varphi(t-2n-k), \varphi(t-l) \rangle. \tag{4.5.10}$$

Because of the orthogonality of $\varphi(t)$ with respect to its integer translates, the inner product in the above equation is equal to $\delta[l - 2n - k]$. Therefore, only the term with $l = 2n - k$ is kept from the second summation. With a change of variable, we can write (4.5.10) as

$$f^{(1)}[n] = \sum_k g_0[k - 2n]\, f^{(0)}[k].$$

With the definition $\tilde{g}_0[n] = g_0[-n]$, we obtain

$$f^{(1)}[n] = \sum_k \tilde{g}_0[2n - k]\, f^{(0)}[k], \tag{4.5.11}$$

that is, the coefficients of the projection onto V_1 are obtained by filtering $f^{(0)}$ with $\tilde{g}_0$ and downsampling by 2. To calculate the projection onto W_1, we use

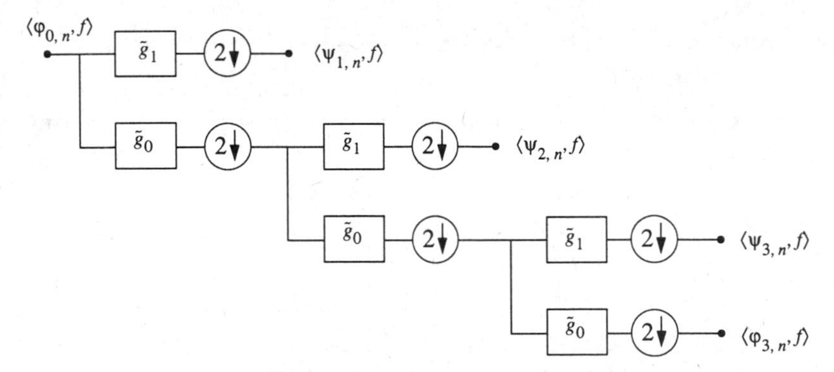

Figure 4.25 Computation of the wavelet series coefficients. Starting with the coefficients $f^{(0)}[n] = \langle \varphi(t-n), f(t) \rangle, n \in \mathcal{Z}$, we obtain the wavelet expansion coefficient by a filter bank algorithm.

the fact that $\psi(t) = \sqrt{2} \sum_k g_1[k] \cdot \varphi(2t-k)$. Calling $d^{(1)}[n]$ the coefficients of the projection onto W_1, or

$$d^{(1)}[n] = \langle \frac{1}{\sqrt{2}} \psi\left(\frac{t}{2} - n\right),\ f(t) \rangle,$$

and using the two-scale equation for $\psi(t)$ as well as the expansion for $f(t)$ given in (4.5.7), we find, similarly to (4.5.9–4.5.11)

$$\begin{aligned} d^{(1)}[n] &= \sum_k \sum_l g_1[k]\ f^{(0)}[l]\ \langle \varphi(t-2n-k), \varphi(t-l) \rangle \\ &= \sum_k \sum_l g_1[k]\ f^{(0)}[l]\ \delta[l-2n-k] \\ &= \sum_l g_1[l-2n]\ f^{(0)}[l] = \sum_l \tilde{g}_1(2n-l)\ f^{(0)}[l], \end{aligned}$$

where $\tilde{g}_1[n] = g_1[-n]$. That is, the coefficients of the projection onto W_1 are obtained by filtering $f^{(0)}$ with $\tilde{g}_1$ and downsampling by 2, exactly as we obtained the projection onto V_1 using $\tilde{g}_0$. Of course, projections onto V_2 and W_2 can be obtained similarly from filtering $f^{(1)}$ and downsampling by 2. Therefore, the projections onto W_m, $m = 1, 2, 3, \ldots$ are obtained from $m-1$ filtering with $\tilde{g}_0[n]$ followed by downsampling by 2, as well as a final filtering by $\tilde{g}_1[n]$ and downsampling. This purely discrete-time algorithm to implement the wavelet series expansion is depicted in Figure 4.25.

A key question is how to obtain an orthogonal projection $\hat{f}(t)$ onto V_0 from an arbitrary signal $f(t)$. Because $\{\varphi(t-n)\}$ is an orthonormal basis for V_0, $\hat{f}(t)$ equals

$$\hat{f}(t) = \sum_n \langle \varphi(t-n), f(t) \rangle\ \varphi(t-n),$$

and $\hat{f}(t) - f(t)$ is orthogonal to $\varphi(t-n), n \in \mathcal{Z}$. Thus, given an initial signal $f(t)$, we have to compute the set of inner products $f^{(0)}(n) = \langle \varphi(t-n), f(t) \rangle$. This, unlike the further decomposition which involves only discrete-time processing, requires continuous-time processing. However, if V_0 corresponds to sufficiently fine resolution compared to the resolution of the input signal $f(t)$, than sampling $f(t)$ will be sufficient. This follows because $\varphi(t)$ is a lowpass filter with an integral equal to 1. If $f(t)$ is smooth and $\varphi(t)$ is sufficiently short-lived, then we have

$$\langle \varphi(t-n), f(t) \rangle \simeq f(n).$$

Of course, if V_0 is not fine enough, one can start with V_{-m} for m sufficiently large so that

$$\langle 2^{\frac{m}{2}} \varphi(2^m t - n), f(t) \rangle \simeq 2^{-m/2} f(2^{-m} n).$$

If $f(t)$ has some regularity (for example, it is continuous), there will be a resolution at which sampling is a good enough approximation of the inner products needed to begin Mallat's algorithm. Generalizations of Mallat's algorithm, which include more general initial approximation problems, are derived in [261] and [296].

4.6 GENERALIZATIONS IN ONE DIMENSION

In this section, we discuss some of the more common generalizations in one dimension, most notably, the biorthogonal and recursive filter cases, as well as wavelets obtained from multichannel filter banks. For treatment of wavelets with rational dilation factors see [16] and [33].

4.6.1 Biorthogonal Wavelets

Instead of orthogonal wavelet families, one can construct biorthogonal ones, that is, the wavelet used for the analysis is different from the one used at the synthesis [58]. Basically, we relax the orthogonality requirement used so far in this chapter. However, we still maintain the requirement that the set of functions $\psi_{m,n}$ or $\tilde{\psi}_{k,l}$ are linearly independent and actually form a basis. In Chapter 5, this requirement will be relaxed, and we will work with linearly dependent sets or frames. Calling $\{\psi_{m,n}(t)\}$ and $\{\tilde{\psi}_{m,n}(t)\}$[7] the families used at synthesis and analysis respectively (m and n stand for dilation and shift) then, in a biorthogonal family, the following relation is satisfied:

$$\langle \psi_{m,n}(t), \tilde{\psi}_{k,l}(t) \rangle = \delta[m-k]\, \delta[n-l]. \tag{4.6.1}$$

[7]Note that here, the " ˜ " does not denote time reversal, but is used for a dual function.

If in addition the family is complete in a given space such as $L_2(\mathcal{R})$, then any function of the space can be written as

$$f(t) = \sum_m \sum_n \langle \psi_{m,n}, f \rangle \, \tilde{\psi}_{m,n}(t) \tag{4.6.2}$$

$$= \sum_m \sum_n \langle \tilde{\psi}_{m,n}, f \rangle \, \psi_{m,n}(t), \tag{4.6.3}$$

since ψ and $\tilde{\psi}$ play dual roles. There are various ways to find such biorthogonal families. For example, one could construct a biorthogonal spline basis by simply not orthogonalizing the Battle-Lemarié wavelet.

Another approach consists in starting with a biorthogonal filter bank and using the iterated filter bank method just as in the orthogonal case. Now, both the analysis and the synthesis filters (which are not just time-reversed versions of each other) have to be iterated. For example, one can use finite-length linear phase filters and obtain wavelets with symmetries and compact support (which is impossible in the orthogonal case).

In a biorthogonal filter bank with analysis/synthesis filters $H_0(z)$, $H_1(z)$, $G_0(z)$, and $G_1(z)$, perfect reconstruction with FIR filters means that (see (3.2.21))

$$G_0(z)H_0(z) + G_0(-z)H_0(-z) = 2 \tag{4.6.4}$$

and

$$H_1(z) = -z^{2k+1} G_0(-z), \tag{4.6.5}$$

$$G_1(z) = z^{-2k-1} H_0(-z) \tag{4.6.6}$$

following (3.2.18), where $\det(H_m(z)) = 2z^{2k+1}$ (we assume noncausal analysis filters in this discussion). Now, given a polynomial $P(z)$ satisfying $P(z) + P(-z) = 2$, we can factor it into $P(z) = G_0(z)H_0(z)$ and use $\{H_0(z), G_0(z)\}$ as the analysis/synthesis lowpass filters of a biorthogonal perfect reconstruction filter bank (the highpass filters follow from (4.6.5–4.6.6).

We can iterate such a biorthogonal filter bank on the lowpass channel and find equivalent iterated filter impulse responses. Note that now, analysis and synthesis impulse responses are not simply time-reversed versions of each other (as in the orthogonal case), but are typically very different (since they depend on $H_0(z)$ and $G_0(z)$, respectively). We can define the iterated lowpass filters as

$$H_0^{(i)}(z) = \prod_{k=0}^{i-1} H_0(z^{2^k}),$$

$$G_0^{(i)}(z) = \prod_{k=0}^{i-1} G_0(z^{2^k}).$$

For the associated limit functions to converge, it is necessary that both $H_0(z)$ and $G_0(z)$ have a zero at $z = -1$ (see Proposition 4.6). Therefore, following (4.6.4), we have that

$$G_0(1)\, H_0(1) = \left(\sum_n g_0[n]\right)\left(\sum_n h_0[n]\right) = 2.$$

That is, we can "normalize" the filters such that

$$\sum_n g_0[n] = \sum_n h_0[n] = \sqrt{2}.$$

This is necessary for the iteration to be well-defined (there is no square normalization as in the orthogonal case). Define

$$\tilde{M}_0(\omega) = \frac{H_0(e^{j\omega})}{\sqrt{2}}, \qquad M_0(\omega) = \frac{G_0(e^{j\omega})}{\sqrt{2}}$$

and the associated limit functions

$$\tilde{\Phi}(\omega) = \prod_{k=1}^{\infty} \tilde{M}_0\left(\frac{\omega}{2^k}\right)$$

$$\Phi(\omega) = \prod_{k=1}^{\infty} M_0\left(\frac{\omega}{2^k}\right)$$

where the former is the scaling function at analysis (within time reversal) and the latter is the scaling function at synthesis. These two scaling functions can be very different, as shown in Example 4.6.

Example 4.6

Consider a biorthogonal filter bank with length-4 linear phase filters. This is a one-parameter family with analysis and synthesis lowpass filters given by ($\alpha \neq \pm 1$):

$$H_0(z) = \frac{1}{\sqrt{2}(\alpha+1)}\,(1 + \alpha z + \alpha z^2 + z^3),$$

$$G_0(z) = \frac{1}{\sqrt{2}(\alpha-1)}\,(-1 + \alpha z^{-1} + \alpha z^{-2} - z^{-3}).$$

In Figure 4.26 we show the iteration of the filter $H_0(z)$ for a range of values α. Looking at the iterated filter for α and $-\alpha$, one can see that there is no solution having both a regular analysis and a regular synthesis filter. For example, for $\alpha = 3$, the analysis filter converges to a quadratic spline function, while the iterated synthesis filter exhibits fractal behavior and no regularity.

In order to derive the biorthogonal wavelet family, we define

$$\tilde{M}_1(\omega) = \frac{H_1(e^{j\omega})}{\sqrt{2}}, \qquad M_1(\omega) = \frac{G_1(e^{j\omega})}{\sqrt{2}}, \tag{4.6.7}$$

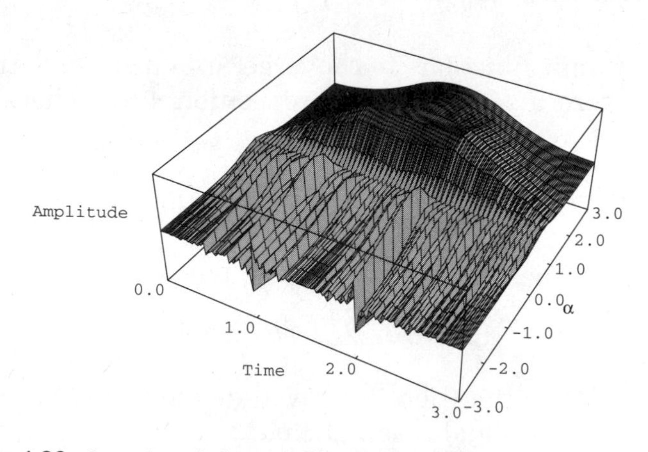

Figure 4.26 Iteration of a lowpass filter with impulse response $c_\alpha \cdot [1, \alpha, \alpha, 1]$ for $\alpha \in [-3 \ldots 3]$. The sixth iteration is shown. For $\alpha = 3$, the iteration converges to a quadratic spline. Note that for $\alpha = 0$, there is no convergence and $\alpha = \pm 1$ does not correspond to a biorthogonal filter bank.

as well as (similarly to (4.4.15))

$$\begin{aligned} \tilde{\psi}(\omega) &= \tilde{M}_1\left(\frac{\omega}{2}\right)\prod_{k=2}^{\infty}\tilde{M}_0\left(\frac{\omega}{2^k}\right), \\ \psi(\omega) &= M_1\left(\frac{\omega}{2}\right)\prod_{k=2}^{\infty}M_0\left(\frac{\omega}{2^k}\right). \end{aligned} \qquad (4.6.8)$$

Note that the regularity of the wavelet is the same as that of the scaling function (we assume FIR filters). Except that we define scaling functions and wavelets as well as their duals, the construction is analogous to the orthogonal case. The biorthogonality relation (4.6.1) can be derived similarly to the orthogonal case (see Proposition 4.4), but using properties of the underlying biorthogonal filter bank instead [58, 319]. As can be seen in the previous example, a difficult task in designing biorthogonal wavelets is to guarantee simultaneous regularity of the basis and its dual.[8] To illustrate this point further, consider the case when one of the two wavelet bases is piecewise linear.

Example 4.7 *Piecewise Linear Biorthogonal Wavelet Bases*

Choose $G_0(z) = 1/2\sqrt{2}\,(z + 2 + z^{-1})$. It can be verified that the associated scaling function $\varphi(t)$ is the triangle function or linear B-spline. Now, we have to choose $H_0(z)$ so that (i) (4.6.4) is satisfied, (ii) $H_0(-1) = 0$, and (iii) $\tilde{\varphi}(t)$ has some regularity. First, choose

$$H_0(z) = \frac{1}{4\sqrt{2}}(-z^2 + 2z + 6 + 2z^{-1} - z^{-2}) = \frac{1}{4\sqrt{2}} \cdot (1+z)(1+z^{-1})(-z + 4 - z^{-1})$$

[8]Regularity of both the wavelet and its dual is not necessary. Actually, they can be very different and still form a valid biorthogonal expansion.

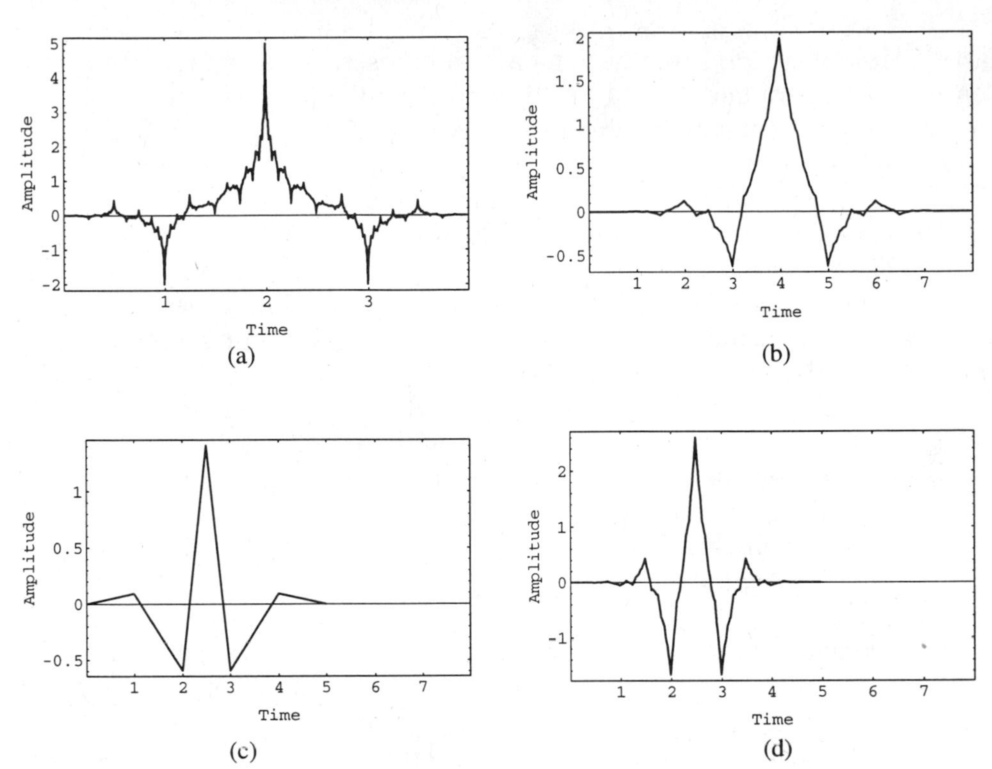

Figure 4.27 Biorthogonal wavelet bases. The scaling function $\varphi(t)$ is the hat function or linear spline (shown in Figure 4.12(b)). (a) Biorthogonal scaling function $\tilde{\varphi}(t)$ based on a length-5 filter. (b) Biorthogonal scaling function $\tilde{\varphi}'(t)$ based on a length-9 filter. (c) Wavelet $\psi'(t)$ which is piecewise linear. (d) Dual wavelet $\tilde{\psi}'(t)$.

which satisfies (i) and (ii) above. As for regularity, we show the iterated filter $H_0^{(i)}(z)$ in Figure 4.27(a) leading to an approximation of $\tilde{\varphi}(t)$. As can be seen, the dual scaling function is very "spiky". Instead, we can take a higher-order analysis lowpass filter, in particular having more zeros at $z = -1$. For example, using

$$H_0'(z) = \frac{1}{64\sqrt{2}}(1+z)^2(1+z^{-1})^2(3z^2 - 18z + 38 - 18z^{-1} + 3z^{-2})$$

leads to a smoother dual scaling function $\tilde{\varphi}'(t)$ as shown in Figure 4.27(b). The wavelet $\psi'(t)$ and its dual $\tilde{\psi}'(t)$ are shown in Figure 4.27(c) and (d). Note that both of these examples are simply a refactorization of the autocorrelation of the Daubechies' filters for $N = 2$ and 3, respectively (see Table 4.2).

Given the vastly different behavior of the wavelet and its dual, a natural question that comes to mind is which of the two decomposition formulas, (4.6.2) or (4.6.3), should be used. If all wavelet coefficients are used, and we are not worried about the speed of convergence of the wavelet series, then it does not matter. However, if approximations are to be used (as in image compression), then the two formulas can exhibit different behavior. First, zero moments of the analyzing wavelet will tend to reduce the number of significant wavelet coefficients (see Section 4.5.2) and thus, one should use the wavelet with many zeros at $\omega = 0$ for the analysis. Since $\tilde{\psi}(\omega)$ involves $H_1(e^{j\omega})$ (see (4.6.7–4.6.8)) and $H_1(z)$ is related to $G_0(-z)$, zeros at the origin for $\tilde{\psi}(\omega)$ correspond to zeros at $\omega = \pi$ for $G_0(e^{j\omega})$. Thus many zeros at $z = -1$ in $G_0(z)$ will give the same number of zero moments for $\tilde{\psi}(\omega)$ and contribute to a more compact representation of smooth signals. Second, the reconstructed signal is a linear combination of the synthesis wavelet and its shifts and translates. If not all coefficients are used in the reconstruction, a subset of wavelets should give a "close" approximation to the signal and in general, smooth wavelets will give a better approximation (for example in a perceptual sense for image compression). Again, smooth wavelets at the synthesis are obtained by having many zeros at $z = -1$ in $G_0(z)$. In practice, it turns out that (4.6.2) and (4.6.3) indeed lead to a different behavior (for example in image compression) and usually, the schemes having smooth synthesis scaling function and wavelet are preferred [14].

This concludes our brief overview of biorthogonal wavelet constructions based on filter banks. For more material on this topic, please refer to [58] (which proves completeness of the biorthogonal basis under certain conditions on the filters), [289] (which discusses general properties of biorthogonal wavelet bases) and [130, 319] (which explores further properties of biorthogonal filter banks useful for designing biorthogonal wavelets).

4.6.2 Recursive Filter Banks and Wavelets with Exponential Decay

In Section 3.2.5, filter banks using recursive or IIR filters were discussed. Just like their FIR counterparts, such filter banks can be used to generate wavelets by iteration [130, 133]. We will concentrate on the orthogonal case, noting that biorthogonal solutions are possible as well.

We start with a valid autocorrelation $P(z)$, that is, $P(z) + P(-z) = 2$, but where $P(z)$ is now a ratio of polynomials. The general form of such a $P(z)$ is given in (3.2.75) and a distinctive feature is that the denominator is a function of z^2. Given a valid $P(z)$, we can take one of its spectral factors. Call this spectral factor $G_0(z)$ and use it as the lowpass synthesis filter in an orthogonal recursive filter bank. The other filters follow as usual (see (3.2.76–3.2.77)) and we assume that there is no additional allpass component (this only increases complexity, but does not improve frequency selectivity).

Assume that $G_0(z)$ has at least one zero at $z = 1$ and define $M_0(w) = 1/\sqrt{2} \cdot G_0(e^{j\omega})$ (thus ensuring that $M_0(0) = 1$). As usual, we can define the iterated filter $G_0^{(i)}(z)$ (4.4.9) and the graphical function $\varphi^{(i)}(t)$ (4.4.10). Assuming convergence of the graphical function, the limit will be a scaling function $\varphi(t)$ just as in the FIR case. The two-scale equation property holds (see (4.4.20)), the only difference being that now, an infinite number of $\varphi(2t-n)$'s are involved.

An interesting question arises: What are the maximally flat IIR filters, or the equivalent of the Daubechies' filters? This question has been studied by Herley, who gave the class of solutions and the associated wavelets [130, 133]. Such maximally flat IIR filters lead to scaling functions and wavelets with high regularity and exponential decay in time domain. Because IIR filters have better frequency selectivity than FIR filters for a given computational complexity, it turns out that wavelets based on IIR filters offer better frequency selectivity as well. Interestingly, the most regular wavelets obtained with this construction are based on very classic filters, namely Butterworth filters (see Examples 2.2 and 3.6).

Example 4.8 *Wavelets based on Butterworth Filters*

The general form of the autocorrelation $P(z)$ of a half-band digital Butterworth filter is given in (3.2.78). Choose $N = 5$ and the spectral factorization of $P(z)$ given in (3.2.79–3.2.80). Then, the corresponding scaling function and wavelet (actually, an approximation based on the sixth iteration) are shown in Figure 4.28. These functions have better regularity (twice differentiable) than the corresponding Daubechies' wavelets but do not have compact support.

The Daubechies' and Butterworth maximally flat filters are two extreme cases to solving for a minimum degree autocorrelation $R(z)$ such that

$$(1+z)^N(1+z^{-1})^N R(z) + (1-z)^N(1-z^{-1})^N R(-z) = 2$$

is satisfied. In the Daubechies' solution, $R(z)$ has zeros only, while in the Butterworth case, $R(z)$ is all-pole. For $N \geq 4$, there are intermediate solutions where $R(z)$ has both poles and zeros and these are described in [130, 133]. The regularity of the associated wavelets is very close to the Butterworth case and thus, better than the corresponding Daubechies' wavelets.

The freedom gained by going from FIR to IIR filters allows the construction of orthogonal wavelets with symmetries or linear phase; a case excluded in the FIR or wavelet with compact support case (except for the Haar wavelet). Orthogonal IIR filter banks having linear phase filters were briefly discussed in Section 3.2.5. In particular, the example derived in (3.2.81–3.2.82) is relevant for wavelet constructions. Take synthesis filters $G_0(z) = A(z^2) + z^{-1}A(z^{-2})$ and $G_1(z) = G_0(-z)$ (similar to (3.2.81)) and $A(z)$ as the allpass given in (3.2.82). Then

$$G_0(z) = \frac{1}{\sqrt{2}} \frac{(1+z^{-1})(49 - 20z^{-1} + 198z^{-2} - 20z^{-3} + 49z^{-4})}{(15 + 42z^{-2} + 7z^{-4})(7 + 42z^{-2} + 15z^{-4})}$$

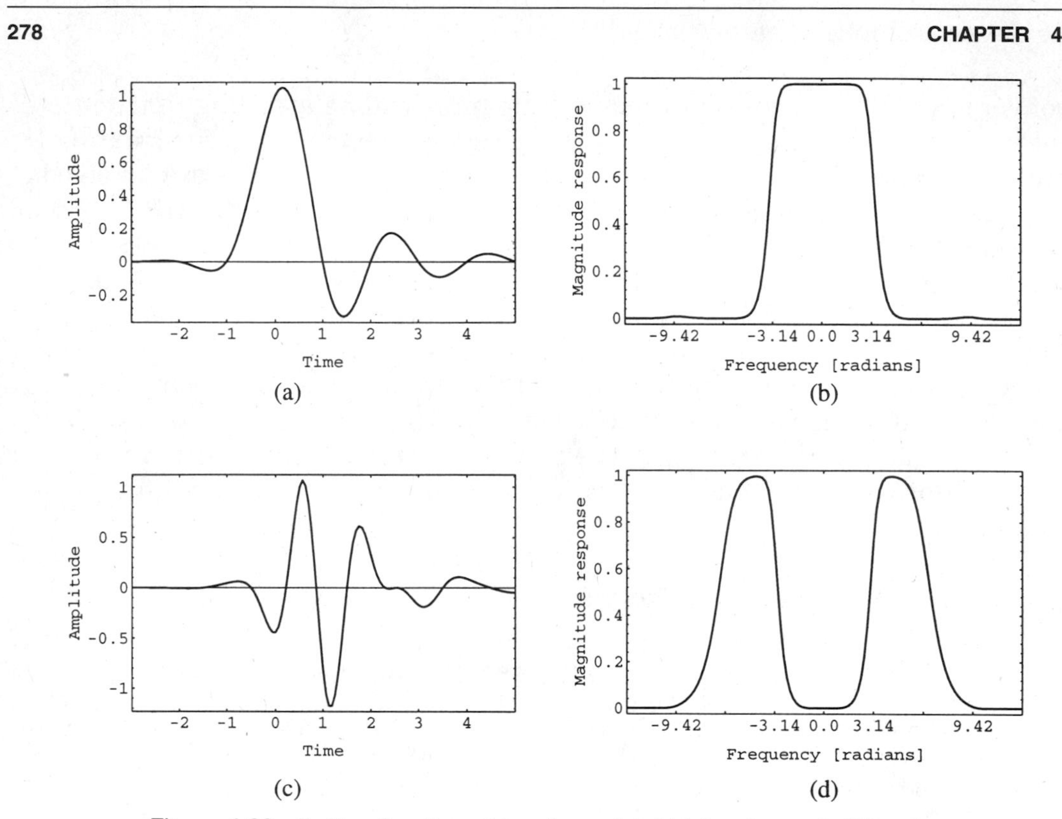

Figure 4.28 Scaling function $\varphi(t)$ and wavelet $\psi(t)$ based on a half-band digital Butterworth filter with five zeros at $w = \pi$. (a) Scaling function $\varphi(t)$. (b) Fourier transform magnitude $\Phi(\omega)$. (c) Wavelet $\psi(t)$. (d) Fourier transform magnitude $\Psi(\omega)$.

has linear phase and five zeros at $z = -1$. It leads, through iteration, to a smooth, differentiable scaling function and wavelet with exponential decay (but obviously noncausal).

4.6.3 Multichannel Filter Banks and Wavelet Packets

Consider the tree-structured filter bank case first and assume that the lowpass filter $g_0[n]$ is regular and orthogonal to its even translates. Thus, there is a limit function $\varphi(t)$ which satisfies a two-scale equation. However, $\varphi(t)$ satisfies also two-scale equations with scale changes by any power of 2 (by iteration). The linear combination is given by the iterated filter $g_0^{(i)}[n]$:

$$\varphi(t) = 2^{i/2} \sum_{k=0}^{L^{(i)}-1} g_0^{(i)}[k]\varphi(2^i t - k).$$

Then, we can design different "wavelet" bases based on iterated low and highpass filters. Let us take a simple example. Consider the following four filters, corresponding to a four-channel filter bank derived from a binary tree:

$$F_0(z) = G_0(z)G_0(z^2) \qquad F_1(z) = G_0(z)G_1(z^2), \tag{4.6.9}$$
$$F_2(z) = G_1(z)G_0(z^2) \qquad F_3(z) = G_1(z)G_1(z^2). \tag{4.6.10}$$

This corresponds to an orthogonal filter bank as we had seen in Section 3.3. Call the impulse responses $f_i[n]$. Then, the following $\varphi(t)$ is a scaling function (with scale change by 4):

$$\varphi(t) = 2\sum_k f_0[k]\ \varphi(4t-k).$$

Note that $\varphi(t)$ is just the usual scaling function from the iterated two-channel bank, but now written with respect to a scale change by 4 (which involves the filter $f_0[k]$). The following three functions are "wavelets":

$$\psi_i(t) = 2\sum_k f_i[k]\varphi(4t-k), \qquad i \in \{1,2,3\}.$$

The set $\{\varphi(t-k), \psi_1(t-l), \psi_2(t-m), \psi_3(t-n), \}$ is orthonormal and $2^j\psi_i(4^jt-l_i), i \in \{1,2,3\}, l_i, j \in \mathcal{Z}$ is an orthonormal basis for $L^2(\mathcal{R})$ following similar arguments as in the classic "single" wavelet case (we have simply expanded two successive wavelet spaces into three spaces spanned by $\psi_i(t), i \in \{1,2,3\}$). Of course, this is a simple variation on the normal wavelet case (note that $\psi_1(t)$ is the usual wavelet). With these methods and the previously discussed concept of wavelet packets in Section 3.3.4 it can be seen how to obtain continuous-time wavelet packets. That is, given any binary tree built with two-channel filter banks, we can associate a set of "wavelets" with the highpass and bandpass channels. These functions, together with appropriate scales and shifts will form orthonormal wavelet packet bases for $L_2(R)$.

The case for general filter banks is very similar [129, 277]. Assume we have a size-N filter bank with a regular lowpass filter. This filter has to be regular with respect to downsampling by N (rather than 2), which amounts (in a similar fashion to Proposition 4.7) to having a sufficient number of zeros at the Nth roots of unity (the aliasing frequencies, see discussion below). The lowpass filter will lead to a scaling function satisfying

$$\varphi(t) = N^{1/2}\sum_k g_0[k]\ \varphi(Nt-k).$$

The $N-1$ functions

$$\psi_i(t) = N^{1/2}\sum_k g_i[k]\varphi(Nt-k), \qquad i = 1,\ldots,N-1,$$

will form a wavelet basis with respect to scale changes by N.

Let us consider the issue of regularity for multichannel filter banks. It is clear that if a regular two-channel filter bank is cascaded a finite number of times in order to obtain wavelet packets (as was done above in (4.6.9–4.6.10)), then regularity of the lowpass filter is necessary and sufficient in order to obtain regular wavelet packets. This follows since the scaling function is the same and the wavelet packets are finite linear combinations of scaling functions. In the more general case of a filter bank with N channels, we have to test the regularity of the lowpass filter $G_0(z)$ with respect to sampling rate changes by N. That is, we are interested in the behavior of the iterated filter $G_0^{(i)}(z)$;

$$G_0^{(i)}(z) = \prod_{k=0}^{i-1} G_0(z^{N^k}), \tag{4.6.11}$$

and the associated graphical function

$$\varphi^{(i)}(t) = N^{i/2} \cdot g_0^{(i)}[n], \qquad \frac{n}{N^i} \leq t < \frac{n+1}{N^i}. \tag{4.6.12}$$

Since the filter $G_0(z)$ is orthogonal with respect to translation by multiples of N, it satisfies (see (3.4.11))

$$\sum_{k=0}^{N-1} G_0(e^{j(\omega+2\pi k/N)})\, G_0(e^{-j(\omega+2\pi k/N)}) = N. \tag{4.6.13}$$

A necessary condition for convergence of the graphical function is that (see Problem 4.15)

$$G_0(e^{j(\omega+2\pi k/N)}) = 0, \qquad k = 1, \ldots, N-1, \tag{4.6.14}$$

that is, $G_0(z)$ has at least one zero at each of the aliasing frequencies $\omega = 2\pi k/N$, $k = 1, \ldots, N-1$. Then, using (4.6.14) in (4.6.13), we see that

$$G_0(1) = \sqrt{N}.$$

Introducing a normalized version of the lowpass filter,

$$M_0(\omega) = \frac{1}{\sqrt{N}} G_0(e^{j\omega})$$

and assuming convergence, it follows that the Fourier transform of the scaling function equals

$$\Phi(\omega) = \prod_{i=1}^{\infty} M_0\left(\frac{\omega}{N^i}\right).$$

A sufficient condition for the convergence of the graphical function (4.6.12) to a continuous function can be derived very similarly to Proposition 4.7. Write

$$M_0(\omega) = \left(\frac{1 + e^{j\omega} + \cdots + e^{j(N-1)\omega}}{N}\right)^K R(\omega)$$

where $K \geq 1$ because of the necessary condition for convergence and call

$$B = \sup_{w \in [0,2\pi]} |R(\omega)|.$$

Then

$$B < N^{K-1} \tag{4.6.15}$$

ensures that the limit $\varphi^{(i)}(t)$ as $i \to \infty$ is continuous (see Problem 4.16).

The design of lowpass filters with a maximum number of zeros at aliasing frequencies (the equivalent of the Daubechies' filters, but for integer downsampling larger than 2) is given in [277]. An interesting feature of multichannel wavelet schemes is that now, orthogonality and compact support are possible simultaneously. This follows from the fact that there exist unitary FIR filter banks having linear phase filters for more than two channels [321]. A detailed exploration of such filter banks and their use for the design of orthonormal wavelet bases with symmetries (for example, a four-band filter bank leading to one symmetric scaling function as well as one symmetric and two antisymmetric wavelets) is done in [275].

The problem with scale changes by $N > 2$ is that the resolution steps are even larger between a scale and the next coarser scale than for the typical "octave-band" wavelet analysis. A finer resolution change could be obtained for rational scale changes between 1 and 2. In discrete time such finer steps can be achieved with filter banks having rational sampling rates [160]. The situation is more complicated in continuous time. In particular, the iterated filter bank method does not lead to wavelets in the same sense as for the integer-band case. Yet, orthonormal bases can be constructed which have a similar behavior to wavelets [33]. A direct wavelet construction with rational dilation factors is possible [16] but the coefficients of the resulting two-scale equation do not correspond to either FIR or IIR filters.

4.7 MULTIDIMENSIONAL WAVELETS

In Chapter 3, we have seen that, driven by applications such as image compression, some of the concepts from the theory of one-dimensional filter banks have been extended to multiple dimensions. Hence, this section can be seen as generalization of both Section 3.6 and the concepts introduced in this chapter.

An easy way to construct two-dimensional wavelets, for example, is to use tensor products of their one-dimensional counterparts. This results, as will be seen later, in one scaling function and three different "mother" wavelets. Since now, scale change is represented by matrices, the scaling matrix in this case will be $2\boldsymbol{I}$, that is, each dimension is dilated by 2. As for multidimensional filter banks, true multidimensional treatment of wavelets offers several advantages.

First, one can still have a diagonal dilation (scaling) matrix and yet design non-separable (irreducible) scaling function and wavelets. Then, the scale change of $\sqrt{2}$, for example, is possible, leading to one scaling function and one wavelet or a true two-dimensional counterpart of the well-known one-dimensional dyadic case. However, unlike for the filter banks, matrices used for dilation are more restricted in that one requires dilation in each dimension. As in one dimension, the powerful connection with filter banks (through the method of iterated filter banks) can be exploited to design multidimensional wavelets. However, the task is more complicated due to incomplete cascade structures and the difficulty of imposing a zero of a particular order at aliasing frequencies. Regularity is much harder to achieve, and up-to-date, orthonormal families with arbitrarily high regularity, have not been found. In the biorthogonal case, transformations of one-dimensional perfect reconstruction filter banks into multidimensional ones can be used to design multidimensional wavelets by iteration.

4.7.1 Multiresolution Analysis and Two-Scale Equation

The axiomatic definition of a multiresolution analysis is easily generalized: The subspaces V_j in (4.2.1) are now subspaces of $\mathcal{R}^m$ and scaling is represented by a matrix $\boldsymbol{D}$. This matrix has to be well-behaved, that is,

$$\begin{aligned} \boldsymbol{D}\mathcal{Z}^m &\subset \mathcal{Z}^m \\ |\lambda_i| &> 1, \qquad \forall i. \end{aligned}$$

The first condition requires $\boldsymbol{D}$ to have integer entries, while the second one states that all the eigenvalues of $\boldsymbol{D}$ must be strictly greater than 1 in order to ensure dilation in each dimension. For example, in the quincunx case, the matrix $\boldsymbol{D}_Q$ from (3.B.2)

$$\boldsymbol{D}_Q = \begin{pmatrix} 1 & 1 \\ 1 & -1 \end{pmatrix}, \tag{4.7.1}$$

as well as

$$\boldsymbol{D}_{Q_1} = \begin{pmatrix} 1 & -1 \\ 1 & 1 \end{pmatrix},$$

are both valid matrices, while

$$\boldsymbol{D}_{Q_2} = \begin{pmatrix} 2 & 1 \\ 0 & 1 \end{pmatrix},$$

is not, since it dilates only one dimension. Matrix $\boldsymbol{D}_Q$ from (4.7.1) is a so-called "symmetry" dilation matrix, used in [164], while $\boldsymbol{D}_{Q_1}$ is termed a "rotation" matrix used in [57]. As will be seen shortly, although both of these matrices represent the same lattice, they are fundamentally different when it comes to constructing wavelets.

For the case obtained as a tensor product, the dilation matrix is diagonal. Specifically, in two dimensions, it is the matrix D_S from (3.B.1)

$$D_s = \begin{pmatrix} 2 & 0 \\ 0 & 2 \end{pmatrix}. \tag{4.7.2}$$

The number of wavelets is determined by the number of cosets of $D\mathcal{Z}^n$, or

$$|\det(D)| - 1 = N - 1,$$

where N represents the downsampling rate of the underlying filter bank. Thus, in the quincunx case, we have one "mother" wavelet, while in the 2×2 separable case (4.7.2), there are three "mother" wavelets ψ_1, ψ_2, ψ_3.

The two-scale equation is obtained as in the one-dimensional case. For example, using D_Q (we will drop the subscript when there is no risk of confusion)

$$\varphi(t) = \sqrt{2} \sum_{n \in \mathcal{Z}^2} g_0[n]\, \varphi(Dt - n),$$

$$\varphi(t_1, t_2) = \sqrt{2} \sum_{n_1, n_2 \in \mathcal{Z}} g_0[n_1, n_2]\, \varphi(t_1 + t_2 - n_1, t_1 - t_2 - n_2).$$

We have assumed that $\sum g_0[n] = \sqrt{2}$.

4.7.2 Construction of Wavelets Using Iterated Filter Banks

Since the construction is similar to the one-dimensional case, we will concentrate on the quincunx dilation matrices by way of example.

Consider again Figure 4.14 with the matrix D_Q replacing upsampling by 2. Then the equivalent low branch after i steps of filtering and sampling by D_Q will be

$$G_0^{(i)}(\omega_1, \omega_2) = \prod_{k=0}^{i-1} G_0\left(\left(D_Q^t\right)^k \begin{pmatrix} \omega_1 \\ \omega_2 \end{pmatrix}\right), \tag{4.7.3}$$

where $G_0^{(0)}(\omega_1, \omega_2) = 1$. Observe here that instead of scalar powers, we are dealing with powers of matrices. Thus, for different matrices, iterated filters are going to exhibit vastly different behavior. Some of the most striking examples are multidimensional generalizations of the Haar basis which were independently discovered by Gröchenig and Madych [123] and Lawton and Resnikoff [172] (see next section).

Now, as in the one-dimensional case, construct a continuous-time "graphical" function based on the iterated filter $g_0^{(i)}[n_1, n_2]$:

$$\varphi^{(i)}(t_1, t_2) = 2^{i/2}\, g_0^{(i)}[n_1, n_2],$$

$$\begin{pmatrix} 1 & 1 \\ 1 & -1 \end{pmatrix}^i \begin{pmatrix} t_1 \\ t_2 \end{pmatrix} \in \begin{pmatrix} n_1 \\ n_2 \end{pmatrix} + [0,1) \times [0,1).$$

Note that these regions are not in general rectangular and specifically in this case, they are squares in even, and diamonds (tilted squares) in odd iterations. Note that one of the advantages of using the matrix $\boldsymbol{D}_Q$ rather than $\boldsymbol{D}_{Q_1}$, is that it leads to separable sampling (diagonal matrix) in every other iteration since $\boldsymbol{D}_Q^2 = 2\boldsymbol{I}$. The reason why this feature is useful is that one can use certain one-dimensional results in a separable manner in even iterations. We are again interested in the limiting behavior of this "graphical" function. Let us first assume that the limit of $\varphi^{(i)}(t_1, t_2)$ exists and is in $L_2(\mathcal{R}^2)$ (we will come back later to the conditions under which it exists). Hence, we define the scaling function as

$$\varphi(t_1, t_2) = \lim_{i \to \infty} \varphi^{(i)}(t_1, t_2), \quad \varphi(t_1, t_2) \in L_2(\mathcal{R}^2). \tag{4.7.4}$$

Once the scaling function exists, the wavelet can be obtained from the two-dimensional counterpart of (4.2.14). Again, the coefficients used in the two-scale equation and the quincunx version of (4.2.14) are the impulse response coefficients of the lowpass and highpass filters, respectively. To prove that the wavelet obtained in such a fashion actually produces an orthonormal basis for $L_2(\mathcal{R}^2)$, one has to demonstrate various facts. The proofs of the following statements are analogous to the one-dimensional case (see Proposition 4.4), that is, they rely on the orthogonality of the underlying filter banks and the two-scale equation property [164]:

(a) $\langle \varphi(\boldsymbol{D}_Q^m \boldsymbol{t} - \boldsymbol{n}), \varphi(\boldsymbol{D}_Q^m \boldsymbol{t} - \boldsymbol{k}) \rangle = 2^{-m} \delta[\boldsymbol{n} - \boldsymbol{k}]$, that is, the scaling function is orthogonal to its translates by multiples of $\boldsymbol{D}_Q^{-m}$ at all scales.

(b) The same holds for the wavelet.

(c) $\langle \varphi(\boldsymbol{t}), \psi(\boldsymbol{t} - \boldsymbol{k}) \rangle$, the scaling function is orthogonal to the wavelet and its integer translates.

(d) Wavelets are orthogonal across scales.

It follows that the set

$$S = \{2^{-m/2} \psi(\boldsymbol{D}^{-m} \boldsymbol{t} - \boldsymbol{n}) \mid m \in \mathcal{Z}, \boldsymbol{n} \in \mathcal{Z}^2, \boldsymbol{t} \in \mathcal{R}^2\},$$

is an orthonormal set. What is left to be shown is completeness, which can be done similarly to the one-dimensional case (see Theorem 4.5 and [71]).

The existence of the limit of $\varphi^{(i)}(t_1, t_2)$ was assumed. Now we give a necessary condition for its existence. Similarly to the one-dimensional case, it is necessary for the lowpass filter of the iterated filter bank to have a zero at aliasing frequencies. This condition holds in general, but will be given here for the case we have been following throughout this section, that is, the quincunx case. The proof of necessity is similar to that of Proposition 4.6.

PROPOSITION 4.9

If the scaling function $\varphi(t_1, t_2)$ exists for some $(t_1, t_2) \in \mathcal{R}^2$, then

$$\sum_{k \in \mathcal{Z}^2} g_0[\boldsymbol{D}_1 \boldsymbol{k} + \boldsymbol{k}_i] = \frac{1}{\sqrt{2}}, \qquad \boldsymbol{k}_0 = \begin{pmatrix} 0 \\ 0 \end{pmatrix}, \; \boldsymbol{k}_1 = \begin{pmatrix} 1 \\ 0 \end{pmatrix}, \tag{4.7.5}$$

or, in other words

$$G_0(1,1) = \sqrt{2}, \quad G_0(-1,-1) = 0.$$

PROOF

Following (4.7.3), one can express the equivalent filter after i steps in terms of the equivalent filter after $(i-1)$ steps as

$$g_0^{(i)}[\boldsymbol{n}] = \sum_k g_0[\boldsymbol{k}]\, g_0^{(i-1)}[\boldsymbol{n} - \boldsymbol{D}^{i-1}\boldsymbol{k}] = \sum_k g_0^{(i-1)}[\boldsymbol{k}]\, g_0[\boldsymbol{n} - D\boldsymbol{k}],$$

and thus

$$g_0^{(i)}[\boldsymbol{D}\boldsymbol{n}] = \sum_k g_0[\boldsymbol{D}\boldsymbol{k}]\, g_0^{(i-1)}[\boldsymbol{n} - \boldsymbol{k}].$$

Using (4.7.4) express $g_0^{(i-1)}$ and $g_0^{(i)}$ in terms of $\varphi^{(i-1)}$ and $\varphi^{(i)}$ and then take the limits (which we are allowed to do by assumption)

$$\varphi(\boldsymbol{D}\boldsymbol{t}) = \sqrt{2} \sum_k g_0[\boldsymbol{D}\boldsymbol{k}]\, \varphi(\boldsymbol{D}\boldsymbol{t}). \tag{4.7.6}$$

Doing now the same for $g_0^{(i)}[\boldsymbol{D}\boldsymbol{n} + \boldsymbol{k}_1]$ one obtains

$$\varphi(\boldsymbol{D}\boldsymbol{t}) = \sqrt{2} \sum_k g_0[\boldsymbol{D}\boldsymbol{n} + \boldsymbol{k}_1]\, \varphi(\boldsymbol{D}\boldsymbol{t}). \tag{4.7.7}$$

Equating (4.7.6) and (4.7.7), one obtains (4.7.5).

Now, a single zero at aliasing frequency is in general not sufficient to ensure regularity. Higher-order zeros have led to regular scaling functions and wavelets, but the precise relationship is a topic of current research.

4.7.3 Generalization of Haar Basis to Multiple Dimensions

The material in this section is based on the work of Lawton and Resnikoff [172], and Gröchenig and Madych [123]. The results are stated in the form given in [123].

Recall the Haar basis introduced at the beginning of this chapter and recall that the associated scaling function is 1 over the interval $[0, 1)$ and 0 otherwise. In other words, this scaling function can be viewed as the characteristic function of the set $Q = [0, 1)$. Together with integer translates, the Haar scaling function "covers" the real line. The idea is to construct analogous multidimensional generalized Haar bases that would have, as scaling functions, characteristic functions of appropriate sets with dilation replaced by a suitable linear transformation.

The approach in [123] consists of finding a characteristic function of a compact set Q that would be the scaling function for an appropriate multiresolution analysis. Then to find the wavelets, one would use the standard techniques. An interesting property of such scaling functions is that they form self-similar tilings of $\mathcal{R}^n$. This is not an obvious feature for some scaling functions of exotic shapes.

The algorithm for constructing a scaling function for multiresolution analysis with matrix dilation $\boldsymbol{D}$ basically states that one takes a set of points belonging to different cosets of the lattice and forms a discrete filter being 1 on these points. The filter is then iterated as explained earlier. If it converges, we obtain an example of a generalized Haar wavelet. For a more formal definition of the algorithm, the reader is referred to [123]. For example, in the quincunx case, the set of points of coset representatives would consist only of two elements (since the quincunx lattice has only two cosets) and its elements would represent the two taps of the lowpass filter. Thus, the corresponding subband schemes would consist of two-tap filters. The algorithm, when it converges, can be interpreted as the iteration of a lowpass filter with only two nonzero taps (each equal to one and being in a different coset) which converges to the characteristic function of some compact set, just as the one-dimensional Haar filter converged to the indicator function of the unit interval.

A very interesting scaling function is obtained when using the "rotation" matrix $\boldsymbol{D}_{Q_1}$ from (4.7.1) and points $\{(0, 0), (1, 0)\}$, that is, the lowpass filter with $g_0[0, 0] = g_0[1, 0] = 1$, and 0 otherwise. Iterating this filter leads to the "twin-dragon" case [190], as given in Figure 4.29. Note that $\varphi(t) = 1$ over the white region and 0 otherwise. The wavelet is 1 and -1 in the white/black regions respectively, and 0 otherwise. Note also how the wavelet is formed by two "scaled" scaling functions, as required by the two-dimensional counterpart of (4.2.9), and how this fractal shape tiles the space.

4.7.4 Design of Multidimensional Wavelets

As we have seen in Section 3.6, the design of multidimensional filter banks is not easy, and it becomes all the more involved by introducing the requirement that the lowpass filter be regular. Here, known techniques will be briefly reviewed, for more details the reader is referred to [57] and [164].

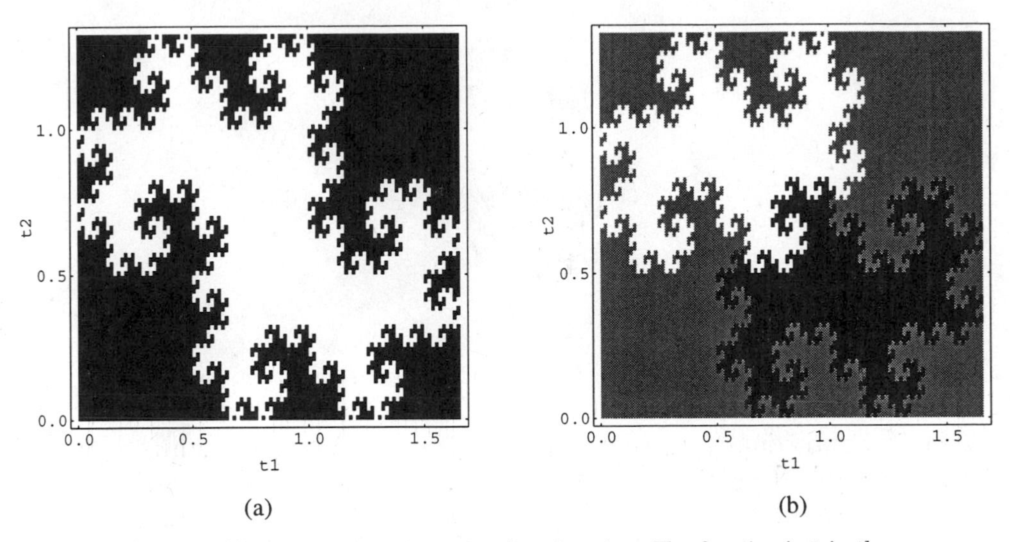

Figure 4.29 (a) The twin-dragon scaling function. The function is 1 in the white area and 0 otherwise. (b) The twin-dragon wavelet. The function is 1 in the white area, −1 in the black area, and 0 otherwise.

Direct Design To achieve perfect reconstruction in a subband system, cascade structures are perfect candidates (see Section 3.6), since beside perfect reconstruction, some other properties such as orthogonality and linear phase can be easily imposed.

Recall that in one dimension, a zero of a sufficiently high order at π would guarantee the desired degree of regularity. Unfortunately, imposing a zero of a particular order in multiple dimensions becomes a nontrivial problem and thus, algebraic solutions can be obtained only for very small size filters.

As an example of direct design, consider again the quincunx case with matrix D_1 and the perfect reconstruction filter pair given in (3.6.7). Thus, the approach is to impose a zero of the highest possible order at (π, π) on the lowpass filter in (3.6.7), that is

$$\left. \frac{\partial^{k-1} H_0(\omega_1, \omega_2)}{\partial^l \omega_1 \partial^{k-l-1} \omega_2} \right|_{(\pi,\pi)} = 0, \qquad \begin{array}{l} k = 1, \ldots, m, \\ l = 0, \ldots, k-1. \end{array}$$

Upon imposing a second-order zero the following solutions are obtained

$$a_0 = \pm\sqrt{3}, \quad a_1 = \pm\sqrt{3}, \quad a_2 = 2 \pm \sqrt{3}, \tag{4.7.8}$$

$$a_0 = \pm\sqrt{3}, \quad a_1 = 0, \quad a_2 = 2 \pm \sqrt{3}. \tag{4.7.9}$$

Note that the filters should be scaled by $(1 - \sqrt{3})/(4\sqrt{2})$. The solution in (4.7.9)

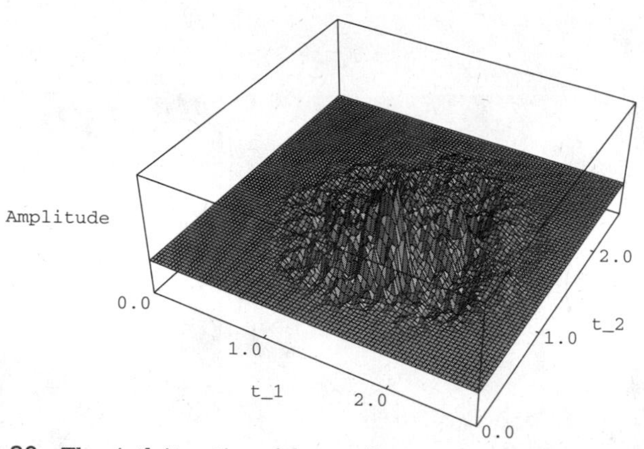

Figure 4.30 The sixth iteration of the smallest regular two-dimensional filter.

is the one-dimensional D_2 filter, while (4.7.8) would be the smallest "regular" two-dimensional filter (actually, a counterpart of D_2). Figure 4.30 shows the fourth iteration of this solution. As can be seen from the plot, the function looks continuous, but not differentiable at some points. As a simple check of continuity, the largest first-order differences of the iterated filter can be computed (in this case, these differences decrease with an almost constant rate — a good indicator that the function is continuous [164]). Recently, a method for checking the continuity was developed [324]. Using this method, it was confirmed that this solution indeed leads to a continuous scaling function and consequently a continuous wavelet.

This method, however, fails for larger size filters, since imposing a zero of a particular order means solving a large system of nonlinear equations (in the orthogonal case). Note, however, that numerical approaches are always possible [163].

One to Multidimensional Transformations Another way to approach the design problem is to use transformations of one-dimensional filters into multidimensional ones in such a way that [164]

(a) Perfect reconstruction is preserved (in order to have a valid subband coding system).

(b) Zeros at aliasing frequencies are preserved (necessary but not sufficient for regularity).

We have already discussed how to obtain perfect reconstruction in Section 3.6. Here, we will concern ourselves only with properties that might be of interest for designing wavelets. If we used the method of separable polyphase components, an advantage is that the zeros at aliasing frequencies carry over into multiple dimensions. As we pointed out in Section 3.6, the disadvantage is that only IIR solutions are possible, and thus we cannot obtain wavelets with compact support. In the McClellan case, however, wavelets with compact support are possible, but not orthonormal ones. For more details on these issues, see [164].

4.8 LOCAL COSINE BASES

At the beginning of this chapter (see Section 4.1.2), we examined a piecewise Fourier series expansion that was an orthogonal local Fourier basis. Unfortunately, because the basis functions were truncated complex exponentials (and thus discontinuous), they achieved poor frequency localization (actually, the time-bandwidth product of the basis functions is unbounded). Because of the Balian-Low Theorem [73], there are no "good" orthogonal bases in the Gabor or windowed Fourier transform case (see Chapter 5). However, if instead of using modulation by complex exponentials, one uses modulation by cosines, it turns out that good orthonormal bases do exist, as will be shown next. This result is the continuous-time equivalent of the modulated lapped orthogonal transforms, seen in Section 3.4.

We will start with a simple case which, when refined, will lead to what Meyer calls "Malvar's wavelets" [193]. Note that, beside this construction, there exists other orthonormal bases with similar properties [61]. Thus, consider the following set of basis functions:

$$\varphi_{j,k}(t) = \sqrt{\frac{2}{L_j}}\, w_j(t) \cos\left[\frac{\pi}{L_j}(k+\frac{1}{2})(t-a_j)\right], \tag{4.8.1}$$

for $k = 0, 1, 2, \ldots$, and $j \in \mathcal{Z}$, a_j is an increasing sequence of real numbers and the window function $w_j(t)$ is centered around the interval $[a_j, a_{j+1}]$. As can be seen, (4.8.1) is the continuous-time counterpart of (3.4.17) seen in the discrete-time case.

4.8.1 Rectangular Window

Let us start with the simplest case and assume that

$$a_j = (j - \frac{1}{2})L, \qquad L_j = a_{j+1} - a_j = L, \tag{4.8.2}$$

while the "window" functions $w_j(t)$ will be restricted as (see Figure 4.31(a))

$$w_j(t) = w(t - jL), \quad w(t) = \frac{1}{\sqrt{2}}, \qquad -L \leq t \leq L.$$

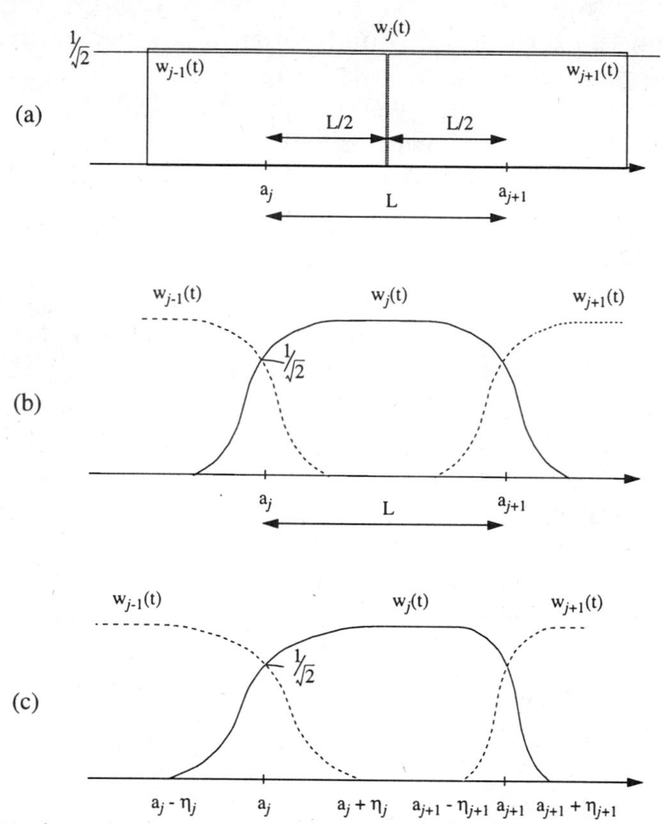

Figure 4.31 Relationship among windows for the local cosine bases. (a) Rectangular window. All windows are the same. (b) Smooth window satisfying the power complementary condition. All windows are the same. (c) General case.

That is, we have rectangular windows which overlap by L with their neighbors, as given in Figure 4.31(a). Thus, the basis functions from (4.8.1) become

$$\varphi_{j,k}(t) \;=\; \frac{1}{\sqrt{L}} \cos\left[\frac{\pi}{L}(k+\frac{1}{2})(t-jL+\frac{L}{2})\right], \qquad (j-1)L \leq t \leq (j+1)L.$$

To prove that this set of functions forms a basis, we have to demonstrate the orthogonality of the basis functions, as well as completeness. Since the proof of completeness is quite involved, we refer the reader to [61] for details (note that in [61], the proof is given for a slightly different set of basis functions, but the idea is the same). As for orthogonality, first note that $\varphi_{j,k}$ and $\varphi_{j',m}$ do not overlap for $j - j' \geq 2$. To prove that $\varphi_{j,k}$ and $\varphi_{j+1,m}$ are mutually orthogonal, write

$$\langle\varphi_{j,k}, \varphi_{j+1,m}\rangle \;=\; \frac{1}{L}\int_{jL}^{(j+1)L} \cos\left[\frac{\pi}{L}(k+\frac{1}{2})(t-jL+\frac{L}{2})\right]$$

$$\times \cos\left[\frac{\pi}{L}(m+\frac{1}{2})(t-(j+1)L+\frac{L}{2})\right]\, dt, \qquad (4.8.3)$$

which, with change of variable $x = t-(j+1)L+L/2$ becomes

$$\langle \varphi_{j,k}, \varphi_{j+1,m}\rangle = \pm\frac{1}{L}\int_{-L/2}^{L/2} \sin\left[\frac{\pi}{L}(k+\frac{1}{2})x\right]\cos\left[\frac{\pi}{L}(m+\frac{1}{2})x\right]\, dx = 0,$$

since the integrand is an odd function of x.

Finally, orthogonality of $\varphi_{j,k}$ and $\varphi_{j,m}$ follows from (again with the change of variable $x = t - jL$)

$$\langle \varphi_{j,k}, \varphi_{j,m}\rangle = \frac{1}{L}\int_{-L}^{L} \cos\left[\frac{\pi}{L}(k+\frac{1}{2})(x+\frac{L}{2})\right]\cos\left[\frac{\pi}{L}(m+\frac{1}{2})(x+\frac{L}{2})\right]\, dx = 0.$$

What we have constructed effectively, is a set of basis functions obtained from the cosines of various frequencies, shifted in time to points jL on the time axis, and modulated by a square window of length $2L$.

4.8.2 Smooth Window

Suppose now that we still keep the regular spacing of L between shifts as in (4.8.2), but allow for a smooth window of length $2L$ satisfying the following (see Figure 4.31(b)):

$$w_j(t) = w(t-jL),\quad w(t) = w(-t),\quad -L \le t \le L,\quad w^2(t)+w^2(L-t) = 1, \quad (4.8.4)$$

and the basis functions are as in (4.8.1) (see Figure 4.31(b)). Note that here, on top of cosines overlapping, we have to deal with the windowing of the cosines. To prove orthogonality, again we will have to demonstrate it only for $\varphi_{j,k}$ and $\varphi_{j+1,m}$, as well as for $\varphi_{j,k}$ and $\varphi_{j,m}$.

By using the same change of variable as in (4.8.3), we obtain that

$$\langle \varphi_{j,k}, \varphi_{j+1,m}\rangle = \pm\frac{2}{L}\int_{-L/2}^{L/2} w(t+\frac{L}{2})\, w(t-\frac{L}{2})\, \sin\left[\frac{\pi}{L}(k+\frac{1}{2})t\right]\cos\left[\frac{\pi}{L}(m+\frac{1}{2})t\right]\, dt.$$

Since $w(t+L/2)w(t-L/2)$ is an even function of t, while the rest is an odd function of t as before, the above inner product is zero. For orthogonality of $\varphi_{j,k}$ and $\varphi_{j,m}$ write

$$\langle \varphi_{j,k}, \varphi_{j,m}\rangle = \frac{2}{L}\int_{-L}^{L} w^2(t)\, \cos\left[\frac{\pi}{L}(k+\frac{1}{2})(t+\frac{L}{2})\right]\cos\left[\frac{\pi}{L}(m+\frac{1}{2})(t+\frac{L}{2})\right]\, dt.$$

Divide the above integral into three parts: from $-L$ to $-L/2$, from $-L/2$ to $L/2$, and from $L/2$ to L. Let us concentrate on the last one. With the change of variable $x = L - t$, it becomes

$$\frac{2}{L}\int_{L/2}^{L} w^2(t)\, \cos\left[\frac{\pi}{L}(k+\frac{1}{2})(t+\frac{L}{2})\right]\cos\left[\frac{\pi}{L}(m+\frac{1}{2})(t+\frac{L}{2})\right]\, dt =$$

$$\frac{2}{L}\int_0^{L/2} w^2(L-x)\,\cos\left[\frac{\pi}{L}(k+\frac{1}{2})(\frac{3}{2}L-x)\right]\,\cos\left[\frac{\pi}{L}(m+\frac{1}{2})(\frac{3}{2}L-x)\right]\,dx.$$

However, since $\cos[(\pi/L)(k+1/2)((3/2)L-x) = -\cos[\pi/L(k+1/2)(x+1/2)]$, we can merge this integral to the second one from 0 to $L/2$. Using the same argument for the one from $-L$ to $-L/2$, we finally obtain

$$\frac{2}{L}\int_{-L/2}^{L/2}\underbrace{(w^2(t)+w^2(L-t))}_{1}\,\cos\left[\frac{\pi}{L}(k+\frac{1}{2})(t+\frac{L}{2})\right]\,\cos\left[\frac{\pi}{L}(m+\frac{1}{2})(t+\frac{L}{2})\right]\,dt = 0.$$

We now see why it was important for the window to satisfy the power complementary condition given in (4.8.4), exactly as in the discrete-time case. Therefore, we have progressed from a rectangular window to a smooth window.

4.8.3 General Window

The final step is to lift the restriction on a_j imposed in (4.8.2) and allow for windows $w_j(t)$ to be different. We outline the general construction [61]. The proofs of orthogonality will be omitted, however, since they follow similarly to the two simpler cases discussed above. They are left as an exercise for the reader (see Problem 4.22). For the proof of completeness, we again refer the reader to [61] (although for a slightly different set of basis functions).

Assume, thus, that we have an increasing sequence of real numbers $a_j, j \in \mathcal{Z}$, $\ldots a_{j-1} < a_j < a_{j+1} \ldots$ We will denote by L_j the distance between a_{j+1} and a_j, $L_j = a_{j+1} - a_j$. We will also assume that we are given a sequence of numbers $\eta_j > 0$ such that $\eta_j + \eta_{j+1} \leq L_j$, $j \in \mathcal{Z}$, which ensures that windows will only overlap with their nearest neighbor. The given windows $w_j(t)$ will be differentiable (possibly infinitely) and of compact support, with the following requirements:

(a) $0 \leq w_j(t) \leq 1$, $w_j(t) = 1$ if $a_j + \eta_j \leq t \leq a_{j+1} - \eta_{j+1}$.

(b) $w_j(t)$ is supported within $[a_j - \eta_j, a_{j+1} + \eta_{j+1}]$.

(c) If $|t - a_j| \leq \eta_j$ then $w_{j-1}(t) = w_j(2a_j - t)$, and $w_{j-1}^2(t) + w_j^2(t) = 1$.

This last condition ensures that the "tails" of the adjacent windows are power complementary. An example of such a window is taking $w_j(t) = \sin[(\pi/2)\theta((t - a_j + \eta_j)/(2\eta_j))]$ for $|t - a_j| \leq \eta_j$, and $w_j(t) = \cos[(\pi/2)\theta((t - a_{j+1} + \eta_{j+1})/\eta_{j+1})]$ for $|t - a_{j+1}| \leq \eta_{j+1}$. Here, $\theta(t)$ is the function we used for constructing the Meyer's wavelet given in (4.3.1), Section 4.3.1. With these conditions, the set of functions as in (4.8.1) forms an orthonormal basis for $L_2(\mathcal{R})$. It helps to visualize the above conditions on the windows as in Figure 4.31(c). Therefore, in this most general case, the window can go anywhere from length $2L$ to length L (being a constant window in this latter case of height 1) and is arbitrary as long as it satisfies the above three conditions.

Let us see what has been achieved. The time-domain functions are local and smooth and their Fourier transforms have arbitrary polynomial decay (depending on the smoothness or differentiability of the window). Thus, the time-bandwidth product is now finite (unlike in the piecewise Fourier series case), and we have a local modulated basis with good time-frequency localization.

APPENDIX 4.A PROOF OF THEOREM 4.5

PROOF

As mentioned previously, what follows is a brief outline of the proof, for more details, refer to [71].

(a) It can be shown that

$$\sum_k [g_0[n-2k]\varphi_{jk} + g_1[n-2k]\psi_{jk}] = \varphi_{j-1,n}.$$

(b) Using this, it can be shown that

$$\sum_n |\langle \varphi_{j-1,n}, f\rangle|^2 = \sum_k |\langle \varphi_{jk}, f\rangle|^2 + \sum_k |\langle \psi_{jk}, f\rangle|^2.$$

(c) Then, by iteration, for all $N \in \mathcal{N}$

$$\sum_n |\langle \varphi_{-N,n}, f\rangle|^2 = \sum_k |\langle \varphi_{Nk}, f\rangle|^2 + \sum_{j=-N}^{N} \sum_k |\langle \psi_{jk}, f\rangle|^2. \tag{4.A.1}$$

(d) It can be shown that

$$\lim_{N\to\infty} \sum_k |\langle \varphi_{Nk}, f\rangle|^2 = 0,$$

and thus the limit of (4.A.1) reduces to

$$\lim_{N\to\infty} |\langle \varphi_{-Nn}, f\rangle|^2 = \lim_{N\to\infty} \sum_{j=-N}^{N} \sum_k |\langle \psi_{jk}, f\rangle|^2. \tag{4.A.2}$$

(e) Concentrating on the left side of (4.A.2)

$$\sum_k |\langle \varphi_{-Nk}, f\rangle|^2 = 2\pi \int |\Phi(2^{-N}\omega)|^2 |F(\omega)|^2 d\omega + R,$$

with $|R| \leq C2^{-3N/2}$ and thus

$$\lim_{N\to\infty} |R| = 0,$$

or

$$\lim_{N\to\infty} \sum_k |\langle \varphi_{-Nk}, f\rangle|^2 = \lim_{N\to\infty} 2\pi \int |\Phi(2^{-N}\omega)|^2 |F(\omega)|^2 d\omega,$$

or again, substituting into (4.A.2)

$$\begin{aligned}\lim_{N\to\infty} \sum_{j=-N}^{N} \sum_k |\langle \psi_{jk}, f\rangle|^2 &= \sum_k |\langle \psi_{jk}, f\rangle|^2, \\ &= \lim_{N\to\infty} 2\pi \int |\Phi(2^{-N}\omega)|^2 |F(\omega)|^2 d\omega.\end{aligned}$$

(f) Finally, the right side of the previous equation can be shown to be

$$\lim_{N\to\infty} 2\pi \int |\Phi(2^{-N}\omega)|^2 |F(\omega)|^2 d\omega = \|f\|^2,$$

and

$$\sum_k |\langle \psi_{jk}, f\rangle|^2 = \|f\|^2,$$

which completes the proof of the theorem.

PROBLEMS

4.1 Consider the wavelet series expansion of continuous-time signals $f(t)$ and assume $\psi(t)$ is the Haar wavelet.

(a) Give the expansion coefficients for $f(t) = 1$, $t \in [0,\ 1]$, and 0 otherwise (that is, the scaling function $\varphi(t)$).

(b) Verify that $\sum_m \sum_n |\langle \psi_{m,n}, f\rangle|^2 = 1$ (Parseval's identity for the wavelet series expansion).

(c) Consider $f'(t) = f(t - 2^{-i})$, where i is a positive integer. Give the range of scales over which expansion coefficients are different from zero.

(d) Same as above, but now $f'(t) = f(t - 1/\sqrt{2})$.

4.2 Consider a multiresolution analysis and the two-scale equation for $\varphi(t)$ given in (4.2.8). Assume that $\{\varphi(t-n)\}$ is an orthonormal basis for V_0. Prove that

(a) $\|g_0[n]\| = 1$

(b) $g_0[n] = \sqrt{2}\ \langle \varphi(2t-n), \varphi(n)\rangle$.

4.3 In a multiresolution analysis with a scaling function $\varphi(t)$ satisfying orthonormality to its integer shifts, consider the two-scale equation (4.2.8). Assume further $0 < |\Phi(0)| < \infty$ and that $\Phi(\omega)$ is continuous in $\omega = 0$.

(a) Show that $\sum_N g_0[n] = \sqrt{2}$.

(b) Show that $\sum_n g_0[2n] - \sum_n g_0[2n+1]$.

4.4 Consider the Meyer wavelet derived in Section 4.3.1 and given by equation (4.3.5). Prove (4.3.6). *Hint:* in every interval $[(2^k\pi)/3, (2^{k+1}\pi)/3]$ there are only two "tails" present.

4.5 A simple Meyer wavelet can be obtained by choosing $\theta(x)$ in (4.3.1) as

$$\theta(x) \ = \ \begin{cases} 0 & x \le 0 \\ x & 0 \le x \le 1 \\ 1 & 1 \le x \end{cases} .$$

(a) Derive the scaling function and wavelet in this case (in Fourier domain).

(b) Discuss the decay in time of the scaling function and wavelet, and compare it to the case when $\theta(x)$ given in (4.3.2) is used.

(c) Plot (numerically) the scaling function and wavelet.

4.6 Consider B-splines as discussed in Section 4.3.2.

(a) Prove (4.3.11) and (4.3.12).

(b) Given that $\beta^{(2N+1)}(t) = \beta^{(N)}(t) * \beta^{(N)}(t)$, prove that

$$b^{(2N+1)}[n] \ = \ \int_{\infty}^{\infty} \beta^{(N)}(t)\ \beta^{(N)}(t-n)\ dt.$$

(This is an alternate proof of (4.3.23).)

(c) Calculate $b^{(2N+1)}[n]$ for $N = 1$ and 2.

4.7 *Battle-Lemarié wavelets:* Calculate the Battle-Lemarié wavelet for the quadratic spline case (see (4.3.26–4.3.27)).

4.8 *Battle-Lemarié wavelets based on recursive filters:* In the orthogonalization procedure of the Battle-Lemarié wavelet (Section 4.3.2), there is a division by $\sqrt{B^{(2N+1)}(\omega)}$ (see (4.3.14), (4.3.17)). Instead of taking a square root, one can perform a spectral factorization of

$B^{(2N+1)}(\omega)$ when $B^{(2N+1)}(\omega)$ is a polynomial in $e^{j\omega}$ (for example, (4.3.16)). For the linear spline case (Section 4.3.2), perform a spectral factorization of $B^{(2N+1)}(\omega)$ into

$$B^{(2N+1)}(\omega) \;=\; R(e^{j\omega}) \cdot R(e^{-j\omega}) \;=\; |R(e^{j\omega})|^2,$$

and derive $\Phi(\omega)$, $\varphi(t)$ (use the fact that $1/R(e^{j\omega})$ is a recursive filter and find the set $\{\alpha_n\}$) and $G_0(e^{j\omega})$. Indicate also $\Psi(\omega)$ in this case.

4.9 Prove that if $g(t)$, the nonorthogonal basis for V_0, has compact support, then $D(\omega)$ in (4.3.20) is a trigonometric polynomial and has a stable (possibly noncausal) spectral factorization.

4.10 *Orthogonality relations of Daubechies' wavelets:* Prove Relations (b) and (c) in Proposition 4.4, namely:
   (a) $\langle \psi(t-n), \psi(t-n')\rangle = \delta[n-n']$ (where we skipped the scaling factor for simplicity)
   (b) $\langle \varphi(t-n), \psi(t-n')\rangle = 0$,

4.11 *Infinite products and the Haar scaling function:*
   (a) Consider the following infinite product:

$$p_k = \prod_{i=0}^{k} a^{b^i} \quad |b| < 1,$$

and show that its limit as $k \to \infty$ is

$$p = \lim_{i\to\infty} p_k = a^{1/(1-b)}.$$

   (b) In Section 4.4.1, we derived the Haar scaling function as the limit of a graphical function, showing that it was equal to the indicator function of the unit interval. Starting from the Haar lowpass filter $G_0(z) = (1+z^{-1})/\sqrt{2}$ and its normalized version $M_0(\omega) = G_0(e^{j\omega})/\sqrt{2}$, show that from (4.4.14),

$$\Phi(\omega) \;=\; \prod_{k=1}^{\infty} M_0\left(\omega/2^k\right) \;=\; e^{-j\omega/2}\frac{\sin(\omega/2)}{\omega/2}.$$

Hint: Use the identity $\cos(\omega) = \sin(2\omega)/2\sin(\omega)$.
   (c) Show, using (4.4.15), that the Haar wavelet is given by

$$\Psi(\omega) \;=\; je^{-j\omega/2}\frac{\sin^2(\omega/4)}{\omega/4}.$$

4.12 Consider the product

$$\Phi^{(i)}(\omega) \;=\; \prod_{k=1}^{i} M_0\left(\frac{\omega}{2^k}\right)$$

where $M_0(\omega)$ is 2π-periodic and satisfies $M_0(0) = 1$ as well as $|M_0(\omega)| \le 1$, $\omega \in [-\pi, \pi]$.
   (a) Show that the infinite product $\Phi^{(i)}(\omega)$ converges pointwise to a limit $\Phi(\omega)$.
   (b) Show that if $M_0(\omega) = 1/\sqrt{2}G_0(e)^{\omega}$ and $G_0(e)^{\omega}$ is the lowpass filter in an orthogonal filter bank, then $|M_0(\omega)| \le 1$ is automatically satisfied and $M_0(0) = 1$ implies $M_0(\pi) = 0$.

4.13 *Maximally flat Daubechies' filters:* A proof of the closed form formula for the autocorrelation of the Daubechies' filter (4.4.34) can be derived as follows (assume $Q = 0$). Rewrite (4.4.32) as

$$P(y) \;=\; \frac{1}{(1-y)^N}\,[1 - y^N\, P(1-y)].$$

Use Taylor series expansion of the first term and the fact that $\deg[P(y)] < N$ (which can be shown using Euclid's algorithm) to prove (4.4.34).

4.14 Given the Daubechies' filters in Table 4.2 or 4.3, verify that they satisfy the regularity bound given in Proposition 4.7. Do they meet higher regularity as well? (you might have to use alternate factorizations or cascades).

4.15 In an N-channel filter bank, show that at least one zero at all aliasing frequencies $2\pi k/N$, $k = 1, \ldots, N-1$, is necessary for the iterated graphical function to converge. *Hint:* See the proof of Proposition 4.6.

4.16 Consider a filter $G_0(z)$ whose impulse response is orthonormal with respect to shifts by N. Assume $G_0(z)$ as K zeros at each of the aliasing frequencies $\omega = 2\pi k/N$, $k = 1, \ldots, N-1$. Consider the iteration of $G_0(z)$ with respect to sampling rate change by N and the associated graphical function (see (4.6.11–4.6.12)). Prove that the condition given in (4.6.15) is sufficient to ensure a continuous limit function $\varphi(t) = \lim_{i\to\infty} \varphi^{(i)}(t)$. *Hint:* The proof is similar to that of Proposition 4.7.

4.17 *Successive interpolation [131]:* Given an input signal $x[n]$, we would like to compute an interpolation by applying upsampling by 2 followed by filtering, and this i times. Assume that the interpolation filter $G(z)$ is symmetric and has zero phase, or $G(z) = g_0 + g_1 z + g_{-1}z^{-1} + g_2 z^2 + g_{-2}z^{-2} + \ldots$

(a) After one step, we would like $y^{(1)}[2n] = x[n]$, while $y^{(1)}[2n+1]$ is interpolated. What conditions does that impose on $G(z)$?

(b) Show that if condition (a) is fulfilled, then after i iterations, we have $y^{(i)}[2^i n] = x[n]$ while other values are interpolated.

(c) Assume $G(z) = 1/2z + 1 + 1/2z^{-1}$. Given some input signal, sketch the output signal $y^{(i)}[n]$ for some small i.

(d) Assume we associate a continuous-time function $y^{(i)}(t)$ with $y^{(i)}[n]$:

$$y^{(i)}(t) = y^{(i)}[n], \quad n/2^i \le t < (n+1)/2^i.$$

What can you say about the limit function $y^{(i)}(t)$ as i goes to infinity and $G(z)$ is as in example (c)? Is the limit function continuous? differentiable?

(e) Consider $G(z)$ to be the autocorrelation of the Daubechies' filters for $N = 2 \ldots 6$, that is, the $P(z)$ given in Table 4.2. Does this satisfy condition (a)? For $N = 2 \ldots 6$, consider the limit function $y^{(i)}(t)$ as i goes to infinity and try to establish the "regularity" of these limit functions (are they continuous, differentiable, etc.?).

4.18 *Recursive subdivision schemes:* Assume that a function $f(t)$ satisfies a two-scale equation $f(t) = \sum_n c_n f(2t-n)$. We can recursively compute $f(t)$ at dyadic rationals with the following procedure. Start with $f^{(0)}(t) = 1$, $-1/2 \le t \le 1/2$, 0 otherwise. In particular, $f^{(0)}(0) = 1$ and $f^{(0)}(1) = f^{(0)}(-1) = 0$. Then, recursively compute

$$f^{(i)}(t) = \sum_n c_n\, f^{(i-1)}(2t-n).$$

In particular, at step i, one can compute the values $f^{(i)}(t)$ at $t = 2^{-i}n, n \in \mathcal{Z}$. This will successively "refine" $f^{(i)}(t)$ to approach the limit $f(t)$, assuming it exists.

(a) Consider this successive refinement for $c_0 = 1$ and $c_1 = c_{-1} = 1/2$. What is the limit $f^{(i)}(t)$ as $i \to \infty$?

(b) A similar refinement scheme can be applied to a discrete-time sequence $s[n]$. Create a function $g^{(0)}(t) = s[n]$ at $t = n$. Then, define

$$g^{(i)}\left(\frac{n}{2^{i-1}}\right) = g^{(i-1)}\left(\frac{n}{2^{i-1}}\right),$$
$$g^{(i)}\left(\frac{2n+1}{2^{i}}\right) = \frac{1}{2}g^{(i-1)}\left(\frac{n}{2^{i-1}}\right) + \frac{1}{2}g^{(i-1)}\left(\frac{n+1}{2^{i-1}}\right).$$

To what function $g(t)$ does this converge in the limit of $i \to \infty$? This scheme is sometimes called bilinear interpolation, explain why.

(c) A more elaborate successive refinement scheme is based on the two-scale equation

$$f(t) = f(2x) + \frac{9}{16}[f(2x+1) + f(2x-1)] - \frac{1}{16}[f(2x+3) + f(2x-3)].$$

Answer parts (a) and (b) for this scheme. (Note: the limit $f(x)$ has no simple closed form expression).

4.19 *Interpolation filters and functions:* A filter with impulse response $g[n]$ is called an interpolation filter with respect to upsampling by 2 if $g[2n] = \delta[n]$. A continuous-time function $f(t)$ is said to have the interpolation property if $f(n) = \delta[n]$. Examples of such functions are the sinc and the hat function.

(a) Show that if $g[n]$ is an interpolation filter and the graphical function $\varphi^{(i)}(t)$ associated with the iterated filter $g^{(i)}[n]$ converges pointwise, then the limit $\varphi(t)$ has the interpolation property.

(b) Show that if $g[n]$ is a finite-length orthogonal lowpass filter, then the only solution leading to an interpolation filter is the Haar lowpass filter (or variations thereof).

(c) Show that if $\varphi(t)$ has the interpolation property and satisfies a two-scale equation

$$\varphi(t) = \sum_n c_n \, \varphi(2t - n),$$

then $c_{2l} = \delta[l]$, that is, the sequence c_n is an interpolation filter.

4.20 Assume a continuous scaling function $\varphi(t)$ with decay $O(1/t^{(1+\epsilon)})$, $\epsilon > 0$, satisfying the two-scale equation

$$\varphi(t) = \sum_n c_n \, \varphi(2t - n).$$

Show that $\sum_n c_{2n} = \sum_n c_{2n+1} = 1$ implies that

$$f(t) = \sum_n \varphi(t - n) = \text{constant} \neq 0.$$

Hint: Show that $f(t) = f(2t)$.

4.21 Assume a continuous and differentiable function $\varphi(t)$ satisfying a two-scale equation

$$\varphi(t) = \sum_n c_n \, \varphi(2t - n)$$

where $\sum_n c_{2n} = \sum_n c_{2n+1} = 1$. Show that $\varphi'(t)$ satisfies a two-scale equation and show this graphically in the case of the hat function (which is differentiable almost everywhere).

4.22 Prove the orthogonality relations for the set of basis functions (4.8.1) in the most general setting, that is, when the windows $w_j(t)$ satisfy conditions (a)–(c) given at the end of Section 4.8.

5

Continuous Wavelet and Short-Time Fourier Transforms and Frames

"Man lives between the infinitely large and the infinitely small."
— Blaise Pascal, *Thoughts*

In this chapter, we consider expansions of continuous-time functions in terms of two variables, such as shift and scale for the wavelet transform, or shift and frequency for the short-time Fourier transform. That is, a one-variable function is mapped into a two-variable function. This representation is redundant but has interesting features which will be studied here. Because of the redundancy, the parameters of the expansion can be discretized, leading to overcomplete series expansions called frames.

Recall Section 2.6.4, where we have seen that one could define the continuous wavelet transform of a function as an inner product between shifted and scaled versions of a single function — the mother wavelet, and the function itself. The mother wavelet we chose was not arbitrary, rather it satisfied a zero-mean condition. This condition follows from the "admissibility condition" on the mother wavelet, which will be discussed in the next section. At the same time, we saw that the resulting transform depended on two parameters — shift and scale, leading to a representation we denote, for a function $f(t)$, by $CWT_f(a,b)$ where a stands for scale and b for shift. Since these two parameters continuously span the real plane (except that scale cannot be zero), the resulting representation is highly redundant.

A similar situation exists in the short-time Fourier transform case (see Section 2.6.3). There, the function is represented in terms of shifts and modulates of a basic window function $w(t)$. As for the wavelet transform, the span of the shift and frequency parameters leads to a redundant representation, which we denote by $STFT_f(\omega, \tau)$ where ω and τ stand for frequency and shift, respectively.

Because of the high redundancy in both $CWT_f(a, b)$ and $STFT_f(\omega, \tau)$, it is possible to discretize the transform parameters and still be able to achieve reconstruction. In the STFT case, a rectangular grid over the (ω, τ) plane can be used, of the form $(m \cdot \omega_0, n \cdot \tau_0)$, $m, n \in \mathcal{Z}$ and with ω_0 and τ_0 sufficiently small $(\omega_0 \tau_0 < 2\pi)$.

In the wavelet transform case, a hyperbolic grid is used instead (with a dyadic grid as a special case when scales are powers of 2). That is, the (a, b) plane is discretized into $(\pm a_0^m, n \cdot a_0^m b_0)$. In this manner, large basis functions (when a_0^m is large) are shifted in large steps, while small basis functions are shifted in small steps. In order for the sampling of the (a, b) plane to be sufficiently fine, a_0 has to be chosen sufficiently close to 1, and b_0 close to 0.

These discretized versions of the continuous transforms are examples of *frames*, which can be seen as overcomplete series expansions (a brief review of frames is given in Section 5.3.2). Reconstruction formulas are possible, but depend on the sampling density. In general, they require different synthesis functions than analysis functions, except in a special case, called a *tight frame*. Then, the frame behaves just as an orthonormal basis, except that the set of functions used to expand the signal is redundant and thus the functions are not independent.

An interesting question is the following: Can one discretize the parameters in the discussed continuous transforms such that the corresponding set of functions is an orthonormal basis? From Chapter 4, we know that this can be done for the wavelet case, with $a_0 = 2$, $b_0 = 1$, and an appropriate wavelet (which is a constrained function). For the STFT, the answer is less obvious and will be investigated in this chapter. However, as a rule, we can already hint at the fact that when the sampling is highly redundant (or, the set of functions is highly overcomplete), we have great freedom in choosing the prototype function. At the other extreme, when the sampling becomes critical, that is, little or no redundancy exists between various functions used in the expansion, then possible prototype functions become very constrained.

Historically, the first instance of a signal representation based on a localized Fourier transform is the Gabor transform [102], where complex sinusoids are windowed with a Gaussian window. It is also called a short-time Fourier transform and has been used extensively in speech processing [8, 226]. A continuous wavelet transform was first proposed by Morlet [119, 125], using a modulated Gaussian as the wavelet (called the Morlet wavelet). Morlet also proposed the inversion formula.[1] The discretization of the continuous transforms is related to the theory of frames, which has been studied in nonharmonic Fourier analysis [89]. Frames of wavelets and short-time Fourier transforms have been studied by Daubechies [72] and an excellent treatment can be found in her book [73] as well, to which we refer for more details. A text that discusses both the continuous wavelet and short-time Fourier transforms is [108]. Several papers discuss these topics as well [10, 60, 99, 293].

Further discussions and possible applications of the continuous wavelet transform can be found in the work of Mallat and coworkers [182, 183, 184] for singularity detection, and in [36, 78, 253, 266] for multiscale signal analysis. Representations involving both scale and modulation are discussed in [185, 291]. Additional material can also be found in edited volumes on wavelets [51, 65, 251].

The outline of the chapter is as follows: The case of continuous transform variables is discussed in the first two sections. In Section 5.1 various properties of the continuous wavelet transform are derived. In particular, the "zooming" property, which allows one to characterize signals locally, is described. Comparisons are made with the STFT, which is presented in Section 5.2. Frames of wavelets and of the STFT are treated in Section 5.3. Tight frames are discussed, as well as the interplay of redundancy and freedom in the choice of the prototype basis function.

5.1 Continuous Wavelet Transform

5.1.1 Analysis and Synthesis

Although the definition of the wavelet transform was briefly introduced in Section 2.6.4, we repeat it here for completeness. Consider the family of functions obtained by shifting and scaling a "mother wavelet" $\psi(t) \in L_2(\mathcal{R})$,

$$\psi_{a,b}(t) = \frac{1}{\sqrt{|a|}}\,\psi\left(\frac{t-b}{a}\right), \tag{5.1.1}$$

where $a, b \in \mathcal{R}$ ($a \neq 0$), and the normalization ensures that $\|\psi_{a,b}(t)\| = \|\psi(t)\|$ (for now, we assume that a can be both positive and negative). In the following, we will assume that the wavelet satisfies the *admissibility condition*

$$C_\psi = \int_{-\infty}^{\infty} \frac{|\Psi(\omega)|^2}{|\omega|} d\omega < \infty, \tag{5.1.2}$$

[1] Morlet proposed the inversion formula based on intuition and numerical evidence. The story goes that when he showed it to a mathematician for verification, he was told: "This formula, being so simple, would be known if it were correct..."

where $\Psi(\omega)$ is the Fourier transform of $\psi(t)$. In practice, $\Psi(\omega)$ will always have sufficient decay so that the admissibility condition reduces to the requirement that $\Psi(0) = 0$ (from (2.4.7–2.4.8)):

$$\int_{-\infty}^{\infty} \psi(t)dt = \Psi(0) = 0.$$

Because the Fourier transform is zero at the origin and the spectrum decays at high frequencies, the wavelet has a bandpass behavior. We now normalize the wavelet so that it has unit energy, or

$$\|\psi(t)\|^2 = \int_{-\infty}^{\infty} |\psi(t)|^2 dt = \frac{1}{2\pi}\int_{-\infty}^{\infty} |\Psi(\omega)|^2 d\omega = 1.$$

As a result, $\|\psi_{a,b}(t)\|^2 = \|\psi(t)\|^2 = 1$ (see (5.1.1)). The continuous wavelet transform of a function $f(t) \in L_2(\mathcal{R})$ is then defined as

$$CWT_f(a,b) = \int_{-\infty}^{\infty} \psi_{a,b}^*(t) f(t)dt = \langle \psi_{a,b}(t), f(t) \rangle. \tag{5.1.3}$$

The function $f(t)$ can be recovered from its transform by the following reconstruction formula, also called *resolution of the identity*:

PROPOSITION 5.1

Given the continuous wavelet transform $CWT_f(a,b)$ of a function $f(t) \in L_2(\mathcal{R})$ (see (5.1.3)), the function can be recovered by:

$$f(t) = \frac{1}{C_\psi}\int_{-\infty}^{\infty}\int_{-\infty}^{\infty} CWT_f(a,b)\, \psi_{a,b}(t)\, \frac{da\, db}{a^2}, \tag{5.1.4}$$

where reconstruction is in the L_2 sense (that is, the L_2 norm of the reconstruction error is zero). This states that any $f(t)$ from $L_2(\mathcal{R})$ can be written as a superposition of shifted and dilated wavelets.

PROOF

In order to simplify the proof, we will assume that $\psi(t) \in L_1$, $f(t) \in L_1 \cap L_2$ as well as $F(\omega) \in L_1$ (or $f(t)$ is continuous) [108]. First, let us rewrite $CWT_f(a,b)$ in terms of the Fourier transforms of the wavelet and signal. Note that the Fourier transform of $\psi_{a,b}(t)$ is

$$\Psi_{a,b}(\omega) = \sqrt{a}e^{-jb\omega}\Psi(a\omega).$$

According to Parseval's formula (2.4.11) given in Section 2.4.2, we get from (5.1.3)

$$\begin{aligned} CWT_f(a,b) = \int_{-\infty}^{\infty} \psi_{a,b}^*(t) f(t)dt &= \frac{1}{2\pi}\int_{-\infty}^{\infty} \Psi_{a,b}^*(\omega)F(\omega)d\omega \\ &= \frac{\sqrt{a}}{2\pi}\int_{-\infty}^{\infty} \Psi^*(a\omega)F(\omega)e^{jb\omega}d\omega. \end{aligned} \tag{5.1.5}$$

Note that the last integral is proportional to the inverse Fourier transform of $\Psi^*(a\omega)F(\omega)$ as a function of b. Let us now compute the integral over b in (5.1.4), which we call $J(a)$,

$$J(a) = \int_{-\infty}^{\infty} CWT_f(a,b)\, \psi_{a,b}(t) db,$$

and substituting (5.1.5)

$$\begin{aligned} J(a) &= \frac{\sqrt{a}}{2\pi} \int_{-\infty}^{\infty} \left(\int_{-\infty}^{\infty} \Psi^*(a\omega)F(\omega)e^{jb\omega} d\omega \right) \psi_{a,b}(t) db \\ &= \frac{\sqrt{a}}{2\pi} \int_{-\infty}^{\infty} \Psi^*(a\omega)F(\omega) \int_{-\infty}^{\infty} \psi_{a,b}(t)e^{jb\omega} db\, d\omega. \end{aligned} \tag{5.1.6}$$

The second integral in the above equation equals (with substitution $b' = (t-b)/a$)

$$\int_{-\infty}^{\infty} \psi_{a,b}(t)e^{jb\omega} db = \frac{1}{\sqrt{a}} \int_{-\infty}^{\infty} \psi\left(\frac{t-b}{a}\right) e^{jb\omega} db$$

$$= \sqrt{a}e^{j\omega t} \int_{-\infty}^{\infty} \psi(b')e^{-j\omega a b'} db' = \sqrt{a}e^{j\omega t}\Psi(a\omega). \tag{5.1.7}$$

Therefore, substituting (5.1.7) into (5.1.6), $J(a)$ becomes equal to

$$J(a) = \frac{|a|}{2\pi} \int_{-\infty}^{\infty} |\Psi(a\omega)|^2 F(\omega) e^{j\omega t} d\omega.$$

We now evaluate the integral in (5.1.4) over a (the integral is multiplied by C_ψ):

$$\int_{-\infty}^{\infty} J(a) \frac{da}{a^2} = \frac{1}{2\pi} \int_{-\infty}^{\infty} \int_{-\infty}^{\infty} F(\omega)e^{j\omega t} \frac{|\Psi(a\omega)|^2}{|a|} d\omega\, da. \tag{5.1.8}$$

Because of the restrictions we imposed on $f(t)$ and $\psi(t)$, we can change the order of integration. We evaluate (use the change of variable $a' = a\omega$)

$$\int_{-\infty}^{\infty} \frac{|\Psi(a\omega)|^2}{|a|} da = \int_{-\infty}^{\infty} \frac{|\Psi(a')|^2}{|a'|} da' = C_\psi, \tag{5.1.9}$$

that is, this integral is independent of ω, which is the key property that makes it all work. It follows that (5.1.8) becomes (this is actually the right side of (5.1.4) multiplied by C_ψ)

$$\frac{1}{2\pi} \int_{-\infty}^{\infty} F(\omega)e^{j\omega t} C_\psi d\omega = C_\psi \cdot f(t),$$

and thus, the inversion formula (5.1.4) is verified almost everywhere. It also becomes clear why the admissibility condition (5.1.2) is required (see (5.1.9)).

If we relax the conditions on $f(t)$ and $\psi(t)$, and require only that they belong to $L_2(\mathcal{R})$, then the inversion formula still holds but the proof requires some finer arguments [73, 108].

There are possible variations on the reconstruction formula (5.1.4) if additional constraints are imposed on the wavelet [75]. We restrict $a \in \mathcal{R}^+$, and if the following modified admissibility condition is satisfied

$$C_\psi = \int_0^{\infty} \frac{|\Psi(\omega)|^2}{|\omega|} d\omega = \int_{-\infty}^{0} \frac{|\Psi(\omega)|^2}{|\omega|} d\omega, \tag{5.1.10}$$

then (5.1.4) becomes

$$f(t) = \frac{1}{C_\psi} \int_0^\infty \int_{-\infty}^\infty CWT_f(a,b)\psi_{a,b}(t)\frac{da\,db}{a^2}.$$

For example, (5.1.10) is satisfied if the wavelet is real and admissible in the usual sense given by (5.1.2).

A generalization of the analysis/synthesis formulas involves two different wavelets; $\psi_1(t)$ for analysis and $\psi_2(t)$ for synthesis, respectively. If the two wavelets satisfy

$$\int_{-\infty}^\infty \frac{|\Psi_1(\omega)||\Psi_2(\omega)|}{|\omega|} d\omega < \infty,$$

then the following reconstruction formula holds [73]:

$$f(t) = \frac{1}{C_{\psi_1,\psi_2}} \int_{-\infty}^\infty \int_{-\infty}^\infty \langle \psi_{1_{a,b}}, f\rangle \psi_{2_{a,b}} \frac{da\,db}{a^2}, \tag{5.1.11}$$

where $C_{\psi_1,\psi_2} = \int (\Psi_1^*(\omega)\Psi_2(\omega)/|\omega|)d\omega$. An interesting feature of (5.1.11) is that $\psi_1(t)$ and $\psi_2(t)$ can have significantly different behavior, as we have seen with biorthogonal systems in Section 4.6.1. For example, $\psi_1(t)$ could be compactly supported but not $\psi_2(t)$, or one could be continuous and not the other.

5.1.2 Properties

The continuous wavelet transform possesses a number of properties which we will derive. Some are closely related to Fourier transform properties (for example, energy conservation) while others are specific to the CWT (such as the reproducing kernel). Some of these properties are discussed in [124]. In the proofs we will assume that $\psi(t)$ is real.

Linearity The linearity of the CWT follows immediately from the linearity of the inner product.

Shift Property If $f(t)$ has a continuous wavelet transform given by $CWT_f(a,b)$, then $f'(t) = f(t-b')$ leads to the following transform:[2]

$$CWT_{f'}(a,b) = CWT_f(a, b-b').$$

This follows since

$$\begin{aligned} CWT_{f'}(a,b) &= \frac{1}{\sqrt{|a|}} \int_{-\infty}^\infty \psi\left(\frac{t-b}{a}\right) f(t-b')dt \\ &= \frac{1}{\sqrt{|a|}} \int_{-\infty}^\infty \psi\left(\frac{t'+b'-b}{a}\right) f(t')dt' = CWT_f(a, b-b'). \end{aligned}$$

[2] In the following, $f'(t)$ denotes the modified function (rather than the derivative).

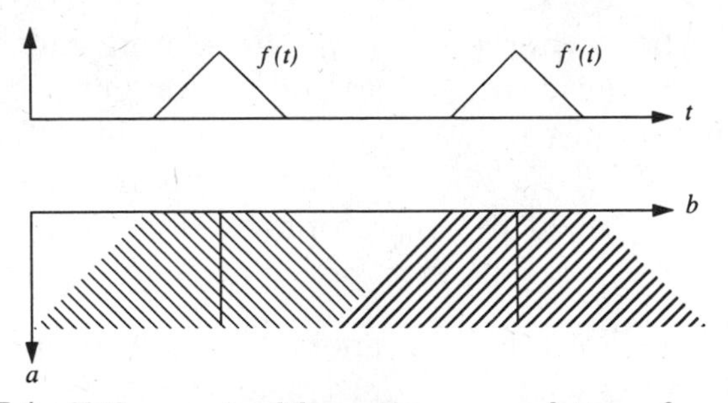

Figure 5.1 Shift property of the continuous wavelet transform. A shift of the function leads to a shift of its wavelet transform. The shading in the (a, b) plane indicates the region of influence.

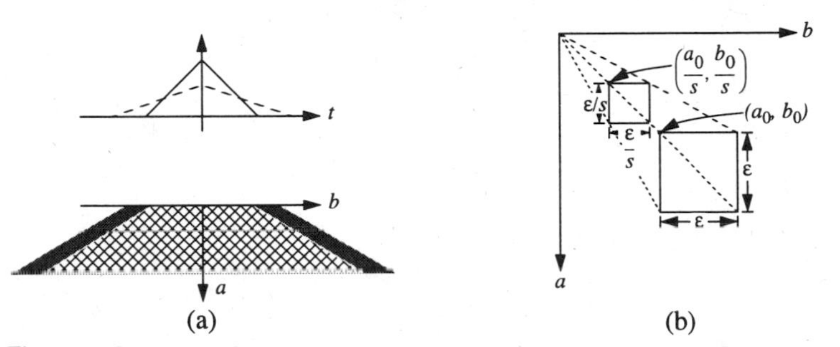

Figure 5.2 The scaling property. (a) Scaling by a factor of 2. (b) Two squares of constant energy in the wavelet-transform plane (after [238]).

This shift invariance of the continuous transform is to be contrasted with the shift variance of the discrete-time wavelet series seen in Chapter 4. Figure 5.1 shows the shift property pictorially.

Scaling Property If $f(t)$ has $CWT_f(a,b)$ as its continuous wavelet transform, then $f'(t) = (1/\sqrt{s})f(t/s)$ has the following transform:

$$CWT_{f'}(a,b) = CWT_f\left(\frac{a}{s}, \frac{b}{s}\right).$$

This follows since

$$\begin{aligned} CWT_{f'}(a,b) &= \frac{1}{\sqrt{|a| \cdot s}} \int_{-\infty}^{\infty} \psi\left(\frac{t-b}{a}\right) f\left(\frac{t}{s}\right) dt \\ &= \sqrt{\frac{s}{|a|}} \int_{-\infty}^{\infty} \psi\left(\frac{st'-b}{a}\right) f(t')dt' = CWT_f\left(\frac{a}{s}, \frac{b}{s}\right). \end{aligned}$$

The scaling property is shown in Figure 5.2(a). We chose $f'(t)$ such that it has the same energy as $f(t)$. Note that an elementary square in the CWT of f', with the upper left corner (a_0, b_0) and width ε, corresponds to an elementary square in the CWT of f with the corner point $(a_0/s, b_0/s)$ and width ε/s, as shown in Figure 5.2(b). That is, assuming a scaling factor greater than 1, energy contained in a given region of the CWT of f is spread by a factor of s in both dimensions in the the CWT of f'. Therefore, we have an intuitive explanation for the measure $(da\ db)/a^2$ used in the reconstruction formula (5.1.4), which weights elementary squares so that they contribute equal energy.

Energy Conservation The CWT has an energy conservation property that is similar to Parseval's formula of the Fourier transform (2.4.12).

PROPOSITION 5.2

Given $f(t) \in L_2(\mathcal{R})$ and its continuous wavelet transform $CWT_f(a,b)$, the following holds:

$$\int_{-\infty}^{\infty} |f(t)|^2 dt = \frac{1}{C_\psi} \int_{-\infty}^{\infty} \int_{-\infty}^{\infty} |CWT_f(a,b)|^2 \frac{da\ db}{a^2}. \tag{5.1.12}$$

PROOF

From (5.1.5) we can write

$$\int_{-\infty}^{\infty} \int_{-\infty}^{\infty} |CWT_f(a,b)|^2 \frac{da\ db}{a^2} = \int_{-\infty}^{\infty} \left(\int_{-\infty}^{\infty} \left| \frac{\sqrt{a}}{2\pi} \int_{-\infty}^{\infty} \Psi^*(a\omega) F(\omega) e^{jb\omega} d\omega \right|^2 db \right) \frac{da}{a^2}.$$

Calling now $P(\omega) = \Psi^*(a\omega)F(\omega)$, we obtain that the above integral equals

$$\begin{aligned} \int_{-\infty}^{\infty} \int_{-\infty}^{\infty} |CWT_f(a,b)|^2 \frac{da\ db}{a^2} &= \int_{-\infty}^{\infty} \left(\int_{-\infty}^{\infty} |\frac{1}{2\pi} \int_{-\infty}^{\infty} P(\omega) e^{jb\omega} d\omega|^2 db \right) \frac{da}{|a|} \\ &= \int_{-\infty}^{\infty} \left(\int_{-\infty}^{\infty} |p(b)|^2 db \right) \frac{da}{|a|} \\ &= \int_{-\infty}^{\infty} \left(\frac{1}{2\pi} \int_{-\infty}^{\infty} |P(\omega)|^2 d\omega \right) \frac{da}{|a|}, \end{aligned} \tag{5.1.13}$$

where we have again used Parseval's formula (2.4.12). Thus, (5.1.13) becomes

$$\int_{-\infty}^{\infty} \left(\frac{1}{2\pi} \int_{-\infty}^{\infty} |\Psi^*(a\omega)|^2 |F(\omega)|^2 d\omega \right) \frac{da}{|a|} = \frac{1}{2\pi} \int_{-\infty}^{\infty} |F(\omega)|^2 \int_{-\infty}^{\infty} \frac{|\Psi(a\omega)|^2}{|a|} da\ d\omega. \tag{5.1.14}$$

The second integral is equal to C_ψ (see (5.1.9)). Applying Parseval's formula again, (5.1.14), and consequently (5.1.13) become

$$\frac{1}{C_\psi} \int_{-\infty}^{\infty} \int_{-\infty}^{\infty} |CWT_f(a,b)|^2 \frac{da\ db}{a^2} = \frac{1}{C_\psi} \cdot \frac{C_\psi}{2\pi} \int_{-\infty}^{\infty} |F(\omega)|^2 d\omega = \int_{-\infty}^{\infty} |f(t)|^2 dt,$$

thus proving (5.1.12).

Again, the importance of the admissibility condition (5.1.2) is evident. Also, the measure $(da\ db)/a^2$ used in the transform domain is consistent with our

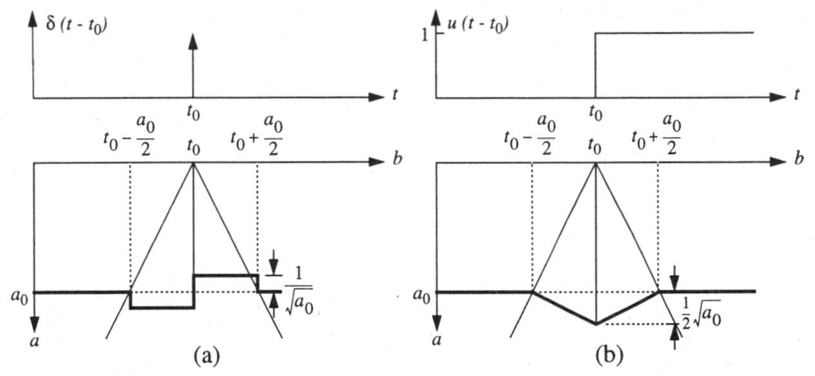

Figure 5.3 Time localization property, shown for the case of a zero-phase Haar wavelet. (a) Behavior of $f(t) = \delta(t - t_0)$. The cone of influence has a width of $a_0/2$ on each side of t_0 and the height is $a_0^{-1/2}$. (b) Behavior for $f(t) = u(t - t_0)$, that is, the unit-step function. The cone of influence is as in part (a), but the height is $-1/2a_0^{1/2}$.

discussion of the scaling property. Scaling by s while conserving the energy will spread the wavelet transform by s in both the dimensions a and b, and thus a renormalization by $1/a^2$ is necessary.

A generalization of this energy conservation formula involves the inner product of two functions in time and in wavelet domains. Then, (5.1.12) becomes [73]

$$\int f^*(t) \cdot g(t) dt = \frac{1}{C_\psi} \int_{-\infty}^{\infty} \int_{-\infty}^{\infty} CWT_f^*(a,b) \cdot CWT_g(a,b) \frac{da\, db}{a^2}, \tag{5.1.15}$$

that is, the usual inner product of the time-domain functions equals, up to a multiplicative constant, the inner product of their wavelet transform, but with the measure $(da\, db)/a^2$.

Localization Properties The continuous wavelet transform has some localization properties, in particular sharp time localization at high frequencies (or small scales) which distinguishes it from more traditional, Fourier-like transforms.

Time Localization Consider a Dirac pulse at time t_0, $\delta(t - t_0)$ and a wavelet $\psi(t)$. The continuous wavelet transform of the Dirac is

$$CWT_\delta(a,b) = \frac{1}{\sqrt{a}} \int \psi\left(\frac{t-b}{a}\right) \delta(t - t_0) dt = \frac{1}{\sqrt{a}} \psi\left(\frac{t_0 - b}{a}\right).$$

For a given scale factor a_0, that is, a horizontal line in the wavelet domain, the transform is equal to the scaled (and normalized) wavelet reversed in time and centered at the location of the Dirac. Figure 5.3(a) shows this localization for

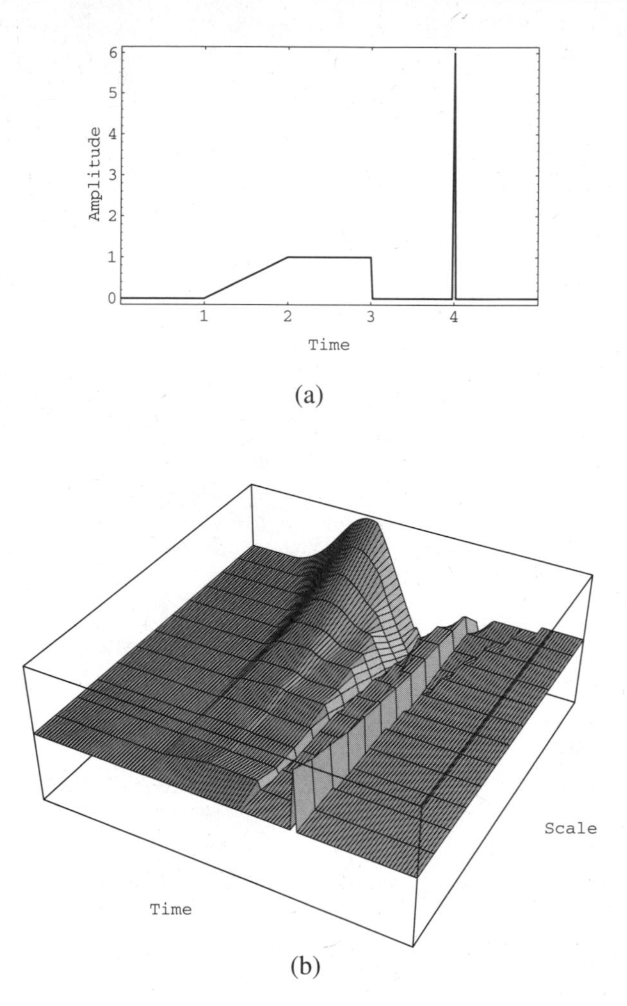

Figure 5.4 Continuous wavelet transform of a simple signal using the Haar wavelet. (a) Signal containing four singularities. (b) Continuous wavelet transform, with small scales toward the front. Note the different behavior at the different singularities and the good time localization at small scales.

the compactly supported Haar wavelet (with zero phase). It is clear that for small a's, the transform "zooms-in" to the Dirac with a very good localization for very small scales. Figure 5.3(b) shows the case of a step function, which has a similar localization but a different magnitude behavior. Another example is given in Figure 5.4 where the transform of a simple synthetic signal with different singularities is shown.

Frequency Localization For the sake of discussion, we will consider the sinc wavelet, that is, a perfect bandpass filter. Its magnitude spectrum is 1 for $|\omega|$

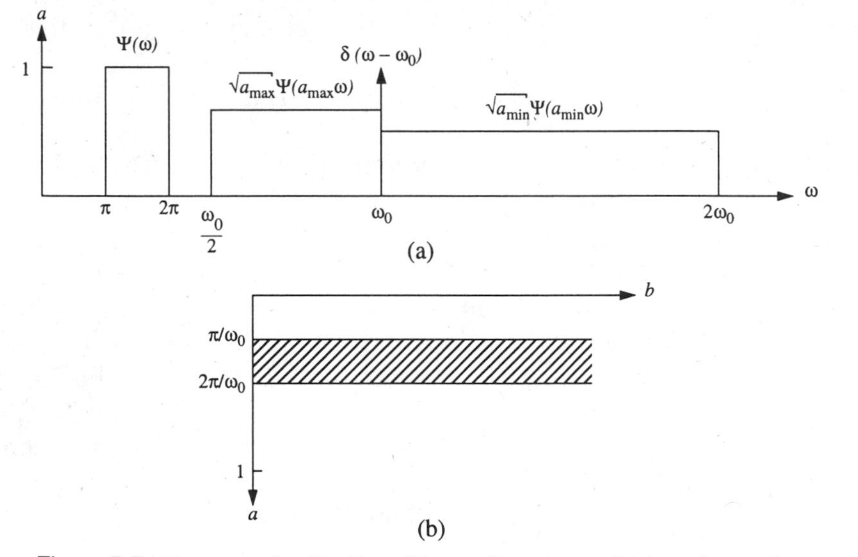

Figure 5.5 Frequency localization of the continuous wavelet transform using a sinc wavelet. (a) Magnitude spectrum of the wavelet and its scaled versions involved in the resolution of a complex sinusoid at ω_0. (b) Nonzero magnitude of the continuous wavelet transform.

between π and 2π. Consider a complex sinusoid of unit magnitude and at frequency ω_0. The highest-frequency wavelet that will pass the sinusoid through, has a scale factor $a_{min} = \pi/\omega_0$ (and a gain of $\sqrt{\pi/\omega_0}$) while the lowest-frequency wavelet passing the sinusoid is for $a_{max} = 2\pi/\omega_0$ (and a gain of $\sqrt{2\pi/\omega_0}$). Figure 5.5(a) shows the various octave-band filters, and Figure 5.5(b) shows the continuous wavelet transform of a sinusoid using a sinc wavelet.

The frequency resolution using an octave-band filter is limited, especially at high frequencies. An improvement is obtained by going to narrower bandpass filters (third of an octave, for example).

Characterization of Regularity In our discussion of time localization (see Figures 5.3 and 5.4), we saw the "zooming" property of the wavelet transform. This allows a characterization of local regularity of signals; a feature which makes the wavelet transform more attractive than the Fourier or local Fourier transform. Indeed, while global regularity of a function can be measured from the decay of its Fourier transform, little can be said about the local behavior. For example, a single discontinuity in an otherwise smooth function will produce an order $1/|\omega|$ decay of its Fourier transform (as an example, consider the step function). The local Fourier transform is able to indicate local regularity within a window, but not more locally. The wavelet transform, because of the zooming property, will isolate the discontinuity from the rest of the function and the behavior of the wavelet transform in the neighborhood of the discontinuity will characterize it.

Consider the wavelet transform of a Dirac impulse in Figure 5.3(a) and of a step function in Figure 5.3(b). In the former case, the absolute value of the wavelet transform behaves as $|a|^{-1/2}$ when approaching the Dirac. In the latter case, it is easy to verify, that the wavelet transform, using a Haar wavelet (with zero phase), is equal to a hat function (a triangle) of height $-1/2 \cdot a_0^{1/2}$ and width from $t_0 - a_0/2$ to $t_0 + a_0/2$. Along the line $a = a_0$, the CWT in 5.3(a) is simply the derivative of the CWT in 5.3(b). This follows from the fact that the CWT can be written as a convolution of the signal with a scaled and time-reversed wavelet. From the differentiation property of the convolution and from the fact that the Dirac is the derivative of the step function (in the sense of distributions), the result follows. In Figure 5.4, we saw the different behavior of the continuous wavelet transform for different singularities, as scale becomes small. A more thorough discussion of the characterization of local regularity can be found in [73, 183] (see also Problem 5.1).

Reproducing Kernel As indicated earlier, the CWT is a very redundant representation since it is a two-dimensional expansion of a one-dimensional function. Consider the space V of square-integrable functions over the plane (a, b) with respect to $(da\, db)/a^2$. Obviously, only a subspace H of V corresponds to wavelet transforms of functions from $L_2(\mathcal{R})$.

PROPOSITION 5.3

If a function $F(a, b)$ belongs to H, that is, it is the wavelet transform of a function $f(t)$, then $F(a, b)$ satisfies

$$F(a_0, b_0) = \frac{1}{C_\psi} \int\int K(a_0, b_0, a, b) F(a, b) \frac{da\, db}{a^2}, \tag{5.1.16}$$

where

$$K(a_0, b_0, a, b) = \langle \psi_{a_0,b_0}, \psi_{a,b} \rangle,$$

is the *reproducing kernel*.

PROOF

To prove (5.1.16), note that $K(a_0, b_0, a, b)$ is the complex conjugate of the wavelet transform of ψ_{a_0,b_0} at (a, b),

$$K(a_0, b_0, a, b) = CWT^*_{\psi_{a_0,b_0}}(a, b), \tag{5.1.17}$$

since $\langle \psi_{a_0,b_0}, \psi_{a,b} \rangle = \langle \psi_{a,b}, \psi_{a_0,b_0} \rangle^*$. Since $F(a, b) = CWT_f(a, b)$ by assumption and using (5.1.17), the right side of (5.1.16) can be written as

$$\frac{1}{C_\psi} \int_{-\infty}^{\infty} \int_{-\infty}^{\infty} K(a_0, b_0, a, b) F(a, b) \frac{da\, db}{a^2}$$

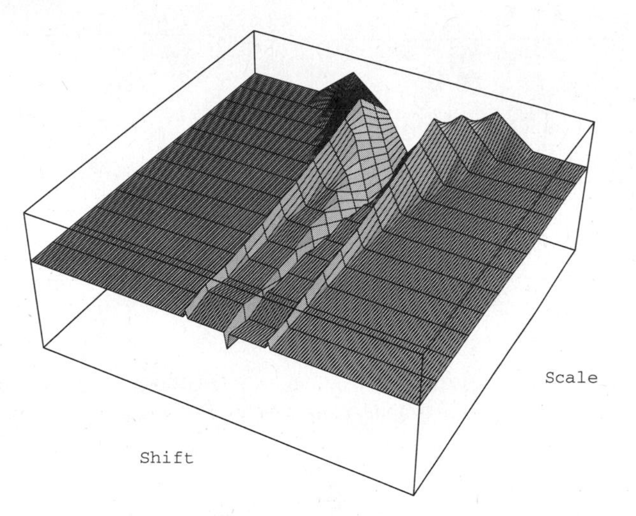

Figure 5.6 Reproducing kernel of the Haar wavelet.

$$\begin{aligned} &= \frac{1}{C_\psi} \int_{-\infty}^{\infty} \int_{-\infty}^{\infty} CWT^*_{\psi_{a_0,b_0}}(a,b) \cdot CWT_f(a,b) \frac{da\, db}{a^2} \\ &= \langle \psi_{a_0,b_0}, f \rangle = CWT_f(a_0,b_0) = F(a_0,b_0), \end{aligned}$$

where (5.1.15) was used to come back to the time domain.

Of course, since $K(a_0, b_0, a, b)$ is the wavelet transform of $\psi_{a,b}$ at location a_0, b_0, it indicates the correlation across shifts and scales of the wavelet ψ.

We just showed that if a two-dimensional function is a continuous wavelet transform of a function, then it satisfies the reproducing kernel relation (5.1.16). It can be shown that the converse is true as well, that is, if a function $F(a,b)$ satisfies (5.1.16), then there is a function $f(t)$ and a wavelet $\psi(t)$ such that $F(a,b) = CWT_f(a,b)$ [238]. Therefore, $F(a,b)$ is a CWT *if and only if* it satisfies the reproducing kernel relation (5.1.16).

An example of a reproducing kernel, that is, the wavelet transform of itself (the wavelet is real), is shown in Figure 5.6 for the Haar wavelet. Note that because of the orthogonality of the wavelet with respect to the dyadic grid, the reproducing kernel is zero at the dyadic grid points.

5.1.3 Morlet Wavelet

The classic example of a continuous-time wavelet analysis uses a windowed complex exponential as the prototype wavelet. This is the *Morlet wavelet*, as first proposed in [119, 125] for signal analysis, and given by

$$\psi(t) = \frac{1}{\sqrt{2\pi}} e^{-j\omega_0 t} e^{-t^2/2}, \qquad (5.1.18)$$

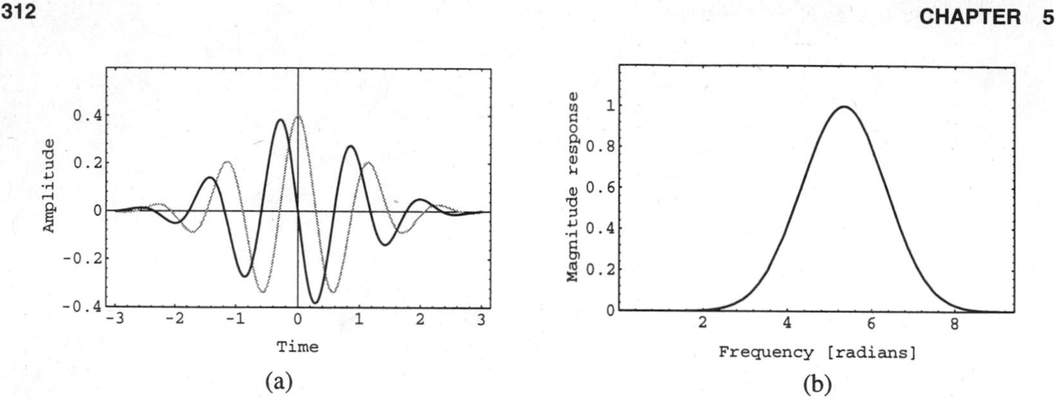

Figure 5.7 Morlet wavelet. (a) Time domain (real and imaginary parts are the continuous and dotted graphs, respectively). (b) Magnitude spectrum.

$$\Psi(\omega) \quad = \quad e^{-(\omega-\omega_0)^2/2}.$$

The factor $1/\sqrt{2\pi}$ in (5.1.18) ensures that $\|\psi(t)\| = 1$. The center frequency ω_0 is usually chosen such that the second maximum of $Re\{\psi(t)\}$, $t > 0$, is half the first one (at $t = 0$). This leads to

$$\omega_0 \quad = \quad \pi\sqrt{\frac{2}{\ln 2}} \quad = \quad 5.336.$$

It should be noted that this wavelet is not admissible since $\Psi(\omega)|_{\omega=0} \neq 0$, but its value at zero frequency is negligible ($\sim 7 \cdot 10^{-7}$), so it does not present any problem in practice. The Morlet wavelet can be corrected so that $\Psi(0) = 0$, but the correction term is very small. Figure 5.7 shows the Morlet wavelet in time and frequency. The latter graph shows that the Morlet wavelet is roughly an octave-band filter. Displays of signal analyses using the continuous-time wavelet transform are often called *scalograms*, in contrast to spectrograms which are based on the short-time Fourier transform.

5.2 Continuous Short-Time Fourier Transform

This transform, also called windowed Fourier or Gabor transform, was briefly introduced in Section 2.6.3. The idea is that of a "localization" of the Fourier transform, using an appropriate window function centered around a location of interest (which can be moved). Thus, as the wavelet transform, it is an expansion along two parameters, frequency and time shift. However, it has a different behavior because of the fixed window size as opposed to the scaled window used in the wavelet transform.

5.2.1 Properties

In the short-time Fourier transform (STFT) case, the functions used in the expansion are obtained by shifts and modulates of a basic window function $w(t)$

$$g_{\omega,\tau}(t) = e^{j\omega t} w(t-\tau). \tag{5.2.1}$$

This leads to an expansion of the form

$$STFT_f(\omega,\tau) = \int_{-\infty}^{\infty} e^{-j\omega t} w^*(t-\tau) f(t) dt = \langle g_{\omega,\tau}(t), f(t)\rangle.$$

There is no admissibility constraint on the window (unlike (5.1.2)) since it is sufficient for the window to have finite energy. It is convenient to choose the window such that $\|w(t)\| = 1$ and we will also assume that $w(t)$ is absolutely integrable, which is the case in practice.

Similarly to the wavelet case, the function $f(t)$ can be recovered, in the L_2 sense, by a double integral

$$f(t) = \frac{1}{2\pi}\int_{-\infty}^{\infty}\int_{-\infty}^{\infty} STFT_f(\omega,\tau) g_{\omega,\tau}(t) d\omega\, d\tau, \tag{5.2.2}$$

where $\|w(t)\| = 1$ was assumed (otherwise, a factor $1/\|w(t)\|^2$ has to be used).

The proof of (5.2.2) can be done by introducing

$$f_A(t) = \frac{1}{2\pi}\int_{-\infty}^{\infty}\int_{-A}^{A} STFT_f(\omega,\tau) g_{\omega,\tau}(t) d\omega d\tau$$

and showing that $\lim_{A\to\infty} f_A(t) = f(t)$ in $L_2(\mathcal{R})$ (see [108] for a detailed proof).

There is also an energy conservation property for the STFT.

PROPOSITION 5.4

Given $f(t) \in L_2(\mathcal{R})$ and its short-time Fourier transform $STFT_f(\omega,\tau)$, the following holds:

$$\|f(t)\|^2 = \frac{1}{2\pi}\int_{-\infty}^{\infty}\int_{-\infty}^{\infty} |STFT_f(\omega,\tau)|^2 d\omega d\tau.$$

PROOF

First, using Parseval's formula, let us write the STFT in Fourier domain as

$$STFT_f(\Omega,\tau) = \int_{-\infty}^{\infty} g^*_{\Omega,\tau}(t) f(t) dt = \frac{1}{2\pi}\int_{-\infty}^{\infty} G^*_{\Omega,\tau}(\omega) F(\omega)\, d\omega, \tag{5.2.3}$$

where

$$G_{\Omega,\tau}(\omega) = e^{-j(\omega-\Omega)\tau} W(\omega-\Omega) \tag{5.2.4}$$

and $W(\omega)$ is the Fourier transform of $w(t)$. Using (5.2.4) in (5.2.3), we obtain

$$\begin{aligned} STFT_f(\Omega,\tau) &= \frac{1}{2\pi} e^{-j\Omega\tau} \int_{-\infty}^{\infty} W^*(\omega-\Omega) F(\omega) e^{j\omega\tau}\, d\omega \\ &= e^{-j\Omega\tau} F^{-1}[W^*(\omega-\Omega)F(\omega)](\tau). \end{aligned}$$

where $F^{-1}[\cdot](\tau)$ is the inverse Fourier transform at τ. Therefore,

$$\begin{aligned}\frac{1}{2\pi}\int_{-\infty}^{\infty}\int_{-\infty}^{\infty}|STFT_f(\Omega,\tau)|^2 d\Omega d\tau &= \frac{1}{2\pi}\int_{-\infty}^{\infty}\left(\int_{-\infty}^{\infty}|F^{-1}[W^*(\omega-\Omega)F(\omega)](\tau)|^2 d\tau\right)d\Omega \\ &= \frac{1}{2\pi}\int_{-\infty}^{\infty}\left(\frac{1}{2\pi}\int_{-\infty}^{\infty}|W^*(\omega-\Omega)F(\omega)|^2 d\omega\right)d\Omega\end{aligned} \tag{5.2.5}$$

where we used Parseval's relation. Interchanging the order of integration (it can be shown that $W^*(\omega-\Omega)F(\omega)$ is in $L_2(\mathcal{R})$), (5.2.5) becomes

$$\int_{-\infty}^{\infty}\frac{1}{2\pi}|F(\omega)|^2\left(\frac{1}{2\pi}\int_{-\infty}^{\infty}|W^*(\omega-\Omega)|^2 d\Omega\right)d\omega = \frac{1}{2\pi}\int_{-\infty}^{\infty}|F(\omega)|^2 d\omega = \|f(t)\|^2$$

where we used the fact that $\|w(t)\|^2 = 1$ or $\|W(\omega)\|^2 = 2\pi$.

5.2.2 Examples

Since the STFT is a local Fourier transform, any classic window that is used in Fourier analysis of signals is a suitable window function. A rectangular window will have poor frequency localization, so smoother windows are preferred. For example, a triangular window has a spectrum decaying in $1/\omega^2$ and is already a better choice. Smoother windows have been designed for data analysis, such as the Hanning window [211]:

$$w(t) = \begin{cases} [1+\cos(2\pi t/T)]/2 & t \in [-T/2, T/2], \\ 0 & \text{otherwise.}\end{cases}$$

The classic window, originally used by Gabor, is the Gaussian window

$$w(t) = \beta e^{-\alpha t^2}, \qquad \alpha, \beta > 0, \tag{5.2.6}$$

where α controls the width, or spread, in time and β is a normalization factor. Its Fourier transform $W(\omega)$ is given by

$$W(\omega) = \beta\sqrt{\frac{\pi}{\alpha}}e^{-\omega^2/4\alpha}.$$

Modulates of a Gaussian window (see (5.2.1)) are often called Gabor functions. An attractive feature of the Gaussian window is that it achieves the best joint time and frequency localization since it meets the lower bound set by the uncertainty principle (see Section 2.6.2).

It is interesting to see that Gabor functions and the Morlet wavelet (see (5.1.18), are related, since they are both modulated Gaussian windows. That is, given a certain α in (5.2.6) and a certain ω_0 in (5.1.18), we have that $\psi_{a,0}(t)$, using the Morlet wavelet, is (we assume zero time shift for simplicity)

$$\psi_{a,0}(t) = \frac{1}{\sqrt{2\pi a}}e^{j\omega_0 t/a}e^{-t^2/2a^2},$$

while $g_{\omega,0}(t)$, using the Gabor window, is

$$g_{\omega,0}(t) = \beta e^{j\omega t} e^{-\alpha t^2},$$

that is, they are equal if $a = 1/\sqrt{2\alpha}$ and $\omega = \omega_0\sqrt{2\alpha}$. Therefore, there is a frequency and a scale at which the Gabor and wavelet transforms coincide. At others, the analysis is different since the wavelet transform uses variable-size windows, as opposed to the fixed-size window of the local Fourier analysis.

This points to a key design question in the STFT, namely the choice of the window size. Once the window size is chosen, all frequencies will be analyzed with the same time and frequency resolutions, unlike what happens in the wavelet transform. In particular, events cannot be resolved if they appear close to each other (within the window spread).

As far as regularity of functions is concerned, one can use Fourier techniques which will indicate regularity estimates within a window. However, it will not be possible to distinguish different behaviors within a window spread. An alternative is to use STFT's with multiple window sizes (see [291] for such a generalized STFT).

5.3 Frames of Wavelet and Short-Time Fourier Transforms

In Chapter 3, we have considered discrete-time orthonormal bases as well as overcomplete expansions. For the latter ones, we pointed out some advantages of relaxing the sampling constraints: As the oversampling factor increases, we get more and more freedom in choosing our basis functions, that is, we can get better filters. In Chapter 4, orthonormal wavelet bases for continuous-time signals were discussed, while at the beginning of this chapter, the continuous-time wavelet and short-time Fourier transforms, that is, very redundant representations, were introduced.

Our aim in this section is to review overcomplete continuous-time expansions called *frames*. They are sets of nonindependent vectors that are able to represent every vector in a given space and can be obtained by discretizing the continuous-time transforms (both wavelet and short-time Fourier transforms). We will see that a frame condition is necessary if we want a numerically stable reconstruction of a function f from a sequence of its transform coefficients (that is, $(\langle \psi_{m,n}, f\rangle)_{m,n\in\mathcal{Z}}$ in the wavelet transform case, and $(\langle g_{m,n}, f\rangle)_{m,n\in\mathcal{Z}}$ in the short-time Fourier transform case).[3] Therefore, the material in this section can be seen as the continuous-time counterpart of overcomplete expansions seen briefly in Section 3.5, as well as a "middle ground" between two extreme cases: Nonredundant orthonormal bases of Chapter 4 and extremely redundant

[3]Round brackets are used to denote sequences of coefficients.

continuous-time wavelet and short-time Fourier transforms at the beginning of this chapter. As in Chapter 3, there will be a trade-off between oversampling and freedom in choosing our basis functions. In the most extreme case, for the short-time Fourier transform frames, the Balian-Low theorem tells us that when critical (Nyquist) sampling is used, it will not be possible to obtain frames with good time and frequency resolutions (and consequently, orthonormal short-time Fourier transform bases will not be achievable with basis functions being well localized in time and frequency). On the other hand, wavelet frames are less restricted and this is one of the reasons behind the excitement that wavelets have generated over the past few years.

A fair amount of the material in this section follows Daubechies's book [73]. For more details and a more rigorous mathematical presentation, the reader is referred to [73], as well as to [26, 72] for more advanced material.

5.3.1 Discretization of the Continuous-Time Wavelet and Short-Time Fourier Transforms

As we have seen previously, the continuous-time wavelet transform employs basis functions given by (5.1.1) where $b \in \mathcal{R}$, $a \in \mathcal{R}^+, a \neq 0$, and the reconstruction formula is based on a double integral, namely the resolution of the identity given by (5.1.4). However, we would like to be able to reconstruct the function from samples taken on a discrete grid. To that end, we choose the following discretization of the scaling parameter a: $a = a_0^m$, with $m \in \mathcal{Z}$ and $a_0 \neq 1$. As for the shift b, consider the following: For $m = 0$, discretize b by taking integer multiples of a fixed b_0 ($b_0 > 0$). The step b_0 should be chosen in such a way that $\psi(t - nb_0)$ will "cover" the whole time axis. Now, the step size b at scale m cannot be chosen independently of m, since the basis functions are rescaled. If we define the "width" of the function, $\Delta_t(f)$, as in (2.6.1), then one can see that the width of $\psi_{a_0^m,0}(t)$ is a_0^m times the width of $\psi(t)$, that is

$$\Delta_t(\psi_{a_0^m,0}(t)) \;=\; a_0^m \Delta_t(\psi(t)).$$

Then, it is obvious that for $\psi_{a,b}(t)$ to "cover" the whole axis at a scale $a = a_0^m$, the shift has to be $b = nb_0 a_0^m$. Therefore, we choose the following discretization:

$$a \;=\; a_0^m, \quad b \;=\; nb_0 a_0^m, \qquad m, n \in \mathcal{Z},\; a_0 > 1,\; b_0 > 0.$$

The discretized family of wavelets is now

$$\psi_{m,n}(t) \;=\; a_0^{-m/2}\psi(a_0^{-m}t - nb_0).$$

As illustrated in Figure 5.8, to different values of m correspond wavelets of different widths: Narrow, high-frequency wavelets are translated by smaller steps in order to "cover" the whole axis, while wider, lower-frequency wavelets

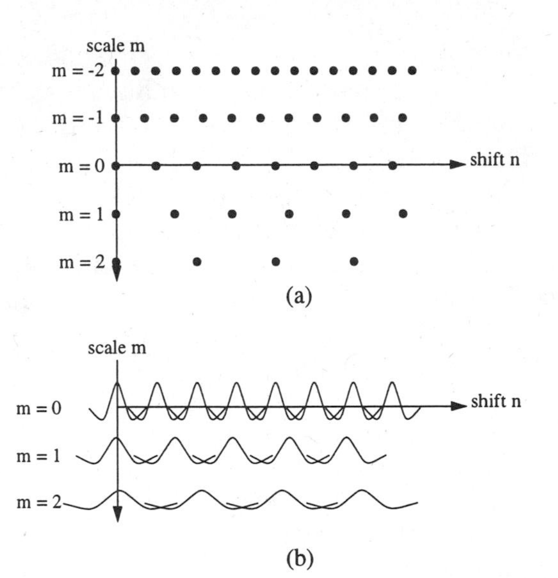

Figure 5.8 By discretizing the values of dilation and shift parameters $a = a_0^m$, $b = nb_0 a_0^m$, one obtains (a) the sampling grid and (b) the corresponding set of functions (the case $a_0 = 2^{1/2}, b_0 = 1$, is shown). To different values of m correspond wavelets of different width: Shorter, high-frequency wavelets are translated by smaller steps, while wider, low-frequency wavelets are translated by larger steps.

are translated by larger steps. For $a_0 = 2,\ b_0 = 1$, we obtain the dyadic case introduced in Chapter 4, for which we know that orthonormal bases exist and reconstruction from transform coefficients is possible.

We would like to answer the following question: Given the sequence of transform coefficients $\langle \psi_{m,n}, f \rangle$, is it possible to reconstruct f in a numerically stable way? In the continuous-parameter case, this is answered by using the resolution of the identity. When the parameters are discretized, there is no equivalent formula. However, in what follows, it will be shown that reconstruction is indeed possible, that is, for certain ψ and appropriate $a_0,\ b_0$, there exist $\tilde{\psi}_{m,n}$ such that the function f can be reconstructed as follows:

$$f = \sum_m \sum_n \langle \psi_{m,n}, f \rangle \tilde{\psi}_{m,n}.$$

It is also intuitively clear that when a_0 is close to one, and b_0 is close to zero, reconstruction should be possible by using the resolution of the identity (since the double sum will become a close approximation to the double integral used in the resolution of the identity). Also, as we said earlier, we know that for some choices of a_0 and b_0 (such as the dyadic case and orthonormal bases in general), reconstruction is possible as well. What we want to explore are the cases in between.

Let us now see what is necessary in order to have a stable reconstruction. Intuitively, the operator that maps a function $f(t)$ into coefficients $\langle \psi_{m,n}, f \rangle$ has to be bounded. That is, if $f(t) \in L_2(\mathcal{R})$, then $\sum_{m,n} |\langle \psi_{m,n}, f \rangle|^2$ has to be finite. Also, no $f(t)$ with $\|f\| > 0$ should be mapped to 0. These two conditions lead to *frame bounds* which guarantee stable reconstruction. Consider the first condition. For any wavelet with some decay in time and frequency, having zero mean, and any choice for $a_0 > 1$, $b_0 > 0$, it can be shown that

$$\sum_{m,n} |\langle \psi_{m,n}, f \rangle|^2 \leq B \, \|f\|^2 \tag{5.3.1}$$

(this just states that the sequence $(\langle \psi_{m,n}, f \rangle)_{m,n}$ is in $l_2(\mathcal{Z}^2)$, that is, the sequence is square-summable [73]). On the other hand, the requirement for stable reconstruction means that if $\sum_{m,n} |\langle \psi_{m,n}, f \rangle|^2$ is small, $\|f\|^2$ should be small as well (that is, $\sum_{m,n} |\langle \psi_{m,n}, f \rangle|^2$ should be "close" to $\|f\|^2$). This further means that there should exist $\alpha < \infty$ such that $\sum_{m,n} |\langle \psi_{m,n}, f \rangle|^2 < 1$ implies $\|f\|^2 \leq \alpha$. Take now an arbitrary f and define $\tilde{f} = \left[\sum_{m,n} |\langle \psi_{m,n}, f \rangle|^2 \right]^{-1/2} f$. Then it is obvious that $\sum_{m,n} |\langle \psi_{m,n}, \tilde{f} \rangle|^2 \leq 1$ and consequently, $\|\tilde{f}\|^2 \leq \alpha$. This is equivalent to

$$A \, \|f\|^2 \leq \sum_{m,n} |\langle \psi_{m,n}, f \rangle|^2, \tag{5.3.2}$$

for some $A = 1/\alpha$. Take now $f = f_1 - f_2$. Then, (5.3.2) means also that the distance $\|f_1 - f_2\|$ cannot be arbitrarily large if $\sum_{m,n} |\langle \psi_{m,n}, f_1 \rangle - \langle \psi_{m,n}, f_2 \rangle|^2$ is small, or, (5.3.2) is equivalent to the stability requirement. Putting (5.3.1) and (5.3.2) together tells us that a numerically stable reconstruction of f from its transform (wavelet) coefficients is possible only if

$$A \, \|f\|^2 \leq \sum_{m,n} |\langle \psi_{m,n}, f \rangle|^2 \leq B \, \|f\|^2.$$

If this condition is satisfied, then the family $(\psi_{m,n})_{m,n \in \mathcal{Z}}$ constitutes a *frame*. When $A = B = 1$, and $|\psi_{m,n}| = 1$, for all m, n, the family of wavelets is an orthonormal basis (what we will call a tight frame with a frame bound equal to 1). These notions will be defined in Section 5.3.2.

Until now, we have seen how the continuous-time wavelet transform can be discretized and what the conditions on that discretized version are so that a numerically stable reconstruction from $(\langle \psi_{m,n}, f \rangle)_{m,n}$ is possible. What about the short-time Fourier transform? As we have seen in Section 5.2, the basis functions are given by (5.2.1). As before, we would like to be able to reconstruct the function from the samples taken on a discrete grid. In the same manner as for the wavelet transform, it is possible to discretize the short-time Fourier transform as follows: In $g_{\omega,\tau}(t) = e^{j\omega t}w(t-\tau)$ choose $\omega = m\omega_0$ and $\tau = nt_0$, with $\omega_0,\ t_0 > 0$ fixed, $m,\ n \in \mathcal{Z}$ so that

$$g_{m,n}(t) \;=\; e^{jm\omega_0 t}w(t - nt_0). \tag{5.3.3}$$

Again, we would like to know whether it is possible to reconstruct a given function f from its transform coefficients $(\langle g_{m,n}, f\rangle)_{m,n}$ in a numerically stable way and again, the answer is positive provided that $g_{m,n}$ constitute a frame. Then, the reconstruction formula becomes

$$\sum_{m,n}\langle g_{m,n}, f\rangle\, \tilde{g}_{m,n} \;=\; f \;=\; \sum_{m,n}\langle \tilde{g}_{m,n}, f\rangle\, g_{m,n},$$

where $\tilde{g}_{m,n}$ are the vectors of the dual frame, and

$$\langle g_{m,n}, f\rangle \;=\; \int e^{-jm\omega_0 t}w^*(t - nt_0)f(t)dt.$$

5.3.2 Reconstruction in Frames

As we have just seen, for numerically stable reconstruction, the vectors used for the expansion have to constitute a frame. Therefore, in this section, we will present an overview of frames, as well as an algorithm to reconstruct f from its transform coefficients. For a more detailed and rigorous account of frames, see [72, 73].

DEFINITION 5.5

A family of functions $(\gamma_j)_{j\in J}$ in a Hilbert space H is called a *frame* if there exist $0 < A \leq B < \infty$, such that, for all f in H,

$$A\, \|f\|^2 \;\leq\; \sum_{j\in J} |\langle\gamma_j, f\rangle|^2 \;\leq\; B\, \|f\|^2, \tag{5.3.4}$$

where, A and B are called *frame bounds*.

If the two frame bounds are equal, the frame is called a *tight* frame. In that case, and if $\|\gamma_j\| = 1$, $A = B$ gives the "redundancy ratio", or the oversampling ratio. If that ratio equals to 1, we obtain the "critical" sampling case, or an orthonormal basis. These observations lead to the following proposition [73]:

PROPOSITION 5.6

If $(\gamma_j)_{j\in J}$ is a tight frame, with frame bound $A = 1$, and if $\|\gamma_j\| = 1$, for all $j \in J$, then the γ_j constitute an orthonormal basis.

Note that the converse is just Parseval's formula. That is, an orthonormal basis is also a tight frame with frame bounds equal to 1.

Since for a tight frame $\sum_{j\in J}|\langle\gamma_j, f\rangle|^2 = A\|f\|^2$, or, $\sum_{j\in J}\langle f,\gamma_j\rangle\langle\gamma_j, g\rangle = A\langle f, g\rangle$, we can say that (at least in the weak sense [73])

$$f = \frac{1}{A}\sum_{j\in J}\langle\gamma_j, f\rangle\,\gamma_j. \tag{5.3.5}$$

This gives us an easy way to recover f from its transform coefficients $\langle\gamma_j, f\rangle$ if the frame is tight. Note that (5.3.5) with $A = 1$ gives the usual reconstruction formula for an orthonormal basis.

A frame, however, (even a tight frame) is not an orthonormal basis; it is a set of nonindependent vectors, as is shown in the following examples.

Example 5.1

Consider $\mathcal{R}^2$ and the redundant set of vectors $\varphi_0 = [1,0]^T$, $\varphi_1 = [-1/2, \sqrt{3}/2]^T$ and $\varphi_2 = [-1/2, -\sqrt{3}/2]^T$ (this overcomplete set was briefly discussed in Example 1.1 and shown in Figure 1.1). Creating a matrix $\boldsymbol{M} = [\varphi_0, \varphi_1, \varphi_2]$, it is easy to verify that

$$\boldsymbol{M}\boldsymbol{M}^T = \frac{3}{2}\boldsymbol{I}$$

and thus, any vector $x \in \mathcal{R}^2$ can be written as

$$x = \frac{2}{3}\sum_{i=0}^{2}\langle\varphi_i, x\rangle\,\varphi_i. \tag{5.3.6}$$

Note that $\|\varphi_i\| = 1$, and thus $3/2$ is the redundancy factor. Also, in (5.3.6), the dual set is identical to the vectors of the expansion. However, this set is not unique, because the φ_i's are linearly dependent. Since $\sum_{i=0}^{2}\varphi_i = 0$, we can choose

$$\tilde{\varphi}_i = \varphi_i + \begin{bmatrix}\alpha\\ \beta\end{bmatrix}$$

and still obtain

$$x = \frac{2}{3}\sum_{i=0}^{2}\langle\tilde{\varphi}_i, x\rangle\,\varphi_i.$$

The particular choice of $\alpha = \beta = 0$ leads to $\tilde{\varphi}_i = \varphi_i$.[4] See Problem 5.5 for a more general version of this example.

Example 5.2

Consider a two-channel filter bank, as given in Chapter 3, but this time with no downsampling (see Section 3.5.1). Obviously, the output is simply

$$\hat{X}(z) = [G_0(z)H_0(z) + G_1(z)H_1(z)]\;X(z).$$

[4] This particular choice is unique, and leads to the *dual frame* (which happens to be identical to the frame in this case).

Suppose now that the two filters $G_0(z)$ and $G_1(z)$ are of unit norm and satisfy

$$G_0(z)G_0(z^{-1}) + G_1(z)G_1(z^{-1}) = 2.$$

Then, setting $H_0(z) = G_0(z^{-1})$ and $H_1(z) = G_1(z^{-1})$ we get

$$\hat{X}(z) = [G_0(z)G_0(z^{-1}) + G_1(z)G_1(z^{-1})]\, X(z) = 2 \cdot X(z). \tag{5.3.7}$$

Write this in time domain using the impulse responses $g_0[n]$ and $g_1[n]$ and their translates. The output of the filter $h_0[n] = g_0[-n]$ at time k equals $\langle g_0[n-k], x[n]\rangle$ and thus contributes $\langle g_0[n-k], x[n]\rangle \cdot g_0[m-k]$ to the output at time m. A similar relation holds for $g_1[n-k]$. Therefore, using these relations and (5.3.7), we can write

$$\hat{x}[m] = \sum_{k=-\infty}^{\infty} \sum_{i=0}^{1} \langle g_i[n-k], x[n]\rangle\, g_i[m-k] = 2 \cdot x[m].$$

That is, the set $\{g_i[n-k]\}, i = 0, 1$, and $k \in \mathcal{Z}$, forms a tight frame for $l_2(\mathcal{Z})$ with a redundancy factor $R = 2$. The redundancy factor indicates the oversampling rate, which is indeed a factor of two in our two-channel, nondownsampled case. The vectors $g_i[n-k], k \in \mathcal{Z}$ are not independent; indeed, there are twice as many than what would be needed to uniquely represent the vectors in $l_2(\mathcal{Z})$. This redundancy, however, allows for more freedom in design of $g_i[k-n]$. Moreover, the representation is now shift-invariant, unlike in the critically sampled case.

What about reconstructing with frames that are not tight? Let us define the *frame operator* Γ from $L_2(\mathcal{R})$ to $l_2(J)$ as

$$(\Gamma f)_j = \langle \gamma_j, f\rangle. \tag{5.3.8}$$

Since $(\gamma_j)_{j\in J}$ constitute a frame, we know from (5.3.4) that $\|\Gamma f\|^2 \leq B\|f\|^2$, that is, Γ is bounded, which means that it is possible to find its adjoint operator Γ^*. Note first that the adjoint operator is a mapping from $l_2(J)$ to $L_2(\mathcal{R})$. Then, $\langle f, \Gamma^* c\rangle$ is an inner product over $L_2(\mathcal{R})$, while $\langle \Gamma f, c\rangle$ is an inner product over $l_2(J)$. The adjoint operator can be computed from the following relation (see (2.A.2))

$$\langle f, \Gamma^* c\rangle = \langle \Gamma f, c\rangle = \sum_{j\in J} \langle \gamma_j, f\rangle^* c_j. \tag{5.3.9}$$

Exchanging the order in the inner product, we get that

$$\sum_{j\in J} \langle \gamma_j, f\rangle^* c_j = \sum_{j\in J} c_j \langle f, \gamma_j\rangle = \langle f, \sum_{j\in J} c_j \gamma_j\rangle. \tag{5.3.10}$$

Comparing the left side of (5.3.9) with the right side of (5.3.10), we find the adjoint operator as

$$\Gamma^* c = \sum_{j\in J} c_j \gamma_j. \tag{5.3.11}$$

From this it follows that:

$$\sum_j \langle \gamma_j, f\rangle \gamma_j = \Gamma^* \Gamma f. \tag{5.3.12}$$

Using this adjoint operator, we can express condition (5.3.4) as (I is the identity operator)

$$A \cdot I \;\leq\; \Gamma^*\Gamma \;\leq\; B \cdot I, \tag{5.3.13}$$

from where it follows that $\Gamma^*\Gamma$ is invertible (see Lemma 3.2.2 in [73]). Applying this inverse $(\Gamma^*\Gamma)^{-1}$ to the family of vectors γ_j, leads to another family $\tilde{\gamma}_j$ which also constitutes a frame. The vectors $\tilde{\gamma}_j$ are given by

$$\tilde{\gamma}_j \;=\; (\Gamma^*\Gamma)^{-1}\gamma_j. \tag{5.3.14}$$

This new family of vectors is called a *dual frame* and it satisfies

$$B^{-1}\|f\|^2 \;\leq\; \sum_{j\in J} |\langle\tilde{\gamma}_j, f\rangle|^2 \;\leq\; A^{-1}\|f\|^2,$$

and the reconstruction formula becomes

$$\begin{aligned}
\sum_{j\in J}\langle\gamma_j, f\rangle\,\tilde{\gamma}_j \;&=\; \sum_{j\in J}\langle\gamma_j, f\rangle\,(\Gamma^*\Gamma)^{-1}\gamma_j \\
&=\; (\Gamma^*\Gamma)^{-1}\sum_{j\in J}\langle\gamma_j, f\rangle\,\gamma_j \\
&=\; (\Gamma^*\Gamma)^{-1}\Gamma^*\Gamma f \\
&=\; f,
\end{aligned}$$

where we have used (5.3.14), (5.3.8) and (5.3.11). Therefore, one can write

$$\sum_{j\in J}\langle\gamma_j, f\rangle\tilde{\gamma}_j \;=\; f \;=\; \sum_{j\in J}\langle\tilde{\gamma}_j, f\rangle\,\gamma_j. \tag{5.3.15}$$

The above relation shows how to obtain a reconstruction formula for f from $\langle\gamma_j, f\rangle$, where the only thing one has to compute is $\tilde{\gamma}_j = (\Gamma^*\Gamma)^{-1}\gamma_j$, given by

$$\tilde{\gamma}_j = \frac{2}{A+B}\sum_{k=0}^{\infty}\left(I - \frac{2}{A+B}\Gamma^*\Gamma\right)^k\gamma_j. \tag{5.3.16}$$

We now sketch a proof of this relation (see [73]) for a rigorous development).

PROOF

If frame bounds A and B are close, that is, if

$$\nabla \;=\; \frac{B}{A} - 1 \ll 1,$$

then (5.3.13) implies that $\Gamma^*\Gamma$ is close to $((A+B)/2)I$, or $(\Gamma^*\Gamma)^{-1}$ is close to $(2/(A+B))I$. This further means that the function f can be written as follows:

$$f \;=\; \frac{2}{A+B}\sum_{j\in J}\langle\gamma_j, f\rangle\,\gamma_j + Rf,$$

where R is given by (use (5.3.12))

$$R = I - \frac{2}{A+B}\Gamma^*\Gamma. \tag{5.3.17}$$

Using (5.3.13) we obtain

$$-\frac{B-A}{B+A}I \leq R \leq \frac{B-A}{B+A}I,$$

and as a result,

$$\|R\| \leq \frac{B-A}{B+A} = \frac{\nabla}{2+\nabla} \leq 1. \tag{5.3.18}$$

From (5.3.17) and using (5.3.18), $(\Gamma^*\Gamma)^{-1}$ can be written as (see also (2.A.1))

$$(\Gamma^*\Gamma)^{-1} = \frac{2}{A+B}(I-R)^{-1} = \frac{2}{A+B}\sum_{k=0}^{\infty} R^k,$$

implying that

$$\tilde{\gamma}_j = (\Gamma^*\Gamma)^{-1}\gamma_j = \frac{2}{A+B}\sum_{k=0}^{\infty} R^k\gamma_j = \frac{2}{A+B}\sum_{k=0}^{\infty}(I - \frac{2}{A+B}\Gamma^*\Gamma)^k\gamma_j. \tag{5.3.19}$$

Note that if B/A is close to one, that is, if ∇ is small, then R is close to zero and convergence in (5.3.19) is fast. If the frame is tight, that is, $A = B$, and moreover, if it is an orthonormal basis, that is, $A = 1$, then $R = I$ and $\tilde{\gamma}_j = \gamma_j$.

We have seen, for example, in the wavelet transform case, that to have a numerically stable reconstruction, we require that $(\psi_{m,n})$ constitute a frame. If $(\psi_{m,n})$ do constitute a frame, we found an algorithm to reconstruct f from $\langle f, \psi_{m,n}\rangle$, given by (5.3.15) with $\tilde{\gamma}_j$ as in (5.3.16). For this algorithm to work, we have to obtain estimates of frame bounds.

5.3.3 Frames of Wavelets and STFT

In the last section, we dealt with abstract issues regarding frames and the reconstruction issue. Here, we will discuss some particularities of frames of wavelets and short-time Fourier transform. The main point of this section will be that for wavelet frames, there are no really strong constraints on $\psi(t), a_0, b_0$. On the other hand, for the short-time Fourier transform, the situation is more complicated and having good frames will be possible only for certain choices of ω_0 and τ_0. Moreover, if we want to avoid redundancy and critically sample the short-time Fourier transform, we will have to give up either good time or good frequency resolution. This is the content of the Balian-Low theorem, given later in this section.

In all the cases mentioned above, we need to have some estimates of the frame bounds in order to compute the dual frame. Therefore, we start with wavelet frames and show that a family of wavelets being a frame imposes the admissibility condition for the "mother" wavelet. We give the result here without proof (for a proof, refer to [73]).

PROPOSITION 5.7

If the $\psi_{m,n}(t) = a_0^{-m/2}\psi(a_0^{-m}t - nb_0),\ m, n \in \mathcal{Z}$ constitute a frame for $L^2(\mathcal{R})$ with frame bounds $A,\ B$, then

$$\frac{b_0 \ln a_0}{2\pi} A \le \int_0^\infty \frac{|\Psi(\omega)|^2}{\omega} d\omega \le \frac{b_0 \ln a_0}{2\pi} B, \tag{5.3.20}$$

and

$$\frac{b_0 \ln a_0}{2\pi} A \le \int_{-\infty}^0 \frac{|\Psi(\omega)|^2}{|\,\omega\,|} d\omega \le \frac{b_0 \ln a_0}{2\pi} B. \tag{5.3.21}$$

Compare these expressions with the admissibility condition given in (5.1.2). It is obvious that the fact that the wavelets form a frame, automatically imposes the admissibility condition on the "mother" wavelet. This proposition will also help us find frame bounds in the case when the frame is tight $(A = B)$, since then

$$A = \frac{2\pi}{b_0 \ln a_0} \int_0^\infty \frac{|\Psi(\omega)|^2}{\omega} d\omega = \frac{2\pi}{b_0 \ln a_0} \int_{-\infty}^0 \frac{|\Psi(\omega)|^2}{|\,\omega\,|} d\omega.$$

Moreover, in the orthonormal case (we use the dyadic case as an example, $A = B = 1,\ b_0 = 1,\ a_0 = 2$)

$$\int_0^\infty \frac{|\Psi(\omega)|^2}{\omega} d\omega = \int_{-\infty}^0 \frac{|\Psi(\omega)|^2}{|\,\omega\,|} d\omega = \frac{\ln 2}{2\pi}.$$

We mentioned previously that in order to have wavelet frames, we need not impose really strong conditions on the wavelet, and the scaling and shift factors. In other words, if $\psi(t)$ is at all a "reasonable" function (it has some decay in time and frequency, and $\int \psi(t)dt = 0$) then there exists a whole arsenal of a_0 and b_0, such that $\{\psi_{m,n}\}$ constitute a frame. This can be formalized, and we refer to [73] for more details (Proposition 3.3.2, in particular). In [73], explicit estimates for frame bounds A, B, as well as possible choices for ψ, a_0, b_0, are given.

Example 5.3

As an example to the previous discussion, consider the so-called Mexican-hat function

$$\psi(t) = \frac{2}{\sqrt{3}} \pi^{-1/4}(1 - t^2)\, e^{-t^2/2},$$

given in Figure 5.9. Table 5.1 gives a few values for frame bounds $A,\ B$ with $a_0 = 2$ and varying b_0. Note, for example, how for certain values of b_0, the frame is almost tight — a so-called "snug" frame. The advantage of working with such a frame is that we can use just the 0th-order term in the reconstruction formula (5.3.16) and still get a good approximation of f. Another interesting point is that when the frame is almost tight, the frame bounds (which are close) are inversely proportional to b_0. Since the frame bounds in this case measure redundancy of the frame, when b_0 is halved (twice as many points

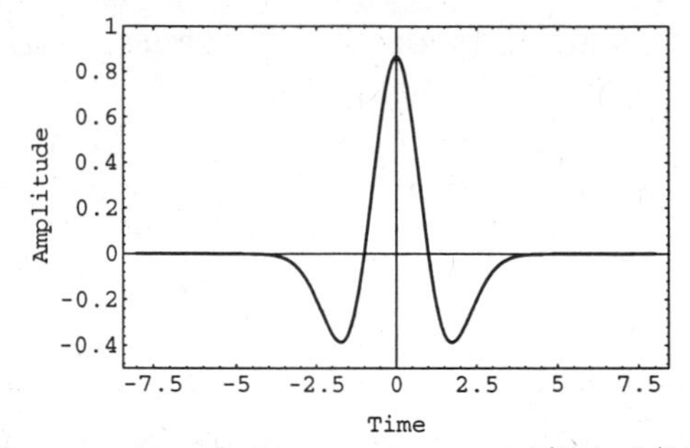

Figure 5.9 The Mexican-hat function $\psi(t) = (2/3^{1/2})\ \pi^{-1/4}(1-t^2)e^{-t^2/2}$. The rotated $\psi(t)$ gives rise to a Mexican hat — thus the name for the function.

Table 5.1 Frame bounds for Mexican-hat wavelet frames with $a_0 = 2$ (from [73]).

b_0	A	B	B/A
0.25	13.091	14.183	1.083
0.50	6.546	7.092	1.083
0.75	4.364	4.728	1.083
1.00	3.223	3.596	1.116
1.25	2.001	3.454	1.726
1.50	0.325	4.221	12.986

on the grid), the frame bounds should double (redundancy increases by two since we have twice as many functions). Note also how for the value of $b_0 = 1.50$, the ratio B/A increases suddenly. Actually, for larger values of b_0, the set $\{\psi_{m,n}\}$ is not even a frame any more, since A is not strictly positive anymore.

Finally, let us say a few words on time-frequency localization properties of wavelet frames. Recall that one of the reasons we opted for the wavelet-type signal expansions is because they allegedly provide good localization in both time and frequency. Let us here, for the sake of discussion, assume that $|\psi|$ and $|\Psi|$ are symmetric. ψ is centered around $t = 0$, and Ψ is centered around $\omega = \omega_0$ (this implies that $\psi_{m,n}$ will be centered around $t = a_0^m n b_0$ and around $\pm a_0^{-m}\omega_0$ in frequency). This means that the inner product $\langle \psi_{m,n}, f\rangle$ represents the "information content" of f near $t = a_0^m n b_0$ and near $\omega_\pm = \pm a_0^{-m}\omega_0$. If the function f is localized (most of its energy lies within $|t| \leq T$ and $\Omega_0 \leq |\omega| \leq \Omega_1$) then only the coefficients $\langle \psi_{m,n}, f\rangle$ for which $(t,\omega) = (a_0^m n b_0, \pm a_0^{-m}\omega_0)$ lies within (or very close) to $[-T,T] \times ([-\Omega_1, -\Omega_0] \cup [\Omega_0, \Omega_1])$ will be necessary for f to be reconstructed up to a good approximation. This approximation property is detailed in [73] (Theorem 3.5.1, in particular).

Let us now shift our attention to the short-time Fourier transform frames.

As mentioned before, we need to be able to say something about the frame bounds in order to compute the dual frame. Then, in a similar fashion to Proposition 5.7, one can obtain a very interesting result, which states that if $g_{m,n}(t)$ (as in (5.3.3)) constitute a frame for $L_2(\mathcal{R})$ with frame bounds A and B, then

$$A \leq \frac{2\pi}{\omega_0 t_0}\|g\|^2 \leq B. \tag{5.3.22}$$

Note how in this case, any tight frame will have a frame bound $A = (2\pi)/(\omega_0 t_0)$ (with $\|g\| = 1$). In particular, an orthonormal basis will require the following to be true:

$$\omega_0 t_0 = 2\pi.$$

Beware, however, that $\omega_0 t_0 = 2\pi$ will not imply an orthonormal basis; it just states that we have "critically" sampled our short-time Fourier transform.[5] Note that in (5.3.22) g does not appear (except $\|g\|$ which can always be normalized to 1), as opposed to (5.3.20), (5.3.21). This is similar to the absence of an admissibility condition for the continuous-time short-time Fourier transform (see Section 5.2). On the other hand, we see that ω_0, t_0 cannot be arbitrarily chosen. In fact, there are no short-time Fourier transform frames for $\omega_0 t_0 > 2\pi$. Even more is true: In order to have good time-frequency localization, we require that $\omega_0 t_0 < 2\pi$. The last remaining case, that of critical sampling, $\omega_0 t_0 = 2\pi$, is very interesting. Unlike for the wavelet frames, it turns out that no critically sampled short-time Fourier transform frames are possible with good time and frequency localization. Actually, the following theorem states just that.

THEOREM 5.8 *(Balian-Low)*

If the $g_{m,n}(t) = e^{j2\pi mt}w(t-n)$, $m, n \in \mathcal{Z}$ constitute a frame for $L_2(\mathcal{R})$, then either $\int t^2|w(t)|^2 dt = \infty$ or $\int \omega^2 |W(\omega)|^2 d\omega = \infty$.

For a proof, see [73]. Note that in the statement of the theorem, $t_0 = 1$, $\omega_0 = 2\pi/t_0 = 2\pi$. Thus, in this case ($\omega_0 t_0 = 2\pi$), we will necessarily have bad localization either in time or in frequency (or possibly both). This theorem has profound consequences, since it also implies that no good short-time Fourier transform orthonormal bases (good meaning with good time and frequency localization) are achievable (since orthonormal bases are necessarily critically sampled). This is similar to the discrete-time result we have seen in Chapter 3, Theorem 3.17. The previous discussion is pictorially represented in Figure 5.10 (after [73]).

A few more remarks about the short-time Fourier transform: First, as in the wavelet case, it is possible to obtain estimates of the frame bounds A, B.

[5] In signal processing terms, this corresponds to the Nyquist rate.

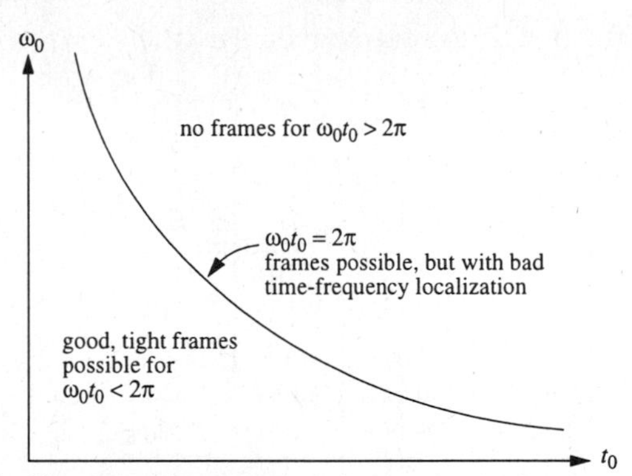

Figure 5.10 Short-time Fourier transform case: no frames are possible for $\omega_0 t_0 > 2\pi$. There exist frames with bad time-frequency localization for $\omega_0 t_0 = 2\pi$. Frames (even tight frames) with excellent time-frequency localization are possible for $\omega_0 t_0 < 2\pi$ (after [73]).

Unlike the wavelet case, however, the dual frame is always generated by a single function $\tilde{w}$. To see that, first introduce the shift operator $Tw(t) = w(t - t_0)$ and the operator $Ew(t) = e^{j\omega_0 t} w(t)$. Then, $g_{m,n}(t)$ can be expressed as

$$g_{m,n}(t) = e^{jm\omega_0 t} w(t - nt_0) = E^m T^n w(t).$$

One can easily check that both T and E commute with $\Gamma^*\Gamma$ and thus with $(\Gamma^*\Gamma)^{-1}$ as well [225]. Then, the dual frame can be found from (5.3.14)

$$\begin{aligned} \text{dual}(g_{m,n})(t) &= (\Gamma^*\Gamma)^{-1} g_{m,n}(t) \\ &= (\Gamma^*\Gamma)^{-1} E^m T^n w(t) \\ &= E^m T^n (\Gamma^*\Gamma)^{-1} w(t) \\ &= E^m T^n \tilde{w}(t), \\ &= \tilde{g}_{m,n}(t). \end{aligned} \tag{5.3.23}$$

To conclude this section, we will consider an example from [73], the Gaussian window, where it can be shown how, as oversampling approaches critical sampling, the dual frame starts to "misbehave."

Example 5.4 *(after [73])*

Consider a Gaussian window

$$w(t) = \pi^{-1/4} e^{-t^2/2}$$

and a special case when $\omega_0 = t_0 = \sqrt{\lambda\, 2\pi}$, or $\omega_0 t_0 = 2\pi\lambda$ (note that $1/\lambda$ gives the oversampling factor). Let us try to find the dual frame. From (5.3.3), recall that (with the Gaussian window)

$$\begin{aligned} g_{m,n}(t) &= e^{jm\omega_0 t} w(t - nt_0) \\ &= \pi^{-1/4} e^{jm\omega_0 t} e^{-(t-nt_0)^2/2}. \end{aligned}$$

Table 5.2 Frame bounds for the Gaussian and $\omega_0 = t_0 = (2\pi\lambda)^{1/2}$, for $\lambda = 0.25, 0.375, 0.5, 0.75, 0.95$ (from [73]).

λ	A	B	B/A
0.250	3.899	4.101	1.052
0.375	2.500	2.833	1.133
0.500	1.575	2.425	1.539
0.750	0.582	2.089	3.592
0.950	0.092	2.021	22.004

Also, since $\tilde{g}_{m,n}(t)$ are generated from a single function $\tilde{w}(t)$ (see (5.3.23)), we will fix $m = n = 0$ and find only $\tilde{w}(t)$ from $g_{0,0}(t) = w(t)$. Then we use (5.3.16) and write

$$\tilde{w}(t) = \frac{2}{A+B}\sum_{k=0}^{\infty}(I - \frac{2}{A+B}\Gamma^*\Gamma)^k w(t). \tag{5.3.24}$$

We will use the frame bounds already computed in [73]. Table 5.2 shows these frame bounds for $\lambda = 0.25, 0.375, 0.5, 0.75, 0.95$, or corresponding $t_0 \cong 1.25, 1.53, 1.77, 2.17, 2.44$. Each of these was taken from Table 3.3 in [73] (we took the nearest computed value). Our first step is to evaluate $\Gamma^*\Gamma w$. From (5.3.12) we know that

$$\Gamma^*\Gamma w \;=\; \sum_m\sum_n \langle g_{m,n}, w\rangle g_{m,n}.$$

Due to the fast decay of functions, one computes only 10 terms on both sides (yielding a total of 21 terms in the summation for m and as many for n). Note that for computational purposes, one has to separate the computations of the real and the imaginary parts. The iteration is obtained as follows: We start by setting $\tilde{w}(t) = w_0(t) = w(t)$. Then for each i, we compute

$$\begin{aligned} w_i(t) &= w_{i-1}(t) - \frac{2}{A+B}\Gamma^*\Gamma w_{i-1}(t), \\ \tilde{w}(t) &= \tilde{w}(t) + w_i(t). \end{aligned}$$

Since the functions decay fast, only 20 iterations were needed in (5.3.24). Figure 5.11 shows plots of $\tilde{w}$ with $\lambda = 0.25, 0.375, 0.5, 0.75, 0.95, 1$. Note how $\tilde{w}$ becomes less and less smooth as λ increases (oversampling decreases). Even so, for all $\lambda < 1$, these dual frames have good time-frequency localization. On the other hand, for $\lambda = 1$, $\tilde{w}$ is not even square-integrable any more and becomes one of the pathological, Baastians' functions [18]. Since in this case $A = 0$, the dual frame function $\tilde{w}$ has to be computed differently. It is given by [225]

$$\tilde{w}_B(t) \;=\; \pi^{7/4}K_0^{-3/2}e^{t^2/2} \sum_{n>|t/\sqrt{2\pi}|-0.5} (-1)^n e^{-\pi(n+0.5)^2},$$

with $K_0 \approx 1.854075$.

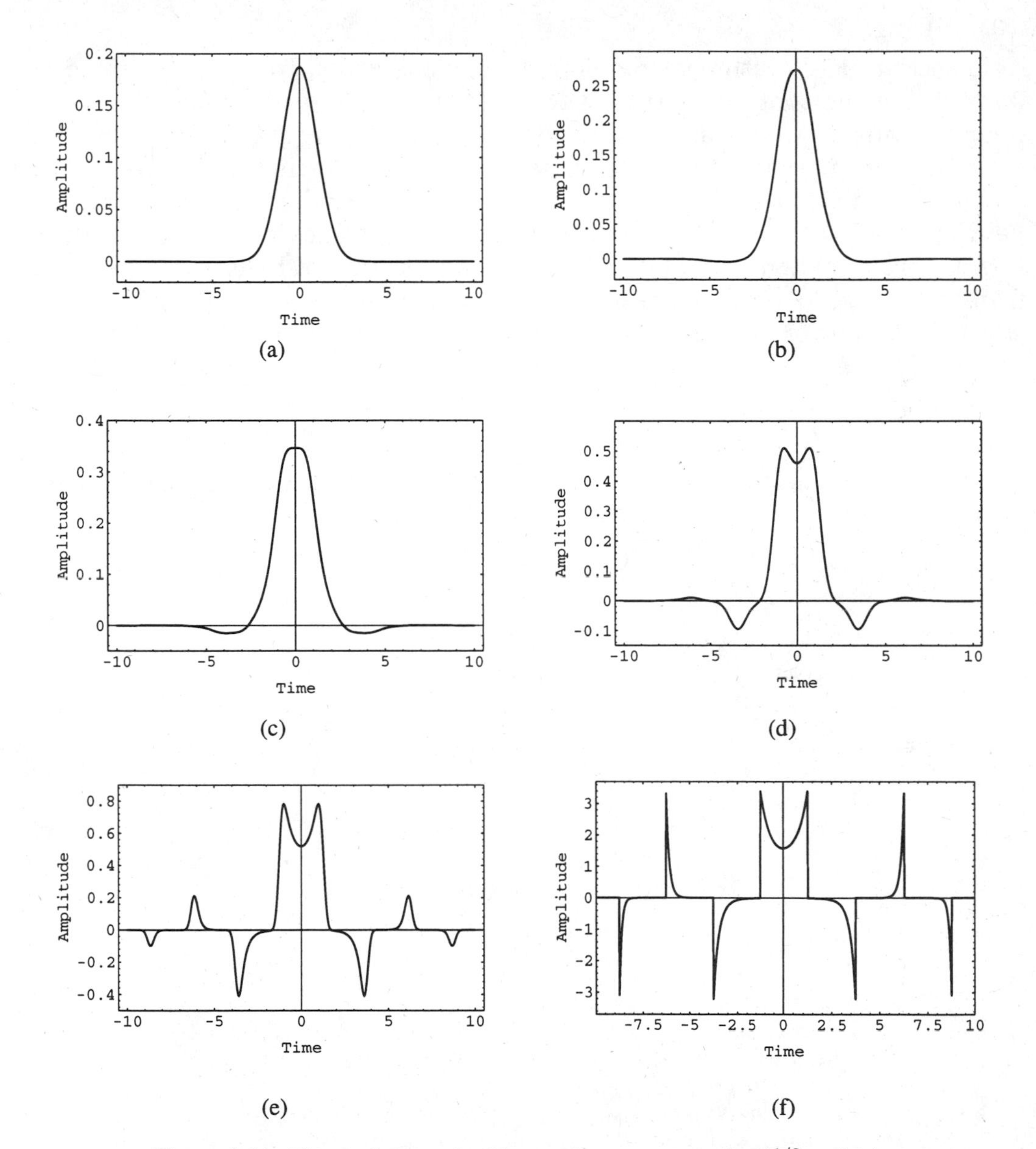

Figure 5.11 The dual frame functions $\tilde{w}$ for $\omega_0 = t_0 = (2\pi\lambda)^{1/2}$ and (a) $\lambda = 0.25$, (b) $\lambda = 0.375$, (c) $\lambda = 0.5$, (d) $\lambda = 0.75$, (e) $\lambda = 0.95$, (f) $\lambda = 1.0$. Note how $\tilde{w}$ starts to "misbehave" as λ increases (oversampling decreases). In fact, for $\lambda = 1$, $\tilde{w}$ is not even square-integrable any more (after [73]).

5.3.4 Remarks

This section dealt with overcomplete expansions called frames. Obtained by discretizing the continuous-time wavelet transform as well as the short-time Fourier transform, they are used to obtain a numerically stable reconstruction of a function f from a sequence of its transform coefficients. We have seen that the conditions on wavelet frames are fairly relaxed, while the short-time Fourier transform frames suffer from a serious drawback given in the Balian-Low theorem: When critical sampling is used, it will not be possible to obtain frames with good time and frequency resolutions. As a result, orthonormal short-time Fourier transform bases are not achievable with basis functions being well localized in time and frequency.

PROBLEMS

5.1 *Characterization of local regularity:* In Section 5.1.2, we have seen how the continuous wavelet transform can characterize the local regularity of a function. Take the Haar wavelet for simplicity.

(a) Consider the function

$$f(t) = \begin{cases} t & 0 \leq t, \\ 0 & t < 0, \end{cases}$$

and show, using arguments similar to the ones used in the text, that

$$CWT_f(a,b) \simeq a^{3/2},$$

around $b = 0$ and for small a.

(b) Show that if

$$f(t) = \begin{cases} t^n & 0 \leq t, \; n = 0, 1, 2 \ldots \\ 0 & t < 0, \end{cases}$$

then

$$CWT_f(a,b) \simeq a^{(2n+1)/2},$$

around $b = 0$ and for small a.

5.2 Consider the Haar wavelet

$$\psi(t) = \begin{cases} 1 & 0 \leq t \leq 1/2, \\ -1 & 1/2 \leq t \leq 1, \\ 0 & \text{otherwise.} \end{cases}$$

(a) Give the expression and the graph of its autocorrelation function $a(t)$,

$$a(t) = \int \psi(\tau)\psi(\tau - t)d\tau.$$

(b) Is $a(t)$ continuous? Derivable? What is the decay of the Fourier transform $A(\omega)$ as $\omega \to \pm\infty$?

5.3 *Nondownsampled filter bank:* Refer to Figure 3.1 without downsamplers.

(a) Choose $\{H_0(z),\ H_1(z),\ G_0(z),\ G_1(z)\}$ as in an orthogonal two-channel filter bank. What is $y[n]$ as a function of $x[n]$? Note: $G_0(z) = H_0(z^{-1})$ and $G_1(z) = H_1(z^{-1})$, and assume FIR filters.

(b) Given the "energy" of $x[n]$, or $\|x\|^2$, what can you say about $\|x_0\|^2 + \|x_1\|^2$? Give either an exact expression, or bounds.

(c) Assume $H_0(z)$ and $G_0(z)$ are given, how can you find $H_1(z)$, $G_1(z)$ such that $y[n] = x[n]$? Calculate the example where

$$H_0(z) = G_0(z^{-1}) = 1 + 2z^{-1} + z^{-2}.$$

Is the solution $(H_1(z),\ G_1(z))$ unique? If not, what are the degrees of freedom? Note: In general, $y[n] = x[n-k]$ would be sufficient, but we concentrate on the zero-delay case.

5.4 *Continuous wavelet transform:* Consider a continuous wavelet transform

$$CWT_f(a,b) = \int_{-\infty}^{\infty} \frac{1}{\sqrt{a}} \psi\left(\frac{t-b}{a}\right) \cdot f(t)dt$$

using a Haar wavelet centered at the origin

$$\psi(t) = \begin{cases} 1 & -\frac{1}{2} \leq t < 0, \\ -1 & 0 \leq t < \frac{1}{2}, \\ 0 & \text{otherwise.} \end{cases}$$

(a) Consider the signal $f(t)$ given by

$$f(t) = \begin{cases} 1 & -\frac{1}{2} \leq t < \frac{1}{2}, \\ 0 & \text{otherwise.} \end{cases}$$

(i) Evaluate $CWT_f(a,b)$ for $a = 1, 1/2, 2$ and all shifts ($b \in \mathcal{R}$).
(ii) Sketch $CWT_f(a,b)$ for all a ($a > 0$) and b, and indicate special behavior, if any (for example, regions where $CWT_f(a,b)$ is zero, behavior as $a \to 0$, anything else of interest).

(b) Consider the case $f(t) = \psi(t)$ and sketch the behavior of $CWT_f(a,b)$, similarly to (ii) above.

5.5 Consider Example 5.1, and choose N vectors φ_i (N odd) for an expansion of $\mathcal{R}^2$, where φ_i is given by

$$\varphi_i = [\cos(2\pi i/N), \sin(2\pi i/N)]^T \qquad i = 0 \ldots N-1.$$

Show that the set $\{\varphi\}$ constitutes a tight frame for $\mathcal{R}^2$, and give the redundancy factor.

5.6 Show that the set $\{\text{sinc}(t - i/N)\}, i \in \mathcal{Z}$ and $N \in \mathcal{N}$, where

$$\text{sinc}(t) = \frac{\sin(\pi t)}{\pi t},$$

forms a tight frame for the space of bandlimited signals (whose Fourier transforms are zero outside $(-\pi, \pi)$. Give the frame bounds and redundancy factor.

5.7 Consider a real $m \times n$ matrix $\boldsymbol{M}$ with $m > n$, rank$(\boldsymbol{m}) = n$ and bounded entries.

(a) Show, given any $\boldsymbol{x} \in \mathcal{R}^n$, that there exist real constants A and B such that

$$0 < A\|\boldsymbol{x}\| \leq \|\boldsymbol{M}x\| \leq B\|\boldsymbol{x}\| < \infty.$$

(b) Show that $\boldsymbol{M}^T\boldsymbol{M}$ is always invertible, and that a possible left inverse of $\boldsymbol{M}$ is given by

$$\tilde{\boldsymbol{M}} = \left(\boldsymbol{M}^T\boldsymbol{M}\right)^{-1}\boldsymbol{M}^T.$$

(c) Characterize all other left inverses of $\boldsymbol{M}$.

(d) Prove that $\boldsymbol{P} = \boldsymbol{M}\tilde{\boldsymbol{M}}$ calculates the orthogonal projection of any vector $\boldsymbol{y} \in \mathcal{R}^m$ onto the range of $\boldsymbol{M}$.

6

Algorithms and Complexity

"... divide each difficulty at hand into as many pieces as possible and as could be required to better solve them."
— René Descartes, *Discourse on the Method*

The theme of this chapter is "divide and conquer." It is the algorithmic counterpart of the multiresolution approximations seen for signal expansions in Chapters 3 and 4. The idea is simple: To solve a large-size problem, find smaller-size subproblems that are easy to solve and combine them efficiently to get the complete solution. Then, apply the division again to the subproblems and stop only when the subproblems are trivial.

What we just said in words, is the key to the fast Fourier transform (FFT) algorithm, discussed in Section 6.1. Other computational tasks such as fast convolution algorithms, have similar solutions.

The reason we are concerned with computational complexity is that the number of arithmetic operations is often what makes the difference between an impractical and a useful algorithm. While considerations other than just the raw numbers of multiplications and additions play an important role as well (such as memory accesses or communication costs), arithmetic or computational complexity is well studied for signal processing algorithms, and we will stay with this point of view in what follows. We will always assume discrete-time data and be mostly concerned with exact rather than approximate algorithms (that is, algorithms that compute the exact result in exact arithmetic).

First, we will review classic digital signal processing algorithms, such as fast convolutions and fast Fourier transforms. Next, we discuss algorithms for multirate signal processing, since these are central for filter banks and discrete-time wavelet series or transforms. Then, algorithms for wavelet series computations are considered, including methods for the efficient evaluation of iterated filters. Even if the continuous wavelet transform cannot be evaluated exactly on a digital computer, approximations are possible, and we study their complexity. We conclude with some special topics, including FFT-based overlap-add/save fast convolution algorithms seen as filter banks.

6.1 Classic Results

We briefly review the computational complexity of some basic discrete-time signal processing algorithms. For more details, we refer to [32, 40, 209, 334].

6.1.1 Fast Convolution

Using transform techniques, the convolution of two sequences

$$c[n] \;=\; \sum_k a[k]\, b[n-k], \tag{6.1.1}$$

reduces to the product of their transforms. If the sequences are of finite length, convolution becomes a polynomial product in transform domain. Taking the z-transform of (6.1.1) and replacing z^{-1} by x, we obtain

$$C(x) \;=\; A(x)\cdot B(x). \tag{6.1.2}$$

Thus, any efficient polynomial product algorithm is also an efficient convolution algorithm.

Cook-Toom Algorithm If $A(x)$ and $B(x)$ are of degree M and N respectively, then $C(x)$ is of degree $M+N$ and has in general $M+N+1$ nonzero coefficients. We are going to use the Lagrange interpolation theorem [32], stating that if we are given a set of $M+N+1$ distinct points α_i, $i=0,\ldots,M+N$, then there exists exactly one polynomial $C(x)$ of degree $M+N$ or less which has the value $C(\alpha_i)$ when evaluated at α_i, and is given by

$$C(x) \;=\; \sum_{i=0}^{M+N} C(\alpha_i)\cdot\left[\frac{\prod_{j\neq i}(x-\alpha_j)}{\prod_{j\neq i}(\alpha_i-\alpha_j)}\right], \tag{6.1.3}$$

where

$$C(\alpha_i) \;=\; A(\alpha_i)\cdot B(\alpha_i), \qquad i=0,\ldots,M+N.$$

Therefore, the Cook-Toom algorithm first evaluates $A(\alpha_i)$, $B(\alpha_i)$, $i=0,\ldots,M+N$, then $C(\alpha_i)$ as in (6.1.2), and finally $C(x)$ as in (6.1.3). Since the α_i's are arbitrary, one can choose them as simple integers and then the evaluation of $A(\alpha_i)$ and $B(\alpha_i)$ can be performed with additions only (however, a very large number of these if M and N grow) or multiplications by integers. Similarly, the reconstruction formula (6.1.3) involves only integer multiplications up to a scale factor (the least common multiple of the denominators). Thus, if one distinguishes carefully multiplications between real numbers (such as the coefficients of the polynomials) and multiplication by integers (or rationals) as interpolation points, one can evaluate the polynomial product in (6.1.2) with $M+N+1$ multiplications only, that is, linear complexity! While this algorithm is impractical for even medium M and N's, it is useful for deriving efficient small size polynomial products, which can then be used in larger problems as we will see.

Example 6.1 *Product of Two Degree-2 Polynomials [32]*

Take $A(x) = a_0 + a_1 x$, $B(x) = b_0 + b_1 x$, and choose $\alpha_0 = 0$, $\alpha_1 = 1$, $\alpha_2 = -1$. Then, according to the algorithm, we first evaluate $A(\alpha_i)$, $B(\alpha_i)$:

$$\begin{aligned} A(0) &= a_0, \quad A(1) = a_0 + a_1, \quad A(-1) = a_0 - a_1, \\ B(0) &= b_0, \quad B(1) = b_0 + b_1, \quad B(-1) = b_0 - b_1, \end{aligned}$$

followed by $C(\alpha_i)$:

$$C(0) = a_0 b_0, \quad C(1) = (a_0 + a_1)(b_0 + b_1), \quad C(-1) = (a_0 - a_1)(b_0 - b_1).$$

We then find the interpolation polynomials and call them $I_i(x)$:

$$I_0(x) = -(x-1)(x+1), \quad I_1(x) = \frac{x(x+1)}{2}, \quad I_2(x) = \frac{x(x-1)}{2}.$$

Finally, $C(x)$ is obtained as

$$C(x) = C(0)I_0(x) + C(1)I_1(x) + C(-1)I_2(x),$$

which could be compactly written as

$$\begin{pmatrix} c_0 \\ c_1 \\ c_2 \end{pmatrix} = \begin{pmatrix} 1 & 0 & 0 \\ 0 & 1/2 & -1/2 \\ -1 & 1/2 & 1/2 \end{pmatrix} \begin{pmatrix} b_0 & 0 & 0 \\ 0 & b_0 + b_1 & 0 \\ 0 & 0 & b_0 - b_1 \end{pmatrix} \begin{pmatrix} 1 & 0 \\ 1 & 1 \\ 1 & -1 \end{pmatrix} \begin{pmatrix} a_0 \\ a_1 \end{pmatrix}.$$

An improvement to this would be if one notes that the highest-order coefficient (in this case c_2) is always obtained as the product of the highest-order coefficients in polynomials $A(x)$ and $B(x)$, that is, in this case $c_2 = a_1 b_1$. Then, one can find a new polynomial $T(x) = C(x) - a_1 b_1 x^2$ and apply the Cook-Toom algorithm on $T(x)$. Thus, with the choice $\alpha_0 = 0$ and $\alpha_1 = -1$, we get

$$\begin{pmatrix} c_0 \\ c_1 \\ c_2 \end{pmatrix} = \begin{pmatrix} 1 & 0 & 0 \\ 1 & -1 & 1 \\ 0 & 0 & 1 \end{pmatrix} \begin{pmatrix} b_0 & 0 & 0 \\ 0 & b_0 - b_1 & 0 \\ 0 & 0 & b_1 \end{pmatrix} \begin{pmatrix} 1 & 0 \\ 1 & -1 \\ 0 & 1 \end{pmatrix} \begin{pmatrix} a_0 \\ a_1 \end{pmatrix}. \tag{6.1.4}$$

The Cook-Toom algorithm is a special case of a more general class of polynomial product algorithms, studied systematically by Winograd [334].

Winograd Short Convolution Algorithms In this algorithm, the idea is to use the Chinese Remainder Theorem [32, 210], which states that an integer $n \in \{0, \ldots, M-1\}$ (where $M = \prod m_i$ and the factors m_i are pairwise coprime) is uniquely specified by its residues $n_i = n \bmod m_i$. The Chinese Remainder Theorem holds for polynomials as well. Thus, a possible way to evaluate (6.1.2) is to choose a polynomial $P(x)$ of degree at least $M + N + 1$, and compute

$$C(x) \;=\; C(x) \bmod P(x) \;=\; A(x) \cdot B(x) \bmod P(x),$$

where the first equality holds because the degree of $P(x)$ is larger than that of $C(x)$, and thus the reduction modulo $P(x)$ does not affect $C(x)$. Factorizing $P(x)$ into its coprime factors, $P(x) = \prod P_i(x)$, one can separately evaluate

$$C_i(x) \;=\; A_i(x) \cdot B_i(x) \bmod P_i(x)$$

(where $A_i(x)$ and $B_i(x)$ are the residues with respect to $P_i(x)$) and reconstruct $C(x)$ from its residues. Note that the Cook-Toom algorithm is a particular case of this algorithm when $P(x)$ equals $\prod(x - \alpha_i)$. The power of the algorithm is that if $P(x)$ is well chosen and factorized over the rationals, then the $P_i(x)$'s can be simple and the reduction operations as well as the reconstruction does not involve much computational complexity. A classic example is to choose $P(x)$ to be of the form $x^L - 1$ and to factor over the rationals. The factors, called cyclotomic polynomials [32], have coefficients $\{1, 0, -1\}$ up to relatively large L's. Note that if $A(x)$ and $B(x)$ are of degree $L - 1$ or less and we compute

$$C(x) \;=\; A(x) \cdot B(x) \bmod (x^L - 1),$$

then we obtain the circular, or, cyclic convolution of the sequences $a[n]$ and $b[n]$:

$$c[n] \;=\; \sum_{k=0}^{L-1} a[k]b[(n-k) \bmod L].$$

Fourier-Domain Computation of Convolution and Interpolation at the Roots of Unity Choosing $P(x)$ as $x^L - 1$ and factoring down to first-order terms leads to

$$x^L - 1 \;=\; \prod_{i=0}^{L-1} (x - W_L^i),$$

where $W_L = e^{-j\,2\pi/L}$. For any polynomial $Q(x)$, it can be verified that

$$Q(x) \bmod (x - a) \;=\; Q(a).$$

Therefore, reducing $A(x)$ and $B(x)$ modulo the various factors of $x^L - 1$ amounts to computing

$$\begin{aligned} A_i(x) \;&=\; A(W_L^i), \\ B_i(x) \;&=\; B(W_L^i), \quad i = 0, \ldots, L-1, \end{aligned}$$

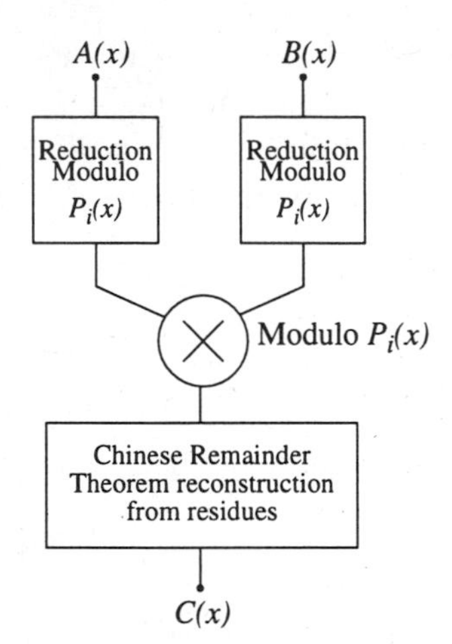

Figure 6.1 Generic fast convolution algorithms. The product $C(x) = A(x) \cdot B(x)$ is evaluated modulo $P(x)$. Particular cases are the Cook-Toom algorithm with $P(x) = \prod(x - \alpha_i)$ and Fourier-domain computation with $P(x) = \prod(x - W_L^i)$ where W_L is the Lth root of unity.

which, according to (2.4.43), is simply taking the length-L discrete Fourier transform of the sequences $a[n]$ and $b[n]$. Then

$$C_i(x) = C(W_L^i) = A(W_L^i) \cdot B(W_L^i), \quad i = 0, \ldots, L-1.$$

The reconstruction is simply the inverse Fourier transform. Of course, this is the convolution theorem of the Fourier transform, but it is seen as a particular case of either Lagrange interpolation or of the Chinese Remainder theorem.

In conclusion, we have seen three convolution algorithms and they all had the generic structure shown in Figure 6.1. First, there is a reduction of the two polynomials involved, then there is a product in the residue domain (which is only a pointwise multiplication if the reduction is modulo first degree polynomials as in the Fourier case) and finally, a reconstruction step concludes the algorithm.

6.1.2 Fast Fourier Transform Computation

The discrete Fourier transform of size N computes (see (2.4.43))

$$X[k] = \sum_{n=0}^{N-1} x[n] \cdot W_N^{nk}, \quad W_N^{nk} = e^{-j\,2\pi/N}. \tag{6.1.5}$$

This is equivalent to evaluating polynomials at the location $x = W_N^k$. Because of the convolution theorem of the Fourier transform, it is clear that a good Fourier transform algorithm will lead to efficient convolution computation.

Let us recall from Section 2.4.8 that the Fourier transform matrix diagonalizes circular convolution matrices. That is, if $\boldsymbol{B}$ is a circulant matrix with first line $(b_0\ b_{N-1}\ b_{N-2} \ldots b_1)$ (the line $i+1$ is a right-circular shift of the line i) then the circular convolution of the sequence $b[n]$ with the sequence $a[n]$ is a sequence $c[n]$ given by

$$\boldsymbol{c} = \boldsymbol{B} \cdot \boldsymbol{a},$$

where the vectors $\boldsymbol{a}$ and $\boldsymbol{c}$ contain the sequences $a[n]$ and $c[n]$, respectively. This can be rewritten, using the convolution theorem of the Fourier transform, as

$$\boldsymbol{c} = \boldsymbol{F}^{-1} \cdot \boldsymbol{\Lambda} \cdot \boldsymbol{F} \cdot \boldsymbol{a},$$

where $\boldsymbol{\Lambda}$ is a diagonal matrix with $\boldsymbol{F} \cdot \boldsymbol{b}$ as the diagonal entries (the vector $\boldsymbol{b}$ contains the sequence $b[n]$). However, unless there is a fast way to compute the matrix-vector products involving $\boldsymbol{F}$ (or $\boldsymbol{F}^{-1}$, which is simply its transpose up to a scale factor), there is no computational advantage in using the Fourier domain for the computation of convolutions.

Several algorithms exist to speed up the product of a vector by the Fourier matrix $\boldsymbol{F}$ which has entries $F_{ij} = W_N^{ij}$ following (6.1.5) (note that rows and columns are numbered starting from 0). We briefly review these algorithms and refer the reader to [32, 90, 209], for more details.

The Cooley-Tukey FFT Algorithm Assume that the length of the Fourier transform is a composite number, $N = N_1 \cdot N_2$. Perform the following change of variable in (6.1.5):

$$\begin{aligned} n &= N_2 \cdot n_1 + n_2, & n_i &= 0, \ldots, N_i - 1, \\ k &= k_1 + N_1 \cdot k_2, & k_i &= 0, \ldots, N_i - 1. \end{aligned} \tag{6.1.6}$$

Then (6.1.5) becomes

$$X[k_1 + N_1 k_2] = \sum_{n_1=0}^{N_1-1} \sum_{n_2=0}^{N_2-1} x[N_2 n_1 + n_2] W_{N_1 N_2}^{(N_2 n_1 + n_2)(k_1 + N_1 k_2)}. \tag{6.1.7}$$

Using the simplifications

$$W_N^{lN} = 1, \quad W_N^{lN_1} = W_{N_2}^{l}, \quad W_N^{lN_2} = W_{N_1}^{l}, \qquad l \in \mathcal{Z},$$

and reordering terms, we can rewrite (6.1.7) as

$$X[k_1 + N_1 k_2] = \sum_{n_2=0}^{N_2-1} W_{N_2}^{n_2 k_2} \left[W_{N_1 N_2}^{n_2 k_1} \cdot \left[\sum_{n_1=0}^{N_1-1} x[N_2 n_1 + n_2] W_{N_1}^{n_1 k_1} \right] \right]. \tag{6.1.8}$$

We recognize:

(a) The right sum as N_2 DFT's of size N_1.

(b) N complex multiplications (by $W_{N_1 N_2}^{n_2 k_1}$).

(c) The left sum as N_1 DFT's of size N_2.

If N_1 and N_2 are themselves composite, one can iterate the algorithm. In particular, if $N = 2^l$ and choosing $N_1 = 2,\ N_2 = N/2$, (6.1.8) becomes

$$X[2k_2] = \sum_{n_2=0}^{N_2-1} W_{N/2}^{n_2 k_2} \cdot (x[n_2] + x[n_2 + N/2]),$$

$$X[2k_2 + 1] = \sum_{n_2=0}^{N_2-1} W_{N/2}^{n_2 k_2} \cdot [W_N^{n_2} \cdot (x[n_2] - x[n_2 + N/2])].$$

Thus, at the cost of $N/2$ complex multiplications (by $W_{N_1 N_2}^{n_2}$) we have reduced the complexity of a size-N DFT to two size-$(N/2)$ DFT's. Iterating $\log_2 N - 1$ times leads to trivial size-2 DFT's and thus, the complexity is of order $N \log_2 N$. Such an algorithm is called a radix-2 FFT and is very popular due to its simplicity and good performance.

The Good-Thomas or Prime Factor FFT Algorithm When performing the index mapping in the Cooley-Tukey FFT (see (6.1.6)), we did not require anything except that N had to be composite. If the factors N_1 and N_2 are coprime, a more powerful mapping based on the Chinese Remainder Theorem can be used [32]. The major difference is that such a mapping avoids the $N/2$ complex multiplications present in the "middle" of the Cooley-Tukey FFT, thus mapping a length-$(N_1 N_2)$ DFT (N_1 and N_2 being coprime) into:

(a) N_1 DFT's of length N_2,

(b) N_2 DFT's of length N_1.

This is equivalent to a two-dimensional FFT of size $N_1 \times N_2$. While this is more efficient than the Cooley-Tukey algorithm, it will require efficient algorithms for lengths which are powers of primes, for which the Cooley-Tukey algorithm can be used. In particular, efficient algorithms for Fourier transforms on lengths which are prime are needed.

Rader's FFT When the length of a Fourier transform is a prime number p, then there exists a permutation of the input and output such that the problem becomes a circular convolution of size $p-1$ (and some auxiliary additions for the frequency zero which is treated separately). While the details are somewhat involved, Rader's method shows that prime-length Fourier transforms can be solved as convolutions and efficient algorithms will be in the generic form we saw in Section 6.1.1 (see the example in (6.1.4)). That is, the Fourier transform matrix $\boldsymbol{F}$ can be written as

$$\boldsymbol{F} = \boldsymbol{CMD}, \tag{6.1.9}$$

where $\boldsymbol{C}$ and $\boldsymbol{D}$ are matrices of output and input additions (which are rectangular) and $\boldsymbol{M}$ is a diagonal matrix containing of the order of $2N$ multiplications.

The Winograd FFT Algorithm We saw that the Good-Thomas FFT mapped a size-$(N_1 N_2)$ Fourier transform into a two-dimensional Fourier transform. Using Kronecker products [32] (see (2.3.2)), we can thus write

$$\boldsymbol{F}_{N_1 \cdot N_2} = \boldsymbol{F}_{N_1} \otimes \boldsymbol{F}_{N_2}. \tag{6.1.10}$$

If N_1 and N_2 are prime, we can use Rader's algorithm to write $\boldsymbol{F}_{N_1}$ and $\boldsymbol{F}_{N_2}$ in the form given in (6.1.9). Finally, using the property of Kronecker products given in (2.3.3) that $(A \otimes B)(C \otimes D) = (A \cdot C) \otimes (B \cdot D)$ (if the products are all well defined), we can rewrite (6.1.10) as

$$\begin{aligned} \boldsymbol{F}_{N_1} \otimes \boldsymbol{F}_{N_2} &= (\boldsymbol{C}_1 \cdot \boldsymbol{M}_1 \cdot \boldsymbol{D}_1) \otimes (\boldsymbol{C}_2 \cdot \boldsymbol{M}_2 \cdot \boldsymbol{D}_2) \\ &= (\boldsymbol{C}_1 \otimes \boldsymbol{C}_2) \cdot (\boldsymbol{M}_1 \otimes \boldsymbol{M}_2) \cdot (\boldsymbol{D}_1 \otimes \boldsymbol{D}_2). \end{aligned}$$

Since the size of $\boldsymbol{M}_1 \otimes \boldsymbol{M}_2$ is of the order of $(2N_1) \cdot (2N_2)$, we see that the complexity is roughly $4N$ multiplications. In general, instead of the $N \log N$ behavior of the Cooley-Tukey FFT, the Winograd FFT has a $C(N) \cdot N$ behavior, where $C(N)$ is slowly growing with N. For example, for $N = 1008 = 7 \cdot 9 \cdot 16$, the Winograd FFT uses 3548 multiplications, while for $N = 1024 = 2^{10}$, the split-radix FFT [90] uses 7172 multiplications. Despite the computational advantage, the complex structure of the Winograd FFT has lead to mixed success in implementations and the Cooley-Tukey FFT is still the most popular fast implementation of Fourier transforms.

Algorithms for Trigonometric Transforms Related to the Fourier Transform Most popular trigonometric transforms used in discrete-time signal processing are closely related to the Fourier transform. Therefore, an efficient way to develop a fast algorithm is to map the computational problem at hand into pre- and post-processing while having a Fourier transform at the center. We will briefly show this for the discrete cosine transform (DCT). The DCT is defined as (see also (7.1.10)-(7.1.11) in Chapter 7)

$$X[k] = \sum_{n=0}^{N-1} x[n] \cos\left(\frac{2\pi(2n+1)k}{4N}\right). \tag{6.1.11}$$

To make it unitary, a factor of $1/\sqrt{N}$ has to be included for $k = 0$, and $\sqrt{2/N}$ for $k \neq 0$, but we skip the scaling since it can be included at the end. If we assume that the transform length N is even, then it can be verified [203] that a simple input permutation given by

$$\begin{aligned} x'[n] &= x[2n], \\ x'[N-n-1] &= x[2n+1], \quad n = 0, \ldots, \frac{N}{2}-1, \end{aligned} \tag{6.1.12}$$

transforms (6.1.11) into

$$X[k] = \sum_{n=0}^{N-1} x'[n] \cos\left(\frac{2\pi(4n+1)k}{4N}\right).$$

This can be related to the DFT of $x'[n]$, denoted by $X'[k]$, in the following manner:

$$X[k] = \cos\left(\frac{2\pi k}{4N}\right) Re[X'[k]] - \sin\left(\frac{2\pi k}{4N}\right) Im[X'[k]].$$

Evaluating $X[k]$ and $X[N-k-1]$ at the same time, it is easy to see that they follow from $X'[k]$ with a rotation by $2\pi k/4N$ [322]. Therefore, the length-N DCT on a real vector has been mapped into a permutation (6.1.12), a Fourier transform of length-N and a set of $N/2$ rotations. Since the Fourier transform on a real vector takes half the complexity of a general Fourier transform [209], this is a very efficient way to compute DCT's. While there exist "direct" algorithms, it turns out that mapping it into a Fourier transform problem is just as efficient and much easier.

6.1.3 Complexity of Multirate Discrete-Time Signal Processing

The key to reduce the complexity in multirate signal processing is a very simple idea: always operate at the slowest possible sampling frequency.

Filtering and Downsampling Convolution followed by downsampling by 2 is equivalent to computing only the even samples of the convolution. Using the polyphase components of the sequences involved (see Section 3.2.1), the convolution (6.1.1)-(6.1.2) followed by downsampling by 2 becomes

$$C_0(x) = A_0(x) \cdot B_0(x) + x \cdot A_1(x) \cdot B_1(x). \tag{6.1.13}$$

This is equivalent to filtering the two independent signals $B_0(x)$ and $B_1(x)$ by the half-length filters $A_1(x)$ and $A_0(x)$ (see Figure 6.2). Because of the independence, the complexity of the two polynomial products in (6.1.13) adds up. Assuming $A(x)$ and $B(x)$ are of odd degree $2M-1$ and $2N-1$, then we have to evaluate two products between polynomials of degree $M-1$ and $N-1$, which takes at least $2(M+N-1)$ multiplications. This is almost as much as the lower bound for the full polynomial product (which is $2(M+$

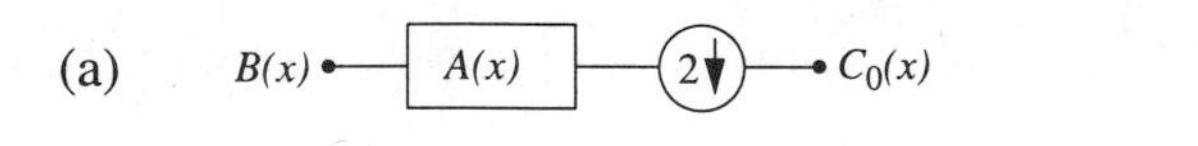

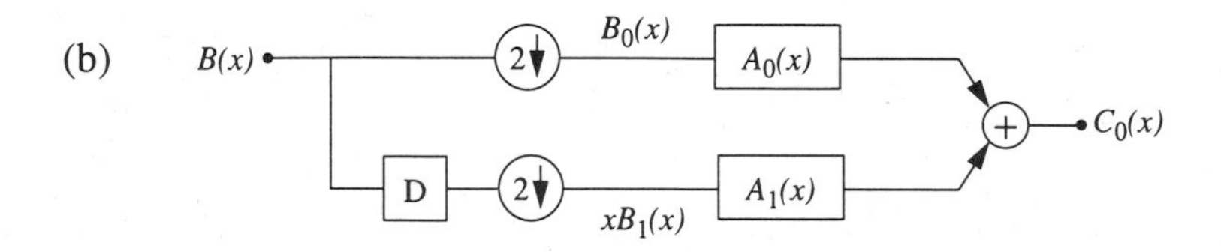

Figure 6.2 Implementation of filtering followed by downsampling by 2. (a) Original system. (b) Decomposition of input into even and odd components followed by filtering with even and odd filters. D stands for a delay by 1.

$N) - 1$ multiplications). If an FFT-based convolution is used, we get some improvement. Assuming that an FFT takes $C \cdot L \cdot \log_2 L$ operations,[1] it takes $2 \cdot C \cdot L \cdot \log_2 L + L$ operations to perform a length-L circular convolution (the transform of the filter is precomputed). Assume a length-N input and a length-N filter and use a length-$2N$ FFT. Direct convolution therefore takes $4 \cdot C \cdot N \cdot (\log_2 N + 1) + 2N$ operations. The computation of (6.1.13) requires two FFT's of size N (for $B_0(x)$ and $B_1(x)$), $2N$ operations for the frequency-domain convolution, and a size-N inverse FFT to recuperate $C_0(x)$, that is, a total of $3 \cdot C \cdot N \cdot \log_2 N + 2N$. This is a saving of roughly 25% over the nondownsampled convolution.

Substantial improvements appear only if straight polynomial products are implemented, since the $4MN$ complexity of the nondownsampled product becomes a $2MN$ complexity for computing the two products in (6.1.13). The main point is that, reducing the size of the polynomial products involved in (6.1.13) might allow one to use almost optimal algorithms, which might not be practical for the full product.

The discussion of the above simple example involving downsampling by 2, generalizes straightforwardly to any downsampling factor K. Then, a polynomial product is replaced by K products with K-times shorter polynomials.

Upsampling and Interpolation The operation of upsampling by 2 followed by interpolation filtering is equivalent to the following convolution:

$$C(x) = A(x) \cdot B(x^2), \qquad (6.1.14)$$

[1] C is a small constant which depends on the particular length and FFT algorithm. For example, the split-radix FFT of a real signal of length $N = 2^n$ requires $2^{n-1}(n-3) + 2$ real multiplications and $2^{n-1}(3n-5) + 4$ real additions [90].

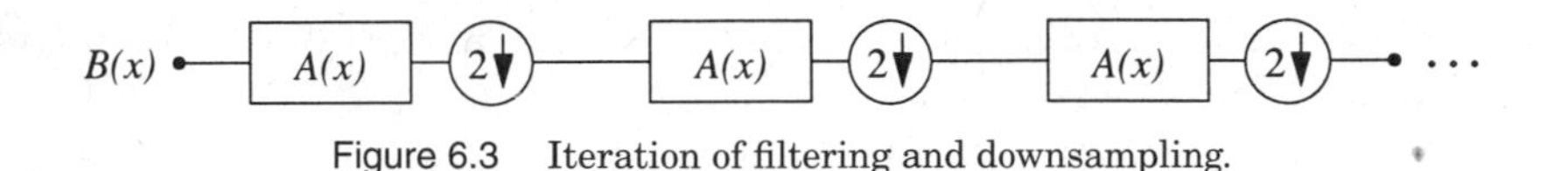

Figure 6.3 Iteration of filtering and downsampling.

where $B(x)$ is the input and $A(x)$ the interpolation filter. Writing $A(x) = A_0(x^2) + x \cdot A_1(x^2)$, the efficient way to compute (6.1.14) is

$$C(x) = B(x^2) \cdot A_0(x^2) + xB(x^2) \cdot A_1(x^2),$$

that is, two polynomial products where each of the terms is approximately of half size, since $B(x^2) \cdot A_0(x^2)$ can be computed as $B(x) \cdot A_0(x)$ and then upsampled (similarly for $B(x^2) \cdot A_1(x^2)$). That this problem seems very similar to filtering and downsampling is no surprise, since they are duals of each other. If one writes the matrix that represents convolution by $a[n]$ and downsampling by two, then its transpose represents upsampling by two followed by interpolation with $\tilde{a}[n]$ (where $\tilde{a}[n]$ is the time-reversed version of $a[n]$). This is shown in a simple three-tap filter example below

$$\begin{pmatrix} \cdots & a[0] & 0 & 0 & \cdots & \cdots \\ 0 & a[2] & a[1] & a[0] & 0 & 0 \\ \cdots & 0 & 0 & a[2] & a[1] & a[0] \end{pmatrix}^T = \begin{pmatrix} \vdots & 0 & 0 \\ a[0] & a[2] & 0 \\ 0 & a[1] & 0 \\ 0 & a[0] & a[2] \\ 0 & 0 & a[1] \\ \vdots & 0 & a[0] \end{pmatrix}.$$

The block diagram of an efficient implementation of upsampling and interpolation is thus simply the transpose of the diagram in Figure 6.2. Both systems have the same complexity, since they require the implementation of two half-length filters ($A_0(x)$ and $A_1(x)$) in the downsampled domain.

Of course, upsampling by an arbitrary factor K followed by interpolation can be implemented by K small filters followed by upsampling, shifts, and summation.

Iterated Multirate Systems A case that appears often in practice, especially around discrete-time wavelet series, is the iteration of an elementary block such as filtering and downsampling as shown in Figure 6.3. An elementary, even if somewhat surprising, result is the following: If the complexity of the first block is C operations/input sample, then the upper bound on the total complexity, irrespective of the number of stages, is $2C$. The proof is immediate, since the second block has complexity C but runs at half sampling rate and similarly, the ith block runs 2^{i-1} times slower than the first one. Thus, the total complexity for K blocks becomes

$$C_{tot} = C + \frac{C}{2} + \frac{C}{4} + \cdots \frac{C}{2^{K-1}} = 2C\left(1 - \frac{1}{2^K}\right) < 2C. \tag{6.1.15}$$

This property has been used to design very sharp filters with low complexity in [236]. While the complexity remains bounded, the delay does not. If the first block contributes a delay D, the second will produce a delay $2D$ and the ith block a delay $2^{i-1}D$. That is, the total delay becomes

$$D_{tot} = D + 2D + 4D + \cdots + 2^{K-1}D = (2^K - 1)D.$$

This large delay is a serious drawback, especially for real-time applications such as speech coding.

Efficient Filtering Using Multirate Signal Processing One very useful application of multirate techniques to discrete-time signal processing has been the efficient computation of narrow-band filters. There are two basic ideas behind the method. First, the output of a lowpass filter can be downsampled, and thus, not all outputs have to be computed. Second, a very long narrow-band filter can be factorized into a cascade of several shorter ones and each of these can be downsampled as well. We will show the technique on a simple example, and refer to [67] for an in-depth treatment.

Example 6.2

Assume we desire a lowpass filter with a cutoff frequency $\pi/12$. Because of this cutoff frequency, we can downsample the output, say by 8. Instead of a direct implementation, we build a cascade of three filters with a cutoff frequency $\pi/3$, each downsampled by two. We call such a filter a third-band filter. Using the interchange of downsampling and filtering property, we get an equivalent filter with a z-transform:

$$H_{equiv}(z) = H(z) \cdot H(z^2) \cdot H(z^4),$$

where $H(z)$ is the z-transform of the third-band lowpass filter. The spectral responses of $H(e^{j\omega})$, $H(e^{j2\omega})$, and $H(e^{j4\omega})$ are shown in Figure 6.4(a) and their product, $H_{equiv}(z)$, is shown in Figure 6.4(b), showing that a $\pi/12$ lowpass filter is realized. Note that its length is approximately equal to $L + 2L + 4L = 7L$, where L is the length of the filter with the cutoff frequency $\pi/3$.

If the filtered signal is needed at the full sampling rate, one can use upsampling and interpolation filtering and the same trick can be applied to that filter as well.

Because of the cascade of shorter filters, and the fact that each stage is downsampled, it is clear that substantial savings in computational complexity are obtained. How this technique can be used to derive arbitrary sharp filters while keeping the complexity bounded is shown in [236].

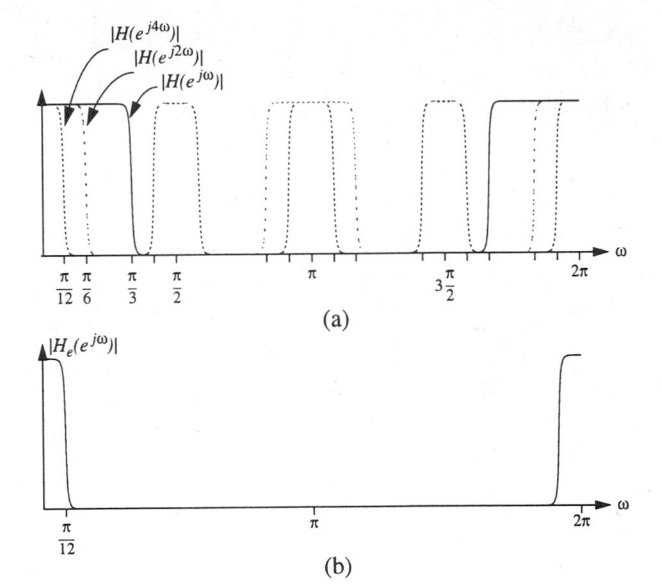

Figure 6.4 Spectral responses of individual filters and the resulting equivalent filter. (a) $|H(e^{j\omega})|$, $|H(e^{j2\omega})|$, $|H(e^{j4\omega})|$. (b) $|H(e^{j\omega})| = |H(e^{j\omega})||H(e^{j2\omega})||H(e^{j4\omega})|$.

6.2 Complexity of Discrete Bases Computation

This section is concerned with the complexity of filter bank related computations. The basic ingredients are the multirate techniques of the previous section, as well as polyphase representations of filter banks.

6.2.1 Two-Channel Filter Banks

Assume a two-channel filter bank with filter impulse responses $h_0[n]$ and $h_1[n]$ of length L. Recall from (3.2.22) in Section 3.2.1, that the channel signals equal

$$\begin{pmatrix} Y_0(z) \\ Y_1(z) \end{pmatrix} = \begin{pmatrix} H_{00}(z) & H_{01}(z) \\ H_{10}(z) & H_{11}(z) \end{pmatrix} \cdot \begin{pmatrix} X_0(z) \\ X_1(z) \end{pmatrix}. \tag{6.2.1}$$

Unless there are special relationships among the filters, this amounts to four convolutions by polyphase filters of length $L/2$ (assuming L even). For comparison purposes, we will count the number of operations for each new input sample. The four convolutions operate at half the input rate and thus, for every two input samples, we compute $4 \cdot L/2$ multiplications and $4((L/2)-1)+2$ additions. This leads to L multiplications and $L-1$ additions/input sample, that is, exactly the same complexity as a convolution by a single filter of size L. If an FFT-based convolution algorithm is used, the transforms of $X_0(z)$ and $X_1(z)$ can be shared for the computation of $Y_0(z)$ and $Y_1(z)$. Assuming again that a length-N FFT uses $C \cdot N \cdot \log_2 N$ operations and that the input signal and the filters are of length L, we get, since we need FFT's of length L to compute the polynomial products in (6.2.1) (which are of size $L/2 \times L/2$):

(a) $2 \cdot C \cdot L \cdot \log_2 L$ operations to get the transforms of $X_0(z)$ and $X_1(z)$,

(b) $4L$ operations to perform the frequency-domain convolutions,

(c) $2 \cdot C \cdot L \cdot \log_2 L$ operations for the inverse FFT's to get $Y_0(z)$ and $Y_1(z)$,

where we assumed that the transforms of the polyphase filters were precomputed. That is, the Fourier-domain evaluation requires

$$4 \cdot C \cdot L \cdot \log_2 L \; + \; 4N \text{ operations},$$

which is of the same order as Fourier-domain computation of a length-L filter convolved with a length-L signal.

In [245], a precise analysis is made involving FFT's with optimized lengths so as to minimize the operation count. Using the split-radix FFT algorithm [90], the number of operations (multiplications plus additions/sample) becomes (for large L)

$$4 \log_2 L \; + \; O(\log \log L), \tag{6.2.2}$$

which is to be compared with $2L-1$ multiplications plus additions for the direct implementation. The algorithm starts to be effective for $L = 8$ and an FFT size of 16, where it achieves around 5 multiplications/point (rather than 8) and leads to improvements by an order of magnitude for large filters such as $L = 64$ or 128. For medium size filters ($L = 6, \ldots, 12$), a method based on fast running convolution is best (see [245] and Section 6.5 below).

Let us now consider some special cases where additional savings are possible.

Linear Phase Filters It is well-known that if a filter is symmetric or antisymmetric, the number of operations can be halved in the direct implementation by simply adding (or subtracting) the two input samples that are multiplied by the same coefficient. This trick can be used in the downsampled case as well, that is, filter banks with linear phase filters require half the number of multiplications, or $L/2$ multiplications/input sample (the number of additions remains unchanged). If the filter length is odd, the polyphase components are themselves symmetric or antisymmetric, and the saving is obvious in (6.2.1).

Certain linear phase filter banks can be written in cascade form [321] (see

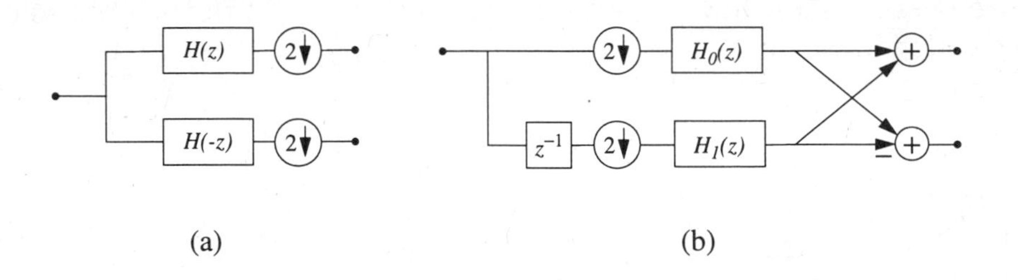

Figure 6.5 Classic QMF filter bank. (a) Initial filter bank. (b) Efficient implementation using polyphase components and a butterfly.

Section 3.2.4). That is, their polyphase matrix is of the form given in (3.2.70):

$$\boldsymbol{H}_p(z) = C \begin{pmatrix} 1 & 1 \\ -1 & 1 \end{pmatrix} \cdot \left[\prod_{i=1}^{K-1} \begin{pmatrix} 1 & 0 \\ 0 & z^{-1} \end{pmatrix} \begin{pmatrix} 1 & \alpha_i \\ \alpha_i & 1 \end{pmatrix} \right].$$

The individual 2×2 symmetric matrices can be written as (we assume $\alpha_i \neq 1$)

$$\begin{pmatrix} 1 & \alpha_i \\ \alpha_i & 1 \end{pmatrix} = \frac{1-\alpha_i}{2} \cdot \begin{pmatrix} 1 & -1 \\ 1 & -1 \end{pmatrix} \begin{pmatrix} \frac{1+\alpha_i}{1-\alpha_i} & 0 \\ 0 & 1 \end{pmatrix} \begin{pmatrix} 1 & -1 \\ 1 & -1 \end{pmatrix}.$$

By gathering the scale factors together, we see that each new block in the cascade structure (which increases the length of the filters by two) adds only one multiplication. Thus, we need order-$(L/2)$ multiplications to compute a new output in each channel, or $L/4$ multiplications/input sample. The number of additions is of the order of L additions/input sample [321].

Classic QMF Solution The classic QMF solution given in (3.2.34)-(3.2.35) (see Figure 6.5(a)), besides using even-length linear phase filters, forces the highpass filter to be equal to the lowpass, modulated by $(-1)^n$. The polyphase matrix is therefore:

$$\boldsymbol{H}_p(z) = \begin{pmatrix} H_0(z) & H_1(z) \\ H_0(z) & -H_1(z) \end{pmatrix} = \begin{pmatrix} 1 & 1 \\ 1 & -1 \end{pmatrix} \cdot \begin{pmatrix} H_0(z) & 0 \\ 0 & H_1(z) \end{pmatrix},$$

where H_0 and H_1 are the polyphase components of the prototype filter $H(z)$. The factorized form on the right indicates that the complexity is halved, and an obvious implementation is shown in Figure 6.5(b). Recall that this scheme only approximates perfect reconstruction when using FIR filters.

Orthogonal Filter Banks As seen in Section 3.2.4, orthogonal filter banks have strong structural properties. In particular, because the highpass is the time-reversed version of the lowpass filter modulated by $(-1)^n$, the polyphase matrix has the following form:

$$\boldsymbol{H}_p(z) = \begin{pmatrix} H_{00}(z) & H_{01}(z) \\ -\tilde{H_{01}}(z) & \tilde{H_{00}}(z) \end{pmatrix}, \tag{6.2.3}$$

where $\tilde{H_{00}}(z)$ and $\tilde{H_{01}}(z)$ are time-reversed versions of $H_{00}(z)$ and $H_{01}(z)$, and $H_{00}(z)$ and $H_{01}(z)$ are the two polyphase components of the lowpass filter. If $H_{00}(z)$ and $H_{01}(z)$ were of degree zero, it is clear that the matrix in (6.2.3) would be a rotation matrix, which can be implemented with three multiplications. It turns out that for arbitrary degree polyphase components, terms can still be gathered into rotations, saving 25% of multiplications (at the cost of 25% more additions) [104]. This rotation property is more obvious in the lattice structure form of orthogonal filter banks [310]. We recall that the two-channel lattice factorizes the paraunitary polyphase matrix into the following form (see (3.2.60)):

$$\boldsymbol{H}_p(z) = \begin{pmatrix} H_{00}(z) & H_{01}(z) \\ H_{10}(z) & H_{11}(z) \end{pmatrix} = \boldsymbol{U}_0 \cdot \left[\prod_{i=1}^{N-1} \begin{pmatrix} 1 & 0 \\ 0 & z^{-1} \end{pmatrix} \boldsymbol{U}_i \right],$$

where filters are of length $L = 2N$ and the matrices $\boldsymbol{U}_i$ are 2×2 rotations. Such rotations can be written as (where we use the shorthand a_i and b_i for $\cos(\alpha_i)$ and $\sin(\alpha_i)$ respectively) [32]

$$\begin{pmatrix} a_i & b_i \\ -b_i & a_i \end{pmatrix} = \begin{pmatrix} 1 & 0 & 1 \\ 0 & 1 & 1 \end{pmatrix} \cdot \begin{pmatrix} a_i + b_i & 0 & 0 \\ 0 & a_i - b_i & 0 \\ 0 & 0 & -b_i \end{pmatrix} \cdot \begin{pmatrix} 1 & 0 \\ 0 & 1 \\ 1 & -1 \end{pmatrix}. \tag{6.2.4}$$

Thus, only three multiplications are needed, or $3N$ for the whole lattice. Since the lattice works in the downsampled domain, the complexity is $3N/2$ multiplications or, since $N = L/2$, $3L/4$ multiplications/input sample and a similar number of additions. A further trick consists in denormalizing the diagonal matrix in (6.2.4) (taking out b_i for example) and gathering all scale factors at the end of the lattice. Then, the complexity becomes $(L/2)+1$ multiplications/input sample. The number of additions remains unchanged.

Table 6.1 summarizes the complexity of various filter banks. Except for the last entry, time-domain computation is assumed. Note that in the frequency-domain computation, savings due to symmetries become minor.

6.2.2 Filter Bank Trees and Discrete-Time Wavelet Transforms

Filter bank trees come mostly in two flavors: the full-grown tree, where each branch is again subdivided, and the octave-band tree, where only the lower branch is further subdivided.

Table 6.1 Number of arithmetic operations/input sample for various two-channel filter banks with length-L filters, where μ and α stand for multiplications and additions, respectively.

Filter bank type	# of μ	# of α
General two-channel filter bank	L	$L-1$
Linear phase filter bank		
direct form	$L/2$	$L-1$
lattice form	$L/4$	L
QMF filter bank	$L/2$	$L/2$
Orthogonal filter bank		
direct form	L	$L-1$
lattice form	$3L/4$	$3L/4$
denormalized lattice	$L/2$	$3L/4$
Frequency-domain computation (assuming large L) [245]	$\log_2 L$	$3\log_2 L$

First, it is clear that techniques used to improve two-channel banks will improve any tree structure when applied to each elementary bank in the tree. Then, specific techniques can be developed to compute tree structures.

Full Trees If an elementary block (a two-channel filter bank downsampled by two) has complexity C_0, then a K-stage full tree with 2^K leaves has complexity $K \cdot C_0$. This holds because the initial block is followed by two blocks at half rate (which contributes $2 \cdot C_0/2$), four blocks at quarter rate and so on. Thus, while the number of leaves grows exponentially with K, the complexity only grows linearly with K.

Let us discuss alternatives for the computation of the full tree structure in the simplest, two-stage case, shown in Figure 6.6(a). It can be transformed into the four-channel filter bank shown in Figure 6.6(b) by passing the second stage of filters across the first stage of downsampling. While the structure is simpler, the length of the filters involved is now of the order of $3L$ if $H_i(z)$ is of degree $L-1$. Thus, unless the filters are implemented in factorized form, this is more complex than the initial structure. However, the regular structure might be preferred in hardware implementations.

Let us consider a Fourier-domain implementation. A simple trick consists of implementing the first stage with FFT's of length N and the second stage with FFT's of length $N/2$. Then, one can perform the downsampling in Fourier domain and then, the forward FFT of the second stage cancels the inverse FFT

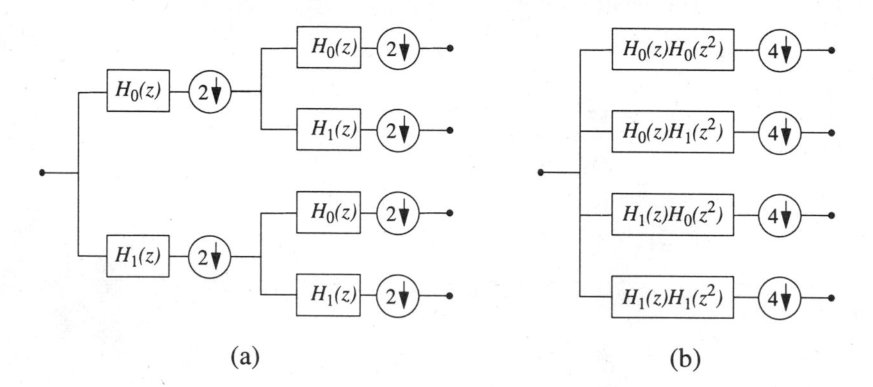

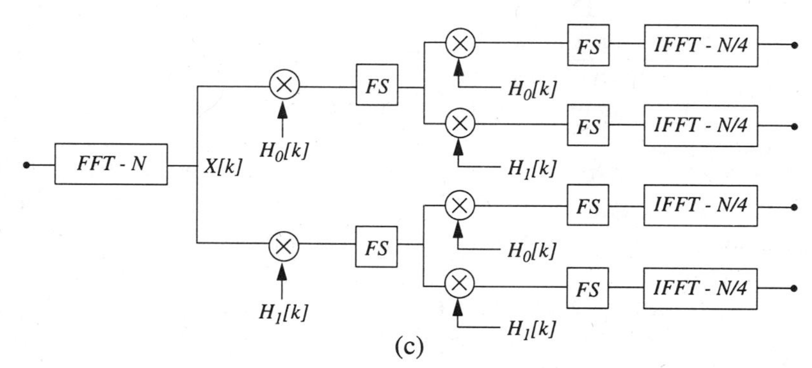

Figure 6.6 Two-stage full-tree filter bank. (a) Initial system. (b) Parallelized system. (c) Fourier-domain computation with implicit cancellation of forward and inverse transforms between stages. FS stands for Fourier-domain downsampling. Note that in the first stage the $H_i[k]$ are obtained as outputs of a size-N FFT, while in the second stage, they are outputs of a size-$N/2$ FFT.

of the first stage. The downsampling in Fourier domain requires $N/2$ additions, since if $X[k]$ is a length-N Fourier transform, the length-$N/2$ Fourier transform of its downsampled version is

$$Y[k] \;=\; \frac{1}{2}\left(X[k] + X\left[k + N/2\right]\right).$$

Figure 6.6(c) shows the algorithm schematically, where, for simplicity, the filters rather than the polyphase components are shown. The polyphase implementation requires to separate even and odd samples in time domain. The even samples are obtained from the Fourier transform $X[k]$ as

$$\begin{aligned} y[2n] &= \sum_{k=0}^{N-1} X[k] W_N^{-2nk} \\ &= \sum_{k=0}^{N/2} (X[k] + X[k+N/2]) W_{N/2}^{-nk}, \end{aligned} \tag{6.2.5}$$

while the odd ones require a phase shift

$$\begin{aligned} y[2n+1] &= \sum_{k=0}^{N-1} X[k] W_N^{-(2n+1)k} \\ &= \sum_{k=0}^{N-1} W_N{}^k (X[k] + X[k+N/2]) W_{N/2}^{-nk}. \end{aligned} \tag{6.2.6}$$

If the next stage uses a forward FFT of size $N/2$ on $y[2n]$ and $y[2n+1]$, the inverse FFT's in (6.2.5) and (6.2.6) are cancelled and only the phase shift in (6.2.6) remains. These complex multiplications can be combined with the subsequent filtering in Fourier domain. Therefore, we have shown how to merge two subsequent stages with only N additions. Note that the length of the FFT's have to be chosen carefully so that linear convolution is computed at each stage. In the case discussed here, $N/2$ (the size of the second FFT) has to be larger than $(3L + L_s - 2)/2$ where L and L_s are the filter and signal lengths, respectively (the factor $1/2$ comes from the fact that we deal with polyphase components).

While this merging improves the computational complexity, it also constrains the FFT length. That is, the length will not be optimal for the first or the second stage, resulting in a certain loss of optimality.

Octave-Band Trees and Discrete-Time Wavelet Series In this case, we can use the property of iterated multirate systems which leads to a complexity independent of the number of stages as seen in (6.1.15). For example, assuming a Fourier-domain implementation of an elementary two-channel bank which uses about $(4 \log_2 L)$ operations/input sample as in (6.2.2), a K-stage discrete-time wavelet series expansion requires of the order of

$$8 \log_2 L \,(1 - 1/2^K) \text{ operations}$$

for long filters implemented in Fourier domain, and

$$4\, L\, (1 - 1/2^K) \text{ operations} \tag{6.2.7}$$

for short filters implemented in time domain. As mentioned earlier, filters of length 8 or more are more efficiently implemented with Fourier-domain techniques.

Of course, the merging trick of inverse and forward FFT's between stages can be used here as well. A careful analysis made in [245] shows that merging of two stages pays off for filter lengths of 16 or more. Merging of more stages is marginally interesting for large filters since it involves very large FFT's, which is probably impractical. Again, fast running convolution methods are best for medium size filters ($L = 6, \ldots, 12$) [245]. Finally, all savings due to special structures, such as orthogonality or linear phase, carry over to tree structures as well.

The study of hardware implementations of discrete-time wavelet transforms is an important topic as well. In particular, the fact that different stages run at different sampling rates makes the problem nontrivial. For a detailed study and various solutions to this problem, see [219].

6.2.3 Parallel and Modulated Filter Banks

General parallel filter banks have an obvious implementation in the polyphase domain. If we have a filter bank with K channels and downsampling by M, we get, instead of (6.2.1), a $K \times M$ matrix times a size-M vector product (where all entries are polynomials). The complexity of straightforward computation is comparable, when $K = M$, to a single convolution since we have M filters downsampled by M. Fourier methods require M forward transforms (for each polyphase component), $K \cdot M$ frequency-domain convolutions, and finally, K inverse Fourier transforms to obtain the channel signals in the time domain.

A more interesting case appears when the filters are related to each other. The most important example is when all filters are related to a single prototype filter through modulation.

The classic example is (see (3.4.13)-(3.4.14) in Section 3.4.3)

$$\begin{aligned} H_i(z) &= H_{pr}(W_N^i z), \quad i = 0, \ldots, N-1, \quad W_N = e^{-j2\pi/N}, && (6.2.8) \\ h_i[n] &= W_N^{-in} h_{pr}[n]. && (6.2.9) \end{aligned}$$

This corresponds to a short-time Fourier or Gabor transform filter bank. The polyphase matrix with respect to downsampling by N has the form shown below (an example for $N = 3$ is given):

$$\begin{aligned} H_p(z) &= \begin{bmatrix} H_{pr_0}(z) & H_{pr_1}(z) & H_{pr_2}(z) \\ H_{pr_0}(z) & W_3 H_{pr_1}(z) & W_3^2 H_{pr_2}(z) \\ H_{pr_0}(z) & W_3^2 H_{pr_1}(z) & W_3 H_{pr_2}(z) \end{bmatrix} \\ &= \boldsymbol{F}_3 \cdot \begin{bmatrix} H_{pr_0}(z) & 0 & 0 \\ 0 & H_{pr_1}(z) & 0 \\ 0 & 0 & H_{pr_2}(z) \end{bmatrix}, \qquad (6.2.10) \end{aligned}$$

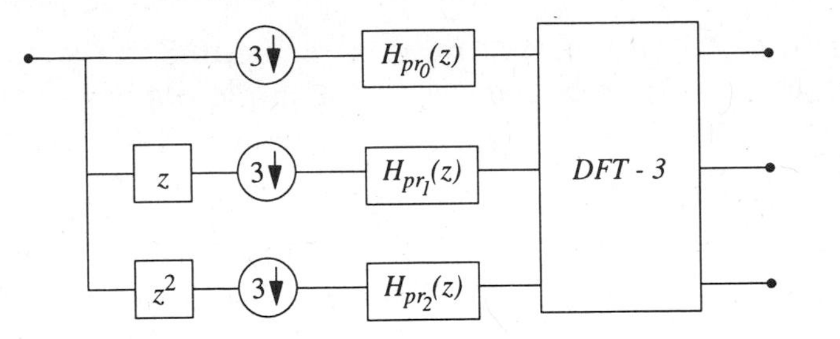

Figure 6.7 Modulated filter bank implemented with an FFT.

where $H_{pr_i}(z)$ is the ith polyphase component of the filter $H_{pr}(z)$ and $\boldsymbol{F}_3$ is the size-3 discrete Fourier transform matrix. The implementation is shown in Figure 6.7. This fast implementation of modulated filter banks using polyphase filters of the prototype filter followed by a fast Fourier transform is central in several applications such as transmultiplexers. This fast algorithm goes back to the early 70's [25]. The complexity is now substantially reduced. The polyphase filters require N-times less complexity than a full filter bank, and the FFT adds an order $N \log_2 N$ operations per N input samples. The complexity is of the order of

$$(2\frac{L}{N} + 2 \cdot \log_2 N) \text{ operations/input sample,} \tag{6.2.11}$$

that is, a substantial reduction over a single, length-L filtering operation. Further reductions are possible by implementing the polyphase filters in frequency domain (reducing the term of order L to $\log_2 L$) and merging FFT's into a multidimensional one [210]. Another important and efficient filter bank is based on cosine modulation. It is sometimes referred to as lapped orthogonal transforms (LOT's) [188] or local cosine bases [63]. Several possible LOT's have been proposed in the literature and are of the general form described in (3.4.17–3.4.18) in Section 3.4.3. Using trigonometric identities, this can be reduced to N polyphase filters followed by a DCT-type of transform of length N (see (6.1.11)). Other LOT's lead to various length-N or length-$2N$ trigonometric transforms, preceded by polyphase filters of length two or larger [187].

6.2.4 Multidimensional Filter Banks

Computational complexity is of particularly great concern in multidimensional systems, since, for example, filtering an $N \times N$ image with a filter of size $L \times L$ requires of the order of $N^2 \cdot L^2$ operations. If the filter is separable, that is, $H(z_1, z_2) = H_1(z_1)H_2(z_2)$, then filtering on rows and columns can be done separately and the complexity is reduced to an order $2N^2L$ operations (N row filterings and N column filterings, each using NL operations).

A multidimensional filter bank can be implemented in its polyphase form, bringing the complexity down to the order of a single nondownsampled convolution, just as in the one-dimensional case. A few cases of particular interest allow further reductions in complexity.

Fully Separable Case When both filters and downsampling are separable, then the system is the direct product of one-dimensional systems. The implementation is done separately over each dimension. For example, consider a two-dimensional system filtering an $N \times N$ image into four subbands using the filters $\{H_0(z_1)H_0(z_2),\ H_0(z_1)H_1(z_2),\ H_1(z_1)H_0(z_2),\ H_1(z_1)H_1(z_2)\}$ each of size $L \times L$ followed by separable downsampling by two in each dimension. This requires N decompositions in one dimension (one for each row), followed by N decompositions in the other, or a total of $2N^2 \cdot L$ multiplications and a similar number of additions. This is a saving of the order of $L/2$ with respect to the nonseparable case. Note that if the decomposition is iterated on the lowpass only (that is, a separable transform), the complexity is only

$$C_{tot} = C + \frac{C}{4} + \frac{C}{16} + \cdots < \frac{4}{3}C,$$

where C is the complexity of the first stage.

Separable Polyphase Components The last example led automatically to separable polyphase components, because in the case of separable downsampling, there is a direct relationship between separability of the filter and its polyphase components [164]. When the downsampling is nonseparable, separable filters yield nonseparable polyphase components in general. Thus, it might be more efficient to compute convolutions with the filters rather than their polyphase components. Finally, one can construct filter banks with separable polyphase components (corresponding to nonseparable filters in the nonseparable downsampling case) having thus an efficient implementation and yielding savings of order $L/2$.

6.3 Complexity of Wavelet Series Computation

The computational complexity of evaluating expansions into wavelet bases is considered in this section, as well as that of related problems such as iterated filters used in regularity estimates of wavelets.

6.3.1 Expansion into Wavelet Bases

Assume a multiresolution analysis structure as defined in Section 4.2. If we have the projection onto V_0, that is, samples $x[n] = \langle \varphi(t-n), x(t) \rangle$, then Mallat's algorithm given in Section 4.5.3, indicates that the expansion onto

W_i, $i = 1, 2, \ldots$ can be evaluated using an octave-band filter bank. Therefore, given the initial projection, the complexity of the wavelet expansion is of order $2L$ multiplications and $2L$ additions/input sample (see (6.2.7)) where L is the length of the discrete-time filter, or equivalently, the order of the two-scale equation. Unless the wavelet $\psi(t)$ is compactly supported, L could be infinite. For example, many of the wavelets designed in Fourier domain (such as the Meyer's and Battle-Lemarié's wavelets) lead to an unbounded L. In general, implementations simply truncate the infinitely long filter and a reasonable approximation is computed with finite computational cost.

A more attractive alternative is to find recursive filters which perform an exact computation at finite computational cost. An example is in the case of spline spaces (see Section 4.3.2), where instead of the usual Battle-Lemarié wavelet, an alternative one can be used which leads to an IIR filter implementation [133, 296].

When we cannot assume to have access to the projection onto V_0, an approximation known as Shensa's algorithm [261] can be used (see Section 4.5.3). It represents, as an initial step, a nonorthogonal projection of the input and the wavelets onto suitable approximation spaces. In terms of computational complexity, Shensa's algorithm involves a prefiltering stage with a discrete-time filter, thus adding an order $2L_p$ number of operations where L_p is the length of the prefilter.

Therefore, the computation of the wavelet series into K octaves requires about

$$2\,L\,(1 - 1/2^K) \;+\; L_p$$

multiplications and a similar number of additions. Of course, applying Fourier transform, the orders L and L_p are reduced to their logarithms. This efficiency for computing in discrete time, a series expansion which normally uses integrals, is one of the main attractive features of the wavelet decomposition.

6.3.2 Iterated Filters

The previous section showed a completely discrete-time algorithm for the computation of the wavelet series. However, underlying this scheme are continuous-time functions $\varphi(t)$ and $\psi(t)$, which often correspond to iterated discrete-time filters. Such iterated filters are usually computed during the design stage of a wavelet transform, so as to verify properties of the scaling function and wavelet such as regularity. Because the complexity appears only once, it is not as important to reduce it as in the computation of the transform itself. However, the algorithms are simple and the computational burden can be heavy especially in multiple dimensions, thus we briefly discuss fast algorithms for iterated filters. Recall from (4.4.9) that we wish to compute

$$G_0^{(i)}(z) = \prod_{k=0}^{i-1} G_0\left(z^{2^k}\right). \tag{6.3.1}$$

For simplicity, we will omit the subscript "0" and will simply call the lowpass filter G. The length of $G^{(i)}(z)$ is equal to

$$L^{(i)} = (2^i - 1)(L-1) + 1.$$

From (6.3.1), the following identities can be verified (Problem 6.5):

$$G^{(i)}(z) = G(z) \cdot G^{(i-1)}(z^2), \tag{6.3.2}$$

$$G^{(i)}(z) = G(z^{2^{i-1}}) \cdot G^{(i-1)}(z), \tag{6.3.3}$$

$$G^{(2^k)}(z) = G^{(2^{k-1})}(z) \cdot G^{(2^{k-1})}(z^{2^{2^{k-1}}}). \tag{6.3.4}$$

The first two relations will lead to recursive algorithms, while the last one produces a doubling algorithm and can be used when iterates which are powers of two are desired. Computing (6.3.2) as

$$G^{(i)}(z) = [G_0(z^2) + z^{-1}G_1(z^2)] \cdot G^{(i-1)}(z^2),$$

where G_0 and G_1 are the two polyphase components of filter G, leads to two products between polynomials of size $L/2$ and $(2^{i-1} - 1)(L-1) + 1$. Calling $O[G^{(i)}(z)]$ the number of multiplications for finding $G^{(i)}(z)$, we get the recursion $O[G^{(i)}(z)] = L \cdot L^{(i-1)} + O[G^{(i-1)}(z)]$. Again, because $G^{(i-1)}(z)$ takes half as much complexity as $G^{(i)}(z)$, we get an order of complexity

$$O\left[G^{(i)}(z)\right] \simeq 2 \cdot L \cdot L^{(i-1)} \simeq 2^i \cdot L^2, \tag{6.3.5}$$

for multiplications and similarly for additions.

For a Fourier-domain evaluation, it turns out that the factorization (6.3.3) is more appropriate. In (6.3.3), we have to compute 2^{i-1} products between polynomials of size L (corresponding to $G(z)$) and of size $L^{(i-1)}/2^{i-1}$ (corresponding to the polyphase components of $G^{(i-1)}(z)$). Now, $L^{(i-1)}/2^{i-1}$ is roughly of size L as well. That is, using direct polynomial products, (6.3.3) takes 2^{i-1} times L^2 multiplications and as many additions, and the total complexity is the same as in (6.3.5). However, using FFT's produces a better algorithm. The $L \times L$ polynomial products require two Fourier transforms of length $2L$ and $2L$ frequency products, or, $L \cdot \log_2 L + 2L$ multiplications using the split-radix FFT. The step leading to $G^{(i)}(z)$ thus uses $2^{(i-1)} \cdot L(\log_2 L + 2)$ multiplications and the total complexity is

$$O\left[G^{(i)}(z)\right] = 2^i \cdot L(\log_2 L + 2)$$

multiplications, and about three times as many additions. This compares favorably to time-domain evaluation (6.3.5). As usual, this is interesting for medium to large L's. It turns out that the doubling formula (6.3.4), which looks attractive at first sight, does not lead to a more efficient algorithm than the ones we just outlined.

The savings obtained by the above simple algorithms are especially useful in multiple dimensions, where the iterates are with respect to lattices. Because multidimensional wavelets are difficult to design, iterating the filter might be part of the design procedure and thus, reducing the complexity of computing the iterates can be important.

6.4 Complexity of Overcomplete Expansions

Often, especially in signal analysis, a redundant expansion of the signal is desired. This is unlike compression applications, where nonredundant expansions are used. As seen in Chapter 5, the two major redundant expansions used in practice are the short-time Fourier (or Gabor) transform, and the wavelet transform. While the goal is to approximate the continuous transforms, the computations are necessarily discrete and amount to computing the transforms on denser grids than their orthogonal counterparts, and this in an exact or approximate manner, depending on the case.

6.4.1 Short-Time Fourier Transform

The short-time Fourier transform is computed with a modulated filter bank as in (6.2.8)-(6.2.9). The only difference is that the outputs are downsampled by $M < N$, and we do not have a square polyphase matrix as in (6.2.10). However, because the modulation is periodic with period N for all filters, there exists a fast algorithm. Compute the following intermediate outputs:

$$x_i[n] = \sum_k h[kN+i] \cdot x[n-kN-i]. \tag{6.4.1}$$

Then, the channel signals $y_i[n]$ are obtained by Fourier transform from the $x_i[n]$'s

$$\boldsymbol{y}[n] = \boldsymbol{F} \cdot \boldsymbol{x}[n],$$

where $\boldsymbol{y}[n] = (y_0[n] \ldots y_{N-1}[n])^T$, $\boldsymbol{x}[n] = (x_0[n] \ldots x_{N-1}[n])^T$, and $\boldsymbol{F}$ is the size $N \times N$ Fourier matrix. The complexity per output vector $\boldsymbol{y}[n]$ is L multiplications and about $L-N$ additions (from (6.4.1)) plus a size-N Fourier transform, or, $(N/2)\log_2 N$ multiplications and three times as many additions. Since $\boldsymbol{y}[n]$ has a rate M times smaller than the input, we get the following multiplicative complexity per input sample (where $K = N/M$ is the oversampling ratio):

$$\frac{1}{M}(L + N\log_2 N) \;=\; K \cdot \left(\frac{L}{N} + \log_2 N\right),$$

that is, K times more than in the critically sampled case given in (6.2.11). The additive complexity is similar (except for a factor of 3 in front of the $\log_2 N$).

Because $M < N$, the polyphase matrix is nonsquare of size $N \times M$ and does not have a structure as simple as the one given in (6.2.10). However, if N is a multiple of M, some structural simplifications can be made.

6.4.2 "Algorithme à Trous"

Mallat's and Shensa's algorithms compute the wavelet series expansion on a discrete grid corresponding to scales $a_i = 2^i$ and shifts $b_{ij} = j \cdot 2^i$ (see Figure 6.8 (a)). We assume $i = 0, 1, 2, \ldots,$ in this discussion. The associated wavelets form an orthonormal basis, but the transform is not shift-invariant, which can be a problem in signal analysis or pattern recognition. An obvious cure is to compute all the shifts, that is, avoid the downsampling (see Figure 6.8(b)). Of course, scales are still restricted to powers of two, but shifts are now arbitrary integers. It is clear that the output at scale a_i is 2^i-times oversampled. To obtain this oversampled transform, one simply finds the equivalent filters for each branch of the octave-band tree which computes the discrete-time wavelet series. This is shown in Figure 6.9. The filter producing the oversampled wavelet transform at scale $a_i = 2^i$ has a z-transform equal to

$$F_i(z) \;=\; H_1\left(z^{2^{i-1}}\right) \cdot \prod_{l=0}^{i-2} H_0\left(z^{2^l}\right).$$

An efficient computational structure simply computes the signals along the tree and takes advantage of the fact that the filter impulse responses are upsampled, that is, nonzero coefficients are separated by 2^k zeros. This lead to the name "algorithme à trous" (algorithm with holes) given in [136]. It is immediately obvious that the complexity of a direct implementation is now $2L$ multiplications and $2(L-1)$ additions/octave and input sample, since each octave requires filtering by highpass and lowpass filters which have L nonzero coefficients. Thus, to compute J octaves, the complexity is of the order of

$$4 \cdot L \cdot J \text{ operations/input sample}$$

that is, a linear increase with the number of octaves. The operations can be moved to Fourier domain to reduce the order L to an order $\log_2 L$ and octaves can be merged, just as in the critically sampled case. A careful analysis of the resulting complexity is made in [245], showing gains with Fourier methods for filters of medium length ($L \geq 9$).

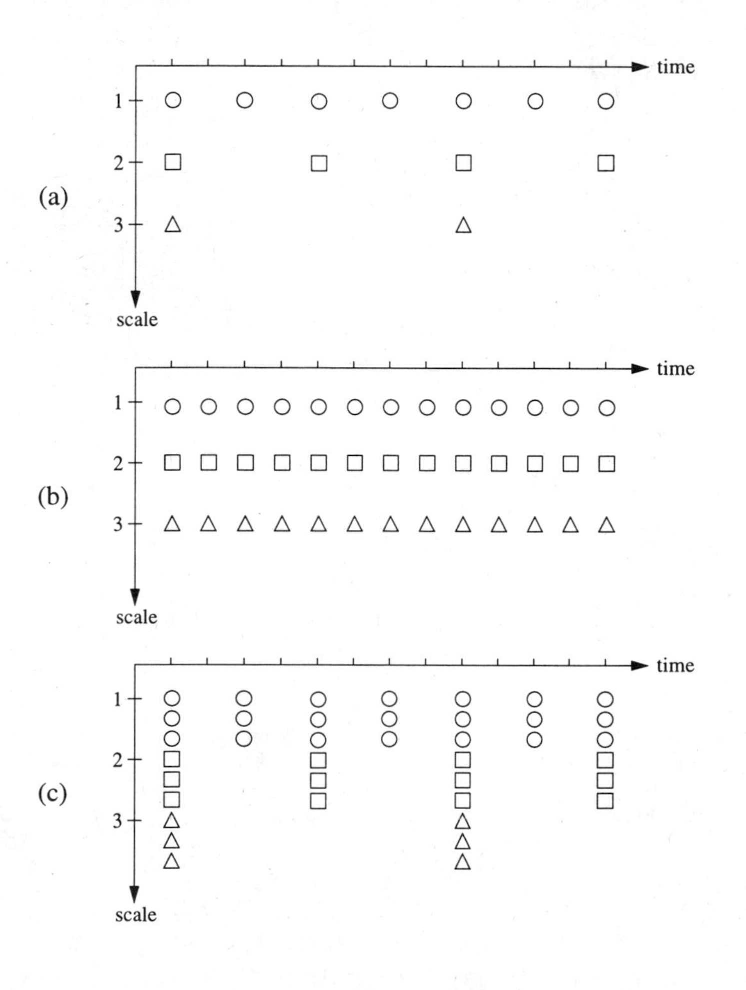

Figure 6.8 Sampling of the time-scale plane. (a) Sampling in the orthogonal discrete-time wavelet series. (b) Oversampled time-scale plane in the "algorithme à trous". (c) Multiple voices/octave. The case of three voices/octave is shown.

6.4.3 Multiple Voices Per Octave

While the above algorithm increased the sampling in time, it remained an "octave by octave" algorithm. Sometimes, finer scale changes are desired. Instead

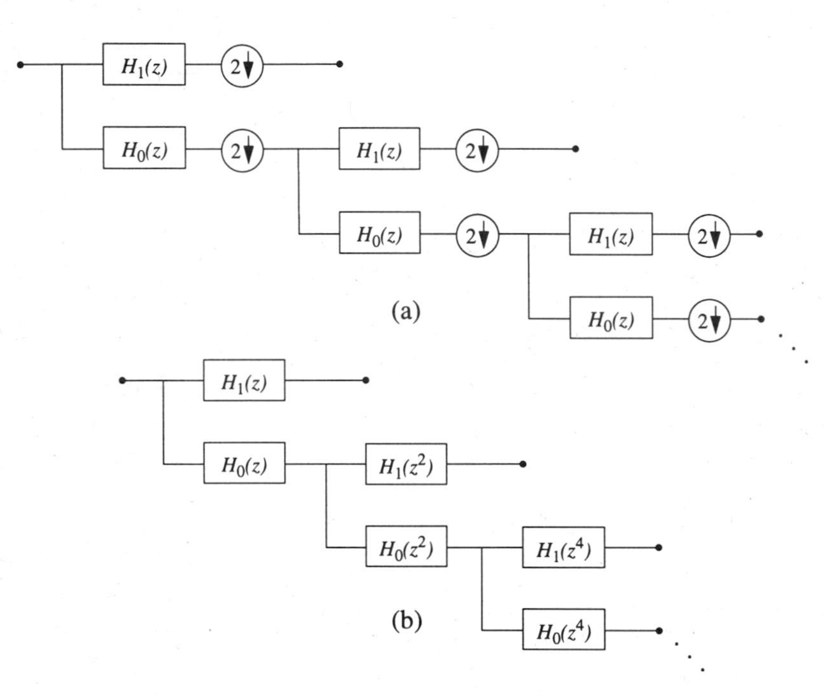

Figure 6.9 Oversampled discrete-time wavelet series. (a) Critically sampled case. (b) Oversampled case obtained from (a) by deriving the equivalent filters and skipping the downsampling. This approximates the continuous-time wavelet transform.

of $a = 2^i$, one uses $a = 2^{j+m/M}$, $m = 0, \ldots, M-1$, which gives M "voices"/octave. Obviously, for $m = 0$, one can use the standard octave by octave algorithm, involving the wavelet $\psi(t)$. To get the scales for $m = 1, \ldots, M-1$, one can use the slightly stretched versions

$$\psi^{(m)}(t) \;=\; 2^{-m/2M}\psi\left(2^{-m/M}t\right), \quad m = 1, \ldots, M-1.$$

The tiling of the time-scale plane is shown in Figure 6.8(c) for the case of three voices/octave (compare this with Figure 6.8(a)). Note that lower voices are oversampled, but the whole scheme is redundant in the first place since one voice would be sufficient. The complexity is M times that of a regular discrete-time wavelet series, if the various voices are computed in an independent manner.

The parameters of each of the separate discrete-time wavelet series have to be computed (following Shensa's algorithm), since the discrete-time filters will not be "scales" of each other, but different approximations. Thus, one has to find the appropriate highpass and lowpass filters for each of the m-voice wavelets. An alternative is to use the scaling property of the wavelet transform. Since

$$\langle x(t), \varphi(at) \rangle = \frac{1}{a} \langle x(t/a), \varphi(t) \rangle,$$

we can start a discrete-time wavelet series algorithm with m signals which are scales of each other; $x_m(t) = 2^{m/2M} x(2^{m/M} t)$, $m = 0, \ldots, M-1$. Again, the complexity is M times higher than a single discrete-time wavelet series. The problem is to find the initial sequence which corresponds to the projection of the $x_m(t)$ onto V_0. One way to do this is given in [300].

Finally, one can combine the multivoice with the "à trous" algorithm to compute a dense grid over scales as well as time. The complexity then grows linearly with the number of octaves and the number of voices, as

$$4 \cdot L \cdot J \cdot M \text{ operations/input sample},$$

where J and M are the number of octaves and voices respectively. This is an obvious algorithm, and there might exist more efficient ways yet to be found.

This concludes our discussion of algorithms for oversampled expansions, which closely followed their counterparts for the critically sampled case.

6.5 Special Topics

6.5.1 Computing Convolutions Using Multirate Filter Banks

We have considered improvements in computing convolutions that appear in filter banks. Now, we will investigate schemes where filter banks can be used to speed up convolutions.

Overlap-Add/Save Computation of Running Convolution When computing the linear convolution of an infinite signal with a finite-length filter using fast Fourier transforms, one has to segment the input signal into blocks. Assume a filter of length L and an FFT of size $N > L$. Then, a block of signal of length $N - L + 1$ can be fed into the FFT so as to get the linear convolution of the signal with the filter. The overlap-add algorithm [32, 209] segments the input signal into pieces of length $N - L + 1$, computes the FFT-based convolution, and adds the overlapping tails of adjacent segments ($L - 1$ outputs spill over to next segments of outputs).

The overlap-save algorithm [32, 209], takes N input samples and computes a circular convolution of which $N - L + 1$ samples are valid linear convolution outputs and $L - 1$ samples are wrap-around effects. These last $L - 1$ samples are discarded, the $N - L + 1$ valid ones kept, and the algorithm moves up by $N - L + 1$ samples.

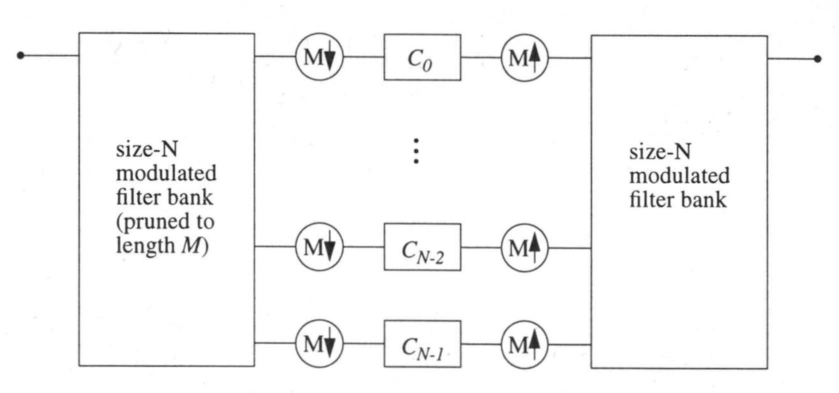

Figure 6.10 Overlap-add algorithm as a filter bank.

Both of these algorithms have an immediate filter bank interpretation [226] which has the advantage of permitting generalizations [317]. We will now focus on the overlap-add algorithm. Computing a size-N FFT with $M = N - L + 1$ nonzero inputs amounts to an analysis filter bank with N channels and downsampling by M. The filters are given by [317]

$$\begin{aligned} H(z) &= z^{M-1} + z^{M-2} + \cdots + z + 1, \\ H_i(z) &= z^{-M+1} \cdot H\left(W_N^i z\right). \end{aligned}$$

In frequency domain, convolution corresponds to pointwise multiplication by the Fourier transform of the filter $c[n]$ given by

$$C_i = \frac{1}{N} \sum_{l=0}^{L-1} W_n^{il} c[l].$$

Finally, the inverse Fourier transform is obtained with upsampling by M followed by filtering with an N-channel synthesis bank where the filters are given by

$$\begin{aligned} G(z) &= 1 + z^{-1} + z^{-2} + \cdots + z^{-N+1}, \\ G_i(z) &= G\left(W_N^i z\right). \end{aligned}$$

The algorithm is sketched in Figure 6.10. The proof that it does compute a running convolution is simply by identification of the various steps with the usual overlap-add algorithm. Note that the system produces a delay of $M - 1$ samples (since all filters are causal), that is

$$Y(z) = z^{-(M-1)} C(z) X(z).$$

A simple generalization consists in replacing the pointwise multiplications by C_i, $i = 0, \ldots, N-1$, by filters $C_i(z)$, $i = 0, \ldots, N-1$. Because the system is linear, we can use the superposition principle and decompose $C_i(z)$ into its components. Call c_{il} the lth coefficient of the ith filter. Now, the set $\{c_{i0}\}$, $i = 0, \ldots, N-1$ produces an impulse response $c_0[n]$ obtained from the inverse Fourier transform of the coefficients c_{i0}. Therefore, because the filters $C_i(z)$ exist in a domain downsampled by M, the set $\{c_{il}\}$ produces an impulse response $c_l[n]$ which is the inverse Fourier transform of c_{il} delayed by $l \cdot M$ samples.

Finally, if $C_i(z)$ is of degree K, the generalized overlap-add algorithm produces a running convolution with a filter of length $(K+1)M$ when $M = L$ and $N = 2M$. Conversely, if an initial filter $c[n]$ is given, one first decomposes it into segments of length M, each of which is Fourier transformed into a set $\{c_{il}\}$. That is, a length-$(K+1)M$ convolution is mapped into N size-$(K+1)$ convolutions, where N is about two times M, and this using size-N modulated filter banks. The major advantage of this method is that the delay is substantially reduced, an issue of primary concern in real-time systems. This is because the delay is of the order of the downsampling M, while a regular overlap-add algorithm would have a delay of the order of $(K+1) \cdot M$.

Table 6.2 gives a comparison of several methods for computing running convolution, highlighting the trade-off between computational complexity and input-output delay, as well as architectural complexity [317].

Short Running Convolution It is well-known that Fourier methods are only worthwhile for efficiently computing convolutions by medium to long filters. If a filter is short, one can use transposition of the short linear convolution algorithms seen in Section 6.1.1 to get efficient running convolutions. For example, the algorithm in (6.1.4) for 2×2 linear convolution, when transposed, computes two successive outputs of a length-2 filter with impulse response $(b_1\ b_0)$, since

$$\begin{pmatrix} b_0 & 0 \\ b_1 & b_0 \\ 0 & b_1 \end{pmatrix}^T = \begin{pmatrix} b_0 & b_1 & 0 \\ 0 & b_0 & b_1 \end{pmatrix}$$

$$= \begin{pmatrix} 1 & 1 & 0 \\ 0 & -1 & 1 \end{pmatrix} \begin{pmatrix} b_0 & 0 & 0 \\ 0 & b_0 - b_1 & 0 \\ 0 & 0 & b_1 \end{pmatrix} \begin{pmatrix} 1 & 1 & 0 \\ 0 & -1 & 0 \\ 0 & 1 & 1 \end{pmatrix}. \quad (6.5.1)$$

The multiplicative complexity is unchanged at three multiplications/two outputs (rather than four), while the number of additions goes up from three to four.

The same generalization we made for overlap-add algorithms works here as well. That is, the pointwise multiplications in (6.5.1) can be replaced by

Table 6.2 Computation of running convolution with a length-32 filter (after [317]). The filter and signal are assumed to be complex.

Method	Delay	Multiplications per point	Architecture
(a) Direct	0	96	Simple
(b) 128-point FFT downsampled by 97	96	15	Complex (128-pt FFT's)
(c) 16-point FFT downsampled by 8 and length-4 channel filters	7	29	Medium (16-pt FFT's)
(d) Same as (c) but with efficient 4-pt convolutions in the channel	31	18.5	Medium (as (c) plus simple short convolution algorithms)

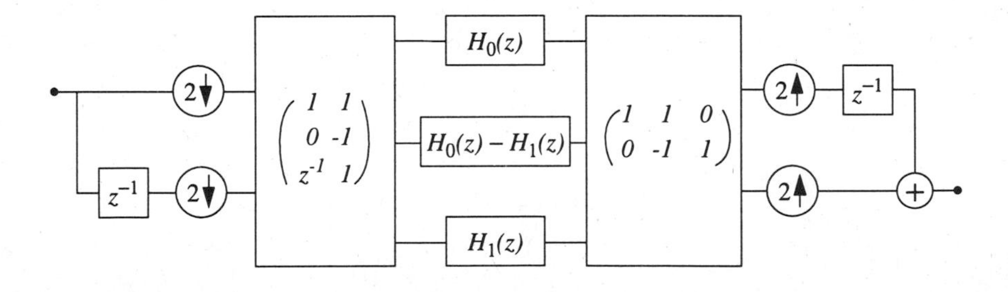

Figure 6.11 Fast running convolution algorithm with channel filters. The input-output relationship equals $H_{tot}(z) = z^{-1}(H_0(z^2) + z^{-1}H_1(z^2))$.

filters in order to achieve longer convolutions. This again is best looked at as a filter bank algorithm, and Figure 6.11 gives an example of equation (6.5.1) with channel filters instead of pointwise multiplications. After a forward polyphase transform, a polyphase matrix (obtained from the rightmost addition matrix in (6.5.1) produces the three channel signals. The channel filters are the polyphase components of the desired filter and their difference. Then, a synthesis polyphase matrix (the left addition matrix from (6.5.1)) precedes an inverse polyphase transform. The transfer matrix between forward and inverse polyphase transform is

$$\begin{aligned} \boldsymbol{T}(z) &= \begin{pmatrix} 1 & 1 & 0 \\ 0 & -1 & 1 \end{pmatrix} \begin{pmatrix} H_0(z) & 0 & 0 \\ 0 & H_0(z) - H_1(z) & 0 \\ 0 & 0 & H_1(z) \end{pmatrix} \begin{pmatrix} 1 & 1 \\ 0 & -1 \\ z^{-1} & 1 \end{pmatrix} \\ &= \begin{pmatrix} H_0(z) & H_1(z) \\ z^{-1}H_1(z) & H_0(z) \end{pmatrix}, \end{aligned}$$

which is pseudocirculant, as required for a time-invariant system [311]. The above $\boldsymbol{T}(z)$ gives the following input-output relationship for the total system

$$H_{tot}(z) = z^{-1}(H_0(z^2) + z^{-1}H_1(z^2)).$$

That is, at the price of a single delay, we have replaced a length-L convolution by three length-$L/2$ convolutions at half-rate, that is, a saving of 25%. This simple example is part of a large class of possible algorithms which have been studied in [198, 199, 317]. Their attractive features are that they are simple, numerically well-conditioned (no approximations are necessary), and the building blocks remain convolutions (for which optimized hardware is available).

6.5.2 Numerical Algorithms

We will briefly discuss an original application of wavelets to numerical algorithms [30]. These algorithms are approximate using exact arithmetic, but arbitrary precision can be obtained. Thus, these are unlike the previous algorithms in this chapter which reduced computations while being exact in exact arithmetic. The idea is that matrices can be compressed just like images! In applications such as iterative solution of large linear systems, the recurrent operation is a very large matrix-vector product which has complexity N^2. If the matrix is the discrete version of an operator which is smooth (except at some singularities), the wavelet transform[2] can be used to "compress" the matrix by concentrating most of the energy into well-localized bands. If coefficients smaller than a certain threshold are set to zero, the transformed matrix becomes sparse. Of course, we now deal with an approximated matrix, but the error can be bounded. Beylkin, Coifman and Rokhlin [30] show that for a large class of operators, the number of coefficients after thresholding is of order N.

[2] Since this will be a matrix operation of finite dimension, we call it a wavelet transform rather than a discrete-time wavelet series.

We will concentrate on the simplest version of such an algorithm. Call $\boldsymbol{W}$ the matrix which computes the orthogonal wavelet transform of a length-N vector. Its inverse is simply its transpose. If we desire the matrix vector product $\boldsymbol{y} = \boldsymbol{M} \cdot \boldsymbol{x}$, we can compute:

$$\boldsymbol{y} \;=\; \boldsymbol{W}^T \cdot (\boldsymbol{W} \cdot \boldsymbol{M} \cdot \boldsymbol{W}^T) \cdot \boldsymbol{W} \cdot \boldsymbol{x}. \tag{6.5.2}$$

Recall that $\boldsymbol{W} \cdot \boldsymbol{x}$ has a complexity of order $L \cdot N$, where L is the filter length and N the size of the vector. The complexity of $\boldsymbol{W} \cdot \boldsymbol{M} \cdot \boldsymbol{W}^T$ is of order $L \cdot N^2$, and thus, (6.5.2) is not efficient if only one product is evaluated. However, if we are in the case of an iterative algorithm, we can compute $\boldsymbol{M}' = \boldsymbol{W} \cdot \boldsymbol{M} \cdot \boldsymbol{W}^T$ once (at a cost of LN^2) and then use $\boldsymbol{M}'$ in the sequel. If $\boldsymbol{M}'$, after thresholding, has order-N nonzero entries, then the subsequent iterations, which are of the form:

$$\boldsymbol{y}' \;=\; \boldsymbol{W}^T \cdot \boldsymbol{M}' \cdot \boldsymbol{W} \cdot \boldsymbol{x}',$$

are indeed of order N rather than N^2. It turns out that the computation of $\boldsymbol{M}'$ itself can be reduced to an order N problem [30]. An interpretation of $\boldsymbol{M}'$ is of interest. Premultiplying $\boldsymbol{M}$ by $\boldsymbol{W}$ is equivalent to taking a wavelet transform of the columns of $\boldsymbol{M}$, while postmultiplying $\boldsymbol{M}$ by $\boldsymbol{W}^T$ amounts to taking a wavelet transform of its rows. That is, $\boldsymbol{M}'$ is the two-dimensional wavelet transform of $\boldsymbol{M}$, where $\boldsymbol{M}$ is considered as an image. Now, if $\boldsymbol{M}$ is smooth, one expects $\boldsymbol{M}'$ to have energy concentrated in some well-defined and small regions. It turns out that the zero moments of the wavelets play an important role in concentrating the energy, as they do in image compression. This short discussion only gave a glimpse of these powerful methods, and we refer the interested reader to [30] and the references therein for more details.

PROBLEMS

6.1 *Toeplitz matrix-vector products:* Given a Toeplitz matrix $\boldsymbol{T}$ of size $N \times N$, and a vector $\boldsymbol{x}$ of size N, show that the product $\boldsymbol{T}\boldsymbol{x}$ can be computed with an order $N \log_2 N$ operations. The method consists in extending $\boldsymbol{T}$ into a circulant matrix $\boldsymbol{C}$. What is the minimum size of $\boldsymbol{C}$, and how does it change if $\boldsymbol{T}$ is symmetric?

6.2 *Block circulant matrices:* A block-circulant matrix of size $NM \times NM$ is like a circulant matrix of size $N \times N$, except that the elements are now blocks of size $M \times M$. For example, given two $M \times M$ matrices $\boldsymbol{A}$ and $\boldsymbol{B}$,

$$\boldsymbol{C} = \begin{bmatrix} \boldsymbol{A} & \boldsymbol{B} \\ \boldsymbol{B} & \boldsymbol{A} \end{bmatrix}$$

is a size $2M \times 2M$ block-circulant matrix. Show that block-circulant matrices are block-diagonalized by block Fourier transforms of size $NM \times NM$ defined as

$$\boldsymbol{F}_{NM}^{B} = \boldsymbol{F}_N \otimes \boldsymbol{I}_M,$$

where $\boldsymbol{F}_N$ is the size-N Fourier matrix, $\boldsymbol{I}_M$ is the size-M identity matrix and $\otimes$ is the Kronecker product (2.3.2).

6.3 The Walsh-Hadamard transform of size $2N$ (N is a power of 2) is defined as

$$W_{2N} = W_2 \otimes W_N,$$

where

$$W_2 = \begin{bmatrix} 1 & 1 \\ 1 & -1 \end{bmatrix},$$

and $\otimes$ is the Kronecker product (2.3.2). Derive an algorithm that uses $N \log_2 N$ additions for a size-N transform.

6.4 *Complexity of MUSICAM filter bank:* The filter bank used in MUSICAM (see also Section 7.2.3) is based on modulation of a single prototype of length 512 to 32 bandpass filters. For the sake of this problem, we assume a complex modulation by W_{32}^{nk}, that is

$$h_k[n] = h_p[n]\, W_{32}^{nk}, \qquad W_{32} = e^{-j2\pi/32},$$

and thus, the filter bank can be implemented using polyphase filters and an FFT (see Section 6.2.3). In a real MUSICAM system, the modulation is with cosines and the implementation involves polyphase filters and a fast DCT, thus it is very similar to the complex case we analyze here. Assuming an input sampling rate of 44.1 kHz, give the number of operations per second required to compute the filter bank.

6.5 *Iterated filters:* Consider

$$H^{(i)}(z) = \prod_{k=0}^{i-1} H\left(z^{2^k}\right) \qquad i = 1, 2, \ldots$$

and prove the following recursive formulas:

$$\begin{aligned} H^{(i)}(z) &= H(z) \cdot H^{(i-1)}(z^2), \\ H^{(i)}(z) &= H(z^{2^{i-1}}) \cdot H^{(i-1)}(z), \\ H^{(2^K)}(z) &= H^{(2^{k-1})}(z) \cdot H^{(2^{k-1})}(z^{2^{2^{k-1}}}). \end{aligned}$$

6.6 *Overlap-add/save filter banks:* Consider a size-4 modulated filter bank downsampled by 2 and implementing overlap-add or save running convolution (see Figure 6.10 for example).

(a) Derive explicitly the analysis and synthesis filter banks.

(b) Derive the channel coefficients. How long can the time-domain impulse response be if the channel coefficients are scalars and the system is LTI?

(c) Implement a filter with a longer impulse response than found in (b) above by using polynomial channel coefficients. Give an example, and verify that the system is LTI.

6.7 Consider a 3-channel analysis/synthesis filter bank downsampled by 2, with filtering of the channels (see Figure 3.18). The filters are given by

$$
\begin{array}{rrr}
H_0(z) = z^{-1}, & H_1(z) = 1 + z^{-1}, & H_2(z) = 1 \\
G_0(z) = 1 - z^{-1}, & G_1(z) = z^{-1}, & G_2(z) = z^{-2} - z^{-1} \\
C_0(z) = F_0(z), & C_1(z) = F_0(z) + F_1(z), & C_2(z) = F_1(z).
\end{array}
$$

Verify that the overall system is shift-invariant and performs a convolution with a filter having the z-transform $F(z) = (F_0(z^2) + z^{-1}F_1(z^2))z^{-1}$.

7

Signal Compression and Subband Coding

"That which shrinks must first expand."
— Lao-Tzu, *Tao Te Ching*

The compression of signals, which is one of the main applications of digital signal processing, uses signal expansions as a major component. Some of these expansions were discussed in previous chapters, most notably discrete-time expansions via filter banks. When the channels of a filter bank are used for coding, the resulting scheme is known as *subband coding*. The reasons for expanding a signal and processing it in transform domain are numerous. While source coding can be performed on the original signal directly, it is usually more efficient to find an appropriate transform. By efficient we mean that for a given complexity of the encoder, better compression is achieved.

The first useful property of transforms, or "generalized" transforms such as subband coding, is their decorrelation property. That is, in the transform domain, the transform coefficients are not correlated, which is equivalent to diagonalizing the autocovariance matrix of the signal, as will be seen in Section 7.1. This diagonalization property is similar to the convolution property (or the diagonalization of circulant matrices) of the Fourier transform as we discussed in Section 2.4.8. However, the only transform that achieves exact diagonalization, the Karhunen-Loève transform, is usually impractical. Many other transforms come close to exact diagonalization and are therefore popular, such as the discrete cosine transform, or, appropriately designed subband or wavelet transforms. The second advantage of transforms is that the new domain is often more appropriate for quantization using perceptual criterions. That is, the transform domain can be used to distribute errors in a way that is less objectionable for the human user. For example, in speech and audio coding, the frequency bands used in subband coding might mimic operations performed in the inner ear and thus one can exploit the reduced sensitivity or even masking between bands. The third advantage of transform coding is that the previous features come at a low computational price. The transform decomposition itself is computed using fast algorithms as discussed in Chapter 6, quantization in the transform domain is often simple scalar quantization, and entropy coding is done on a sample-by-sample basis.

Together, these advantages produced successful compression schemes for speech, audio, images and video, some of which are now industry standards (32 Kbits/sec subband coding for high-quality speech [192], AC [34, 290], PAC [147], and MUSICAM for audio [77, 279], JPEG for images [148, 327], MPEG for video [173, 201]).

It is important to note that the signal expansions on which we have focused so far are only one of the three major components of such compression schemes. The other two are quantization and entropy coding. This three part view of compression will be developed in detail in Section 7.1, together with the strong interaction that exists among them. That is, in a compression context, there is no need for designing the "ultimate" basis function system unless adequate quantization and entropy coding are matched to it. This interplay, while fairly obvious, is often insufficiently stressed in the literature. Note that this section is a review and can be skipped by readers familiar with basic signal compression.

Section 7.2 concentrates on one-dimensional signal compression, that is, speech and audio coding. Subband methods originated from speech compression research, and for good reasons: Dividing the signal in frequency bands imitates the human auditory system well enough to be the basis for a series of successful coders.

Section 7.3 discusses image compression, where transform and subband/wavelet methods hold a preeminent position. It turns out that representing images at multiple resolutions is a desirable feature in many systems using image compression such as image databases, and thus, subband or wavelet methods are a popular choice. We also discuss some new schemes which contain wavelet decompositions as a key ingredient.

Section 7.4 adds one more dimension and discusses video compression. While straight linear transforms have been used, they are outperformed by methods using a combination of motion based modeling and transforms. Again, a multiresolution feature is often desired and will be discussed.

Section 7.5 discusses joint source-channel coding using multiresolution source decompositions and matched channel coding. It turns out that several upcoming applications, such as digital broadcasting and transmission over

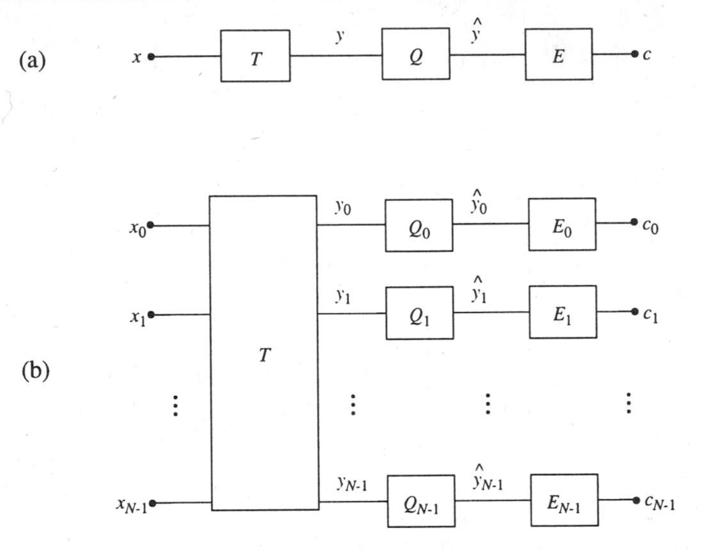

Figure 7.1 Compression system based on linear transformation. The linear transform $\boldsymbol{T}$ is followed by quantization ($\boldsymbol{Q}$) and entropy coding ($\boldsymbol{E}$). The reconstruction is simply $\hat{\boldsymbol{x}} = \boldsymbol{T}^{-1}\hat{\boldsymbol{y}}$. (a) Global view. (b) Multichannel case with scalar quantization and entropy coding.

highly varying channels such as wireless channels or channels corresponding to packet-switched transmission, are improved by using multiresolution techniques.

7.1 Compression Systems Based on Linear Transforms

In this section, we will deal with compression systems, as given in Figure 7.1(a). The linear transformation (T) is the first step in the process which includes quantization (Q) and entropy coding (E). Quantization introduces nonlinearities in the system and results in loss of information, while entropy coding is a reversible process. A system as given in Figure 7.1 is termed an open-loop system, since there is no feedback from the output to the input. On the other hand, a closed-loop system, such as the DPCM (see Figure 7.5), includes the quantization in the loop. We mostly concentrate on open-loop systems, because of their close connection to signal expansions. Following Figure 7.1, we start by discussing various linear transforms with an emphasis on the optimal Karhunen-Loève transform, followed by quantization, and end up briefly describing entropy coding methods. We try to emphasize the interplay among these three parts, as well as indicate the importance of perceptual criterions in designing the overall system. Our discussion is based on the excellent text by Gersho and Gray [109], to which we refer for more details. This chapter uses results from statistical signal processing, which are reviewed in Appendix 7.A.

Let us here define the measures of quality we will be using. First, the *mean square error* (MSE), or, *distortion*, equals

$$D = E(\sum_{i=0}^{N-1} |x_i - \hat{x}_i|^2), \tag{7.1.1}$$

where x_i are the input values and $\hat{x}_i$ are the reconstructed values. For a zero-mean input, the signal-to-noise ratio (SNR) is given by

$$SNR = 10 \log_{10} \frac{\sigma^2}{D}, \tag{7.1.2}$$

where D is as given in (7.1.1) and σ^2 is the input variance. The peak signal-to-noise ratio (SNR_p) is defined as [138]

$$SNR_p = 10 \log_{10} \frac{M^2}{D}, \tag{7.1.3}$$

where M is the maximum peak-to-peak value in the signal (typically 256 for 8-bit images). Distortion measures based on squared error have shortcomings when assessing the quality of a coded signal such as an image. An improved distortion measure is a perceptually weighted mean square error. Even better are distortion models which include masking. These distortion metrics are signal specific, and some of them will be discussed in conjunction with practical compression schemes in later sections.

7.1.1 Linear Transformations

Assume a vector $\boldsymbol{x}[n] = (x[n], x[n+1], \ldots x[n+N-1]\,)^T$ of N consecutive samples of a real wide-sense stationary random process (see Appendix 7.A). Typically, these samples are correlated and independent coding of the samples is inefficient. The idea is to apply a linear transform[1] so that the transform coefficients are decorrelated. While there is no general formal result that guarantees more efficient compression by decorrelation, it turns out in practice (and for certain cases in theory) that scalar quantization of decorrelated transform coefficients is more efficient than direct scalar quantization of the samples.

Since we assumed that the process is wide-sense stationary and we will be dealing only with the second-order statistics, we do not need to keep the index n for $\boldsymbol{x}[n]$ and can abbreviate it simply as $\boldsymbol{x}$. From now on, we will assume that the process is zero-mean and thus its autocorrelation and autocovariance are

[1]This can also be seen as a discrete-time series expansion. However, since it is usually implemented as a matrix block transform we will adhere to the compression literature's convention and call it a transform.

the same, that is, $K[n,m] = R[n,m]$. The autocovariance matrix of the input vector $\boldsymbol{x}$ is

$$\boldsymbol{K}_x = E(\boldsymbol{x} \cdot \boldsymbol{x}^T).$$

Again, since the process is wide-sense stationary and zero-mean, $K[n,m] = K[n-m] = R[n-m]$ (see Appendix 7.A). Therefore, the matrix $\boldsymbol{K}_x$ has the following form:

$$\boldsymbol{K}_x = \begin{pmatrix} R[0] & R[1] & \dots & R[N-1] \\ R[1] & R[0] & \dots & R[N-2] \\ \vdots & \vdots & \vdots & \vdots \\ R[N-1] & R[N-2] & \dots & R[0] \end{pmatrix}.$$

This matrix is Toeplitz, symmetric (see Section 2.3.5), and nonnegative definite since all of its eigenvalues are greater or equal to zero (this holds in general for autocorrelation matrices). Consider now the transformed vector $\boldsymbol{y}$,

$$\boldsymbol{y} = \boldsymbol{T}\boldsymbol{x}, \tag{7.1.4}$$

where $\boldsymbol{T}$ is an $N \times N$ unitary matrix which thus satisfies $\boldsymbol{T}^T\boldsymbol{T} = \boldsymbol{T}\boldsymbol{T}^T = \boldsymbol{I}$. Then the autocovariance of $\boldsymbol{y}$ is

$$\begin{aligned} \boldsymbol{K}_y &= E(\boldsymbol{y}\boldsymbol{y}^T) = E(\boldsymbol{T}\boldsymbol{x}\boldsymbol{x}^T\boldsymbol{T}^T) = \boldsymbol{T}E(\boldsymbol{x}\boldsymbol{x}^T)\boldsymbol{T}^T \\ &= \boldsymbol{T}\boldsymbol{K}_x\boldsymbol{T}^T. \end{aligned} \tag{7.1.5}$$

Karhunen-Loève Transform We would like to obtain uncorrelated transform coefficients. Recall that for each two coefficients to be uncorrelated, their covariance has to be zero (see Appendix 7.A). Thus, we are looking for a diagonal $\boldsymbol{K}_y$. For that to hold, $\boldsymbol{T}$ has to be chosen with its rows equal to the eigenvectors of $\boldsymbol{K}_x$. Call $\boldsymbol{v}_i$ the eigenvector (normalized to unit norm) of $\boldsymbol{K}_x$ associated with the eigenvalue λ_i, that is, $\boldsymbol{K}_x\boldsymbol{v}_i = \lambda_i\boldsymbol{v}_i$, and choose the following ordering for the λ_i's:

$$\lambda_0 \geq \lambda_1 \geq \cdots \geq \lambda_{N-1} \geq 0, \tag{7.1.6}$$

where the last inequality holds because $\boldsymbol{K}_x$ is nonnegative definite. Moreover, since $\boldsymbol{K}_x$ is symmetric, there is a complete set of orthonormal eigenvectors (see Section 2.3.2). Take $\boldsymbol{T}$ as

$$\boldsymbol{T} = [\boldsymbol{v}_0\ \boldsymbol{v}_1\ \dots\ \boldsymbol{v}_{N-1}]^T, \tag{7.1.7}$$

then, from (7.1.5),

$$\boldsymbol{K}_y = \boldsymbol{T} \cdot \boldsymbol{K}_x \cdot \boldsymbol{T}^T = \boldsymbol{T} \cdot \boldsymbol{T}^T \cdot \boldsymbol{\Lambda} = \boldsymbol{\Lambda}, \tag{7.1.8}$$

where $\boldsymbol{\Lambda}$ is a diagonal matrix with $\Lambda_{ii} = \lambda_i = \sigma_i^2 = y_i^2$, $i = 0, \dots, N-1$. The transform defined in (7.1.7) which achieves decorrelation as shown in (7.1.8) is the *discrete-time Karhunen-Loève* (KLT) or *Hotelling transform* [109, 138]. The following approximation result is intuitive:

PROPOSITION 7.1

If only k out of the N transform coefficients are kept, then the coefficients $y_0, \ldots, y_{k-1}$ will minimize the MSE between $\boldsymbol{x}$ and its approximation $\hat{\boldsymbol{x}}$.

Although the proof of this result follows from general orthonormal expansions results given in Chapter 2, we describe it here for completeness.

PROOF

Following (7.1.1), the MSE is equal to

$$D = E\left(\sum_{i=0}^{N-1}(x_i - \hat{x}_i)^2\right) = E((\boldsymbol{x} - \hat{\boldsymbol{x}})^T \cdot (\boldsymbol{x} - \hat{\boldsymbol{x}})) = E((\boldsymbol{y} - \hat{\boldsymbol{y}})^T \cdot (\boldsymbol{y} - \hat{\boldsymbol{y}})), \quad (7.1.9)$$

where the last equality follows from the fact that $\boldsymbol{T}$ is a unitary transform, that is, the MSE is conserved between transform and original domains. Keeping only the first k coefficients means that $\hat{y}_i = y_i$ for $i = 0, \ldots, k-1$ and $\hat{y}_i = 0$, for $i = k, \ldots, N-1$. Then the MSE equals

$$D_k = E\left(\sum_{i=0}^{N-1}(y_i - \hat{y}_i)^2\right) = \frac{1}{N}\sum_{i=k}^{N-1} y_i^2 = \frac{1}{N}\sum_{i=k}^{N-1}\lambda_i,$$

and this is obviously smaller or equal to any other set of $N-k$ coefficients because of the ordering in (7.1.6). Recall here that the assumption of zero mean still holds.

Another way to say this is that the first k coefficients contain most of the energy of the transformed signal. This is the "energy packing" property of the Karhunen-Loève transform. Actually, among all unitary transforms, the KLT is the one that packs most energy into the first k coefficients.

There are two major problems with the KLT, however. First, the KLT is signal dependent, since it depends on the autocovariance matrix. Second, it is computationally complex, since no structure can be assumed for $\boldsymbol{T}$, and no fast algorithm can be used. This leads to an order N^2 operations for applying the transform.

Discrete Cosine Transform Due to the discussed problems, various approximations to the KLT have been proposed. These approximations usually have fast algorithms for efficient implementation. The most successful is the *discrete cosine transform* (DCT), which calculates the vector $\boldsymbol{y}$ from $\boldsymbol{x}$ as

$$y_0 = \frac{1}{\sqrt{N}}\sum_{n=0}^{N-1} x_n, \quad (7.1.10)$$

$$y_k = \sqrt{\frac{2}{N}}\sum_{n=0}^{N-1} x_n \cos\left(\frac{2\pi(2n+1)k}{4N}\right), \quad k = 1, \ldots, N-1. \quad (7.1.11)$$

The DCT was developed [2] as an approximation for the KLT of a first-order Gauss-Markov process with a large positive correlation coefficient ρ ($\rho \to 1$). In this case, $\boldsymbol{K}_x$ is of the following form (assuming unit variance and zero mean)

$$\boldsymbol{K}_x = \begin{bmatrix} 1 & \rho & \rho^2 & \rho^3 & \cdots \\ \rho & 1 & \rho & \rho^2 & \cdots \\ \rho^2 & \rho & 1 & \rho & \cdots \\ \vdots & \vdots & \vdots & \vdots & \ddots \end{bmatrix}.$$

For large ρ's, the DCT approximately diagonalizes $\boldsymbol{K}_x$. Actually, the DCT (as well as some other transforms) is asymptotically equivalent to the KLT of an arbitrary wide-sense stationary process when the block size N tends to infinity [294]. It should be noted that even if the assumptions do not hold exactly (images are not first-order Gauss-Markov), the DCT has proven to be a robust approximation to the KLT, and is used in several standards for speech, image and video compression as we shall see.

The DCT also has shortcomings. One must block the input stream in order to perform the transform and this blocking is quite arbitrary. The block boundaries often create not only loss of compression (correlation across the boundaries is not removed) but also annoying blocking effects. This is one of the reasons for using lapped transforms and subband or wavelet coding schemes. However, the goal of these generalized transforms is the same, namely, to create decorrelated outputs from a correlated input stream, and then to quantize the outputs separately.

Discussion We recall that decorrelation leads to independence only if the input is Gaussian (see Appendix 7.A). Also, even independent random variables are better quantized as a block (or as a vector) than as independent scalars, due to sphere packing gains (see discussion of vector quantization in Section 7.1.2). However, the complexity of doing so is high, and thus, scalar quantization is often preferred. It will be shown below, after a discussion of quantization and bit allocation, that the KLT is the optimal linear transformation (under certain assumptions) among block transforms. The performance of subband coding will also be analyzed.

The major point is that all these schemes are unitary transformations on the input and thus, if $\hat{\boldsymbol{x}}$ and $\hat{\boldsymbol{y}}$ are the approximate versions of $\boldsymbol{x}$ and $\boldsymbol{y}$, respectively, we always have (similarly to (7.1.9))

$$\|\boldsymbol{x} - \hat{\boldsymbol{x}}\| = \|\boldsymbol{y} - \hat{\boldsymbol{y}}\|. \tag{7.1.12}$$

Note that nonorthogonal systems (such as linear phase biorthogonal filter banks) are usually designed to almost satisfy (7.1.12). If they do not, there

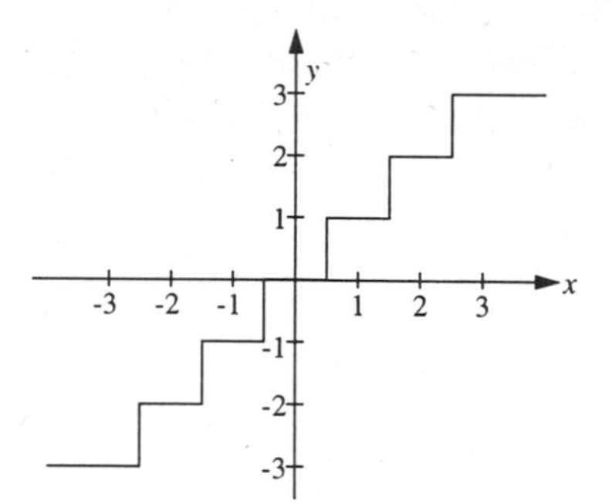

Figure 7.2 Uniform scalar quantizer with $N = 7$ and $\Delta = 1$. The decision levels $\{x_i\}$ are $\{-5/2, -3/2, -1/2, 1/2, 3/2, 5/2\}$ and the outputs $\{y_i\}$ are $\{-3, -2, -1, 0, 1, 2, 3\}$.

is a risk that small errors in the transform domain are magnified after reconstruction. The key problem now is to design the set of quantizers so as to minimize $E(\|\boldsymbol{y} - \hat{\boldsymbol{y}}\|)$.

7.1.2 Quantization

While we deal with discrete-time signals in this chapter, the sample values are real numbers, that is, continuously distributed in amplitude. In order to achieve compression, we need to map the real value of samples into a discrete set, or *discrete alphabet*. This process of mapping the real line into a countable discrete alphabet is called *quantization*. In practical situations, the sample values are mapped into a finite alphabet. An excellent treatment of quantization can be found in [109]. In its simplest form, each sample is individually quantized, which is called *scalar quantization*. A more powerful method consists in quantizing several samples at once, which is referred to as *vector quantization*. Also, one can quantize the difference between a signal and a suitable prediction of it, and this is called *predictive quantization*. We would like to stress here that the results on optimal quantization for a given signal are well-known, and can be found in [109, 143].

Scalar Quantization An example of a scalar quantizer is shown in Figure 7.2. The input range is divided into intervals $I_i = (x_{i-1}, x_i]$ (a partition of the real line) and the output value y_i is typically chosen in the interval I_i. The set $\{y_i\}$ is called the *codebook* and y_i the *codewords*. For the simple, uniform quantizer shown in Figure 7.2, the intervals are of the form $(i-1/2, i+1/2]$ and $y_i = i$. Note that the number of intervals is finite. Thus, there are two unbounded intervals which correspond to what is called "overload" regions of the quantizer, that is, for $x < -5/2$ and $x > 5/2$. Given that the number of intervals is N, there are N output symbols. Thus, $R = \lceil \log_2 N \rceil$ bits are needed to represent the output of the quantizer, and this is called the *rate*. The operation of selecting the interval is sometimes called *coding*, while assigning the output value y_i for the interval I_i is called *decoding*. Thus, we have a two-step process

$$(x_{i-1}, x_i] \underbrace{\longrightarrow}_{\text{coder}} i \underbrace{\longrightarrow}_{\text{decoder}} y_i.$$

The performance of a quantizer is measured as the distance between the input and the output, and typically, the squared error is used:

$$d(x, \hat{x}) = |x - \hat{x}|^2.$$

Given an input distribution, worst case or more often average distortion is measured. Thus, the MSE is

$$D = E(|x - \hat{x}|^2) = \sum_i \int_{x_{i-1}}^{x_i} (x - y_i)^2 f_X(x) dx, \tag{7.1.13}$$

where $f_X(x)$ is the probability density function (pdf) of x. For example, assume a uniform input pdf and a bounded input with N intervals, then uniform quantization with intervals of width Δ and $y_i = (x_i + x_{i-1})/2$ leads to an MSE equal to

$$D = \frac{\Delta^2}{12}. \tag{7.1.14}$$

The derivation of (7.1.14) is left as an exercise (see Problem 7.1). The error due to quantization is called quantization noise:

$$e[n] = \hat{x}[n] - x[n],$$

if x and $\hat{x}$ are the input and the output of the quantizer, respectively. While $e[n]$ is a deterministic function of $x[n]$, it is often modeled as a noise process which is uncorrelated to the input, white and with a uniform sample distribution. This is called an additive noise model, since $\hat{x}[n] = x[n] + e[n]$. While this is clearly an approximation, it is a fair one in the case of high-resolution uniform quantization (when Δ is much smaller than the standard deviation σ of the input signal and N is large).

Uniform quantization, while not optimal for nonuniform input pdf's, is very simple and thus often used in practice. One design parameter, besides the quantization step Δ, is the number of intervals, or the boundaries which correspond to the overload region. Usually, they are chosen as a multiple of the standard deviation σ of the input pdf (typically, 4 σ away from the mean). Given constant boundaries a and b, then $\Delta = (b - a)/N$. Thus, Δ decreases as $1/N = 1/2^R$ where R is the number of bits of the quantizer. The distortion D is of the form (following (7.1.14))

$$D = \frac{\Delta^2}{12} = \frac{(b-a)^2}{12N^2} = \sigma^2 2^{-2R} = C \cdot 2^{-2R}, \tag{7.1.15}$$

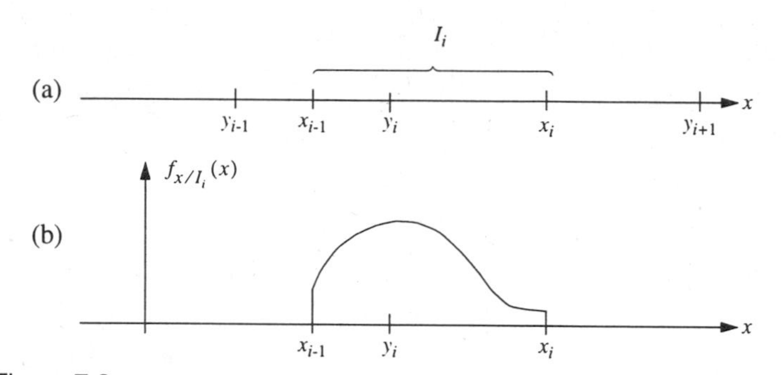

Figure 7.3 Optimality conditions for scalar quantizers. (a) Nearest neighbor condition. (b) Centroid condition.

since $\sigma^2 = (b-a)^2/12$ for uniform input pdf. In general, C is a function of σ^2 and depends on the distribution. This means that the SNR goes up by 6 dB for every additional bit in the quantizer. To see that, add a bit to R, $R' = R+1$. Then

$$D' = C \cdot 2^{-2(R+1)} = C \cdot 2^{-2R} \cdot 2^{-2}.$$

The new SNR' equals (use (7.1.2))

$$SNR' = 10 \log_{10} 4 \frac{\sigma^2}{C2^{-2R}} = SNR + 10 \log_{10} 4 \simeq SNR + 6 \text{ dB}.$$

When the pdf is not uniform, optimal quantization will not be uniform either. An optimal MSE quantizer is one that minimizes D in (7.1.13) for a given number of output symbols N. For a quantizer to be MSE optimal, it has to satisfy the following two necessary conditions [109]:

(a) *Nearest neighbor condition* For a given set of output levels, the optimal partition cells are such that an input is assigned to the nearest output level. For MSE minimization, this leads to the midpoint decision level between every two adjacent output levels.

(b) *Centroid condition* Given a partition of the input, the optimal decoding levels with respect to the MSE are the centroids of the intervals, that is, $y_i = E(x \mid x \in I_i)$.

Note that such a quantizer is not necessarily optimal for compression since it does not take into account entropy coding.[2] The two conditions are sketched in Figure 7.3. Both conditions are intuitive, and can be used to verify optimality of a quantizer or actually design an optimal one. This is done in the Lloyd

[2]A suitable modification, called entropy constrained quantization, takes entropy into account in the design of the quantizer.

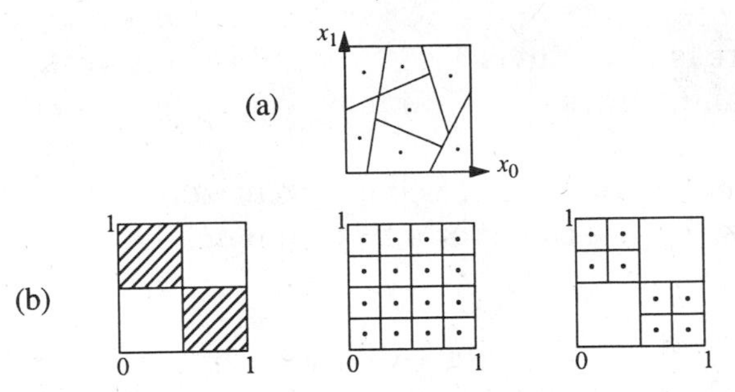

Figure 7.4 Vector quantization. (a) Example of a regular vector quantizer in two dimensions. (b) Comparison of scalar and vector quantizations. On the left, a two-dimensional probability density function is shown. It equals 2 in shaded areas and 0 otherwise. Note that x_0 and x_1 have uniform marginal distributions. For a given distortion, in the middle, optimal scalar (separable) quantization is shown, with 4.0 bits, or, 2.0 bits/sample. For the same distortion, on the right, vector quantization is shown, with 3.0 bits, or, 1.5 bits/sample.

algorithm, which iteratively improves a codebook for a given pdf and a number of codewords N (the pdf can be given analytically or through measurements). Starting with some initial codebook $\{y_i^{(0)}\}$, it alternates between

(a) Given $\{y_i^{(n)}\}$, find the partition $\{x_i^{(n)}\}$, based on the nearest neighbor condition.

(b) Given $\{x_i^{(n)}\}$, find the next $\{y_i^{(n+1)}\}$, satisfying the centroid condition.

and stops when $D^{(n)}$ is only marginally improved. The resulting quantizer is called a Lloyd-Max quantizer.

The above discussion assumed quantization of a continuous variable into a discrete set. Often, a discrete input set of size M has to be quantized into a set of size $N < M$. A "discrete" version of the Lloyd algorithm, which uses the same necessary conditions (nearest neighbor and centroid), can then be used.

While the above method yields quantizers with minimum distortion for a given codebook size, entropy coding was not considered. We will see that if entropy coding is used after quantization, a uniform quantizer can actually be attractive.

Vector Quantization While *vector quantization* (VQ) [109, 120] is much more than just a generalization of scalar quantization to multiple dimensions, we will only look at it in this restricted way in our brief treatment. Figure 7.4(a) shows a regular vector quantizer for a two-dimensional variable. Note that the partition of the square is into convex[3] regions and the separation into regions is performed using straight lines (in N dimensions, these would be hyperplanes of dimension $N-1$).

There are several advantages of vector quantizers over scalar quantizers. For the sake of discussion, we consider a two-dimensional case, but it obviously generalizes to N dimensions.

(a) *Packing gain* Even if two variables are independent, there is gain in quantizing them together. The reason is that there exist better partitions of the space then the rectangular partition obtained when we separately scalar quantize each variable. For example, in two dimensions, it is well-known that hexagonal tiling achieves a smaller MSE than the square tiling for the quantization of uniformly distributed random variables, given a certain density. The packing gain increases with dimensionality.

(b) *Removal of linear and nonlinear dependencies* While linear dependencies could be removed using a linear transformation, VQ also removes nonlinear dependencies. To see this, let us consider the classic example shown in Figure 7.4(b). The two-dimensional probability density function equals 2 in shaded areas and 0 otherwise. Because the marginal distributions are uniform, scalar quantization of each variable is uniform. Vector quantization "understands" the dependency, and only allocates partitions where necessary. Thus, instead of 4.0 bits, or, 2.0 bits/sample for the scalar quantization, we obtain 3.0 bits, or, 1.5 bits/sample for the vector quantization, reducing the bit rate by 25% while keeping the same distortion (see Figure 7.4(b)).

(c) *Fractional bit rate* At low bit rates, choosing between 1.0 bits/sample or 2.0 bits/sample is a rather crude choice. By quantizing several samples together and allocating an integer number of bits to the group, fractional bit rates can be obtained.

For a vector quantizer to be MSE optimal, it has to satisfy the same two conditions we have seen for scalar quantizers, namely:

(a) The nearest neighbor condition.

(b) The centroid condition.

A codebook satisfying these two necessary conditions is locally optimal (small

[3]Convex means that if two points x and y belong to one region, then all the points on the straight line connecting x and y will belong to the same region as well.

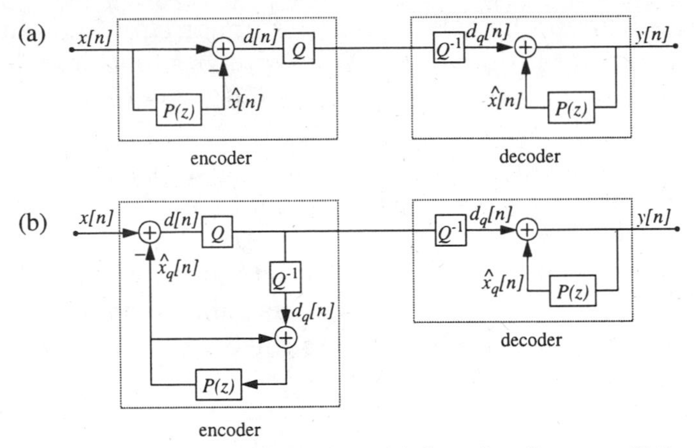

Figure 7.5 Predictive quantization. (a) Open-loop linear predictive quantization. (b) Closed-loop predictive quantization or differential pulse code modulation (DPCM).

perturbations will not decrease D) but is usually not globally optimal. The design of VQ codebooks is thus a sophisticated technique, where a good initial guess is crucial and is followed by an iterative procedure. For escaping local minimums, stochastic relaxation is used. For details, we refer to [109].

A drawback of VQ is its complexity, which limits the size of vectors that can be used. One solution is to structure the codebook so as to simplify the search of the best matching vector, given the input. This is achieved with tree-structured VQ. Another approach is to use linear transforms (including subband or wavelet transforms) and apply VQ to the relevant transform coefficients. Finally, lattice VQ uses multidimensional lattices as a partition, allowing large vectors with reasonable complexity, since lattice VQ is the equivalent of uniform quantization in multiple dimensions.

Predictive Quantization An important and useful technique is when, instead of quantizing the samples $x[n]$ of the signal to be compressed, one quantizes the difference between a prediction $\hat{x}[n]$ and $x[n]$, or $d[n] = x[n] - \hat{x}[n]$ [109, 143]. Obviously, if the prediction is accurate, $d[n]$ will be small. In other words, for a given number of quantization levels, the quantization error will decrease as compared to a straight quantization of $x[n]$. Prediction is usually linear and based on a finite number of past samples. An example is shown in Figure 7.5(a), where $P(z)$ is a strictly causal filter, $P(z) = a_1 z^{-1} + a_2 z^{-2} + \cdots + a_L z^{-L}$. That is, $x[n]$ is predicted based on a linear combination of L past samples, $\{x[n-1], \ldots, x[n-L]\}$. Furthermore, $1 - P(z)$ is chosen to be minimum phase so that its inverse, used in the decoder, is a stable filter. Given a predictor order and a stationary input signal, the best linear prediction filter that minimizes the variance of $d[n]$ is found by solving a set of linear equations involving the autocorrelation matrix of the signal (the Yule-Walker equations).

An interesting alternative is closed-loop predictive quantization or *differential pulse code modulation* (DPCM), as shown in Figure 7.5(b). In the absence of quantization, DPCM is equivalent to the open-loop predictive quantization in Figure 7.5(a). An important feature here is that since we are predicting $x[n]$ based on its past quantized values $\hat{x}_q[k]$, $k = n - L, \ldots, n - 1$, we can generate the same $\hat{x}[n]$ at the decoder side from these past values $\hat{x}_q[k]$. The idea is that in the decoder, we can add back exactly what was subtracted in the encoder and thus, the error made on the signal is equal to the error made when quantizing the difference signal. In other words, since

$$d[n] = x[n] - \hat{x}_q[n],$$

and

$$y[n] = d_q[n] + \hat{x}_q[n],$$

we get that

$$E(\,|x[n] - y[n]|^2\,) = E(\,|d[n] - d_q[n]|^2\,),$$

where $x[n]$ and $y[n]$ are the input and output of the DPCM, while $d[n]$ and $d_q[n]$ are the prediction error and its quantized version, respectively.

An important figure of merit of the above closed-loop predictive quantization is the closed-loop prediction gain. It is defined as the ratio of the variances of the input and of the prediction error,

$$G = \frac{\sigma_x^2}{\sigma_d^2}.$$

Note that when the quantization is coarse, this can be quite different from the open-loop prediction gain, which is the equivalent relation but with the prediction as in Figure 7.5(a). For practical reasons, the predictor $P(z)$ in the closed-loop case is usually chosen as in the open-loop case, that is, we are using the predicted coefficients that are optimal for the true past L samples of the signal.

A further improvement involves adaptive prediction, and can be used both in the open-loop and in the closed-loop cases. The predictor is updated every K samples based on the local signal characteristics and sent to the decoder as side information.

Linear predictive quantization is used successfully in speech and image compression (both in the open-loop and closed-loop forms). In video, a special form of adaptive DPCM, over time, involves motion-based prediction called motion compensation, which is discussed in Section 7.4.2.

Bit Allocation Looking back at the transform coding diagram in Figure 7.1, the obvious question is: How do we choose the quantizers for the various transform coefficients? This is a classical resource allocation problem, where one tries to maximize (or minimize) a cost function which describes the quality of approximation under the constraint of finite resources, that is, a given number of bits that can be used to code the signal. Let us first recall an important fact: The total squared error between the input and the output is the sum of individual errors because the transform is unitary. To see that, call $\boldsymbol{x}$ and $\hat{\boldsymbol{x}}$ the input and reconstructed input, respectively. Then $\boldsymbol{y}$ and $\hat{\boldsymbol{y}}$ will be the input and the output of the quantizer. That is,

$$\boldsymbol{y} = \boldsymbol{T}\boldsymbol{x}, \quad \hat{\boldsymbol{x}} = \boldsymbol{T}^T\hat{\boldsymbol{y}},$$

where the last equation holds since the transform $\boldsymbol{T}$ is unitary, that is, $\boldsymbol{T}^T\boldsymbol{T} = \boldsymbol{T}\boldsymbol{T}^T = \boldsymbol{I}$. Then the total distortion is

$$\begin{aligned} D &= E((\boldsymbol{x}-\hat{\boldsymbol{x}})^T \cdot (\boldsymbol{x}-\hat{\boldsymbol{x}})) = E((\boldsymbol{y}-\hat{\boldsymbol{y}})^T \cdot \boldsymbol{T}\boldsymbol{T}^T \cdot (\boldsymbol{y}-\hat{\boldsymbol{y}})) \\ &= E((\boldsymbol{y}-\hat{\boldsymbol{y}})^T \cdot (\boldsymbol{y}-\hat{\boldsymbol{y}})) = E\left(\sum_{i=0}^{N-1} (y_i - \hat{y}_i)^2\right) = \sum_{i=0}^{N-1} D_i, \end{aligned}$$

where D_i is the expected squared error of the ith coefficient. Then, the bit allocation problem is to minimize

$$D = \sum_{i=0}^{N-1} D_i, \tag{7.1.16}$$

while satisfying the bit budget

$$\sum_{i=0}^{N-1} R_i \le R, \tag{7.1.17}$$

where R is the total budget and R_i the number of bits allocated to the ith coefficient. A dual situation appears when a maximum allowable distortion is given and the rate has to be minimized. Before considering specific allocation procedures, we will discuss some aspects of optimal solutions.

The fundamental trade-off in quantization is between rate (number of bits used) and distortion (approximation error) and is formalized as rate-distortion theory [28, 121]. A rate-distortion function for a given source specified by a statistical model precisely indicates the possible trade-off. While rate-distortion bounds are usually not closely met in practice, implementable systems have a similar behavior. Figure 7.6(a) shows a possible rate-distortion function as well as points reached by a practical system (called an operational rate-distortion curve). Note that the true rate-distortion function is convex, while the operational one is not necessarily.

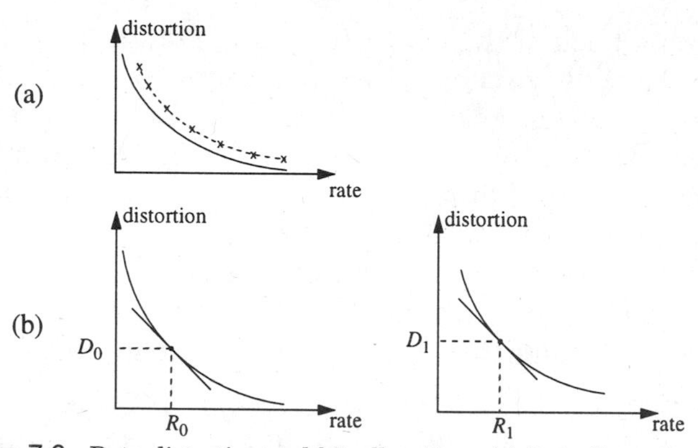

Figure 7.6 Rate distortion and bit allocation. (a) Rate-distortion curve for a statistically described source (solid line) and an operational rate-distortion curve (dashed line) based on a set of quantizers. (b) Constant-slope solution for an optimal allocation between two sources having the above rate-distortion curves.

For example, for high-resolution scalar quantization, the distortion D_i is related to the rate R_i as (see (7.1.15))

$$D_i(R_i) \simeq C_i \, \sigma_i^2 \, 2^{-2R_i}, \tag{7.1.18}$$

where C_i is a constant depending on the pdf of the quantized variable (for example, in the case of a zero-mean Gaussian variable, $C_i = \sqrt{3}\pi/2$).

Returning to our initial problem as stated in (7.1.16) and (7.1.17), we will consider a two-variable case for illustration. Assume we separately code two variables x_0 and x_1, each having a given rate-distortion function. A key property we assume is that both rate and distortion are additive. This is, for example, the case in transform coding if the coefficients are independent. How shall we allocate bits to each variable so as to minimize distortion? It is important to note that in a rate-distortion problem, we have to consider both rate and distortion in order to be optimal. Since the two dimensions are not related (one is bits and the other is MSE), we use a new cost function L combining the two through a positive Lagrange multiplier λ:

$$\begin{aligned} L &= D + \lambda \cdot R, \\ L_i &= D_i + \lambda \cdot R_i, \qquad i = 0, 1, \end{aligned}$$

where $L = L_0 + L_1$. Finding a minimum of L (which now depends on λ) amounts to finding minimums for each L_i (because the costs are additive). Writing distortion as a function of rate, $D_i(R_i)$, and taking the derivative to find a minimum, we get

$$\frac{\partial L_i}{\partial R_i} = \frac{\partial D_i(R_i)}{\partial R_i} + \lambda = 0,$$

that is, the slope of the rate-distortion function is equal to $-\lambda$, for $i = 0, 1$ and $\partial D_0(R_0)/\partial R_0 = \partial D_1(R_1)/\partial R_1 = -\lambda$. Uniqueness follows from the convexity of the rate-distortion curves. Thus, for a solution to be optimal, the set of chosen rates R_0 and R_1 have to correspond to constant-slope points on their respective rate-distortion curves [262], as shown in Figure 7.6(b). This solution is also very intuitive. Consider what would happen if (R_0, D_0), (R_1, D_1) did not have the same slope, and suppose that λ_0 is much steeper than λ_1. We assume we are within the budget R, that is, $R = R_0 + R_1$. Increase now the rate R_0 by ϵ. Since we need to stay within the budget, we have to decrease the rate of R_1 by the same amount. In the process, we have decreased the distortion D_0 and increased the distortion D_1. However, since we assumed that the first slope is steeper, it actually paid off to do this since we remained with the same budget while decreasing the overall distortion. Repeating the process, we move closer and closer to the optimal solution. Once we reach the point where both slopes are the same, we do not gain anything by moving further.

A constant-slope solution is obtained for any fixed value of R. To enforce the constraint (7.1.17) exactly, one has to search over all slopes λ until the budget is met and then we have an optimal solution that satisfies the constraints. In practice, the exact functions $D_i(R_i)$ might not be known, but one can still use similar ideas on operational rate-distortion curves [262]. The main point of our discussion was to indicate the philosophy of the approach: Based on rate-distortion curves, find operating points that satisfy an optimality criterion and search until the budget constraint is satisfied as well.

When high-resolution quantization approximations can be used, it is possible to give closed-form allocation expressions. Assume the N sources have the same type of distribution but different variances. Then $D_i(R_i)$ is given in (7.1.18) with a fixed constant $C_i = C$. Taking the derivative, it follows that:

$$\frac{\partial D_i(R_i)}{\partial R_i} = C' \cdot \sigma_i^2 \cdot 2^{-2R_i},$$

with $C' = -2\ln 2 \cdot C$. The constant-slope solution, that is, $\partial D_i(R_i)/\partial R_i = -\lambda$, forces the rates to be of the following form:

$$R_i = \alpha + \log_2 \sigma_i.$$

Since we also have the budget constraint (7.1.17),

$$\sum R_i = N \cdot \alpha + \sum_{i=0}^{N-1} \log_2 \sigma_i = R,$$

we find

$$\alpha = \frac{R}{N} - \frac{1}{N} \cdot \sum_{i=0}^{N-1} \log_2 \sigma_i,$$

and

$$R_i = \frac{R}{N} + \log_2 \sigma_i - \frac{1}{N}\sum_{i=0}^{N-1} \log_2 \sigma_i = \bar{R} + \log_2 \frac{\sigma_i}{\rho}, \tag{7.1.19}$$

where $\bar{R} = R/N$ is the mean rate and ρ is the geometric mean of the variances

$$\rho = \left(\prod_{i=0}^{N-1} \sigma_i\right)^{1/N}.$$

Note that each quantizer has the same average distortion

$$\begin{aligned} D_i &= C \cdot \sigma_i^2 2^{-2R_i} = C \cdot \sigma_i^2 2^{-2(\bar{R} + \log_2 \sigma_i/\rho)} \\ &= C \cdot \sigma_i^2 \cdot 2^{-2\bar{R}} 2^{2\log_2(\rho/\sigma_i)} = C \cdot \rho^2 \cdot 2^{-2\bar{R}}. \end{aligned} \tag{7.1.20}$$

The result of this allocation procedure is intuitive, since the number of quantization levels allocated to the ith quantizer,

$$2^{R_i} = \frac{2^{\bar{R}}}{\rho} \cdot \sigma_i,$$

is simply proportional to the standard deviation or spread of the variable x_i. The allocation (7.1.19) can be modified for nonidentically distributed random variables and weighted errors (the ith error is weighted by W_i in the total distortion). In this case σ_i^2, in the allocation problem, is replaced by $C_i \cdot W_i \cdot \sigma_i^2$, leading to the appropriate modification of (7.1.19).

The problem with the above allocation procedure is that the resulting rates are noninteger and even worse, small variances can lead to negative allocations. Both problems can be tackled by starting with the solution given by (7.1.19) and forcing nonnegative integer allocations (this might lead to slight suboptimality, however).

The next algorithm [109] tackles the problem directly by allocating one bit at a time to the quantizer where it is most needed. It is a "greedy" algorithm and not optimal, but leads to good solutions. Call $R_i[n]$ the number of bits allocated to quantizer i at the nth iteration of the algorithm. Then, the algorithm iterates over n until all bits have been allocated and at each step, allocates the next bit to the quantizer j which has maximum distortion with the current allocation,

$$D_j(R_j[n]) \geq D_i(R_i[n]), \quad i \neq j.$$

That is, the next bit is allocated to where it is most needed. Since D_i can be given in analytical form or measured on a training set, this algorithm is easily applicable. More sophisticated algorithms, optimal or near optimal, are based on Lagrange methods applied to arbitrary rate-distortion curves [262].

Coding Gain Now that we have discussed quantization and bit allocation, we can return to our study of transform coding and see what advantage is obtained by doing quantization in the transform domain (see Figure 7.1).

First, recall that the Karhunen-Loève transform leads to uncorrelated variables with variance λ_i (see (7.1.8)). Assume that the input to the transform is zero-mean Gaussian with variance σ_x^2, and that fine quantization is used. This leads us to Proposition 7.2.

PROPOSITION 7.2 *Optimality of Karhunen-Loève Transform*

Among all block transforms and at a given rate, the Karhunen-Loève transform will minimize the expected distortion.

PROOF

After the KLT with optimal scalar quantization and bit allocation, the total distortion for all N channels is (following (7.1.20)),

$$D_{KLT} = N \cdot C \cdot 2^{-2\bar{R}} \cdot \rho^2 = N \cdot C \cdot 2^{-2\bar{R}} \left(\prod_{i=0}^{N-1} \lambda_i \right)^{1/N}, \tag{7.1.21}$$

where $C = \sqrt{3}\pi/2$ (see (7.1.18)). Since the determinant of a matrix is equal to the product of its eigenvalues, the last term is equal to $(\det(\boldsymbol{K}_x))^{1/N}$ where $\boldsymbol{K}_x$ is the autocovariance matrix (assuming zero mean, $\boldsymbol{K}_x = \boldsymbol{R}_x$). To prove the optimality of the KLT, we need the following inequality for the determinant of an autocorrelation matrix of N zero-mean variables with variances σ_i^2 [109]:

$$\det(\boldsymbol{R}_x) \leq \prod_{i=0}^{N-1} \sigma_i^2, \tag{7.1.22}$$

with equality *if and only if* $\boldsymbol{R}_x$ is diagonal. It turns out that the more correlated the variables are, the smaller the determinant.

Consider now an arbitrary orthogonal transform, with transform variables having variance σ_i^2. The distortion is

$$D_T = N \cdot C \cdot 2^{-2\bar{R}} \left(\prod_{i=0}^{N-1} \sigma_i \right)^{1/N}.$$

Because of (7.1.22) and the fact that the determinant is conserved by unitary transforms, this is greater or equal than

$$D_T \geq N \cdot C \cdot 2^{-2\bar{R}} \det(\boldsymbol{R}_x)^{1/N}.$$

Since the KLT achieves a diagonal $\boldsymbol{R}_x$, then the equality is reached by the KLT following (7.1.21). This proves that if the input to the transform is Gaussian and the quantization is fine, the KLT is optimal among all unitary transforms.

What is the gain we just obtained? If the samples are directly quantized, the distortion will be

$$D_{PCM} = N \cdot C \cdot 2^{-2\bar{R}} \cdot \sigma_x^2, \tag{7.1.23}$$

(where PCM stands for pulse code modulation, that is, sample-by-sample quantization) and the coding gain due to optimal transform coding is

$$\frac{D_{PCM}}{D_{KLT}} = \frac{\sigma_x^2}{\left(\prod_{i=0}^{N-1}\sigma_i^2\right)^{1/N}} = \frac{1/N\sum_{i=0}^{N-1}\sigma_i^2}{\left(\prod_{i=0}^{N-1}\sigma_i^2\right)^{1/N}}, \qquad (7.1.24)$$

where we used the fact that $N \cdot \sigma_x^2 = \sum \sigma_i^2$. Recalling that the variances σ_i^2 are the eigenvalues of $\boldsymbol{R}_x$, it follows that the coding gain is the ratio of the arithmetic and geometric means of the eigenvalues of the autocorrelation matrix (under the zero-mean assumption). The lower bound on the gain is 1, which is attained only if all eigenvalues are identical.

Subband coding, being a generalization of transform coding, has a similar behavior. If the input is Gaussian, the channel signals are Gaussian as well. If the filters are ideal bandpass filters, the channels will be decorrelated. In any case, the distortion resulting from optimally allocating $R = N \cdot \bar{R}$ bits across N channels with variances σ_i^2 is, as in the usual transform case

$$D_{SBC} = N \cdot C \cdot 2^{-2\bar{R}} \cdot \rho^2,$$

where ρ is the geometric mean of the subband variances. Using (7.1.23) for direct quantization we get, similarly to (7.1.24), the subband coding gain as

$$\frac{D_{PCM}}{D_{SBC}} = \frac{1/N\sum_{i=0}^{N-1}\sigma_i^2}{\left(\prod_{i=0}^{N-1}\sigma_i^2\right)^{1/N}},$$

where the σ_i^2's are the subband variances. That is, if the spectrum is far from being flat, there will be a large coding gain in subband methods. This is to be expected, since it becomes possible to match the spectral characteristics of the signal very closely, unlike in a sample-domain quantization. It is worthwhile to note that when the number of channels grows to infinity, both transform and subband coding achieve the theoretical performance of predictive coding with infinitely long predictor [143].

The obvious question is of course how do transform and subband coding compare? The ratio of D_{KLT} and D_{SBC} is:

$$\frac{D_{KLT}}{D_{SBC}} = \frac{\rho_{KLT}^2}{\rho_{SBC}^2},$$

that is, the one with the smaller geometric mean wins. Qualitatively, the one with the larger spread in variances will achieve better coding gain. The exact comparison thus requires measurements of variances in specific transforms (such as the DCT) versus filter banks (of finite length rather than ideal ones).

While the above considerations use some idealized assumptions, the concept holds true in general: The wider the variations between the component signals (transform coefficients or subbands), the higher the potential for coding gain. More about the above can be found in [5, 220, 273, 292, 295].

7.1.3 Entropy Coding

The last step in transform coding as shown in Figure 7.1 is entropy coding. Similarly to the first step, it is reversible and thus, there is no approximation problem as in quantization. After quantization, the variables take values drawn from a finite set $\{a_i\}$. The idea is to find a reversible mapping M to a new set $\{b_i\}$ such that the average number of bits/symbol is minimized. A historical example is the Morse code which assigns short codes to the letters that appear frequently in the English language while reserving long codes to less frequent ones. The parameters in searching for the mapping M are the probabilities of occurrence of the symbols a_i, $p(a_i)$. If the quantized variable is stationary, these probabilities are fixed, and a fixed mapping such as Huffman coding can be used. If the probabilities evolve over time, more sophisticated adaptive methods such as adaptive arithmetic coding can be used. Such mappings will transform fixed-length codewords into variable-length ones, creating a variable-length bit stream. If a constant bit rate channel is used, buffering has to smooth out variations so as to accommodate the fixed-rate channel.

Huffman Coding Given an alphabet $\{a_i\}$ of size M and its associated probabilities of occurrence $p(a_i)$, the goal is to find a mapping $b_i = F(a_i)$ such that the average length $l(b_i)$ is minimized:

$$E(l(b_i)) = \sum_{i=0}^{M-1} p(a_i) l(b_i). \tag{7.1.25}$$

We also require that a sequence of b_i's should be uniquely decodable (note that invertibility of F is not sufficient). This last requirement puts an extra constraint on the codewords b_i, namely, no codeword is allowed to be a prefix to another one. Then, the stream of b_i's can be uniquely decoded by sequentially removing codewords b_i. The lower bound of the expected length (7.1.25) is given by the entropy of the set $\{a_i\}$

$$H_a = -\sum_{i=0}^{M-1} p(a_i) \log_2(p(a_i)). \tag{7.1.26}$$

Huffman's construction elegantly meets the prefix condition while coming quite close to the entropy lower bound. The design is guided by the following property

Table 7.1 Symbols, probabilities and resulting possible Huffman codewords where $H_a = 2.28$ bits and $E[l(b_i)] = 2.35$ bits. First, the symbols are merged going from (a) to (e). Then, the codewords are assigned going from (e) to (a).

a_i	$p(a_i)$	b_i
0	0.40	0
1	0.20	100
2	0.15	101
3	0.10	110
4	0.10	1110
5	0.05	1111

(a)

a_i	$p(a_i)$	b_i
0	0.40	0
1	0.20	100
2	0.15	101
4 + 5	0.15	111
3	0.10	110

(b)

a_i	$p(a_i)$	b_i
0	0.40	0
3+(4+5)	0.25	11
1	0.20	100
2	0.15	101

(c)

a_i	$p(a_i)$	b_i
0	0.40	0
1 + 2	0.35	10
3+(4+5)	0.25	11

(d)

a_i	$p(a_i)$	b_i
(1+2) + (3+(4+5))	0.60	1
0	0.40	0

(e)

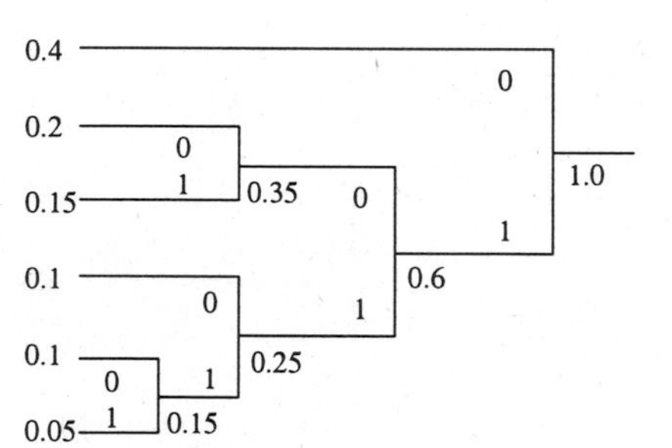

Figure 7.7 Huffman code derived from a binary tree and corresponding to the symbol probabilities given in Table 7.1.

of optimum binary prefix codes: The two least probable symbols have codewords of equal length which differ only in the last symbol.

The design of the Huffman code is best looked at as growing a binary tree from the leaves up to the root. The codeword will be the sequence of zeros and ones encountered as going from the root to the leaf corresponding to the desired symbol. Start with a list of the probabilities of the symbols. Then, take the two least probable symbols and make them two nodes with branches (labeled "0" and "1") to a common node which represents a new symbol. The new symbol has a probability which is the sum of the two probabilities of the merged symbols. The new list of symbols is now shorter by one. Iterate until only one symbol is left. The codewords can now be read off along the branches of the binary tree. Note that at every step, we have used the property of optimum binary prefix codes so that the two least probable symbols were of equal length and had the same prefix.

Example 7.1 *Huffman Coding*

An example is given in Figure 7.7 where a Huffman tree is shown for the symbol probabilities given in Table 7.1(a). Let us first consider only the first two columns of each of the tables. We start from left to right and in Table 7.1(a) choose the two symbols with the lowest probabilities, that is, 4 and 5, and merge them. We then reorder the symbols in the decreasing order, and form Table 7.1(b). Now the process is repeated, joining symbols 3 and $(4 + 5)$. After a couple more steps, we obtain the final Table 7.1(e). Now we start assigning codewords, going from right to left. Thus, 0.6 gets a "1", and 0.4 gets a "0". Then we split 0.6, and assign "10" to 0.35, and "11" to 0.25. The final result of the whole procedure is given in Table 7.1(a) and Figure 7.7.

Note that we call Huffman coding optimal when the average length $E(l(b_i))$ given in (7.1.25) reaches the theoretical lower bound given by the entropy (7.1.26), which is possible only if the symbol probabilities are powers of two. This is a limitation of Huffman coding, which can be surmounted by using arithmetic coding. It is more complicated to implement and, in its simplest form, it also requires a priori knowledge of symbol probabilities. If the source matches the probabilities used to design the arithmetic coder, then the rate approaches the entropy arbitrarily closely for long sequences. See [24] and [109] for more details.

Adaptive Entropy Coding While the above approaches come close to the entropy of a known stationary source, they fail if the source is not well-known or changes significantly over time. A possible solution is to estimate the probabilities on the fly (by counting occurrences of the symbols at both the encoder and decoder) and modify the Huffman code accordingly. While this seems complicated at first sight, it turns out that only minor modifications are necessary, since only a single probability is affected by an entering symbol [105, 109].

Arithmetic coding can be modified as well, in order to estimate probabilities on the fly. This adaptive version is known as a Q-coder [221]. Finally, Ziv-Lempel coding [342] is an elegant lossless coding technique which uses no a priori probabilities. It builds up a dictionary of encountered subsequences in such a way that the decoder can build the same dictionary. Then, the encoder sends only the index to an encountered entry. The dictionary size is fixed and the index uses a fixed number of bits. Thus, the Ziv-Lempel coding maps variable-size input sequences into fixed-size codewords, a dual of the Huffman code. The only limitation of the Ziv-Lempel code is its fixed-size dictionary, which leads to loss in performance when very long sequences are encoded. No new entries can be created once the dictionary is full and the remainder of the sequence has to be coded with the current entries. Modifications of the basic algorithm allow for dictionary updates. Note that since there are many variations on this theme, we refer to [24] for a thorough discussion.

Run-Length Coding Another important lossless coding technique is *run-length coding* [138]. It is useful when a sequence of samples consists of stretches of zeros followed by small packs of nonzero samples (this is typically encountered in subband image coding at the outputs of the highpass channels after uniform quantization with a dead zone, as in Section 7.3.3). It is thus advantageous to encode the length of the stretch of zeros, to then encode the values of the nonzero samples and then an indicator of the start of another run of zeros. Of course, both the length of runs and the nonzero values can be entropy coded.

7.1.4 Discussion

So far we have separately considered the three building blocks of a transform coder as depicted in Figure 7.1. Some interaction between the transform and the quantization was discussed when proving the optimality of the KLT. Including entropy coding after quantization can change the way quantization should be done. In the high-rate, memoryless[4] case, uniform quantization followed by entropy coding turns out to be better than using nonuniform quantization and fixed codewords [109]. However, this leads to variable-rate schemes and thus requires buffering when fixed-rate channels are used. This is done with a finite-size buffer, which has a nonzero probability of overflow. Therefore, a buffer control algorithm is needed. This usually means moving to coarser quantization when the buffer is close to overflow and finer quantization in the underflow case. Obviously, in the overflow control case, there is a loss in performance in such variable-rate schemes. The size of the buffer is limited for cost reasons, but also because of the delay it produces in a real-time transmission case.

Our discussion has focused on MSE-based coding, but we indicated that it extends readily to weighted MSE. Such weights are usually based on perceptual criterions [141, 142], and will be discussed later. We note that certain "tricks" such as the dead zone quantizers used in image compression (uniform quantizers with a zone around zero larger than the step size that maps to the origin) are heuristics derived from experiments that are not optimal in the sense discussed so far, but which produce visually more pleasing images.

[4]Memoryless means that the output value at a present time depends only on the present input value and not on any past or future values.

7.2 Speech and Audio Compression

In this section, we consider the use of signal expansions for one-dimensional signal compression. Subband methods are successful for medium compression of speech [68, 94, 103, 192], and high quality compression of audio [34, 77, 147, 267, 279, 290, 333]. At other rates (for example, low bit rate speech compression) different methods are used, which we will briefly indicate as well.

7.2.1 Speech Compression

Production-Model Based Compression of Speech A particularity of speech is that a good production model can be identified. The vocal cords produce an excitation function which can be roughly classified into voiced (pulse-train like) and unvoiced (noise-like) excitation. The vocal tract, mouth, and lips act as a filter on this excitation signal. Therefore, very high compression systems for speech are based on identifying the parameters of this speech production model. Typically, linear prediction is used to identify a linear filter of a certain order which will whiten the speech signal (this is therefore the inverse filter of the speech production model). Then, the residual signal is analyzed to decide if the speech was voiced or unvoiced, and in the former case, to identify the pitch. Such an analysis is done on a segment-by-segment basis. It reduces the original speech signal to a small set of parameters: voiced/unvoiced decision plus pitch value in the voiced case and filter coefficients (up to 16 typically). At the decoder, the speech is synthesized following the production model and using the parameters identified at the encoder. As to be expected, this approach leads to very high compression factors. Speech sampled at 8 kHz with 8 bits/sample, that is, at 64 Kbits/sec, is compressed down to as low as 2.4 Kbits/sec with adequate intelligibility but some lack of naturalness [141]. At 8 to 16 Kbits/sec, sophisticated versions of linear predictive coders achieve what is called "toll quality," that is, they can be used on public telephone networks. Instead of simple voiced/unvoiced excitation, these higher-quality coders use a codebook from which the best excitation function is chosen. An important advantage of linear predictive coding (LPC) of speech is that low delay is achievable.

High-Quality Speech Compression Certain applications require speech compression with better than telephone quality (for example, audio conferencing). This is often called wideband speech [141] since the sampling rate is raised from 8 kHz to 14 kHz. Because of the desire for high quality, more attention is focused on the perception process, since the goal is to attain a perceptually transparent coding. That is, masking patterns of the auditory system are taken advantage of, so as to place quantization noise in the least sensitive regions of the spectrum. In that sense, wideband speech coding is similar to audio coding,

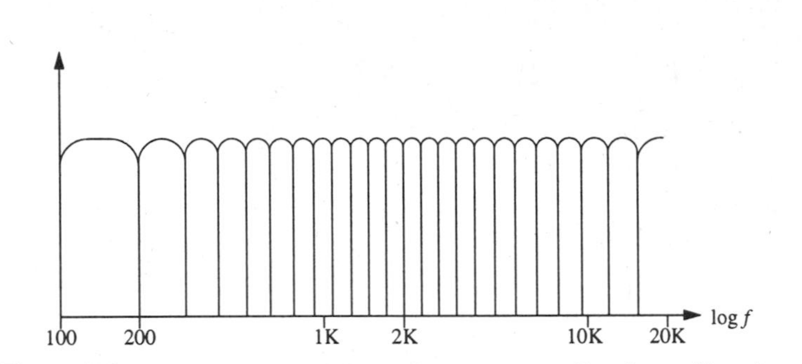

Figure 7.8 Critical bands of the auditory system. Bandpass filters' magnitude response on a logarithmic frequency axis.

and we defer the discussion of masking to the next section. One difference, however, is the delay constraint which is stringent for real-time interactive speech compression, while being relaxed in the audio compression case, since the latter is usually performed off line.

7.2.2 High-Quality Audio Compression

Perceptual Models The auditory system is often modeled as a filter bank in a first approximation. This filter bank is based on critical bands [254], as shown in Figure 7.8 and Table 7.2. The key features of such a spectral view of hearing are [146]:

(a) A constant relative bandwidth behavior of the filter (see Figure7.8).

(b) Masking properties of dominant sounds over weaker ones within a critical band and over nearby bands, as given by a spreading function.

The critical bands can be seen as pieces of the spectrum that are considered as an entity in the auditory process. For example, a sine wave centered in a given critical band will mask noise in this band, but not outside. While the masking properties are very complex and only partly understood, the basic concepts can be successfully used in an audio compression system.

Unlike in the case of speech compression, there is no source model for general audio signals. However, there is a good perceptual model of the auditory process, which can be used for achieving better compression through *perceptual coding* [141].

Perceptual Coders A perceptual coder for transparent coding of audio will attempt to keep quantization noise just below the level where it would become noticeable. Quantization noise within a critical band has to be controlled and an easy way to do that is to use a subband or transform coder. Also, permissible quantization noise levels have to be calculated and this is based on some form

Table 7.2 Critical bands of the auditory system, which are of constant bandwidth at low frequencies (below 500 Hz) and of constant relative bandwidth at high frequencies.

Band number	Lower edge (Hz)	Center (Hz)	Upper edge (Hz)	BW (Hz)
1	0	50	100	100
2	100	150	200	100
3	200	250	300	100
4	300	350	400	100
5	400	450	510	110
6	510	570	630	120
7	630	700	770	140
8	770	840	920	150
9	920	1000	1080	160
10	1080	1170	1270	190
11	1270	1370	1480	210
12	1480	1600	1720	240
13	1720	1850	2000	280
14	2000	2150	2320	320
15	2320	2500	2700	380
16	2700	2900	3150	450
17	3150	3400	3700	550
18	3700	4000	4400	700
19	4400	4800	5300	900
20	5300	5800	6400	1100
21	6400	7000	7700	1300
22	7700	8500	9500	1800
23	9500	10500	12000	2500
24	12000	13500	15500	3500
25	15500	19500		

of spectral analysis of the input. Therefore, a generic perceptual coder for audio is as depicted in Figure 7.9. Note that one can use the analysis filter bank as a spectrum analyzer or calculate a separate spectrum estimation. Usually, the two are integrated for computational reasons.

A filter bank implementing critical bands exactly, is computationally unfeasible. Instead, some approximation is attempted that has roughly a logarithmic behavior, with an initial octave-band filter bank, but uses short-time Fourier-like banks within the octaves to get finer analysis at reasonable computational cost. A possible example is shown in Figure 7.10, where LOT stands for

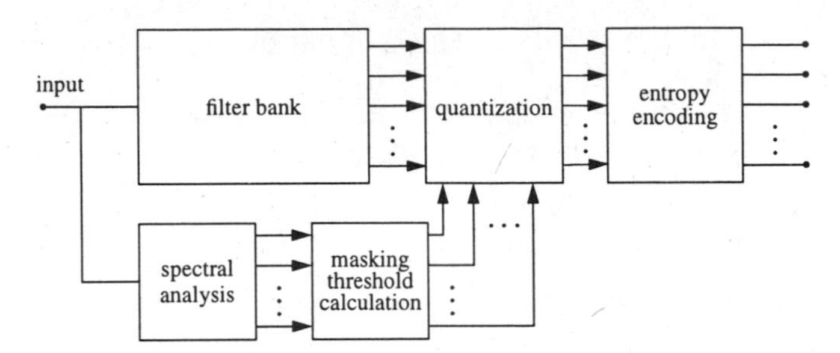

Figure 7.9 Generic perceptual coder for high-quality audio compression (after [146]).

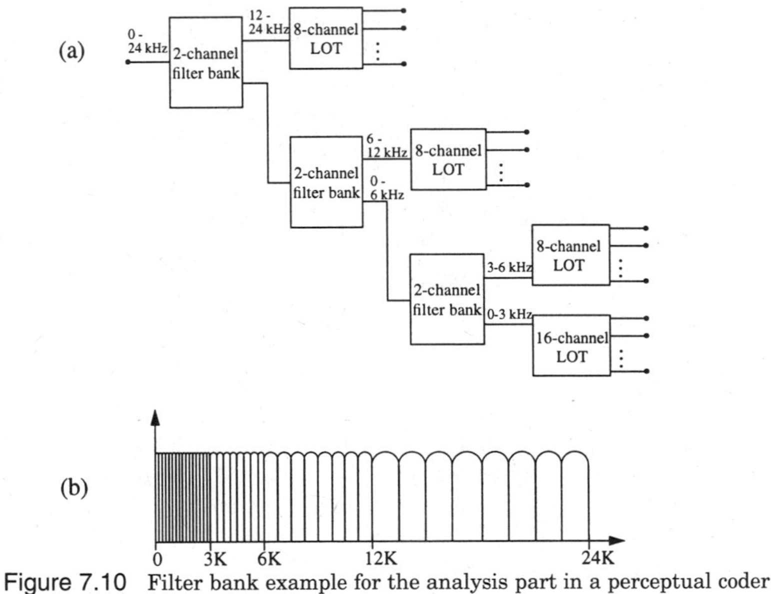

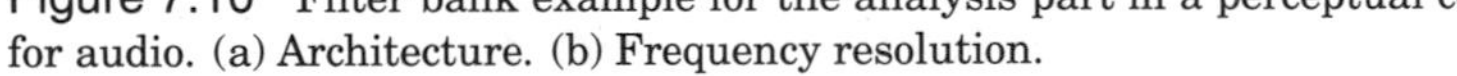

Figure 7.10 Filter bank example for the analysis part in a perceptual coder for audio. (a) Architecture. (b) Frequency resolution.

lapped orthogonal transforms and also refers to cosine-modulated filter banks[5] (Section 3.4.3). Recently, Princen has proposed to use nonuniform modulated filter banks [227]. They are near perfect reconstruction and since they are a straightforward extension of the cosine-modulated filter banks, they are computationally efficient. High-quality audio coding usually does not have to meet delay constraints and thus the delay due to the filter bank is not a problem. Typically, very long filters are used in order to get excellent band discrimination, and to avoid aliasing as much as possible since aliasing is perceptually

[5]Note that this filter bank is known under many names, such as LOT, MLT, MDCT, TDAC, Princen & Bradley filter bank, cosine modulated filter bank [188, 229, 228].

very disturbing in audio.

The next step consists of estimating the masking thresholds within the bands. Typically, a fast Fourier transform is performed in parallel with the filter bank. Based on the signal energy and spectral flatness within a critical band, the maximum tolerable quantization noise level can be estimated. Typically, single tones can be identified, their associated masking function derived, and thus, the allowable quantization steps follow. Bands which have amplitudes below this maximum step can be disregarded altogether. For a detailed description of the perceptual threshold calculations, refer to [145]. Note that this quantization procedure is quite different from an MSE-based approach as discussed in Section 7.1.2, where only the variances within bands mattered. Sometimes, the perceptual and MSE approaches are combined. A first pass allocates an initial number of bits so as to satisfy the minimum perceptual requirements, while a second pass distributes remaining bits according to the usual MSE criterions.

The quantization and bit allocation is recalculated for every new segment of the input signal, and sent as side information to the decoder. Because entropy coding is used on the quantized subband samples, the bit stream has to be buffered if fixed rate transmission is intended. Note that not all systems use entropy coding (for example, MUSICAM does not).

7.2.3 Examples

Various applications such as digital audio broadcasting (DAB) require CD-quality audio (44.1 kHz sampling and 16 bits/sample). This lead to the development of medium compression, high-quality standards for audio coding.

MUSICAM Probably the most well-known audio coding algorithm is MUSICAM (Masking-pattern Universal Subband Integrated Coding and Multiplexing) [77, 279], used in the MPEG-I standard, and thus frequently referred to as MPEG audio [38]. It is also conceptually the simplest coder. This system uses a 32-band uniform filter bank, obtained by modulation of a 512-tap prototype lowpass filter. The magnitude response of this filter bank is shown in Figure 7.11. One reason for choosing such a filter bank is that it has a reasonable computational complexity since it can be implemented with a polyphase filter followed by a fast transform (see Section 6.2). Another reason is its smaller delay when compared to a tree-structured filter bank.

In parallel to the filter bank, a fast Fourier transform is used for spectral estimation. Based on the power spectrum, a masking curve is calculated, an example of which is shown in Figure 7.12. Quantization noise is then allocated in the various subbands according to the masking function. This

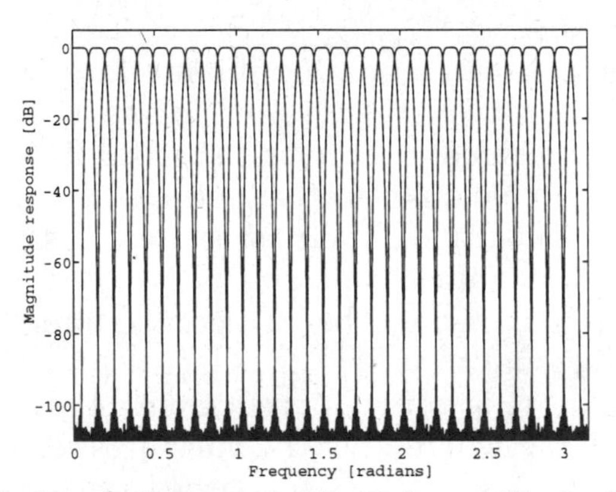

Figure 7.11 Magnitude response of the 32-channel filter bank used in MUSICAM. The prototype is a length 512 window, and cosine modulation is used to get the 32 modulated filters.

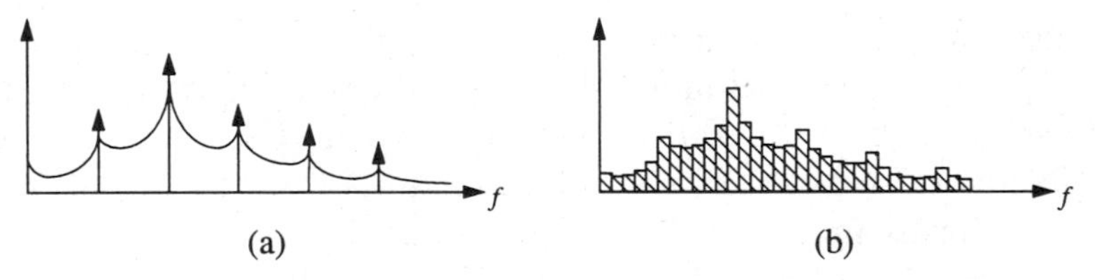

Figure 7.12 Example of quantization based on psychoacoustics. (a) Line spectrum and associated masking function. (b) Quantization noise in the 32 subbands of MUSICAM taking advantage of masking.

allocation is done on a small block of subband samples (typically 12). The maximum value within a block, called *scale factor*, and the quantization step, based on masking, are calculated for each block. They are transmitted as side information, together with the quantized samples. MUSICAM does not use entropy coding, the quantized values are sent (almost) directly.

The resulting system compresses audio signals of about 700 Kbits/sec (44.1 kHz, 16 bit samples) down to around 128 Kbits/sec, without audible impairments [77, 279]. When used on stereo signals, it leads to a bit rate of 256 Kbits/sec.

PAC Coder An interesting coder for high-quality compression of audio is the PAC (Perceptual Audio Coder) coder [147]. In its stereo version, it has been proposed for digital audio broadcasting as well as for a nonbackward compatible MPEG-II audio compression system.

The coder has the basic blocks that are typical of many perceptual coders, given in Figure 7.9. The signal goes through a filter bank and a perceptual model. Then the outputs of the filter bank and the perceptual model are fed into PCM quantization, Huffman coding and rate control.

The filter bank is based on the cosine modulated banks presented in Section 3.4.3, with window switching. The psychoacoustic analysis provides a noise threshold for L (Left), R (Right), S (Sum) and D (Difference) channels, where $S = L + R$ and $D = L - R$. One feature of the PAC algorithm is that it is adaptive in time and frequency since, in each frequency band, it sends either the (L, R) or (S, D) signals, depending on which one is more efficient.

This coder provides transparent or near-transparent quality coding at 192 Kbits/sec/stereo pair, and high-quality coding at 128 Kbits/sec/stereo pair.

AC System Two well-known algorithms for high-quality audio compression are the AC-2 and AC-3 algorithms, coming from Dolby [34, 290]. They have both stereo and five-channel, surround system, versions.

The AC-2 version exploits both the time-domain and frequency-domain psychoacoustic models. It uses a time-frequency division scheme, achieving a trade-off between time and frequency resolutions, on a signal-dependent basis. This is achieved by selecting the optimal transform block length for each 10ms analysis interval. The filter bank is based again on the cosine-modulated filter bank [229, 228]. This coder operates at a variety of bit rates ranging from 64-192 Kbits/sec/channel. The 128 Kbits/sec/channel AC-2 version has been selected for use in a new multichannel NTSC compression system [34].

As can be seen from the above three examples, filter bank methods had a substantial impact on audio compression systems. Note that sophisticated time-frequency analysis is a key component.

7.3 Image Compression

Multiresolution techniques are most naturally applied to images, where notions such as resolution and scale are very intuitive. Multiresolution techniques have been used in computer vision for tasks such as object recognition and motion estimation as well as in image compression, with pyramid [41] and subband coding [111, 314, 337]. An important feature of such image compression techniques is their successive approximation property: As higher frequencies are added (which is equivalent to more bands in subband coding or, difference signals in pyramids), higher-resolution images are obtained. Note that multiresolution successive approximation corresponds to the human visual system which helps the multiresolution techniques in terms of perceptual quality. Transform coding also has a successive approximation property (see the discussion on the Karhunen-Loève transform in Section 7.1.1) and is thus part of this broad class of techniques which are characterized by multiresolution approximations. In short, besides good compression capabilities, these schemes allow partial decoding of the coded version which lead to usable subresolution approximations.

We start by discussing the standard image compression schemes, which are based on block transforms such as the discrete cosine transform (DCT) or overlapping block transforms such as the lapped orthogonal transform. This leads naturally to a description of the current image compression standard based on the DCT, called JPEG [148, 327], indicating some of the constraints of a "real-world" compression system.

We continue by discussing pyramid coding, which is a very simple but flexible image coding method. A detailed treatment of subband/wavelet image coding follows. Several important issues pertaining to the choice of the filters, the decomposition structure, quantization and compression are discussed and some examples are given.

Following these standard coding algorithms, we describe some more recent and sometimes exploratory compression schemes which use multiresolution as an ingredient. These include image compression methods based on wavelet maximums [184], and a method using adaptive wavelet packets [15, 233]. We also discuss some recent work on a successive approximation method for image coding using subband/wavelet trees [259], quantization error analysis in a subband system [331], joint design of quantization and filtering for subband coding [162], and nonorthogonal subband coding [200].

Note that in all experiments, we use the standard image *Barbara*, with 512×512 pixels and 8-bit gray-scale values (see Figure 7.13). For comparison purposes, we will use the peak signal-to-noise ratio (SNR_p) given by (7.1.3).

7.3.1 Transform and Lapped Transform Coding of Images

We have introduced block transforms in Section 3.4.1, and while they are a particular case of filter banks (with filter length L equal to the downsampling factor N), they are usually considered separately. Their importance in practical image coding applications is such that a detailed treatment is justified. As we mentioned in audio coding examples, lapped orthogonal transforms are also filter bank expansions since they use modulated filter banks with filters of length typically twice the downsampling factor, or $L = 2N$. They have been introduced as an extension of block transforms in order to solve the problem of blocking in transform coding. Because of this close relationship between block transforms and lapped transforms, quantization and entropy coding for both schemes are usually very similar. A text on transform coding of images is [54], and lapped transform coding is treated in [188].

Block Transforms Recall that unitary block transforms of size $N \times N$ are defined by N orthonormal basis vectors, that is, the transform matrix $\boldsymbol{T}$ has these basis

Figure 7.13 Standard image used for the image compression experiments, called *Barbara*. The size is 512×512 pixels and 8 bits/pixel.

vectors as its rows (see Section 3.4.1 and (7.1.4)). For two-dimensional signals, one usually takes a separable transform which corresponds to the Kronecker product of $\boldsymbol{T}$ with itself,

$$\boldsymbol{T}_{2D} = \boldsymbol{T} \otimes \boldsymbol{T}.$$

In other words, this separable transform can be evaluated by taking one-dimensional transforms along the rows and columns of a block $\boldsymbol{B}$ of an image. This can be written as:

$$\boldsymbol{B}_T = \boldsymbol{T}\boldsymbol{B}\boldsymbol{T}^T,$$

where the first product corresponds to transforming the columns, while the second product computes the transform on rows of the image block. Many transforms have been proposed for the coding of images. Besides the DCT given in (7.1.10–7.1.11), the sine, slant, Hadamard and Haar transform are common candidates, the last two mainly because of their low computational complexity (only additions and subtractions are involved). All of the transforms have fast, $O(N \log N)$ algorithms, as opposed to the optimal KLT which has $O(N^2)$ complexity and is signal dependent. The performance of the DCT in image compression is sufficiently close to that of the KLT as well as superior to other transforms so that it has become the standard transform. Figure 7.14 shows the 8×8 DCT transform of the original image. Note the two representations shown. In part (a), we display the transform of each block of the image, while part (b) has gathered all coefficients of the same frequency into a block. This latter representation is simply a subband interpretation of the DCT; for example, the lowest left corner is the output of a filter which takes the average of 8×8 blocks. The similarity of this representation with subband-decomposed images is obvious. Note that for quantization and entropy coding purposes, the representation (a) is preferred.

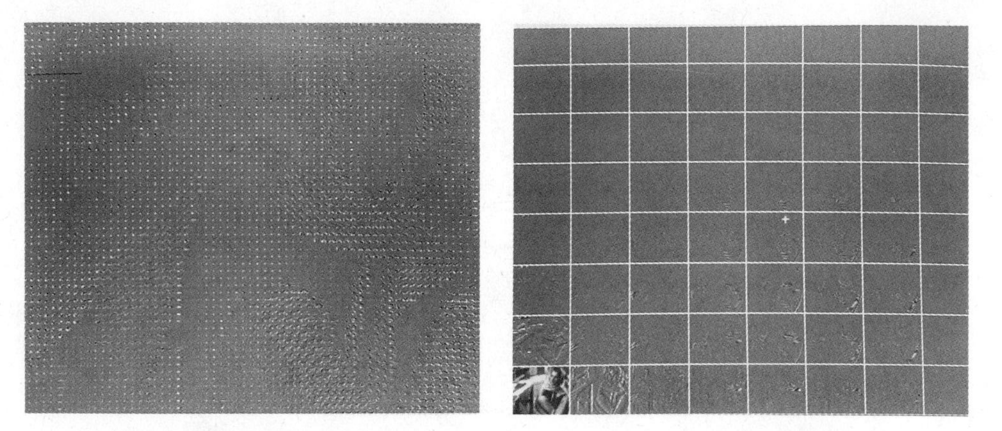

Figure 7.14 8×8 DCT transform of the original image. On the left is the usual block-by-block representation and on the right is the reordering of the coefficients so that same frequencies appear together (subband interpretation of DCT). The lowest frequency is in the lower left corner.

The quantization in the DCT domain is usually scalar and uniform. The lowest two-dimensional frequency component, called the DC coefficient, is treated with particular care. According to (7.1.10), it corresponds to the local average of the block. Mismatches between blocks often lead to the feared blocking effect, that is, the boundaries between the blocks become visible, a visually annoying artifact. Because the DC coefficient has the highest energy, a fine scalar quantization leads to a large entropy. Also, as can be seen in Figure 7.14(b), there is still high correlation among DC coefficients (it resembles the original image). Therefore, predictive quantization, such as the DPCM, of the DC coefficients is often used to increase compression without increasing distortion.

The choice of the quantization steps for the various coefficients of the DCT is a classic bit-allocation problem, since distortion and rate are additive. However, perceptual factors are very important and careful experiments lead to quantization matrices which take into account the visibility of errors (besides the variance and entropy of the coefficients). While this has the flavor of a weighted MSE bit-allocation method, it relies heavily on experimental results. An example quantization matrix, showing the quantizer step sizes used for various DCT coefficients in JPEG, is given in Table 7.3 [148]. What is particularly important is the relative size of the steps, because within a certain range one can scale this quantization matrix, that is, multiply all step sizes by a scale factor greater or smaller than one in order to reduce or increase the bit rate, respectively. This scale factor is very useful for adaptive quantization, where the bit allocation is made between blocks which have various energy levels.

Table 7.3 Example of a quantization matrix as used in DCT transform coding in JPEG [148]. The entries are the step sizes for the quantization of the coefficient (i, j). Note that the relative step sizes are what is critical, since the whole matrix can be multiplied by an overall scale factor. The lowest frequency or DC coefficient is in the upper left corner.

16	11	10	16	24	40	51	61
12	12	14	19	26	58	60	55
14	13	16	24	40	57	69	56
14	17	22	29	51	87	80	62
18	22	37	56	68	109	103	77
24	35	55	64	81	194	113	92
49	64	78	87	103	121	120	101
72	92	95	98	121	100	103	99

Then, one can think of this scale factor as a "super" quantizer step and the goal is to choose the sequence of scale factors that will minimize the total distortion given a certain budget. Each block has its rate-distortion function and thus, the scale factors can be chosen according to the constant-slope rule described in Section 7.1.2. Sometimes, scale factors are fixed for a number of blocks (called macro-block) in order to reduce the overhead.

Of course, bit allocation is done by taking entropy coding into account, which we describe next. As in subband coding, higher frequency coefficients have lower energy and thus have high probability to be zero after quantization. In particular, the conditional probability of a high-frequency coefficient to be zero, given that its predecessors are zero, is close to one. Therefore, there will be runs of zeros, in particular up to the terminal coefficient. To take better advantage of this phenomenon in a two-dimensional transform, an ordering of the coefficients called *zig-zag scanning* is used (see Figure 7.15(a)). Very often, a long stretch of zeros terminates the sequence (see Figure 7.15(b)) and then an "end of block" (EOB) can be sent instead. The nonzero values and the run lengths are entropy coded (typically using Huffman or arithmetic codes).

Note that DCT coding is used not only on images, but also in video coding. While the same principles are used, specific quantization and entropy coding schemes have to be developed, as will be seen in Section 7.4.2.

The coding of color images is performed on a component-by-component basis, that is, after transformation into an appropriate color space such as the luminance and two chrominance components. The components are coded individually with a lesser weighting of the errors in the chrominance components.

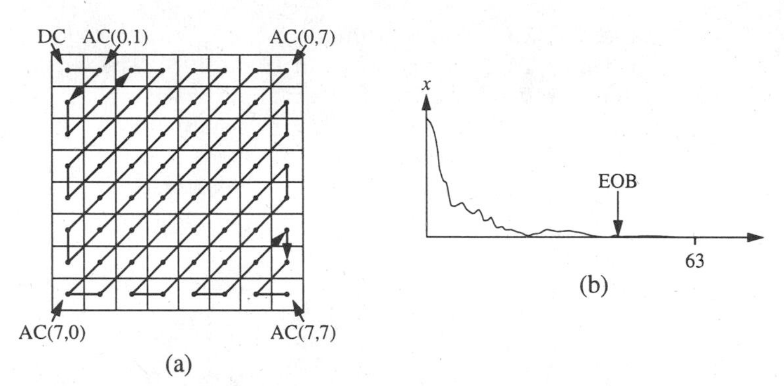

Figure 7.15 Zig-zag scanning of 8×8 DCT coefficients. (a) Ordering of the coefficients. DC stands for the average or constant component, while AC stands for the higher frequencies. (b) Typical sequence of quantized and zig-zag scanned DCT coefficients.

Overlapping Block Transforms Lapped orthogonal transforms (see also Section 3.4.1) were developed specifically to solve the blocking problem inherent to block transforms. Rather than having a hard transition from one block to the next, they smooth out the boundary with an overlapping window [44, 188, 189].

For image coding applications, the LOT basis functions are designed so as to resemble the DCT basis functions and thus, the behavior of lapped orthogonal transform coefficients is very similar to that of DCT coefficients. That is, the DCT quantization and entropy coding strategies will work well in LOT encoding of images as well.

While it is true that blocking effects are reduced in LOT compressed images, other artifacts tend to appear, such as increased ringing around edges due to longer basis functions. Because the blocking effect with the LOT is reduced, one can use more channels, that is, larger blocks, (16×16), and achieve better compression.

The LOT represents an elegant extension of the DCT, however, it has not yet been successful in dislodging it. One of the reasons is that the improvements are not sufficient to justify the increase in complexity. While the LOT has a fast, $O(N \log N)$ algorithm, the structure is more involved since blocks now interact with neighbors. While this small increase in complexity is not much of a problem in software, it has made LOT's less attractive in VLSI implementations so far.

Example: JPEG Image Coding Standard To describe a transform coding example, we will discuss the JPEG industry standard [148, 327]. While it is not the most sophisticated transform coder, its simplicity and good performance (for the type of imagery and bit rate it has been designed for) made it very popular. The availability of special purpose hardware implementing JPEG at high rates (such as 30 frames per second) has further imposed this standard both in still image and in intraframe video compression (see the next section).

An important point is that the JPEG image compression standard specifies only the decoder, thus allowing for possible improvements of the encoder. The JPEG standard comprises several options or modes of operation [327]:

(a) Sequential encoding: block-by-block encoding in scan order.

(b) Progressive encoding: geared at progressive transmission, or successive approximation. To achieve higher-resolution pictures, it uses either more and more DCT coefficients, or more and more bits/coefficient.

(c) Hierarchical encoding: a lower-resolution image is encoded first, upsampled and interpolated to predict the full resolution and the difference or prediction error is encoded with one of the other JPEG versions. This is really a pyramidal coder as will be seen in Section 7.3.2 which uses JPEG on the difference signal.

(d) Lossless encoding: this mode actually does not use the DCT, but predictive encoding based on a causal neighborhood of three samples.

We will only discuss the sequential encoding mode in its simplest version which is called the *baseline JPEG coder*. It uses a size 8×8 DCT, which was found to be a good compromise between coding efficiency (large blocks) and avoidance of blocking effects (small blocks). This holds true for the typical imagery and bit rates for which JPEG is designed, such as the 512×512 *Barbara* image compressed to 0.5 bits/pixel. Note that other types of imagery might use other DCT sizes.

The input is assumed to be 8 bits (typical for regular images) or 12 bits (typical for medical images). Colors are separately treated. After the DCT transform, the quantization uses a carefully designed set of uniform quantizers. Their step sizes are stored in a quantization table, where each entry is an integer belonging to the set $\{1, \ldots, 255\}$. An example was shown in Table 7.3. Quantization is performed by rounding the DCT coefficient divided by the step size to the nearest integer. At the decoder, this rounded value is simply multiplied by the step size. Note that the quantization tables are based on visual experiments, but since they can be specified by the user, they are not part of the standard.

Zig-zag scanning follows quantization and finally entropy coding is performed. First, the DC coefficient (the average of 64 samples) is differentially encoded, that is, $\Delta_l = DC_l - DC_{l-1}$ is entropy coded. This removes some of the

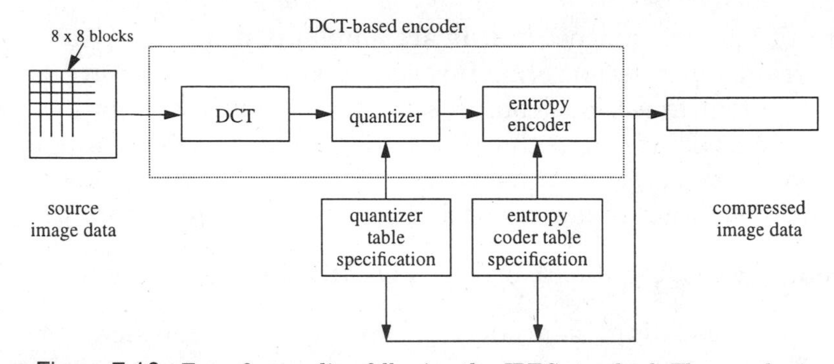

Figure 7.16 Transform coding following the JPEG standard. The encoder is shown. The decoder performs entropy decoding, inverse quantization and an inverse DCT (after [327]).

correlation left between DC coefficients of adjacent blocks. Then, the sequences of remaining DCT coefficients is entropy coded. Because of the high probability of stretches of consecutive zeros, run-length coding is used. A symbol pair (L, A) specifies the length of the run (0 to 15) and the amplitude range (number of bits, $0, \ldots, 10$) of the following nonzero value. Then follows the nonzero value (which has the previously specified number of bits). For example, $(15, 7)$ would mean that we have 15 zeros followed by a number requiring seven bits.

Runs longer than 15 samples simply use a value A equal to zero, signifying continuation of the run, and the pair $(0, 0)$ stands for end of block (no more nonzero values in this block). Finally, the pairs (L, A) are Huffman coded with a table specified by the user (default tables are suggested, but can be replaced). The nonzero values following a run of zeros are now so-called variable-length integers specified by the preceding value A. These are not Huffman coded because of insufficient gain in view of the complexity.

The decoder now operates as follows: Based on the Huffman coding table, it entropy decodes the incoming bit stream, and using the quantization table, it "dequantizes" the transform domain values. Finally, an inverse DCT is applied to reconstruct the image.

Figure 7.16 schematically shows a JPEG encoder. An example of the *Barbara* image coded with the baseline JPEG algorithm is shown in Figure 7.17 at the rate of 0.5 bits/pixel and $SNR_p = 28.26$ dB.

7.3.2 Pyramid Coding of Images

A simple, yet powerful image representation scheme for image compression is the pyramid scheme of Burt and Adelson [41] (see Section 3.5.2). From an original image, derive a coarse approximation, for example, by lowpass filtering and downsampling. Based on this coarse version, predict the original

Figure 7.17 Example of a transform-coded *Barbara* using the JPEG standard. The image has 512 × 512 pixels, the target rate is 0.5 bits/pixel and $SNR_p = 28.26$ dB.

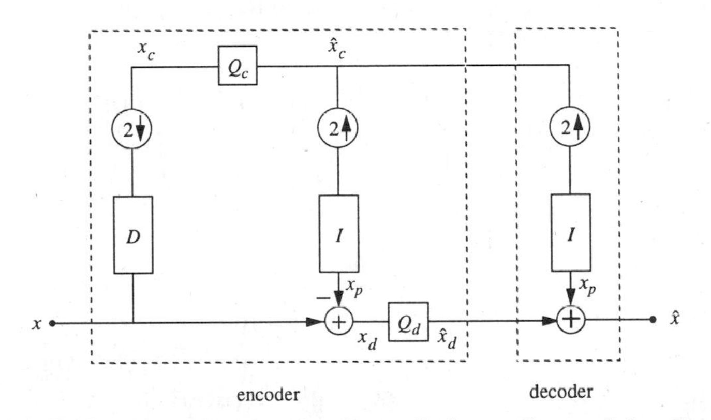

Figure 7.18 One-step pyramid coding. Both encoding and decoding are shown. Note that only the quantization of the difference signal contributes to the reconstruction error. D stands for deriving a coarse version, and I stands for interpolation.

(by upsampling and filtering) and calculate the difference as the prediction error. Instead of the original image, one can compress the coarse version and the prediction error. If the prediction is good (which will be the case for most natural images which have a lowpass characteristic), the error will have a small variance and can thus be well compressed. Of course, the process can be iterated on the coarse version. Figure 7.18 shows such a pyramid scheme. Note how perfect reconstruction, in absence of quantization of the difference signal, is simply obtained by adding back at the decoder the prediction which was subtracted at the encoder.

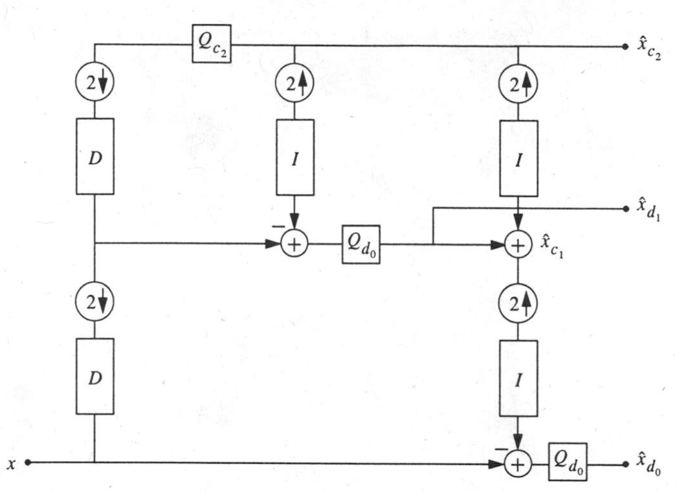

Figure 7.19 Quantization noise feedback in a two-step pyramid. Only the encoder is shown. Note that a decoder is part of the encoder in order to make predictions based on quantized versions only.

Quantization Noise Refer to Figure 7.18. Because the prediction $\boldsymbol{x}_p$ is based on the quantized coarse version $\hat{\boldsymbol{x}}_c$ (rather than $\boldsymbol{x}_c$ itself), the only source of quantization error in the reconstructed signal is the one due to the quantizer Q_d. Since $\hat{\boldsymbol{x}}_d = \boldsymbol{x}_d + \boldsymbol{e}_d$ where $\boldsymbol{e}_d$ is the error due to the quantizer Q_d, we find that

$$\hat{\boldsymbol{x}} = \hat{\boldsymbol{x}}_d + \boldsymbol{x}_p = \boldsymbol{x}_d + \boldsymbol{e}_d + \boldsymbol{x}_p = \boldsymbol{x} + \boldsymbol{e}_d,$$

where we used the fact that $\boldsymbol{x} = \boldsymbol{x}_d + \boldsymbol{x}_p$ in a pyramid coder. This is important if one is interested in the maximum error introduced by coding. In the pyramid case, it will simply be the maximum error of the quantizer Q_d (typically half the largest quantization interval). The property holds also for multilevel pyramids if one uses quantization error feedback [303]. As can be seen from Figure 7.19, the trick is to use only quantized coarse versions in the prediction of a finer version. Thus, the same prediction can be obtained in the decoder as well and the source of quantization noise can be limited to the last quantizer Q_{d_0}. Note that quantizer error feedback requires the reconstruction of $\hat{\boldsymbol{x}}_{c_1}$ in the encoder, and is thus more complex than an encoder without feedback and adds encoding delay.

Decimation and Interpolation Operators In Figures 7.18 and 7.19, we used boxes labeled D and I to denote operators that derive the coarse version and interpolate the fine version, respectively. While these operators are often linear filters, as in the original Burt and Adelson scheme [41], nothing prohibits the use of nonlinear operators [9]. While such generalized operators have not been often used so far, they represent a real potential for pyramid coding. For example, sophisticated methods based on edges could be used to get very rough coarse versions, as long as the prediction reduces the variance of the difference signal sufficiently.

Another attractive feature of this freedom in choosing the operators is that visually pleasing coarse versions are easy to obtain. This is because the filters used for decimation and interpolation, unlike in the subband case, are unconstrained. Typically, zero-phase FIR filters are used where medium lengths already achieve good lowpass behavior and visually good looking coarse versions.

Oversampling A drawback of pyramid coding is the implicit oversampling. Assume we start with an $N \times N$ image. After one step, we have an $N/2 \times N/2$ coarse version, but also an $N \times N$ difference image. If the scheme is iterated we have the following number of samples:

$$N^2(1 + \frac{1}{4} + \frac{1}{4^2} + \cdots) \leq \frac{4}{3}N^2,$$

as was given in (3.5.4). This oversampling of up to 33% has often been considered as a drawback of pyramid coding (in one dimension, the overhead is 100% and thus a real problem). However, it does not prohibit efficient coding a priori and the other attractive features such as the control of quantization noise, quality of coarse pictures, and robustness counterbalance the oversampling problem.

Bit Allocation The problem of allocating bits to the various quantizers is tricky in pyramid coders, especially when quantization noise feedback is present. The reason is that the independence assumption used in the optimal bit allocation algorithm derived in Section 7.1.2 does not hold. Consider Figure 7.18 and assume a choice of quantizers for Q_c and Q_d. Because the choice for Q_c influences the prediction x_p and thus the variable to be quantized x_d, there is no independence between the choices for Q_c and Q_d. For example, increasing the step size of Q_c not only increases the distortion of $\hat{x}_c$, but also of $\hat{x}_d$ (since its variance will probably increase). Thus, in the worst case, one might have to search all possible pairs of quantizers for x_c and x_d and find the best performing pair given a certain bit budget. It is clear that this search grows exponentially as the number of levels increases, since we have K^l possible l-tuples of quantizers, where K is the number of quantizers at every level and l is the number of levels. Even if quantization error feedback is not used, there is a complication because the total error squared is not the sum of the errors e_c and e_d squared (see (7.1.16)), since the pyramid decomposition is not unitary (unless an ideal lowpass filter is assumed). A discussion of dependent quantization and its application to pyramid coding can be found in [232].

7.3.3 Subband and Wavelet Coding of Images

The generalization of subband decomposition to multiple dimensions is straightforward, especially in the separable case [314]. The application to compression of images has become popular [1, 111, 265, 330, 332, 335, 337]. The nonseparable multidimensional case, using quincunx [314] or hexagonal downsampling [264], as well as directional decompositions [19, 287], has also found applications in image compression. Recently, using filters specifically designed for regularity, methods closely related to subband coding have been proposed under the name of wavelet coding [14, 79, 81, 101, 176, 244, 260, 341]. The main difference with pyramid coding, discussed in Section 7.3.2, is that we have a critically sampled scheme and often an orthogonal decomposition. The price paid is more constrained filters in the decomposition, which leads to poorer coarse resolution pictures in general. In what follows, we discuss various forms of subband and wavelet compression schemes tailored to images.

Separable Decompositions We will call separable decompositions those which use separable downsampling. Usually, they also use separable filters (but this is not necessary). When both downsampling and filters are separable, the implementation is very efficient since it can be done on rows and columns separately, at least at each stage of the decomposition.

While being constrained, separable systems are often favored because of their computational efficiency with separable filters, since size-$N \times N$ filters lead to order N rather than N^2 operations/input sample (see Section 6.2.4). Conceptually, separable systems are also much easier to implement since they are cascades of one-dimensional systems. However, from the fact that the two-dimensional filters are products of one-dimensional filters, it is clear that only rectangular pieces of the spectrum can be isolated.

Nonseparable Decompositions Recall that coding gain in subband coding was maximized when the variances in the channels were as different as possible (see Section 7.1.2). If one assumes that images have a power spectrum that is roughly rotationally-invariant and decreases with higher frequencies, then it is clear that separable systems are not best suited for isolating a lowpass channel containing most energy and having highpass channels with low energy. A better solution is found by opting for nonseparable systems. The two most important systems for image processing are based on the quincunx [314] and hexagonal downsamplings [264], for two- and four-channel subband coding systems, respectively. Quincunx and hexagonal sublattices of $\mathcal{Z}^2$ are shown in Figure 7.20, together with the more conventional separable sublattice. They correspond to

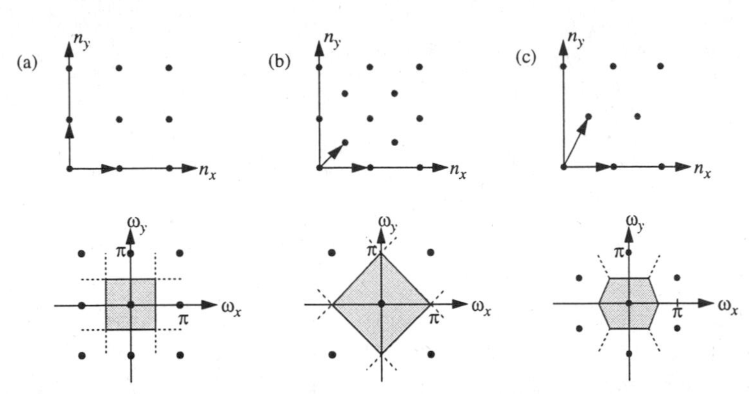

Figure 7.20 Sublattices of $\mathcal{Z}^2$ and shapes of possible ideal lowpass filters (corresponding to the Voronoi cell of the dual lattice, which is indicated as well). (a) Separable sublattice $\boldsymbol{D}_S$. (b) Quincunx $\boldsymbol{D}_Q$. (b) Hexagonal $\boldsymbol{D}_H$.

integer linear combinations of the columns of the following matrices[6]:

$$\boldsymbol{D}_S = \begin{pmatrix} 2 & 0 \\ 0 & 2 \end{pmatrix}, \quad \boldsymbol{D}_Q = \begin{pmatrix} 2 & 1 \\ 0 & 1 \end{pmatrix}, \quad \boldsymbol{D}_H = \begin{pmatrix} 2 & 1 \\ 0 & 2 \end{pmatrix},$$

where the sampling density is reduced by a factor of four for the separable sampling, two for the quincunx sampling (see also Appendix 3.B) and by a factor of four for the hexagonal sampling. The repeated spectrums in Fourier domain due to downsampling appear on the dual lattice, which is given by the transposed inverse of the lattice matrix. Also shown in Figure 7.20 are possible ideal lowpass filters that will avoid aliasing when downsampling to these sublattices. If, as we said, images have circularly symmetric power spectrums that decrease with higher frequencies, then the quincunx lowpass filter will retain more of the original signal's energy than a separable lowpass filter (which would be one-dimensional since the downsampling is by two). Using the same argument, the hexagonal lowpass filter is then better than the corresponding lowpass filter in a separable system with downsampling by two in each dimension. Thus, these nonseparable systems, while being more difficult to design and more complex to implement, represent a better match to usual image spectrums.

Furthermore, the simple quincunx case has the following perceptual advantage: The human visual system is more accurate in horizontal and vertical high frequencies than along diagonals. The lowpass filter in Figure 7.20(b) conserves horizontal and vertical frequencies, while it cuts off diagonals to half of their original range. This is a good match to the human eye and often, the highpass channel (which is complementary to the lowpass channel) can

[6]Recall from Appendix 3.B, that a given sampling lattice may have infinitely many matrix representations.

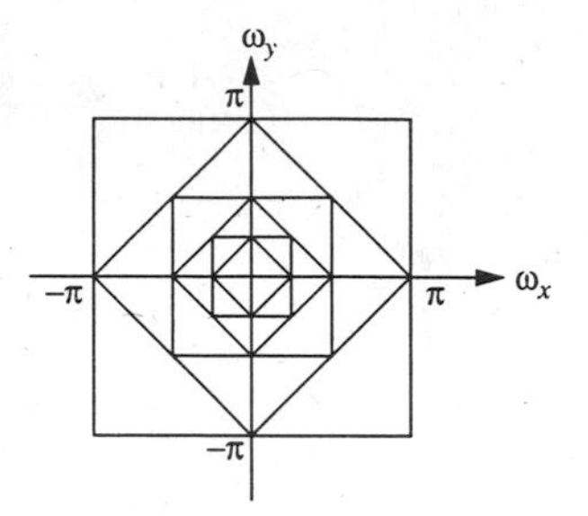

Figure 7.21 Frequency decomposition of iterated quincunx scheme.

be disregarded altogether. That is, a compression by a factor of two can be achieved with no visible degradation. Such preprocessing has been used in intraframe coding of HDTV [12]. The above quincunx scheme is often iterated on the lowpass channel, leading to a frequency decomposition as shown in Figure 7.21. This actually corresponds to a two-dimensional nonseparable wavelet decomposition [164] and has been used for image compression [14].

The hexagonal system, besides having a fairly good approximation to a circularly symmetric lowpass, has three directional channels which can be used to detect directional edges [264]. However, the goal of an isotropic analysis is only approximated, since the horizontal and vertical directions are not treated in the same manner (see Figure 7.20(c)). Therefore, it is not clear if the added complexity of a nonseparable four-channel system based on the hexagonal sublattice is justified for coding purposes.

Choice of Filters Unlike in audio compression, the filters for image subband coding do not need high out-of-band rejection. Instead, a number of other constraints have to be satisfied.

Linear phase In regular image filtering, the need for linear phase is well-known since without linear phase, the phase distortion around edges is very visible. Therefore, the use of linear phase filters in subband coding has been often advocated [14]. Recall from Section 3.2.4, that in two-band FIR systems, linear phase and orthogonality are mutually exclusive and this carries over to four-band separable systems which are most often used in practice.

However, the case for linear phase is not as obvious as it seems at first sight. For example, in the absence of quantization, the phase of the filters has no bearing since the system has perfect reconstruction. This argument carries over for fine quantization as well. In the case of coarse quantization, the situation is more complex. One scenario is to consider the highpass channel as being set to zero. Look at the two impulse responses of this system. Nonlinear phase systems lead to nonsymmetric responses, but so do some of the linear phase systems. Only if the filters meet additional constraints do the two impulse responses remain symmetric. Note also, that for computational purposes, linear phase is more convenient because of the symmetry of the filters.

Note that orthogonal FIR filters of sufficient length can be made almost linear phase by appropriate factorization of their autocorrelation function. Also, there are nonseparable orthogonal filters with linear phase. Finally, by resorting the IIR filters, one can have both linear phase and orthogonality, and such noncausal IIR filters can be used in image processing without problems since we are dealing with finite-length input signals.

Orthogonality Orthogonal filters implement a unitary transform between the input and the subbands. The usual features of unitary transforms hold, such as conservation of energy. In particular, the total distortion is the sum of the subband distortions, or:

$$D = \sum_i D_i, \tag{7.3.1}$$

and the total bit rate is the sum of all the subband's bit rates. Therefore, optimal bit-allocation algorithms which assume additivity of bit rate and distortion can be used (see Section 7.1.2). In thc nonorthogonal case, (7.3.1) does not hold, and thus, these bit allocation algorithms cannot be used directly. It should be noted that well designed linear phase FIR filter banks (that is, with good out-of-band rejection) are often close to being orthogonal and thus satisfy (7.3.1) approximately.

Filter size Good out-of-band rejection or high regularity require long filters. Besides their computational complexity, long filters are usually avoided because they tend to spread coding errors. For example, sharp edges introduce distortions because high-frequency channels are coarsely quantized. If the filters are long (and usually their impulse response has several sign changes), this causes an annoying artifact known as ringing around edges. Therefore, filters used in audio subband compression, such as length-32 filters, are too long for image compression. Instead, shorter "smooth" filters are preferred. Sometimes both their impulse and their step response are considered from a perceptual point of view [167]. The step response is important since edges in images will generate step responses at least in some of the channels. Highly oscillating step responses will require more bits to code, and coarse quantization will produce oscillations which are related to the step response. As can already be seen from this short discussion, there is an intertwining between the choice of filters and the type of quantization that follows. However, it is clear that the frequency-domain criterions used in audio (sharp cut-off, strong out-of-band rejection) have little meaning in the image compression context, where time-domain arguments such as ringing, are more important.

Regularity An orthogonal filter with a certain number of zeros at the aliasing frequency (π in the two-channel case) is called regular if its iteration tends to a continuous function (see Section 4.4). The importance of this property for coding is potentially twofold when the decomposition is iterated. First, the presence of many zeroes at the aliasing frequency can improve the coding gain and second, compression artifacts might be less objectionable. To investigate the first effect, Rioul [243] compared the compression gain for filters of varying regularity used in a wavelet coder, or octave-band subband coder, with four stages. The experiment included bit allocation, quantization, and entropy coding and is thus quite realistic. The results are quite interesting: Some regularity is desired (the performance with no regularity is poor) and higher regularity improves compression further (but not substantially).

As for the compression artifacts, the following argument shows that the filters should be regular when an octave-band decomposition is used: Assume a single quantization error in the lowpass channel. This will add an error to the reconstructed signal which depends only on the equivalent — iterated lowpass filter. If the iterated filter is smooth, this will be less noticeable than if it is a highly irregular function (even though both contribute the same MSE). Note also that the lowest band is upsampled 2^{i-1} times (where i is the number of iterations) and thus, the iterated filter's impulse response is shifted by large steps, making irregular patterns in the impulse response more visible.

In the case of biorthogonal systems such as linear phase FIR filter banks, one is often faced with the case where either the analysis or the synthesis is regular, but not both. In that case, it is preferable to use the regular filter at the synthesis, by the same argument as above. Visually, an irregular analysis is less noticeable than an irregular synthesis, as can be verified experimentally.

When the decomposition is not iterated, regularity is of little concern. A typical example is the lapped orthogonal transform, that is, a multi-channel filter bank which is applied only once.

Frequency selectivity What is probably the major criterion in audio subband filter design is of much less concern in image compression. Aliasing, which is a major problem in audio, is much less disturbing in images [331]. The desire for short filters limits the frequency selectivity as well. One advantage of frequency selectivity is that perceptual weighting of errors is easier, since errors will be confined to the band where they occur.

In conclusion, subband image coding requires relatively short and smooth filters, with some regularity if the decomposition is iterated.

Quantization of the Subbands There are basically two ways to approach quantization of a subband-decomposed image: Either the subbands are quantized independently of each other, or dependencies are taken into account.

Independent quantization of the subbands While the subbands are only independent if the input is a Gaussian random variable and the filters decorrelate the bands, the independence assumption is often made because it makes the system much simpler. Different tree structures will produce subbands with different behaviors, but the following facts usually hold:

(a) The lowest band, being a lowpass and downsampled version of the original, has a behavior much like the original image. That is, traditional quantization methods used for images can be applied here as well, such as DPCM [337] or even transform coding [174, 285].

(b) The highest bands have negligible energy and can usually be discarded with no noticeable loss in visual quality.

(c) Except along edges, little correlation remains within higher bands. Because of the directional filtering, the edges are confined to certain directions in a given subband. Also, the probability density function of the pixel values peaks in zero and falls off very rapidly. While it is often modeled as a Laplacian distribution, it is actually falling off more rapidly. It is more adequately fitted with a generalized Gaussian pdf with faster decay than the Laplacian pdf [329].

Besides the lowband compression, which uses known image coding methods, the bulk of the compression is obtained by appropriate quantization of the high bands. The following quantizers are typically used:

(a) Lloyd quantizers fitted to the distribution of the particular band to be quantized. Tables of such Lloyd quantizers for generalized Gaussian pdf's and decay values of interest for image subbands can be found in [329].

(b) Uniform quantizers with a so-called dead zone which maps a region around the origin to zero (typically of twice the step size used elsewhere). Such dead zone quantizers have proven useful because they increase compression substantially with little loss of visual quality, since they tend to eliminate what is essentially noise in the subbands [111].

Because entropy coding is used after quantization, uniform quantizers are nearly optimal [285]. Thus, since uniform quantizers are much easier to implement than Lloyd quantizers, the former are usually chosen, unless the variable rate associated with entropy codes has to be avoided. Note that vector quantization could be used in the subbands, but its complexity is usually not worthwhile since there is little dependence between pixels anyway.

An important consideration is the relative perceptual importance of various subbands. This leads to a weighting of the MSE in various subbands. This weighting function can be derived through perceptual experiments by finding the level of "just noticeable noise" in various bands [252]. As expected, high bands tolerate more noise because the human visual system becomes less sensitive at high frequencies. Note that more sophisticated models would include masking as well.

Quantization across the bands Looking at subband decomposed images, it is clear that the bands are not independent. A typical example is the representation of a vertical edge. It will be visible in the lowpass image and appears in every band that contains horizontal highpass filtering. It has thus been suggested to use vector quantization across the bands instead of in the bands [329, 332]. While there is some gain in doing so, there is also the following problem: Because the subbands are downsampled versions of the original, we have a shift-variant system. Thus, small shifts can produce changes in the subband signals which reduce the correlation. That is, while visually the edge is "preserved", the exact values in the various bands depend strongly on the location and are thus difficult to predict from band to band. In Section 7.3.4, we will see schemes which, by using an approach that does not rely on vector quantization but simply on local energy, can make use of some dependence between bands.

It should be noted that the straightforward vector quantization across bands is easiest when equal-size subbands are used. In the case of an octave-band decomposition, the vector should use pixels at each level that correspond to the same region of the original signal. That is, the number of pixels should be inversely proportional to scale. The comparison of vector quantization for equally-spaced bands and octave-spaced bands is shown in Figure 7.22 for the one-dimensional case for simplicity.

Bit Allocation For bit allocation between the bands, one can directly use the procedures developed in Section 7.1.2, at least if the filters are orthogonal. Then, the total distortion is the sum of the subbands distortions, and the total rate is the sum of rates for the various bands. In the nonorthogonal case, the distortion is not additive, but can be approximated as such.

The typical allocation problem is the following: For each channel i, one has a choice from a set of quantizers $\{q_{i,j}\}$. Choosing a given quantizer $q_{i,j}$ will produce a distortion $d_{i,j}$ and a rate $r_{i,j}$ for channel i (one can use weighted distortion as well). The problem is to find which combination of quantizers in the

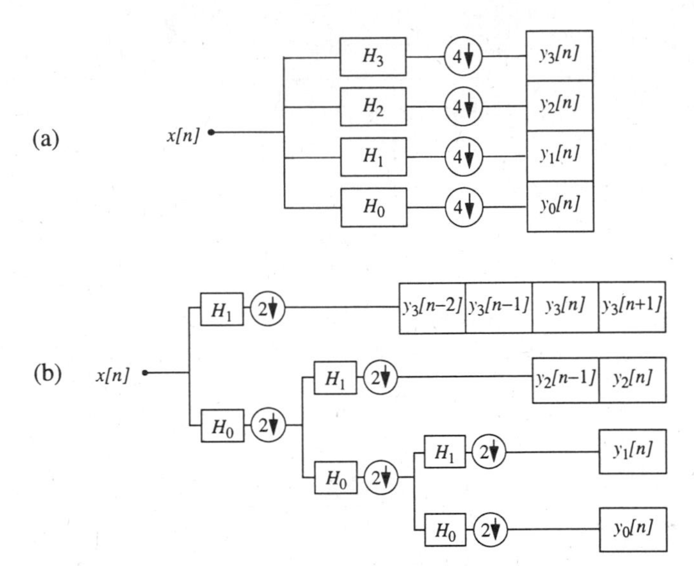

Figure 7.22 Vector quantization across the bands in subband decomposition. (a) Uniform decomposition. (b) Octave-band, or, wavelet decomposition. Note that the number of samples in the various bands corresponds to a fixed region of the input signal.

various channels will produce the minimum squared error while satisfying the budget constraint. The optimal solution is found using the constant-slope solution as described in Section 7.1.2. The pairs $(d_{i,j}, r_{i,j})$, that is, the operational rate-distortion curves can be measured over a set of representative images and then used as a fixed allocation. The problem is that, when applied to a particular image, the budget might not be met. On the other hand, given an image to be coded, one can measure the operational rate-distortion curves and use the constant-slope allocation procedure. This will guarantee an optimal solution, but is computationally expensive. Finally, one can use allocations based on probability density functions, in which case it is often sufficient to measure the variance of a particular channel in order to find its allocation (see (7.1.19) for example). Note that the rates used in the allocation procedure are after entropy coding.

Entropy Coding Substantial reductions in rate, especially in the case of uniform quantizers, is obtained by entropy coding quantized samples or groups of samples. Any of the techniques discussed in Section 7.1.3 can be used, such as Huffman coding. Since Huffman codes are only within one bit of the true entropy [109], they tend to be inefficient for small alphabets. Thus, codewords from small alphabets are gathered into groups and vector Huffman coded (see [285]). Another option is to use vector quantization to group samples [256].

LL, HL	LH, HL	HH, HL	HL, HL
LL, HH	LH, HH	HH, HH	HL, HH
LL, LH	LH, LH	HH, LH	HL, LH
LL, LL	LH, LL	HH, LL	HL, LL

Figure 7.23 Uniform subband decomposition of an image into 16 subbands. The spectral decomposition and ordering of the channels is shown. The first two letters correspond to horizontal filtering and the last two to vertical filtering. LH, for example, means that a lowpass is used in the first stage and a highpass in the second. The ordering is such that frequencies increase from left to right and from bottom to top.

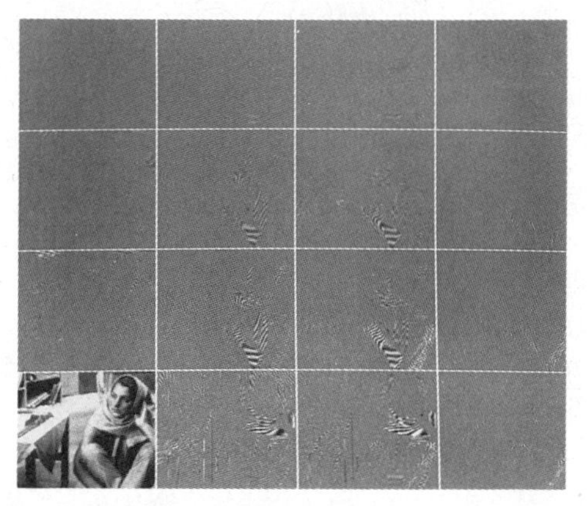

Figure 7.24 Uniform subband decomposition of the *Barbara* image. The ordering of the subbands is given in Figure 7.23.

Because higher bands tend to have large amounts of zeros (especially after deadzone quantizers), run-length coding and an end of block symbol can be used to increase compression substantially.

Examples Two typical coding examples will be described in some detail. The first is a uniform separable decomposition. The second is an octave-band or constant relative bandwidth decomposition (often called a wavelet decomposition).

Uniform decomposition By using a separable decomposition into four bands and iterating it once, we obtain 16 subbands as shown in Figure 7.23. The resulting subband images are shown in Figure 7.24. The filters used are linear phase length-12 QMF's [144] and the image was symmetrically extended before filtering. The variances of the samples in the bands are shown in Table 7.4. We

Table 7.4 Variances in the various bands of a uniform decomposition (defined as in Figure 7.23).

	LL	LH	HH	HL
HL	0.58959	0.86237	1.77899	0.88081
HH	2.87483	6.71625	8.56729	3.25402
LH	23.5474	33.4055	60.9195	14.8490
LL	2711.45	56.0058	52.5202	13.9685

Table 7.5 Step sizes for the quantizers in the various bands (as defined in Figure 7.23), for a target rate of 0.5 bits/pixel. The lowest band was JPEG coded, and the step size corresponds to the quality factor (QF) used in JPEG.

	LL	LH	HH	HL
HL	9.348	8.246	8.657	22.318
HH	8.400	10.161	8.887	13.243
LH	6.552	7.171	10.805	16.512
LL	QF-89	8.673	11.209	15.846

code the lowest subband (LL,LL) with JPEG (see Section 7.3.1). For all other bands, we use uniform quantization with a dead zone of twice the step size used elsewhere. Using a set of step sizes, one can derive rate-distortion curves by measuring the entropy of the resulting quantized channels. A true operational rate-distortion curve would have to include run-length coding and actual entropy coding. Based on these rate-distortion curves, one can perform an optimal constant-slope bit allocation, that is, one can choose the optimal quantizer step sizes for the various bands. The step sizes for a budget of 0.5 bits/pixel are listed in Table 7.5. A set of Huffman codes and run-length codes are designed for each subband channel. Note that the special symbol "start of run" (SR) is entropy coded as any other nonzero pixel. Altogether, one obtains the final rate of 0.497 bits/pixel (the difference in rate comes from the fact that bit allocation was based on entropy measures). Then, the coded image has SNR_p of 30.38 dB. Figure 7.27 (top row) shows the compressed *Barbara* image and a detail at the same rate.

Octave-band decomposition Instead of uniformly decomposing the spectrum of the image, we iterate a separable four-band decomposition three times. The resulting split of the spectrum is shown in Figure 7.25, together with the subband images in Figure 7.26. Here, we used the Daubechies' maximally flat

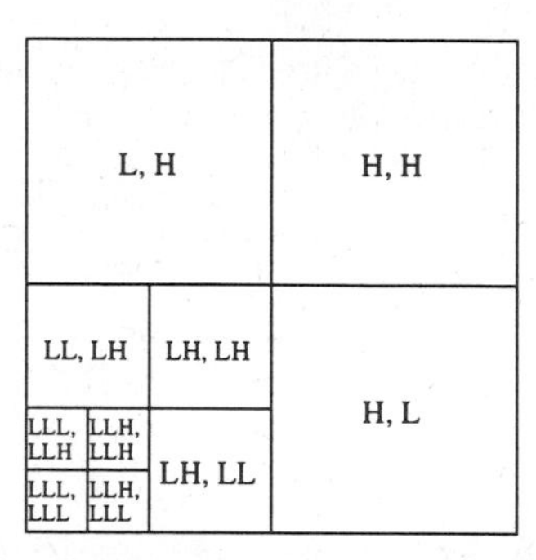

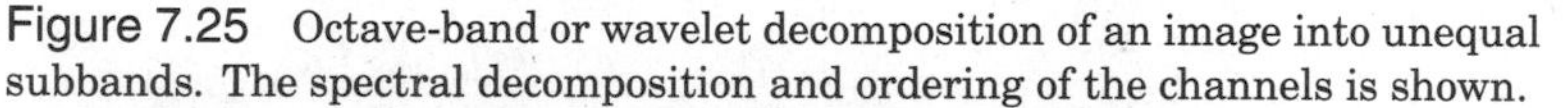
Figure 7.25 Octave-band or wavelet decomposition of an image into unequal subbands. The spectral decomposition and ordering of the channels is shown.

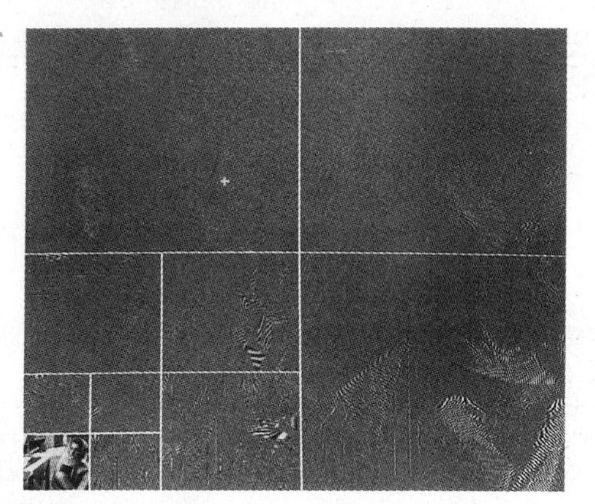

Figure 7.26 Subband images corresponding to the spectral decomposition shown in Figure 7.25.

orthogonal filters of length 8. At the boundaries, we used periodic extension. The variances in the bands are shown in Table 7.6. Histograms of pixel values of the bands are similar to the ones in a uniform decomposition. Because the lowest band (LLL, LLL) is small enough (64×64 pixels), we use scalar quantization on it as on all other bands. Again, uniform quantizers with double-sized dead zone are used and rate-distortion curves are derived for bit-allocation purposes. The resulting step sizes for the target bit rate of 0.5 bits/pixel are given in Table 7.7.

The development of entropy coding (including run-length coding for higher bands) is similar to the uniform-decomposition case discussed earlier. The final rate is 0.499 bits/pixel, with SNR_p of 29.21 dB. The coded image and a detail are shown in Figure 7.27 (bottom row). Note that there is little difference between the uniform and the octave-band decomposition results.

We would like to emphasize that the above examples are "textbook ex-

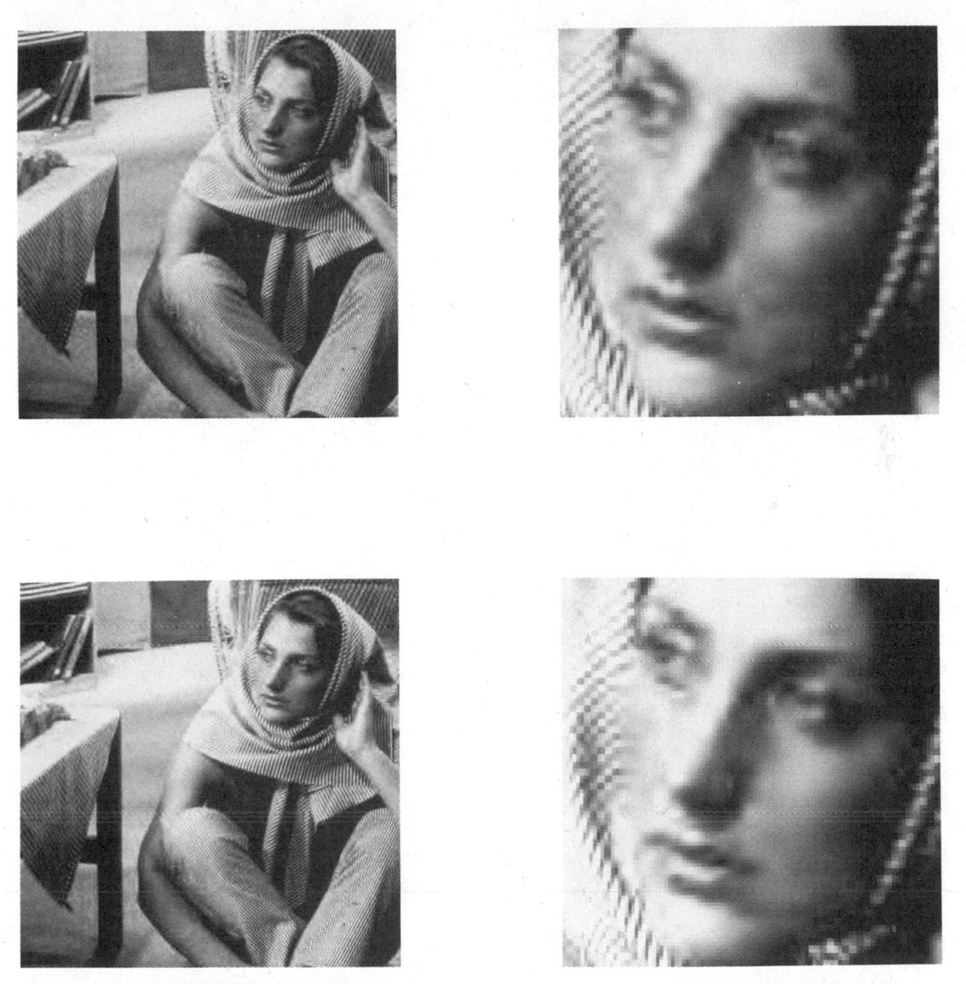

Figure 7.27 Compression results on *Barbara* image. Top left: Subband coding in 16 uniform bands at 0.4969 bits/pixel and $SNR_p = 30.38$ dB. Top right: Detail of top left. Bottom left: Octave-band or wavelet compression at 0.4990 bits/pixel and $SNR_p = 29.21$ dB. Bottom right: Detail of bottom left.

amples" for illustration purposes. For example, no statistics over large sets of images were taken and thus, the entropy coders might perform poorly for a substantially different image. The aim was more to demonstrate the ingredients used in a subband/wavelet image coder.

State of the art coders, which can be found in the current literature, improve substantially the results shown here. Major differences with respect to the simple coders we discussed so far are the following:

Table 7.6 Variances in the different bands of an octave-band decomposition (defined as in Figure 7.25).

Band	Variance
LLL,LLL	2559.8
LLH,LLL	60.7
LLL,LLH	43.8
LLH,LLH	21.2
LH,LL	55.4
LL,LH	24.5
LH,LH	33.7
H,L	141.4
L,H	15.2
H,H	16.2

Table 7.7 Step sizes for uniform quantizer in the octave subband or wavelet decomposition of Figure 7.25, for a target rate of 0.5 bits/pixel.

Band	Step size
LLL,LLL	5.21
LLH,LLL	3.69
LLL,LLH	4.42
LLH,LLH	4.08
LH,LL	8.42
LL,LH	9.22
LH,LH	7.45
H,L	17.23
L,H	22.05
H,H	21.57

(a) *Vector quantization* can be used in the subbands, such as lattice vector quantization [13].

(b) *Adaptive entropy coding* is used to achieve immunity to changes in image statistics.

(c) *Adaptive quantization in the subbands* can take care of busy versus non-busy regions.

(d) *Dependencies across scales,* either by vector quantization or prediction of structures across scales, are used to reduce the bit rate [176, 222, 259].

(e) *Perceptual tuning* using band sensitivity, background luminance level and masking of noise due to high activity can improve the visual quality [252].

The last point — perceptual models for subband compression, is where most gain can be obtained.

With these various fine tunings, good image quality for a compressed version of a 512×512 original image such as *Barbara* can be obtained in the range of 0.25 to 0.5 bits/pixel. Note that the complexity level is still of the same order as the coders we presented and is comparable in order of magnitude to a DCT coder such as JPEG.

7.3.4 Advanced Methods in Subband and Wavelet Compression

The discussion so far has focused on standard methods. Below, we describe some more recent algorithms which are both of theoretical and practical interest.

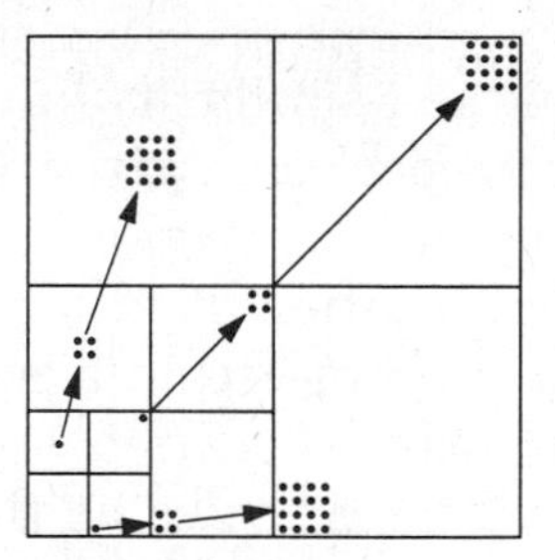

Figure 7.28 Zero-tree structure on an octave-band decomposed image. Three possible trees in different bands are shown.

Zero-Tree Based Compression From looking at subband pictures such as those in Figures 7.24 or 7.26, it is clear that there are some dependencies left among the bands, as well as within the bands. Also, for natural images with decaying spectrums, it is unlikely to find significant high-frequency energy if there is little low-frequency energy in the same spatial location. These observations lead to the development of an entropy coding method specifically tailored to octave-band or wavelet coding. It is based on a data structure called a zero tree [176, 260], which is the analogous to zig-zag scanning and the end of block (EOB) symbol used in the DCT.

The idea is to define a tree of zero symbols which starts at a root which is also zero. Therefore, this root can be labeled as an "end of block". A few such zero trees are shown in Figure 7.28. Because the tree grows as powers of four, a zero tree allows us to disregard many insignificant symbols at once. Note also that a zero tree gathers coefficients that correspond to the same spatial location in the original image.

Zero trees have been combined with bit plane coding in an elegant and efficient compression algorithm due to Shapiro [260, 259]. It incorporates nicely many of the key ideas presented in this section and demonstrates the effectiveness of wavelet based coding. The resulting algorithm is called *embedded zero-tree wavelet* (EZW) algorithm. Embedded means that the encoder can stop encoding at any desired target rate. Similarly, the decoder can stop decoding at any point resulting in the image that would have been produced at the rate of the truncated bit stream. This compression method produces excellent results without requiring a priori knowledge of the image source, without prestored tables of codebooks, and without training.

The EZW algorithm uses the discrete-time wavelet transform decomposition where at each level i the lowest band is split into four more bands: LL_{i+1}, LH_{i+1}, HL_{i+1}, and HH_{i+1}. In simulations in [260], six levels are used with length-9 symmetric filters given in [1].

The second important ingredient is that the absence of significance across scales is predicted by exploiting self-similarity inherent in images. A coefficient x is called *insignificant* with respect to a given threshold T, if $|x| < T$. The assumption is that if x is insignificant, then all of its descendents of the same orientation in the same spatial location at all finer scales are insignificant as well. We call a coefficient at a coarse scale a *parent*. All coefficients at the next finer scale at the same spatial location and of similar orientation are *children*. All coefficients at all finer scales at the same spatial location and of similar orientation are *descendents*. Although there exist counterexamples to the above assumption, it holds true most of the time. Then, one can make use of it, and code such a parent as a zero-tree root (*ZTR*), thereby avoiding to code all its descendants. When the assumption is not true, that is, the parent is insignificant but down the tree, there exists a significant descendant, then such a parent will be coded as an isolated zero (IZ). To code the coefficients, Shapiro uses four symbols, *ZTR*, IZ, POS for a positive significant coefficient, and *NEG* for a negative significant one. In the highest bands which do not have any children, IZ and *ZTR* are merged into a zero symbol (Z). The order in which the coefficients are scanned is of importance as well. It is performed so that no child is scanned before its parent. Thus, one scans bands LL_N, HL_N, LH_N, HH_N, and moves on to the scale $(N-1)$ scanning HL_{N-1}, LH_{N-1}, HH_{N-1}, until reaching the starting scale HL_1, LH_1, HH_1. This scanning pattern orders the coefficients in the order of importance, allowing for embedding.

The next step is successive approximation quantization. It entails keeping at all times two lists: the *dominant* list and the *subordinate* list. The dominant list contains the coordinates of those coefficients that have not yet been found to be significant. The subordinate list contains the magnitudes of those coefficients that have been found to be significant. The process is as follows: We decide on the initial threshold T_0, (for example, it could be half of the positive range of the coefficients) and start with the dominant pass where we evaluate each coefficient in the scanning order described above to be one of the four symbols *ZTR*, IZ, POS and NEG. Then we cut the threshold in half obtaining T_1 and add another bit of precision to the magnitudes on the list of coefficients known to be significant, that is, the subordinate list. More precisely, we assign the symbols 0 and 1 depending whether the refinement leaves the reconstruction of a coefficient in the upper or lower half of the previous bin. We reorder the coefficients in the decreasing order and go onto the dominant pass again with the threshold T_1. Note that now those coefficients that have been found to be significant during a previous pass are set to zero so that they do not preclude a possibility of finding a zero tree. The process then alternates between these two passes until some stopping condition is met, such as that the bit budget is exhausted. Finally, the symbols are losslessly encoded using

Table 7.8 An example of a 3-level discrete-time wavelet transform of an 8 × 8 image.

5	11	5	6	0	3	-4	4
2	-3	6	-4	3	6	3	6
3	0	-3	2	3	-2	0	4
-5	9	-1	47	4	6	-2	2
9	-7	-14	8	4	-2	3	2
15	14	3	-12	5	-7	3	9
-31	23	14	-13	3	4	6	-1
63	-34	49	10	7	13	-12	7

adaptive arithmetic coding.

Example 7.2 *EZW Example from [260]*

Let us consider a simple example given in [260]. We assume that we are given an 8×8 image whose 3-level discrete-time wavelet transform is given in Table 7.8. Since the largest coefficient is 63, the initial threshold is $T_0 = 32$.

We start in the scanning order as we explained before. 63 is larger than 32 and thus gets POS. -34 is larger than 32 in absolute value and gets NEG. We go onto -31 which is smaller in absolute value than 32. However, going through its tree, which consists of bands LH_2 and LH_1, we see that it is not a root of a zero tree due to a large value of 47. Therefore its assigned symbol is IZ. We continue with 23 and establish that it is a root of a zero tree comprising bands HH_2 and HH_3. We continue the process in the scanning order, except that we skip all those coefficients for which we have previously established that they belong to a zero tree. The result of this procedure is given in Table 7.9.

After we have scanned all available coefficients, we are ready to go onto the first subordinate pass. We commence by halving the threshold, to obtain $T_1 = 16$ as well as quantization intervals. The resulting intervals are now $[32, 48)$ and $[48, 64)$. The first significant value, 63, obtains a 1, and is reconstructed to 56. The second one, -34, gets a 0 and is reconstructed to -40, 49 gets a 1 and is reconstructed to 56, and finally, 47 gets a 0 and is reconstructed to 40. We then order these values in the decreasing order of reconstructed values, that is, $(63, 49, 34, 47)$. If we want to continue the process, we start the second dominant pass with the threshold of 16. We first set all significant values from the previous pass to zero, in order to be able to identify zero trees. In this pass, we establish that -31 in LH_3 is NEG and 23 in HH_3 is POS. All the other coefficients are then found to be either zero tree roots or zeros. We add to the list of significant coefficients 31 and 23 and halve the quantization intervals, to obtain, $[16, 24)$, $[24, 32)$, $[32, 40)$, $[40, 48)$, $[48, 56)$, and $[56, 64)$. At the end of this pass, the revised list is $(63, 49, 47, 34, 31, 23)$, while the reconstructed list is $(60, 52, 44, 36, 28, 20)$. This process continues until, for example, the bit budget is met.

Adaptive Decomposition Methods In our discussions of subband and wavelet coding of images, we have seen that both full-tree decompositions and octave-band tree decompositions are used. A natural question is: Why not use arbi-

Table 7.9 The first dominant pass through the coefficients.

Subband	Coefficient	Symbol	Reconstruction
LL_3	63	POS	48
HL_3	-34	NEG	-48
LH_3	-31	IZ	0
HH_3	23	ZTR	0
HL_2	49	POS	48
HL_2	10	ZTR	0
HL_2	14	ZTR	0
HL_2	-13	ZTR	0
LH_2	15	ZTR	0
LH_2	14	IZ	0
LH_2	-9	ZTR	0
LH_2	-7	ZTR	0
HL_1	7	Z	0
HL_1	13	Z	0
HL_1	3	Z	0
HL_1	4	Z	0
LH_1	-1	Z	0
LH_1	47	POS	48
LH_1	-3	Z	0
LH_1	-2	Z	0

trary binary-tree decompositions, and in particular, choose the best binary tree for a given image? This is exactly what the best basis algorithm of Coifman, Meyer, Quake and Wickerhauser [62, 64] attempts. Start with a collection of bases given by all binary subband coding trees of a given depth, called wavelet packets (see Section 3.3.4). From a full tree, the best basis algorithm uses dynamic programming to prune back to the best tree, or equivalently, the best basis.

In [233], the best basis algorithm was modified so as to be optimal in an operational rate-distortion sense, that is, for compression. Assume we choose a certain tree depth K, and for each node of the tree, a set of quantizers. Thus, given an input signal, we can evaluate an operational rate-distortion curve for each node of the binary tree. Then, we can prune the full tree based on operational rate distortion. Specifically, we introduce a Lagrange multiplier λ (as we did in bit allocation, see Section 7.1.2) and compute a cost $L(\lambda) = D + \lambda R$ for a root r and its two children c_1 and c_2. This is done at points of constant slope $-\lambda$. Then, if

$$L_r(\lambda) \; < \; L_{c_1}(\lambda) + L_{c_2}(\lambda),$$

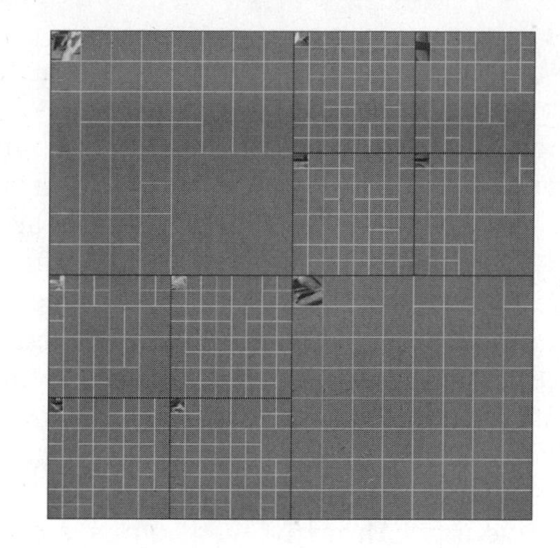

Figure 7.29 Simultaneous space and frequency splitting of the *Barbara* image using the double-tree algorithm. Black lines correspond to spatial segmentations, while white lines correspond to frequency splits.

we can prune the children and keep the root, otherwise, we keep the children. The comparison is made at constant-slope points (of slope λ) on the respective rate-distortion curves. Going up the tree in this fashion will result in an optimal binary tree for the image to be compressed. Note that in order to apply the Lagrange method, we assumed independence of the nodes, an assumption that might be violated (especially for deep trees).

An extension of this idea consists of considering not only frequency divisions (obtained by a subband decomposition) but also splitting of the signal in time, so that different wavelet packets can be used for different portions of the time-domain signal (see also Figure 3.13). This is particularly useful if the signal is nonstationary. The solution consists in jointly splitting in time and frequency using a double-tree algorithm [132, 230] (one tree for frequency and another for time splitting). Using dynamic programming and an operational rate-distortion criterion, one can obtain best time and frequency splittings. This algorithm was applied to image compression in [15]. An example of space and frequency splitting of the *Barbara* image is shown in Figure 7.29, showing that large regions with similar characteristics are gathered into blocks, while busy regions get split into many smaller blocks. Over each of these blocks, a specific wavelet packet is used.

Methods Based on Wavelet Maximums Since edges are critical to image perception [168], there is a strong motivation to find a compression scheme that contains edges as critical information. This is done in Mallat and Zhong's algorithm [184] which is based on wavelet maximums representations. The idea is to decompose the image using a redundant representation which approximates the continuous wavelet transform at scales which are powers of two. This can be done using nondownsampled octave-band filter banks. Because there is no downsampling, the decomposition is shift-invariant. If the highpass filter is designed as an edge detector (such as the derivative of a Gaussian), then we will have edges represented at all scales by some local maximums or minimums. Because the representation is redundant, keeping only these maximums/minimums still allows good reconstruction of the original using an iterative procedure (based on alternating projections onto convex sets [29, 70, 184]). While this is an interesting approach, it turns out that coding the edges is expensive. Also, textures are not easily represented and need separate treatment. Finally, the computational burden, even for reconstruction only, is heavy due to the iterative algorithm involved. Thus, such an approach needs further research in order to fully assess its potential as an image compression method.

Quantization Error Analysis in a Subband System In compression schemes we have seen so far, the approach has been to first design the linear transform and then find the best quantization and entropy coding strategies possible. The problem of analyzing the system as a whole, although of significant theoretical and practical importance, has not been addressed by many authors. One of the few works on the topic is due to Westerink, Biemond and Boekee [331]. The authors use the optimal scalar quantizer to quantize the subbands — Lloyd-Max. For that particular quantizer, it can be shown that (see, for example, [143])

$$\sigma_y^2 = \sigma_x^2 - \sigma_q^2, \tag{7.3.2}$$

where $\sigma_q^2, \sigma_x^2, \sigma_y^2$ are the variances of the quantization error, the input and output signals, respectively. Consider now a so-called "gain plus additive noise" linear model for this quantizer. Its input/output relationship is given by

$$\boldsymbol{y} = \alpha \boldsymbol{x} + \boldsymbol{r}$$

where $\boldsymbol{x}, \boldsymbol{y}$ are the input/output of the quantizer,[7] $\boldsymbol{r}$ is the additive noise term, and α is the gain factor ($\alpha \leq 1$). The main advantage of this model is that, by choosing

$$\alpha = 1 - \frac{\sigma_q^2}{\sigma_x^2}, \tag{7.3.3}$$

the additive noise will not be correlated with the signal and (7.3.2) will hold. In other words, to fit the model to our given quantizer, (7.3.3) must be satisfied. Note also, that the additive noise term is not correlated with the output signal.

[7]Bold letters denote random variables.

The authors in [331] then incorporate this model into a QMF system (where the filters are designed to cancel aliasing, as given in (3.2.34–3.2.35)). That is, each of the two channel signals are quantized, use a gain factor α_i, and generate an additive noise $\boldsymbol{r}_i$. Consequently, the error at the output of the system can be written as the sum of the error terms

$$E(z) = E_Q(z) + E_S(z) + E_A(z) + E_R(z),$$

where

$$\begin{aligned} E_Q(z) &= \frac{1}{2}[H^2(z) - H^2(-z) - 2]\, X(z), \\ E_S(z) &= \frac{1}{2}[(\alpha_0 - 1)H^2(z) - (\alpha_1 - 1)H^2(-z)]\, X(z), \\ E_A(z) &= \frac{1}{2}(\alpha_0 - \alpha_1)\, H(z)\, H(-z)\, X(-z), \\ E_R(z) &= H(z)R_0(z^2) - H(-z)R_1(z^2). \end{aligned}$$

Note that here, z^2 in $R_i(z^2)$ appears since the noise component passes through the upsampler. This breakdown into different types of errors allows one to investigate their influence and severity. Here, E_Q denotes the QMF (lack of perfect reconstruction) error, E_S is the signal error (term with $X(z)$), E_A is the aliasing error (term with $X(-z)$), and E_R is the random error. Note that only the random error E_R is uncorrelated with the signal. The QMF error is insignificant and can be disregarded. Aliasing errors become negligible if filters of length 12 or more are used. Finally, the signal error determines the sharpness while the random error is most visible in flat areas of the image.

Joint Design of Quantization and Filtering in a Subband System Let us now extend the idea from the previous section into more general subband systems. The surprising result is that by changing the synthesis filter bank according to the quantizer used, one can cancel all signal-dependent errors [162]. In other words, the reconstructed signal error will be of only one type, that is, random error, uncorrelated to the signal.

The idea is to use a general subband system with Lloyd-Max quantization and see whether one can eliminate certain types of errors. Note that here, no assumptions are made about the filters, that is, filters (H_0, H_1) and (G_0, G_1) do not constitute a perfect reconstruction pair. Assume, however, that given (H_0, H_1), we find (T_0, T_1) such that the system is perfect reconstruction. Then, it can be shown that if the synthesis filters are chosen as

$$G_0(z) = \frac{1}{\alpha_0} T_0(z), \qquad G_1(z) = \frac{1}{\alpha_1} T_1(z),$$

where α_i are the gain factors of the quantizer models, all errors depending on $X(z)$ and $X(-z)$ are cancelled and the only remaining error is the random error

$$E(z) = E_R(z) = \frac{1}{\alpha_0}T_0(z)R_0(z^2) + \frac{1}{\alpha_1}T_1(z)R_1(z^2),$$

where $R_i(z)$ are the noise terms appearing in the linear model. In other words, by appropriate choice of synthesis filters, the only remaining error is uncorrelated to the signal. The potential benefit of this approach is that one has to deal only with a random, noise-like error at the output, which can then be alleviated with an appropriate noise removal technique. Note, however, that the random error has been boosted by dividing the terms by $\alpha_i \leq 1$. For more details, see [162].

Nonorthogonal Subband Coding Most of the subband coding literature uses orthogonal filters, since otherwise the squared norm of the quantization error would not be preserved leading to a possibly large reconstruction error. If nonorthogonal transforms are used, they are usually very close to the orthogonal ones [14].

Moulin in [200] shows that the fact that nonorthogonal transforms do not perform well when compared to orthogonal ones, is due to an inappropriate formulation of the coding problem, rather than to the use of the nonorthogonal transform itself.

Let us recall how the usual subband decomposition/reconstruction is performed. We have an image x, going through the analysis stage H, to produce subband images

$$\boldsymbol{y} = \boldsymbol{H}\boldsymbol{x}.$$

The next step is to compute a quantized image $\hat{\boldsymbol{y}}$,

$$\hat{\boldsymbol{y}} = Q(\boldsymbol{y}).$$

Finally, we reconstruct the image as

$$\hat{\boldsymbol{x}} = \boldsymbol{G}\hat{\boldsymbol{y}},$$

where the system is perfect or near-perfect reconstruction. Moulin, instead, suggests the following: Find $\hat{\boldsymbol{y}}$ that minimizes the squared error at the output

$$E(\hat{\boldsymbol{y}}_{opt}) = \|\boldsymbol{G}\hat{\boldsymbol{y}}_{opt} - \boldsymbol{x}\|^2,$$

where $\hat{\boldsymbol{y}}_{opt}$ belongs to the set of all possible quantized images. Due to this constraint, the problem becomes a discrete optimization problem and is solved using a numerical relaxation algorithm. Experiments on images show significant visual as well as MSE improvement. For more details, refer to [200].

7.4 Video Compression

Digital video compression has emerged as an area of intense research and development activity recently. This is due to the demand for new video services such as high-definition television, the maturity of the compression techniques, and the availability of technology to implement state of the art coders at reasonable costs. Besides the large number of research papers on video compression, good examples of the increased activity in the field are the standardization efforts such as MPEG [173, 201] (the Moving Pictures Experts Group of the International Standardizations Organization). While the video compression problem is quite different from straight image coding, mainly because of the presence of motion, techniques successful with images are often part of video coding algorithms as well. That is, signal expansion methods are an integral part of most video coding algorithms and are used in conjunction with motion based techniques.

This section will discuss both signal expansion and motion based methods used for moving images. We start by describing the key problems in video compression, one of which is compatibility between standards of various resolutions and has a natural answer in multiresolution coding techniques. Standard motion compensated video compression is described next, as well as the use of transforms for coding the prediction error signal. Then, pyramid coding of video, which attempts to get the best of subband and motion based techniques, is discussed. Subband or wavelet decomposition techniques in three dimensions are presented, indicating both their usefulness and their shortcomings. Finally, the emerging MPEG standard is discussed.

Note that by *intraframe coding* we will denote video coding techniques where each frame is coded separately. On the other hand, *interframe coding* will mean that we take the time dimension and the correlation between frames into account.

7.4.1 Key Problems in Video Compression

Video is a sequence of images, that is, a three-dimensional signal. A number of key features distinguishes video compression from being just a multidimensional extension of previously discussed compression methods. Moreover, the data rates are several orders of magnitude higher than those in speech and audio (for example, digital standard television uses more than 200 Mbits/sec, and high-definition television more than 1 Gbits/sec).

Motion Models in Video The presence of structures related to motion in the video signal indicates ways to achieve high compression by using model based processing. That is, instead of looking at the three-dimensional video signal as

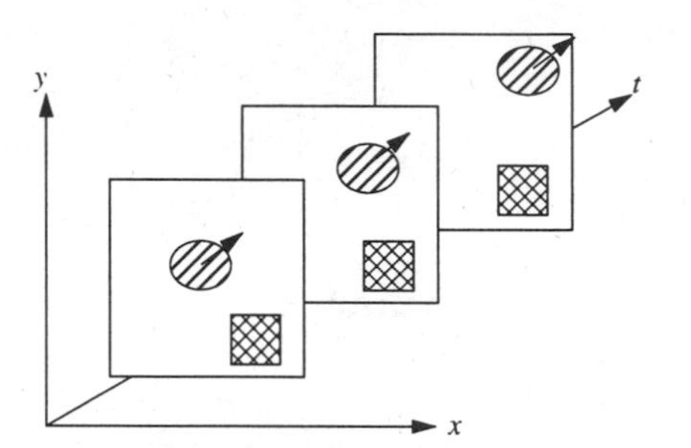

Figure 7.30 Moving objects in a video sequence. One object is still — zero motion, whereas the other has a purely translational motion.

simply a sequence of images, one knows that very often, future images can be deduced from the past ones by some simple transformation such as translation. This is shown schematically in Figure 7.30, where two objects appear in front of a uniform background, one being still (no motion) and the other moving (simple, translational motion). It is clear that a compact description of this scene can be obtained by describing the first image and then indicating only how the objects move in subsequent images. It turns out that most video scenes are well described by such motion models of objects, as well as global modifications such as zooms and pans. Of course, a number of problems have to be addressed such as occlusion or uncovering of background due to an object's movement. Overall, the motion based approaches in video processing have been very successful [207]. Note that motion is an "image-domain" phenomenon, since we are looking for displacements of image features. Thus, many of the motion estimation algorithms are of a correlative nature. An example is the block matching algorithm, which searches for local correlation maximums between successive images.

A Transform-Domain View Assume the following simplified view of video: a single object has a translational motion in front of a black background. One can verify that the three-dimensional Fourier transform is zero except on a plane orthogonal to the motion vector and passing through the origin. The values on the plane are equal to the two-dimensional Fourier transform of the object. That is, motion simply tilts the Fourier transform of a still object. It seems therefore attractive to code the moving object in Fourier space, where the coding would reduce to coding of the object's Fourier transform and the direction of the plane. This idealized view has lead to various proposals for video coding which would first include an appropriate transform domain approximating Fourier space (such as a subband division) and then locate the region where the energy is mostly concentrated (corresponding to the tilted plane of the object). It would then disregard other Fourier components to achieve compression. While such an approach seems attractive at first sight, it has some shortcomings.

First, real video scenes do not match the model. The background, which has an "untilted" Fourier transform, gets covered and uncovered by the moving object, creating spurious frequencies. Then, there are usually several moving objects with different motions, thus several tilted planes would be necessary. Finally, most of the transforms proposed (such as N-band subband division where N is not a large integer for complexity reasons) partition the spectrum coarsely and thus, they cannot approximate the tilted plane very well.

Since coding the spectrum requires coding of one image (or its two-dimensional spectrum) plus the direction of the tilted plane, staying in the sequence domain will perform just as well. Note also that motion is easier to analyze in the image plane rather than the Fourier domain. The argument is simple; compare two images where an object has moved. In the image plane, it is a localized phenomenon described by a single motion vector, while in spectral domain, it results in a different phase shift of every Fourier component.

The Perceptual Point of View Just as in coding of speech or images, the ultimate judge of quality is the human observer. Therefore, spatio-temporal models of the human visual system (HVS) are important. These turn out to be more complex than for static images, especially because of spatio-temporal masking phenomena related to motion. If one considers sensitivity to spatio-temporal gratings (sinusoids with an offset and various frequencies in all three dimensions), then the eye has a lowpass/bandpass characteristic [207]. The sensitivity is maximum at medium spatial and temporal frequencies, falls off slightly at low frequencies, and falls off rapidly toward high frequencies (note that the sensitivity function is not separable in space and time). Finally, sinusoids separated by more than an octave in spatial frequency are treated in an independent manner.

Masking does occur, but it is a very local effect and cannot be well modeled in the frequency domain. This masking is both spatial (reduced sensitivity at sharp transitions) and temporal (reduced sensitivity at scene changes). The perception of motion is a complex phenomenon and psychophysical results are only starting to be applicable to coding. One effect is clear and intuitive however: The perception of a moving object depends on if it is tracked by the eye or not. While in the latter case, the object could be blurred without noticeable effect, in the former, the object will be perceived as accurately as if it were still. Since it cannot be predicted if the viewer will or will not follow the object, one cannot increase compression of moving objects by blurring them. This somewhat naive approach has sometimes been suggested in conjunction with three-dimensional frequency-domain coding methods, but does not work, since more often than not, the interest of the viewer is in the moving object.

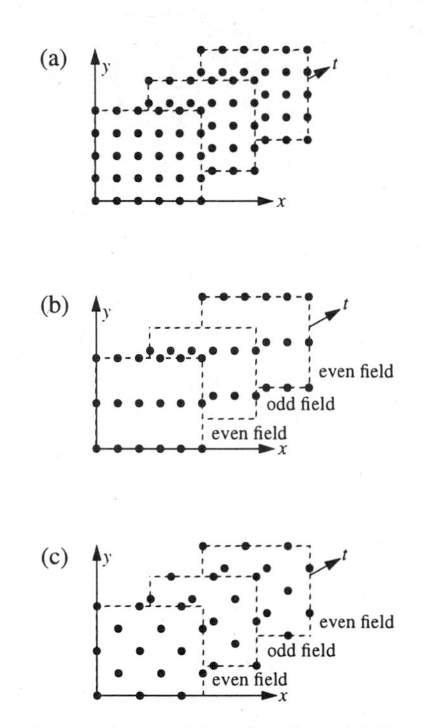

Figure 7.31 Scanning modes used in television. (a) Progressive scanning, which corresponds to the ordinary rectangular lattice. (b) Interlaced scanning, which samples alternately even and odd lines. It corresponds to the quincunx lattice in the (vertical, time)-plane. (c) Face centered orthorhombic (FCO) lattice, which is the true three-dimensional downsampling by two of the rectangular lattice.

Progressive and Interlaced Scanning When thinking of sampling a three-dimensional signal, the most natural sampling lattice seems to be the rectangular lattice, as shown in Figure 7.31(a). The scanning corresponding to this lattice is called *progressive scanning* in television cameras and displays. However, for historical and technological reasons, a different sampling called *interlaced scanning* is often used. It corresponds to a quincunx lattice in the (vertical, time)-plane and its shifted versions along the horizontal axis, as shown in Figure 7.31(b). The name *interlaced* comes from the fact that even and odd lines are scanned alternately. A set of even or odd lines is called a *field*, and two successive fields form a *frame*.

While interlacing complicates a number of signal processing tasks such as motion estimation, it represents an interesting compromise between space and time resolutions for a given number of sampling points in a space-time volume. Typically, high frequencies in both vertical and time dimensions cannot be represented, but this loss in resolution is not very noticeable. Progressive scanning would have to reduce the sampling rate by two in either dimension in Figure 7.31(a) to achieve the same density as in Figure 7.31(b), which is more noticeable than to resort to interlacing.

An even better compromise would be obtained with the face-centered orthorhombic (FCO) lattice [165], which is the true generalization of the two-dimensional quincunx lattice to three dimensions (see Figure 7.31(c)). Then, only frequencies which are high in all three dimensions simultaneously are lost, and these are not well perceived anyway. However, for technological reasons, FCO is less attractive than interlaced scanning. Of course, in the various sampling schemes discussed above, one can always construct counter examples that lose resolution, in particular when tracked by the human observer (for example, objects with high frequency patterns moving in a worst case direction). However, these counter examples are unlikely in real world imagery, particularly for interlaced and even more for FCO scanning.[8]

Compatibility In three-dimensional imagery such as television and movies, the issue of compatibility between various standards, or at least easy transcoding, has become a central issue. For many years, progressive scanning used in movies and interlaced scanning used in television and video had an uneasy coexistence, just as the 50 Hz frame rate for television in Europe versus 60 Hz frame rate for television in US and Japan. Some ad hoc techniques were used to transcode from one standard to another, such as the so-called 2/3 pull-down to go from 24 Hz progressively scanned movies to 60 Hz interlaced video.

The advent of digital television with its potential for higher quality, as well as the development of new formats (usually referred to as high definition television or, HDTV) has pushed compatibility to the forefront of current concerns.

Conceptually, multiresolution techniques form an adequate framework to deal with compatibility issues [323]. For example, standard television can be seen as a subresolution of high definition television (although this is a very rough approximation), but with added problems such as different aspect ratios (the ratio of width and height of the picture). However, there are two basic problems which make the problem difficult:

Sublattice property Unless the lower-resolution scanning standard is a sublattice of the higher-resolution one, it cannot be used directly as a subresolution signal in a multiresolution scheme such as a subband coder. Consider the following two examples in Figure 7.32.

[8]The famous backward turning wagon wheels in movies provide an example of aliasing in progressive scanning which could only be avoided by blurring in time.

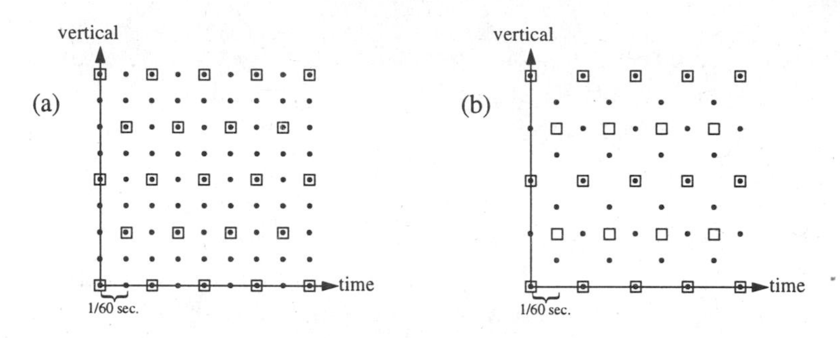

Figure 7.32 Sublattice property for compatibility (the (vertical, time)-plane is shown). The "•" represents the original lattice, and the squares the sparser lattice. (a) 1024 × 1024 progressive, 60 Hz versus 512 × 512 interlaced, 60 Hz. The sublattice property is verified. (b) 1024 × 1024 interlaced, 60 Hz versus 512 × 512 interlaced, 60 Hz. The sublattice property is not verified.

First, take as full resolution a 1024 × 1024 progressive sequence at 60 Hz, with a 512 × 512 interlaced sequence at 60 Hz as subresolution (note that 60 Hz is the frame and field rate in the progressive and interlaced case, respectively). The latter exists on a sublattice of the former, namely, by downsampling by two in the horizontal and vertical dimension, followed by quincunx downsampling in the (vertical, time)-plane (see Figure 7.32(a)).

The second example starts with a 1024 × 1024 interlaced sequence at 60Hz and one would like to obtain a 512 × 512 interlaced one at 60Hz as well (see Figure 7.32(b)). Half of the points have to be interpolated, since the latter scanning is not a sublattice of the former. It can still be used as a coarse resolution in a pyramid coder, but cannot be used as one of the channels in subband coding.

Compatibility as an overconstraint Sometimes, it is stated that all video services from videotelephone to HDTV should be embedded in one another, somewhat like Russian dolls. That is, the whole video hierarchy can be progressively built up from the simplest to the most sophisticated. However, the successive refinement property is a constraint with a price [93] and a complete refinement property with some stringent bit rates requirements (for example, videotelephone at 64 Kbits/sec, standard television at 5 Mbits/sec and HDTV at 20 Mbits/sec) is quite constrained and might not lead to the best quality pictures. This is because each of the individual rates is a difficult target in itself, and the combination thereof can be an overconstrained problem.

While we will discuss compatibility issues and use multiresolution techniques as a possible technique to address the problems, we want to point out that there is no panacea. Each case of compression with compatibility requirement has to be carefully addressed essentially from scratch.

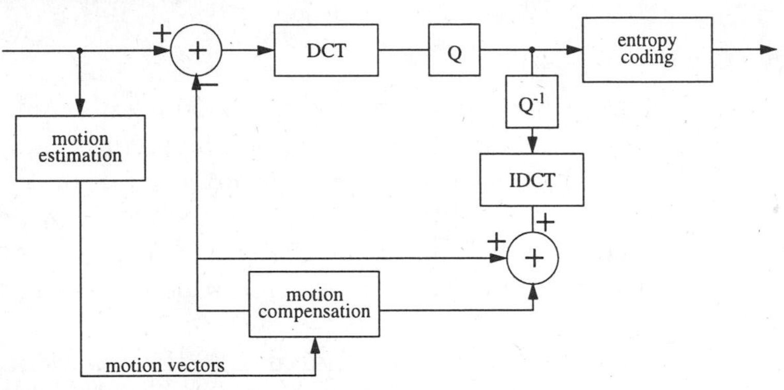

Figure 7.33 Hybrid motion-compensated predictive DCT coding.

7.4.2 Motion-Compensated Video Coding

As discussed above, motion models allow a compact description of moving imagery and motion prediction permits high compression. Typically, a future frame is predicted from past frames using local motion information. That is, a particular $N \times N$ block of the current frame to be coded is predicted as a displaced $N \times N$ block from the previous reconstructed frame and the prediction error is compressed using techniques such as transform coding. The decoder can construct the same prediction and add it to the decoded prediction error. Such a scheme is essentially an adaptive DPCM over the time dimension, where the predictor is based on motion estimation. Figure 7.33 shows such a scheme, which is called *hybrid motion-compensated predictive DCT video coding* and is part of several standard coding algorithms [177].

As can be seen in Figure 7.33, the prediction error is compressed using the DCT, even though there is little correlation left in the prediction error on average.

Note also that the DCT could be replaced by another expansion such as subbands (see Figure 7.39(b)). Because of its resemblance to a standard coder, the approach will work. However, because motion compensation is done on a block-by-block basis (for example, in block matching motion compensation), there can be a block structure in the prediction error. Thus, choosing a DCT of the same block size is a natural expansion, while taking an expansion that crosses the boundaries could suffer from that blocking structure (which creates artificially high frequencies). It should not be forgotten, however, that the bulk of the compression comes from the motion compensation loop using accurate motion estimates and thus, replacing the DCT by a LOT or a discrete wavelet transform can improve the performance, but not dramatically.

7.4.3 Pyramid Coding of Video

The difficulty of including motion in three-dimensional subband coding will be discussed shortly. It turns out that it is much easier to include motion in pyramid coding, due to the fact that the prediction or interpolation from low resolution to full resolution (see Figure 7.18) can be an arbitrary predictor [9], such as a motion based one. This is a general idea which can be used in various forms for video compression and we will describe a particular scheme as an example.

This video compression scheme was studied in [301, 302, 303]. Consider a progressive video sequence and its subresolutions, obtained by spatial filtering and downsampling as well as frame skipping over time. Note that filtering over time would create so-called "double images" when there is motion and thus straight downsampling in time is preferable. This is shown schematically in Figure 7.34(a), where the resolution is decreased by a factor of two in each dimension between one level of the pyramid and the next. Now we apply the classic pyramid coding scheme, which consists of the following:

(a) Coding the low resolution..

(b) Predicting the higher resolution based on the coded low resolution.

(c) Taking the difference between the predicted and the true higher resolution, resulting in the prediction error.

(d) Coding the prediction error.

While these steps could be done in the three dimensions at once, it is preferable to separate the spatial and temporal dimensions. First, the spatial dimension is interpolated using filtering and then the temporal dimension is interpolated using motion-based interpolation. This is shown in Figure 7.34(b). Following each interpolation step, the prediction error is computed and coded and this coded value is added to the prediction before going to the next step. Because at each step, we use coded versions for our prediction, we have a pyramid scheme with quantization noise feedback, as was described in Figure 7.19. Therefore, there is only one source of error, namely the compression of the last prediction error.

The oversampling inherent in pyramid coding is not a problem in the three-dimensional case, since, following (3.5.4), we have a total number of samples which increases only as

$$(1 + \frac{1}{8} + \frac{N}{8^2} + \cdots)N \; < \; \frac{8}{7}N,$$

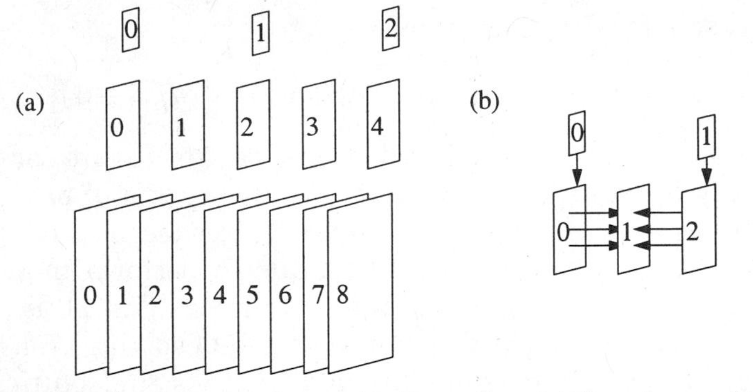

Figure 7.34 Spatio-temporal pyramid video coding. (a) Three layers of the pyramid, corresponding to three resolutions. (b) Prediction of the higher resolution. The spatial resolution is interpolated first (using linear filtering) and then the temporal resolution is increased using motion interpolation.

or at most 14%, since every coarser level has only 1/8th the number of samples of its predecessor.

The key technique in the spatio-temporal pyramid scheme is the motion interpolation step, which predicts a frame from its two neighbors based on motion vectors. Assume the standard rigid-object and pure translational motion model [207]. If we denote the intensity of a pixel at location $\boldsymbol{r} = (x, y)$ and time t by $I(\boldsymbol{r}, t)$, we are looking for a mapping $\boldsymbol{d}(\boldsymbol{r}, t)$ such that we can write:

$$I(\boldsymbol{r}, t) = I(\boldsymbol{r} - \boldsymbol{d}(\boldsymbol{r}, t), t - 1).$$

If motion is not changing over time, we also have:

$$I(\boldsymbol{r}, t) = I(\boldsymbol{r} + \boldsymbol{d}(\boldsymbol{r}, t), t + 1).$$

The goal is to find the function $\boldsymbol{d}(\boldsymbol{r}, t)$, that is, estimate the motion. This is a standard estimation procedure, where some simplifying assumptions are made (such as constant motion over a neighborhood). Typically, for a small block $\boldsymbol{b}$ in the current frame, one searches over a set of possible motion vectors such that the sum of squared differences,

$$\sum_{\boldsymbol{r} \in \boldsymbol{b}} |I(\boldsymbol{r}, t) - \hat{I}(\boldsymbol{r}, t)|^2, \tag{7.4.1}$$

is minimized, where

$$\hat{I}(\boldsymbol{r}, t) = I(\boldsymbol{r} - \boldsymbol{d}_{\boldsymbol{b}}, t - 1), \tag{7.4.2}$$

corresponds to a block in the previous frame displaced by $\boldsymbol{d}_{\boldsymbol{b}}$ (the motion for the block under consideration in the current frame). It is best to actually perform a symmetric search by considering the past (as in (7.4.2)), the future ((7.4.2) with sign reversals for d_b), and the average,

$$\hat{I}(r,t) = \frac{1}{2}[I(r - d_b, t-1) + I(r + d_b, t+1)],$$

and then to choose the best match. Choosing past or future for the interpolation is especially important for covering and uncovering of background due to moving objects, as well as in case of abrupt changes (scene changes).

Interestingly, a very successful technique to perform motion estimation (that is, finding the displacement d_b that minimizes (7.4.1)) is based on multiresolution or successive approximation. Instead of solving (7.4.1) directly, one solves a coarse version of the same problem, refines the solution (by interpolating the motion vector field), and uses this new field as a starting point for a new, finer search. This is not only computationally less complex, but also more robust in general [31, 302]. It is actually a regularization of the motion estimation problem.

As an illustration of this video coding scheme, a few representative pictures are shown. First, Figure 7.35 shows the successive refinement of the motion vector field, which starts with a sparse field on a coarse version and refines it to a fine field on the full-resolution image. In Figure 7.36, we show the resulting spatial and temporal prediction error signals. As can be seen, the spatial prediction error has higher energy than the temporal one, which shows that temporal interpolation based on motion is quite successful (actually, this sequence has high frequency spatial details, which cannot be well predicted from the coarse resolution).

A point to note is that the first subresolution sequence (which is downsampled by 2 in each dimension) is of good visual quality and could be used for a compatible coding scheme. This coding scheme was implemented for high quality coding of HDTV with a compatible subchannel and it performed well at medium compression (of the order of 10-15 to 1) with essentially no visible degradation [301, 303].

7.4.4 Subband Decompositions for Video Representation and Compression

Decompositions for Representation We will discuss here two ways of sampling video by 2; the first, using quincunx sampling along (vertical, time)-dimensions and the second, true three-dimensional sampling by 2, using the FCO sampling lattice.

Quincunx sampling for scanning format conversions We have outlined previously the existence of different scanning standards (such as interlaced and progressive) as well as the desire for compatibility. A simple technique to deal with these problems is to use perfect reconstruction filter banks to go back and forth between

Figure 7.35 Multiresolution motion vector fields used in the interpolation. Each corresponds to a layer in the pyramid, with coarse (top left), medium (top right) and fine (bottom) resolutions.

progressive and interlaced scanning, as shown in Figure 7.37 [320]. This is achieved by quincunx downsampling the channels in the (vertical, time)-plane. Properly designed filter pairs (either orthogonal or biorthogonal solutions) lead to a lowpass channel that is a usable interlaced sequence, while the original sequence can be perfectly recovered when using both the lowpass and highpass channels in the reconstruction. This is a compatible solution in the following sense: A low-quality receiver would only decode the lowpass channel and thus show an interlaced sequence, while a high-quality receiver would synthesize a full resolution progressive sequence based on both the lowpass and the highpass

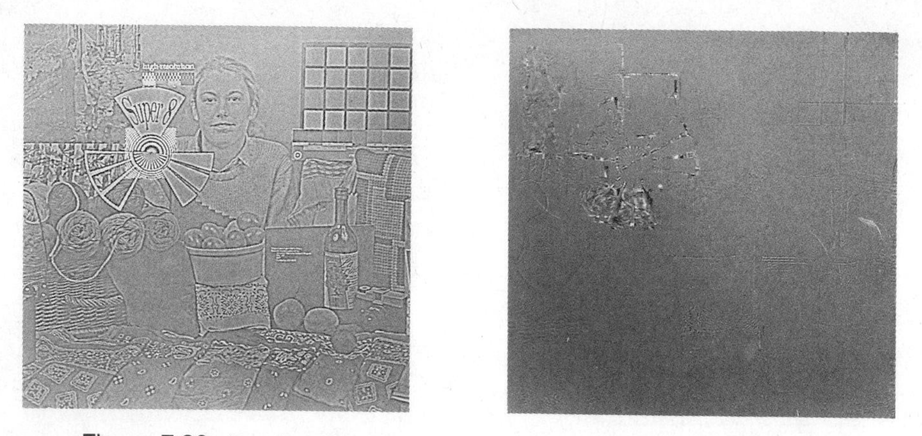

Figure 7.36 Results of spatio-temporal coding of video (after [301]). The spatial (left) and temporal (right) prediction errors are shown. The reconstruction (not shown) is indistinguishable from the original at the rate used in this experiment (around 1.0 bits/pixel).

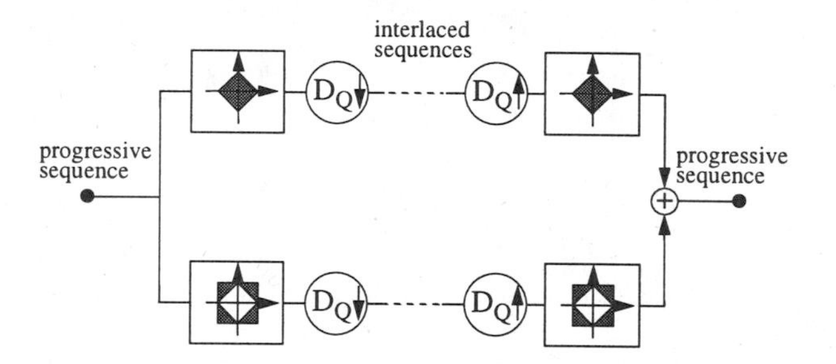

Figure 7.37 Progressive to interlaced conversion using a two-channel perfect reconstruction filter bank with quincunx downsampling.

channels.

If one starts with an interlaced sequence, one can obtain a progressive sequence by quincunx downsampling. Thus, an interlaced sequence can be broken into lowpass and highpass progressive sequences, again allowing perfect reconstruction when perfect reconstruction filter banks are used. This is a very simple, linear technique to produce a deinterlaced sequence (the lowpass signal) as well as a helper signal (the highpass signal) from which to reconstruct the original signal. While more powerful, motion based techniques can produce better results, the above technique is attractive because of its low complexity and the fact that no motion model needs to be assumed.

Perfect reconstruction filter banks for these applications, in particular having low complexity, have been designed in [320]. Both orthogonal and biorthogonal solutions are given. As an example, we give the two-dimensional impulse responses of a simple linear phase filter pair,

$$h_0[n_1,n_2] = \begin{pmatrix} & & -1 & & \\ & -2 & 4 & -2 & \\ -1 & 4 & 28 & 4 & -1 \\ & -2 & 4 & -2 & \\ & & -1 & & \end{pmatrix}, \qquad h_1[n_1,n_2] = \begin{pmatrix} & 1 & \\ 1 & -4 & 1 \\ & 1 & \end{pmatrix}, \tag{7.4.3}$$

which are lowpass and highpass filters, respectively. Since it is a biorthogonal pair, the synthesis filters (if the above are used for analysis) are obtained by modulation with $(-1)^{(n_1+n_2)}$ and thus, the roles of lowpass and highpass are reversed (see also Problem 7.7).

FCO sampling for video representation We mentioned previously that using the FCO lattice (depicted in Figure 7.31(c)) might produce visually more pleasing sequences if a data reduction by two is needed. This is due in part to the fact that an ideal lowpass in the FCO case would retain more of the energy of the original signal than the corresponding quincunx lowpass filter. Actually, assuming that the original signal has a spherically uniform spectrum, and that the ideal lowpass filters are Voronoi regions both in the quincunx and the FCO cases, the quincunx lowpass would retain 84.3% of the original spectrum, while the FCO lowpass would retain 95.5% of the original spectrum [165].

To evaluate the gain of processing a video signal with a true three-dimensional scheme when a data rate reduction of two is needed, we can use a two-channel perfect reconstruction filter bank [165]. The sampling matrix is

$$\boldsymbol{D}_{FCO} = \begin{pmatrix} 1 & 0 & 1 \\ -1 & -1 & 1 \\ 0 & -1 & 0 \end{pmatrix},$$

and the perfect reconstruction filter pair is a generalization of the above diamond-shaped quincunx filters to three dimensions. To compare the low bands obtained in this manner, they are interpolated back to the original lattice, since we cannot observe the FCO output directly. Upon observing the result, the conclusion is that FCO produces visually more pleasing sequences. For more detail, see [165].

Three-Dimensional Subband Decomposition for Compression A straightforward generalization of separable subband decomposition to three dimensions is shown in Figure 7.38, with the separable filter tree shown in part (a) and slicing of the spectrum given in part (b) [153]. In general, most of the energy will be contained in the band that has gone through lowpass filtering in all three directions thus iterating the decomposition on this band is most natural. This is

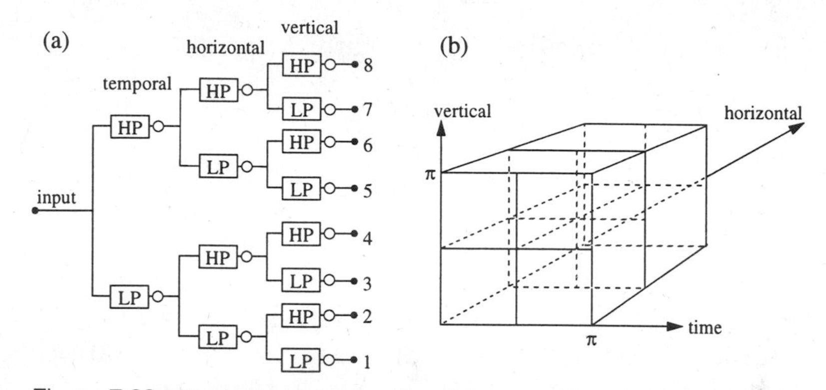

Figure 7.38 Three-dimensional subband decomposition of video. (a) Separable filter bank tree. LP and HP stand for lowpass and highpass filtering, respectively, and the circle indicates downsampling by two. (b) Slicing of the three-dimensional spectrum.

actually a three-dimensional discrete-time wavelet decomposition and is used in [153, 224]. Such three-dimensional decompositions work best for isotropic data, such as tomographic images used in medical imaging or multispectral images used in satellite imagery. In that case, the same filters can be used in each dimension, together with the same compression strategy (at least as a first approximation).

As we said, in video sequences, time should be treated differently from the spatial dimensions. Typically, only very short filters are used along time (such as Haar filters given in (3.1.2) and (3.1.17)) since long filters will smear motion in the lowpass channel and create artificial high frequencies in the highpass channel. If one looks at the output of a three-dimensional subband decomposition, one can note that the lowpass version is similar to the original and the only other channel with substantial energy is the one containing a highpass filter over time followed by lowpass filters in the two spatial dimensions. This channel contains energy every time there is substantial motion and can be used as a motion indicator.

While motion-compensated methods can outperform subband decompositions over time, recently, there have been some promising results [223, 286]. Also, it is a simple, low-complexity method and can easily be used in a joint source-channel coding environment because of the natural ordering in importance of the subbands [323]. Subband representation is also very convenient for hierarchical decomposition and coding [35] and has been used for compression of HDTV [336].

Motion and Subband Coding Intuitively, instead of lowpass and highpass filtering along the time axis, one should filter along the direction of motion instead.

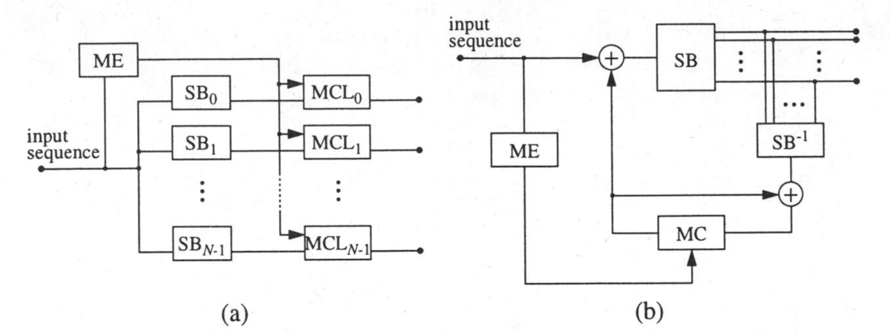

Figure 7.39 Motion-compensated subband coding. SB: subband, ME: motion estimation, MC: motion compensation, MCL: motion-compensation loop. (a) Motion compensation of each subband. (b) Subband decomposition of the motion-compensated prediction error.

Then, motion itself would not create artificial high frequencies as it does in straight three-dimensional subband coding. This view, although conceptually appealing, is difficult to translate into practice, except in very limited cases (such as panning, which corresponds to a single translational motion). In general, there are different motion trajectories as well as covering and uncovering of background by moving objects. Thus, subband decomposition along motion trajectories is not a practical approach (see [167] for further discussions on this topic).

Instead, one has to go back to more traditional motion-compensation techniques and see how they fit into a subband coding framework or, conversely, how subband coding can be used within a motion-compensated coder [110]. Consider inclusion of motion compensation into a subband decomposition. That is, instead of processing the time axis using Haar filters, we use a motion-compensation loop in each of the four spatial bands. One advantage is that the four channels are now treated in an independent fashion. While this scheme should perform better than the straight three-dimensional decomposition, it also has a number of drawbacks. First, motion compensation requires motion estimation. If it is done in the subbands, it is less accurate than the motion estimates obtained from the original sequence. Also, motion estimation in the high frequency subbands will be difficult. Thus, motion estimation should probably be done on the original sequence and the estimates then used in each band after proper rescaling (see Figure 7.39(a)). One of the attractive features of the original scheme, namely that motion processing is done in parallel and at a lower resolution, is thus partly lost, since motion estimation is now shared. Moreover, it is hard to perform motion compensation in the high frequency subbands, since they mostly consist of edge information and thus slight motion errors lead to large prediction errors.

Table 7.10 Comparison of subband and pyramid coding of video. N is the number of channels in the subband decomposition and δ is the quantizer step size.

Method	Subband	Pyramid
Oversampling	0%	< 14%
Maximum coding error	$\sqrt{N}\delta$	δ
Subchannel quality	Limited	Good
Inclusion of motion	Difficult	Easy
Nonlinear processing	Difficult	Easy
Model-based processing	Difficult	Easy
Encoding delay	Moderate	Large

As can be been from the above discussion, motion compensation in the subbands is not easy. An intuitive explanation is the following: motion, that is, translation of objects, is a sequence-domain phenomenon. Going to a subband domain is similar to going into frequency domain, but there, translation is a complex phenomenon, with different phase factors at different frequencies. This shows that motion estimation and compensation is more difficult in the subband domain than in the original sequence domain.

Consider the alternative of using subband decomposition within a motion-compensated coder, as shown in Figure 7.39(b). The subband decomposition is used to decompose the prediction error signal spatially and replaces simply the DCT which is usually present in such a hybrid motion-compensated DCT coder. This approach was discussed in Section 7.4.2, where we indicated its feasibility, but also some of its possible shortcomings.

Comparison of Subband and Pyramid Coding for Video Because both subband and pyramid coding of video are three-dimensional multiresolution decompositions, it is natural to compare them. A slight disadvantage of pyramid over subband coding is the oversampling; however, it is small in this three-dimensional case. Also, the encoding delay is larger in pyramid coding than in subband coding. On all other counts, pyramid coding turns out to be advantageous when compared to subband coding, a somewhat astonishing fact considering the simplicity of the pyramid approach. First, there is an easy control of quantization error, using the quantization error feedback and this leads to a tight bound on a maximum possible error, unlike in transform or subband coders. Second, the inclusion of motion, which we discovered to be difficult in subband coding, is very simple in a pyramidal scheme, as demonstrated in the spatio-temporal scheme discussed previously. The quality of a compatible subchannel is limited in a subband scheme due to the constrained filters that are used. In the pyramid case, however, the freedom on the filters involved both before downsampling and for interpolation can be used to obtain visually pleasing coarse resolutions as well as good quality interpolated versions, a useful feature for compatibility. The above comparison is summarized in Table 7.10.

7.4.5 Example: MPEG Video Compression Standard

Just as in image compression, where several key ideas led to the JPEG standard (see Section 7.3.1), the work on video compression led to the development of a successful standard called MPEG [173, 201]. Currently, MPEG comes in two versions, namely a "coarse" version called MPEG-I (for noninterlaced television at 30 frames/second, and a compressed bit rate of the order of 1 Mbits/sec) and a "finer" version named MPEG-II (for 60 fields/sec regular interlaced television, and a compressed bit rate of 5 to 10 Mbits/sec). The principles used in both versions are very similar and we will concentrate on MPEG-I in the following. What makes MPEG both interesting and powerful is that it combines several of the ideas discussed in image and video compression earlier in this chapter. In particular, it uses both hybrid motion-compensated predictive DCT coding (for a subset of frames) and bidirectional motion interpolation (as was discussed in the context of video pyramids). But first, it segments the infinite sequence of frames into temporal blocks called group of pictures (GOP). A GOP typically consists of 15 frames (that is, half a second of video). The first frame of a GOP is coded using standard image compression and no prediction from the past frames (this decouples the GOP from the past and allows one to decode a GOP independently of other GOP's). This intraframe coded image — *I-frame*, is used as the start frame of a motion-compensation loop which predicts every N-th frame in the GOP where N is typically two or three. The predicted frames (*P-frames)* are then used together with the I-frame in order to interpolate the $N-1$ intermediate frames (called *B-frames* because the interpolation is bidirectional) between the P-frames. A GOP, the various frame types, and their dependencies are shown in Figure 7.40.

Both the intraframe and the various prediction errors (corresponding to the difference between the true frame and its prediction either from the past or from its neighbors in the P and B case, respectively) are compressed using a JPEG-like standard (DCT, quantization with an appropriate quantization matrix, and zigzag scanning with entropy coding). One important difference, however, is that the quantization matrix can be scaled by a multiplicative factor and this factor is sent as overhead. This allows a coarse form of adaptive quantization if desired.

A key for good compression performance is good motion estimation/prediction. In particular, motion can be estimated at different accuracies (motion

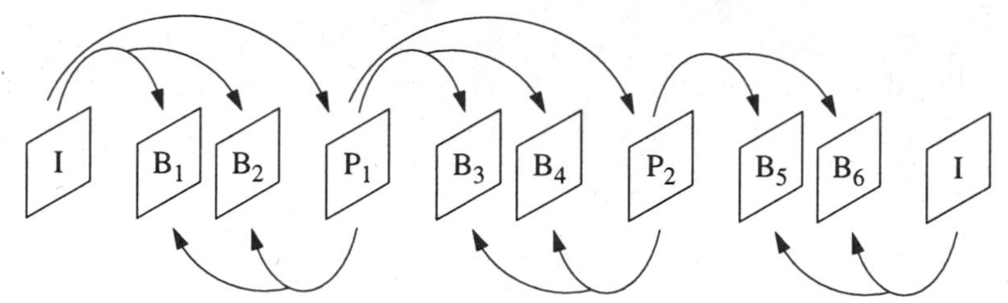

Figure 7.40 A group of pictures (GOP) in the MPEG video coding standard. I, P, and B stand for intra, predicted and bidirectionally interpolated frames, respectively. There are nine frames in this GOP, with two B-frames between every P-frame. The arrows show the dependencies between frames.

by integer pixel distances, or finer, subpixel accuracy). Of course, finer motion information increases the overhead to be sent to the decoder, but typically, the reduction in prediction error justifies this finer motion estimation and prediction. For example, it is common to use half-pixel accuracy motion estimation in MPEG.

7.5 Joint Source-Channel Coding

The source coding methods we have discussed so far are used in order to transport information (such as a video sequence) over a channel with limited capacity (such as a telephone line which can carry up to 20 Kbits/sec). In many situations, source coding can be performed separately from channel coding, which is known as the *separation principle* of source and channel coding. For example, in a point-to-point transmission using a known, time-invariant channel such as a telephone line, one can design the best possible channel coding method to approach channel capacity, that is, achieve a rate R in bits/sec such that $R \leq C$ where C is the channel capacity [258]. Then, the task of the source compression method is to reduce the bit rate so as to match the rate of the channel.

However, there exist other situations where a separation principle cannot be used. In particular, when the channel is time-varying and there is a delay constraint, or when multiple channels are present as in broadcast or multicast, it can be advantageous to jointly design the source and channel coding so that, for example, several transmission rates are possible.

The development of such methods is beyond the scope of this book. As an example, the case of multiple channels falls into a well studied branch of information theory called multiuser information theory [66]. Instead, we will show several examples indicating how multiresolution source coding fits naturally into joint source-channel coding methods. In all these examples, the transmission, or channel coding, uses a principle we call *multiresolution transmission* and can be seen as the dual of multiresolution source coding.

Multiresolution transmission is based on the idea that a transmission system can operate at different rates, depending on the channel conditions, or that certain bits will be better protected than others in case of adverse channel conditions. Such a behavior of the transmission system can be achieved using different techniques, depending on the transmission media. For example, unequal error protection codes can be used, thus making certain bits more robust than others in the case of a noisy channel. The combination of such a transmission scheme with a multiresolution source coder is very natural. The multiresolution source coder segments the information into a part which reconstructs a coarse, first approximation of the signal (such as the lowpass channel in a subband coder) as well as a part which gives the additional detail signal (typically, the higher frequencies). The coarse approximation is now sent using the highly protected bits and has a high probability of arriving successfully, while the detail information will only arrive if the channel condition is good. The scheme generalizes to more levels of quality in an obvious manner. This intuitive matching of successive approximation of the source to different transmission rates, depending on the quality of the channel, is called *multiresolution joint source-channel coding*.

7.5.1 Digital Broadcast

As a first example, we consider digital broadcast. This is a typical instance of a multiuser channel, since a single emitter sends to many users, each with a different channel. One can of course design a digital communication channel that is geared to the worst case situation, but that is somewhat of a waste for the users with better channels. For simplicity, consider two classes of users U_1 and U_2 having "good" and "bad" channels, with capacities $C_1 > C_2$, respectively. Then, the idea is to superimpose information for the users with the good channel on top of the information that can be received by the users with the bad channel (which can also be decoded by the former class of users) [66]. Interestingly, this simple idea improves the joint capacity of both classes of users over simply multiplexing between the two channels (sending information at rate $R_1 \leq C_1$ to U_1 part of the time, and then at rate $R_2 \leq C_2$ to U_1 and U_2 the rest of the time). See Figure 7.41(a) for a graphical description of the joint capacity region and Figure7.41(b) for a typical constellation used in digital transmission where information for the users with better channels is superimposed over information which can be received by both classes of users. Now, keeping our multiresolution paradigm in mind, it is clear that we can send coarse signal information to both classes of users, while superposing detail information that can be taken by the users with the good channel. In [231], a digital broadcast system for HDTV was designed using these principles, including multiresolution video

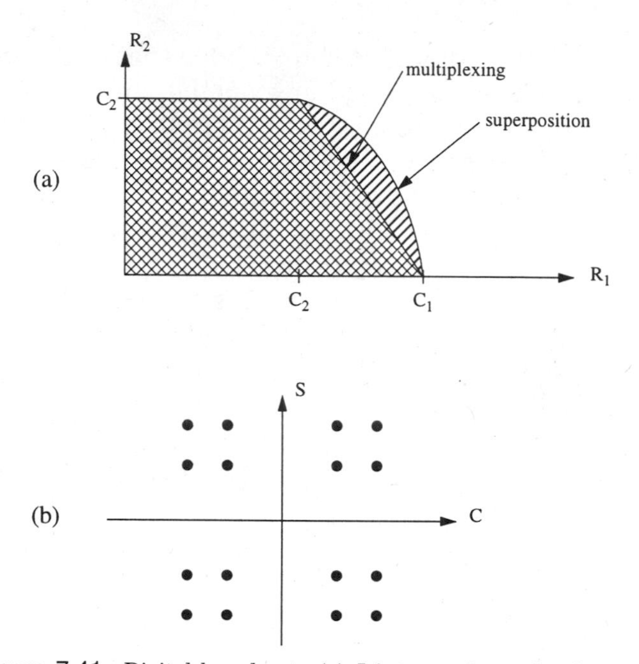

Figure 7.41 Digital broadcast. (a) Joint capacity region for two classes of users with channel capacities C_1 and C_2, respectively, and $C_1 > C_2$. Any point on or below the curves is achievable, but superposition outperforms multiplexing. (b) Example of a signal constellation (showing amplitudes of cosine and sine carriers in a digital communication system) using superposition of information. As can be seen, there are four clouds at four points each. When the channel is good, 16 points can be distinguished, (or four bits of information), while under adverse conditions, only the clouds are seen (or two bits of information).

coding [301] and multiresolution transmission with graceful degradation (using constellations similar to the one in Figure 7.41(b)).

The principles just described can be used for transmission over unknown time-varying channels. Instead of transmitting assuming the worst case channel, one can superpose information decodable on a better channel, in case the channel is actually better than worst case. On average, this will be better than simply assuming worst case all the time. As an example, consider a wireless channel without feedback. Because of the changing location of the user, the channel can vary greatly, and the worst case channel can be very poor. Superposition allows delivery of different levels of quality, depending on how good the reception actually is. When there is feedback (as in two-way wireless communication), then one can use a channel coding optimized for the current channel (see [114]). The source coder then has to adapt to the current transmission rate, which again is easy to achieve using multiresolution source coding. A study of wireless video transmission using a two resolution video source coder can be found in [157].

7.5.2 Packet Video

Another example of application of multiresolution coding for transmission is found in real-time services such as voice and video over asynchronous transfer mode (ATM) networks. The problem is that packet transmission can have greatly varying delays as well as packet losses. However, it is possible to protect certain packets (for example, using priorities). Again, the natural idea is to use multiresolution source coding and put the coarse approximation into high priority so that it will almost surely be received [154]. The detail information is carried with lower priority packets and will only arrive when the network has enough resources to carry them. Such an approach can lead to substantial improvements over nonprioritized transmission [107]. In video compression, this approach is often called layered coding, with the layers corresponding to different levels of approximation (typically, two layers are used) and different layers having different protections for transmission.

This concludes our brief overview of multiresolution methods for joint source and channel coding. It can be argued that because of increasing interconnectivity and heterogeneity, traditional fixed-rate coding and transmission will be replaced by flexible multiresolution source coding and multiple or variable-rate transmission. For an interface protocol allowing such flexible interconnection, see [127]. The main advantage is the added flexibility, which will allow users with different requirements to be interconnected through a mixture of possible channels.

APPENDIX 7.A Statistical Signal Processing

Very often, a signal has some statistical characteristics of which we can take advantage. A full blown treatment of statistical signal processing requires the study of stochastic processes [122, 217]. Here, we will only consider elementary concepts and restrict ourselves to the discrete-time case.

We start by reviewing random variables and then move to random processes. Consider a real-valued random variable X over $\mathcal{R}$ with distribution P_X. The distribution $P_X(A)$ indicates the probability that the random variable X takes on a value in A, where A is a subset of the real line. The *cumulative distribution function* (cdf) F_X is defined as

$$F_X(\alpha) = P_X(\{x|x \leq \alpha\}), \qquad \alpha \in \mathcal{R}.$$

The *probability density function* (pdf) is related to the cdf (assume that F_X is differentiable) as

$$f_X(\alpha) = \frac{dF_X(\alpha)}{d\alpha}, \qquad \alpha \in \mathcal{R},$$

and thus

$$F_X(\alpha) = \int_{-\infty}^{\alpha} f_X(x)dx, \qquad \alpha \in \mathcal{R}.$$

A vector random variable $\boldsymbol{X}$ is a collection of k random variables $(X_0, \ldots, X_{k-1})$, with a cdf $F_{\boldsymbol{X}}$ given by

$$F_{\boldsymbol{X}}(\boldsymbol{\alpha}) = P_{\boldsymbol{X}}(\{\boldsymbol{x}|x_i \leq \alpha_i, i = 0, 1, \ldots, k-1\}),$$

where $\boldsymbol{\alpha} = (\alpha_0, \ldots, \alpha_{k-1})$. The pdf is obtained, assuming differentiability, as

$$f_{\boldsymbol{X}}(\boldsymbol{\alpha}) = \frac{\partial^k}{\partial\alpha_0, \partial\alpha_1, \ldots, \partial\alpha_{k-1}} F_{\boldsymbol{X}}(\alpha_0, \alpha_1, \ldots, \alpha_{k-1}).$$

A key notion is *independence* of random variables. A collection of k random variables is independent *if and only if* the joint pdf has the form

$$f_{X_0 X_1 \cdots X_{k-1}}(x_0, x_1, \ldots, x_{k-1}) = f_{X_0}(x_0) \cdot f_{X_1}(x_1) \cdots f_{X_{k-1}}(x_{k-1}). \tag{7.A.1}$$

In particular, if each random variable has the same distribution, then we have an *independent and identically distributed* (iid) random vector.

Intuitively, a discrete-time *random process* is the infinite-dimensional generalization of a vector random variable. Therefore, any finite subset of random variables from a random process is a vector random variable.

Example 7.3 *Jointly Gaussian Random Process*

An important class of vector random variables is the Gaussian vector random variable of dimension k. To define its pdf, we need a length-k vector $\boldsymbol{m}$ and a positive definite matrix $\boldsymbol{\Lambda}$ of size $k \times k$. Then, the k-dimensional Gaussian pdf is given by

$$f(\boldsymbol{x}) = (2\pi)^{-k/2}(\det \boldsymbol{\Lambda})^{-1/2} e^{-(\boldsymbol{x}-\boldsymbol{m})^T \boldsymbol{\Lambda}^{-1}(\boldsymbol{x}-\boldsymbol{m})/2}, \qquad \boldsymbol{x} \in \mathcal{R}^k \tag{7.A.2}$$

Note how, for $k = 1$ and $\boldsymbol{\Lambda} = \sigma^2$, this reduces to the usual Gaussian (normal) distribution

$$f(x) = \frac{1}{\sqrt{2\pi\sigma^2}} \cdot e^{-(x-m)^2/2\sigma^2}, x \in \mathcal{R},$$

of which (7.A.2) is a k-dimensional generalization.

A discrete-time random process is jointly Gaussian if all finite subsets of samples $\{X_{n_0}, X_{n_1}, \ldots, X_{n_{k-1}}\}$ are Gaussian random vectors. Thus, a Gaussian random process is completely described by $\boldsymbol{m}$ and $\boldsymbol{\Lambda}$, which are called the mean and covariance as we will see.

For random variables as for random processes, a fundamental concept is that of *expectation,* defined as

$$E(X) = \int_{-\infty}^{\infty} x f_X(x)\, dx.$$

Expectation is a linear operator, that is, given two random variables X and Y, we have $E(aX + bY) = aE(X) + bE(Y)$. The expectation of products of random variables leads to the concept of *correlation*. Given two random variables X and Y, their correlation is $E(XY)$. They are uncorrelated if

$$E(XY) = E(X)\, E(Y).$$

From (7.A.1) we see that independent variables are uncorrelated (but uncorrelatedness is not sufficient for independence). Sometimes, the "centralized" correlation, or *covariance*, is used, namely

$$\begin{aligned} \text{cov}(X, Y) &= E((X - E(X))(Y - E(Y))) \\ &= E(XY) - E(X)E(Y), \end{aligned}$$

from which it follows that two random variables are uncorrelated *if and only if* their covariance is zero. The *variance* of X, denoted by σ_X^2, equals $\text{cov}(X, X)$, that is,

$$\sigma_X^2 = E((X - E(X))^2),$$

and its square root σ_X is called the standard deviation of X. Higher-order moments are obtained from $E(X^k), k > 2$. The above functions can be extended to random processes. The autocorrelation function of a process $\{X_n, n \in \mathcal{Z}\}$, is defined by

$$R_X[n, m] = E(X_n\, X_m), \qquad n, m \in \mathcal{Z},$$

and the autocovariance function is

$$\begin{aligned} K_X[n, m] &= \text{cov}(X_n, X_m) \\ &= R_X[n, m] - E(X_n)E(X_m), \qquad n, m \in \mathcal{Z}. \end{aligned}$$

An important class of processes are *stationary random processes*, for which the probabilistic behavior is constant over time. In particular, the following then hold:

$$E(X_n) = E(X), \qquad n \in \mathcal{Z}, \tag{7.A.3}$$

$$\sigma_{X_n}^2 = \sigma_X^2, \qquad n \in \mathcal{Z}. \tag{7.A.4}$$

By the same token, all other moments are independent of n. Also, correlation and covariance depend only on the difference $(n - m)$, or

$$R_X[n, m] = R_X[n - m], \qquad n, m \in \mathcal{Z}, \tag{7.A.5}$$

$$K_X[n, m] = K_X[n - m], \qquad n, m \in \mathcal{Z}. \tag{7.A.6}$$

While stationarity implies that the full probabilistic description is time-invariant, nth-order stationarity means that distributions and expectations involving n samples are time-invariant. The case $n = 2$, which corresponds to (7.A.3–7.A.6) is called *wide-sense stationarity*. An important property of Gaussian random processes is that if they are wide-sense stationary, then they are also strictly stationary.

Often, we are interested in filtering a random process by a linear time-invariant filter with impulse response $h[n]$. That is, the output equals $Y[n] = \sum_{k=-\infty}^{\infty} h[k]\, X[n-k]$. Note that $Y[.]$ and $X[.]$ denote random variables and are thus capitalized, while $h[.]$ is a deterministic value. We will assume a stable and causal filter. The expected value of the output is

$$E(Y[n]) = E(\sum_{k=0}^{\infty} h[k]X[n-k]) = \sum_{k=0}^{\infty} h[k]E(X[n-k]) = \sum_{k=0}^{\infty} h[k]m_{n-k}, \qquad (7.A.7)$$

where m_l is the expected value of X_l. Note that if the input is wide-sense stationary, that is, $E(X_n) = E(X)$ for all n, then the output has a constant expected value equal to $E(X)\sum_{k=0}^{\infty} h[k]$. It can be shown that the covariance function of the output depends also only on the difference $n - m$ (as in (7.A.5)) and thus, filtering by a linear time-invariant system conserves wide-sense stationarity (see Problem 7.9).

When considering filtered wide-sense stationary processes, it is useful to introduce the *power spectral density function* (psdf), which is the discrete-time Fourier transform of the autocorrelation function

$$S_X(e^{j\omega}) = \sum_{n=-\infty}^{\infty} R_X[n]\, e^{-j\omega n}.$$

Then, it can be shown that the psdf of the output process after filtering with $h[n]$ equals

$$S_Y(e^{j\omega}) = \left|H(e^{j\omega})\right|^2 S_X(e^{j\omega}), \qquad (7.A.8)$$

where $H(e^{j\omega})$ is the discrete-time Fourier transform of $h[n]$. Note that when the input is uncorrelated, that is, $R_X[n] = E^2(X)\delta[n]$, then the output autocorrelation is simply the autocorrelation of the filter, or $R_Y[n] = E^2(X)\langle h[k], h[k+n]\rangle$, as can be seen from (7.A.8). If we define the crosscorrelation function

$$R_{XY}[m] = E(X[n]\, Y[n+m]),$$

then its Fourier transform leads to

$$S_{XY}(e^{j\omega}) = H(e^{j\omega})\, S_X(e^{j\omega}). \qquad (7.A.9)$$

Again, when the input is uncorrelated, this can be used to measure $H(e^{j\omega})$.

An important application of filtering is in linear estimation. The simplest linear estimation problem is when we have two random variables X and Y, both with zero mean. We wish to find an estimate $\hat{X}$ of the form $\hat{X} = \alpha Y$ from the observation Y, such that the mean square error (MSE) $E((X - \hat{X})^2)$ is minimized. It is easy to verify that

$$\alpha = \frac{E(XY)}{E(Y^2)},$$

minimizes the expected squared error. One distinctive feature of the MSE estimate is that the estimation error $(X - \hat{X})$ is orthogonal (in expected value) to the observation Y, that is,

$$E((X - \hat{X})Y) = E((X - \alpha Y)Y) = E(XY) - \alpha E(Y^2) = 0.$$

This is known as the *orthogonality principle*: The best linear estimate in the MSE sense is the orthogonal projection of X onto the span of Y. It follows that the minimum MSE is

$$E((X - \hat{X})^2) = E(X^2) - \alpha^2 E(Y^2),$$

because of orthogonality of $(X - \hat{X})$ and Y. This geometric view follows from the interpretation of $E(XY)$ as an inner product and thus $E(X^2)$ is the squared length of the vector X. Similarly, orthogonality of X and Y is seen as $E(XY) = 0$. Based on this powerful geometric point of view, let us tackle a more general linear estimation problem. Assume two zero-mean jointly wide-sense stationary processes $\{X[n]\}$ and $\{Y[n]\}$. We want to estimate $X[n]$ from $Y[n]$ using a filter with the impulse response $h[n]$, that is

$$\hat{X}[n] = \sum_k h[k]\, Y[n-k], \qquad (7.A.10)$$

in such a way that $E((X[n] - \hat{X}[n])^2)$ is minimized. The range of k is restricted to a set K (for example, $k \geq 0$ so that only $y[n], y[n-1], \ldots$ are used). The orthogonality principle states that the optimal solution will satisfy

$$E((X[n] - \hat{X}[n])Y[k]) = 0, \qquad k \in K.$$

Using (7.A.10), we can rewrite the orthogonality condition as

$$\begin{aligned} & E(X[n]Y[k]) - E(\sum_i h[i]Y[n-i]Y[k]) \\ = \; & R_{XY}[n,k] - \sum_i h[i] R_Y[n-i,k] \\ = \; & R_{XY}[n-k] - \sum_i h[i] R_Y[n-k-i], \qquad k \in K, \end{aligned}$$

where we used wide-sense stationarity in $R_{XY}[n,k] = R_{XY}[n-k]$. Replacing $n-k$ by l, we get

$$R_{XY}[l] \;=\; \sum_i h[i]\, R_Y[l-i], \qquad n-l \in K. \tag{7.A.11}$$

In particular, when there is no restriction on the set of samples $\{Y[n]\}$ used for the estimation, that is $K = \mathcal{Z}$, then we can take the Fourier transform of (7.A.11) to find

$$H(e^{j\omega}) \;=\; \frac{S_{xy}(e^{j\omega})}{S_y(e^{j\omega})},$$

which is the optimal linear estimator. Note that this is in general a noncausal filter. Finding a causal solution ($K = (-\infty, n]$) is more involved [122], but the orthogonality principle is preserved.

This concludes our brief overview of statistical signal processing. One more topic, namely the discrete-time Karhunen-Loève transform, is discussed in the main text, in Section 7.1, since it lays the foundation for transform-based signal compression.

PROBLEMS

7.1 For a uniform input pdf, as well as uniform quantization, prove that the distortion between the input and the output of the quantizer is given by (7.1.14), that is

$$D = \frac{\Delta^2}{12},$$

where Δ is the quantizer step size $\Delta = (b-a)/N$, a, b are the boundaries of the input, and N is the number of intervals.

7.2 *Coding gain as a function of number of channels:* Consider the coding gain of an ideal filter bank with N channels (see Section 7.1.2).

(a) Construct a simple example where the coding gain for a 2-channel system is bigger than the coding gain for a 3-channel system. *Hint:* Construct a piecewise constant power spectrum for which the 2-channel system is better matched than the 3-channel system.

(b) For the example constructed above, show that a 4-channel system outperforms both the 2- and 3-channel systems.

7.3 Consider the coding gain (see Section 7.1.2) in an ideal subband coding system with N channels (the filters used are ideal bandpass filters). Start with the case $N = 2$ before looking at the general case.

(a) Assume that the power spectrum of the input signal $|X(e^{j\omega})|^2$ is given by

$$|X(e^{j\omega})|^2 = 1 - \frac{\omega}{\pi} \qquad |\omega| \leq \pi.$$

Give the coding gain as a function of N.

(b) Same as above, but with

$$|X(e^{j\omega})|^2 = e^{-\alpha\omega} \qquad |\omega| \leq \pi.$$

Give the coding gain as a function of N and α, and compare to (a).

7.4 *Huffman and run-length coding:* A stream of symbols has the property that stretches of zeros are likely. Thus, one can use code the length of the stretch of zeros, after a special "start of run" (SR) symbol.

(a) Assume there are runs of lengths 1 to 8, with probabilities:

Length	1	2	3	4	5	6	7	8
Probability	1/2	1/4	1/8	1/16	1/32	1/64	1/128	1/128

Design a Huffman code for the run lengths. How close does it come to the entropy?

(b) There are 8 nonzero symbols, plus the start of run symbols, with probabilities:

Symbol	± 1	± 2	± 3	± 4	SR
Probability	0.2	0.15	0.075	0.05	0.05

Design a Huffman code for these symbols. How close does it come to the entropy?

(c) As an example, take a typical sequence, including stretches of zeros, and encode it, then decode it, with your Huffman code (small example). Can you decode your bit stream?

(d) Give the average compression of this run-length and Huffman coding scheme.

7.5 Consider a pyramid coding scheme as discussed in Section 7.3.2. Assume a one-dimensional signal and an ideal lowpass filter both for coarse-to-fine and fine-to-coarse resolution change.

(a) Assume an exponentially decaying power spectrum

$$|X(e^{j\omega})|^2 = e^{-3\omega/\pi} \qquad |\omega| < \pi.$$

Derive the variances of the coarse and the difference channels.

(b) Assume now that the coarse channel is quantized before being interpolated and used as a prediction. Assume an additive noise model, with variance $c\Delta^2$ where Δ is the quantizer step. Give the variance of the difference channel (which now depends on Δ, or the number of bits allocated to the coarse channel).

(c) Investigate experimentally the bit allocation problem in a pyramid coder using a quantized coarse version for the prediction. That is, generate some correlated random process (for example, first-order Markov with high correlation) and process it using pyramid coding. Allocate part of the bit budget to the coarse version, and the rest for the difference signal. Discuss the two limiting cases, that is, zero bits to the coarse version and all the bits for the coarse version.

7.6 Consider the embedded zero tree wavelet (EZW) transform algorithm discussed in Section 7.3.4, and study a one-dimensional version.

(a) Assume a one-dimensional octave-band filter bank and define a zero tree for this case. Compare to the two-dimensional case. Discuss if the dominant and subordinate passes of the EZW algorithm have to be modified, and if so, how.

(b) One can define a zero tree for arbitrary subband decomposition trees (or wavelet packets). In which case is the zero tree most powerful?

(c) In the case of a full tree subband decomposition in two dimensions (for example, of depth 3, leading to 64 channels), compare the zero tree structure with zig-zag scanning used in DCT.

7.7 *Progressive to interlaced conversion:*

(a) Verify that the filters given in (7.4.3) form a perfect reconstruction filter bank for quincunx downsampling and give the reconstruction filters as well.

(b) Show that cascading the quincunx decomposition twice on a progressive sequence (on the vertical-time dimension) yields again a progressive sequence, with an intermediate interlaced sequence. Use the downsampling matrix

$$\boldsymbol{D} = \begin{pmatrix} 1 & 1 \\ -1 & 1 \end{pmatrix}$$

7.8 Consider a two-channel filter bank for three-dimensional signals (progressive video sequences) using FCO downsampling (see Section 7.4.4).

(a) Consider a lowpass filter

$$H_0(z_1, z_2, z_3) = \frac{1}{\sqrt{2}}(1 + z_1 z_2 z_3),$$

and a highpass filter

$$H_1(z_1, z_2, z_3) = H_0(-z_1, -z_2, -z_3).$$

Show that this corresponds to an orthogonal Haar decomposition for FCO downsampling.

(b) Give the output of a two-channel analysis/synthesis system with FCO downsampling as a function of the input, the aliased version, and the filters.

7.9 *Filtering of wide-sense stationary processes:* Consider a wide-sense stationary process $\{x[n]\}$ and its filtered version $y[n] = \sum_k h[k]x[n-k]$, where $h[k]$ is a stable and causal filter.

(a) In Appendix 7.A, we saw that the mean of $\{y[n]\}$ is independent of n (see below Equation (7.A.7)). Show that the covariance function of $\{y[n]\}$, $K_Y[n,m] = \text{cov}(y[n] \cdot y[m])$ is a function of $(n-m)$ only, and given by

$$K_Y[k] = \sum_{n=0}^{\infty}\sum_{m=0}^{\infty} h[n]\, h[m]\, K_X[k-(n-m)]$$

(b) Prove (7.A.9) in time domain, or assuming zero-mean input,

$$K_{XY}[m] = \sum_{h=0}^{\infty} h[k]\, K_X[m-k].$$

(c) Consider now one-sided wide-sense stationary processes, which can be thought of as wide-sense stationary processes that are "turned on" at time 0. Consider filtering of such processes by causal FIR and IIR filters, respectively. What can be said about $E(Y[n])$ $n \geq 0$ in these cases?

Projects: The following problems are computer-based projects with an experimental flavor. Access to adequate data (images, video) is helpful.

7.10 *Coding gain and R(d) optimal filters for subband coding:* Consider a two-band perfect reconstruction subband coder with orthogonal filters in lattice structure. As an input, use a first-order Markov process with high correlation ($\rho = 0.9$). For small filter lengths ($L = 4, 6$ or so), optimize the lattice coefficients so as to maximize coding gain or minimize first-order entropy after uniform scalar quantization. Find what filter is optimal, and try for fine and coarse quantization steps.
Use optimal bit allocation between the two channels, if possible. The same idea can be extended to Lloyd-Max quantization, and to logarithmic trees. This project requires some experience with coding algorithms. For relevant literature, see [79, 109, 244, 295].

7.11 *Pyramids using nonlinear operators:* One of the attractive features of pyramid coding schemes over critically sampled coding schemes is that nonlinear operators can be used. The goal of the project is to investigate the use of median filters (or some other nonlinear operators) in a pyramidal scheme.
The results could be theoretical or experimental. The project requires image processing background. For relevant literature, see [41, 138, 303, 323].

7.12 *Motion compensation of motion vectors:* In video coding, motion compensation is used to predict a new frame from reconstructed previous frames. Usually, a sparse set of motion vectors is used (such as one per 8×8 block), and thus, sending motion vectors contributes little to the bit rate overhead. An alternative scheme could use a dense motion vector field in order to reduce the prediction error. In order to reduce the overhead, predict the motion vector field, since it is usually not changing radically in time within a video scene. Thus, the aim of the project is to treat the motion vector field as a sequence (of vectors), and find a meta-motion vector field to predict the actual motion vector field (for example, per block of 2×2 motion vectors).

This project requires image/video processing background. For more literature on motion estimation, see [138, 207].

7.13 *Adaptive Karhunen-Loève transform:* The Karhunen-Loève transform is optimal for energy packing of stationary processes, and under certain conditions, for transform coding and quantization of such processes. However, if the process is nonstationary, compression might be improved by using an adaptive transform. An interesting solution is an overhead free transform which is derived from the coded version of the signal, based on some estimate of local correlations.
The goal of the project is to explore such an adaptive transform on some synthetic nonstationary signals, as well as on real signals (such as speech).
This project requires good signal processing background. For more literature, see [143].

7.14 *Three-dimensional wavelet coding:* In medical imaging and remote sensing, one often encounters three-dimensional data. For example, multispectral satellite imagery consists of many spectral band images. Develop a simple three-dimensional coding algorithm based on the Haar filters, and iteration on the lowpass channel. This is the three-dimensional equivalent of the octave-band subband coding of images discussed in Section 7.3.3. Apply your algorithm to real imagery if available, or generate synthetic data with a lowpass nature.

Bibliography

[1] E. H. Adelson, E. Simoncelli, and R. Hingorani. Orthogonal pyramid transforms for image coding. In *Proc. SPIE*, volume 845, pages 50–58, Cambridge, MA, October 1987.

[2] N. Ahmed, T. Natarajan, and K. R. Rao. Discrete cosine transform. *IEEE Trans. on Computers*, 23:88–93, January 1974.

[3] A. Akansu and R. Haddad. *Multiresolution Signal Decomposition*. Academic Press, New York, 1993.

[4] A. N. Akansu, R. A. Haddad, and H. Caglar. The binomial QMF-wavelet transform for multiresolution signal decomposition. *IEEE Trans. on Signal Processing*, 41(1):13–19, January 1993.

[5] A. N. Akansu and Y. Liu. On signal decomposition techniques. *Optical Engr.*, 30:912–920, July 1991.

[6] A. Aldroubi and M. Unser. Families of wavelet transforms in connection with Shannon's sampling theory and the Gabor transform. In C. K. Chui, editor, *Wavelets: A Tutorial in Theory and Applications*, pages 509–528. Academic Press, New York, 1992.

[7] A. Aldroubi and M. Unser. Families of multiresolution and wavelet spaces with optimal properties. *Numer. Functional Anal. and Optimization*, 14:417–446, 1993.

[8] J. B. Allen. Short term spectral analysis, synthesis, and modification by discrete Fourier transform. *IEEE Trans. Acoust., Speech, and Signal Proc.*, 25:235–238, June 1977.

[9] D. Anastassiou. Generalized three-dimensional pyramid coding for HDTV using nonlinear interpolation. In *Proc. of the Picture Coding Symp.*, pages 1.2–1–1.2–2, Cambridge, MA, March 1990.

[10] J. C. Anderson. A wavelet magnitude analysis theorem. *IEEE Transactions on Signal Processing, Special Issue on Wavelets and Signal Processing*, 41(12):3541–3542, December 1993.

[11] R. Ansari. Two-dimensional IIR filters for exact reconstruction in tree-structured sub-band decomposition. *Electr. Letters*, 23(12):633–634, June 1987.

[12] R. Ansari, H. Gaggioni, and D.J. LeGall. HDTV coding using a nonrectangular subband decomposition. In *Proc. SPIE Conf. on Vis. Commun. and Image Proc.*, pages 821–824, Cambridge, MA, November 1988.

[13] M. Antonini, M. Barlaud, and P. Mathieu. Image coding using lattice vector quantization of wavelet coefficients. In *Proc. IEEE Int. Conf. Acoust., Speech, and Signal Proc.*, pages 2273–2276, Toronto, Canada, May 1991.

[14] M. Antonini, M. Barlaud, P. Mathieu, and I. Daubechies. Image coding using wavelet transform. *IEEE Trans. Image Proc.*, 1(2):205–220, April 1992.

[15] K. Asai, K. Ramchandran, and M. Vetterli. Image representation using time-varying wavelet packets, spatial segmentation and quantization. *Proc. of Conf. on Inf. Science and Systems*, March 1993.

[16] P. Auscher. Wavelet bases for $L^2(R)$ with rational dilation factors. In B. Ruskai et al., editor, *Wavelets and Their Applications*. Jones and Bartlett, Boston, MA, 1992.

[17] P. Auscher, G. Weiss, and M. V. Wickerhauser. Local sine and cosine bases of Coifman and Meyer and the construction of smooth wavelets. In C. K. Chui, editor, *Wavelets: A Tutorial in Theory and Applications*. Academic Press, New York, 1992.

[18] M. J. Baastians. Gabor's signal expansion and degrees of freedom of a signal. *Proc. IEEE*, 68:538–539, 1980.

[19] R. H. Bamberger and M. J. T. Smith. A filter bank for the directional decomposition of images: Theory and design. *IEEE Trans. Signal Proc.*, 40(4):882–893, April 1992.

[20] M. Basseville, A. Benveniste, K. C. Chou, S. A. Golden, R. Nikoukhah, and A. S. Willsky. Modeling and estimation of multiresolution stochastic processes. *IEEE Transactions on Information Theory, Special Issue on Wavelet Transforms and Multiresolution Signal Analysis*, 38(2):766–784, March 1992.

[21] G. Battle. A block spin construction of ondelettes. Part I: Lemarié functions. *Commun. Math. Phys.*, 110:601–615, 1987.

[22] G. Battle. A block spin construction of ondelettes. Part II: the QFT connection. *Commun. Math. Phys.*, 114:93–102, 1988.

[23] V. Belevitch. *Classical Network Synthesis*. Holden Day, San Francisco, CA, 1968.

[24] T. C. Bell, J. G. Cleary, and J. H. Witten. *Text Compression*. Prentice-Hall, Englewood Cliffs, NJ, 1990.

[25] M. G. Bellanger and J. L. Daguet. TDM-FDM transmultiplexer: Digital polyphase and FFT. *IEEE Trans. Commun.*, 22(9):1199–1204, September 1974.

[26] J. J. Benedetto. Irregular sampling and frames. In C. K. Chui, editor, *Wavelets: A Tutorial in Theory and Applications*. Academic Press, New York, 1992.

[27] J. J. Benedetto and M. W. Frazier, editors. *Wavelets: Mathematics and Applications*. CRC Press, Boca Raton, 1994.

[28] T. Berger. *Rate Distortion Theory*. Prentice-Hall, Englewood Cliffs, NJ, 1971.

[29] Z. Berman and J. S. Baras. Properties of the multiscale maxima and zero-crossings representations. *IEEE Transactions on Signal Processing, Special Issue on Wavelets and Signal Processing*, 41(12):3216–3231, December 1993.

[30] G. Beylkin, R. Coifman, and V. Rokhlin. Fast wavelet transforms and fast algorithms. In

Y. Meyer, editor, *Wavelets and Applications*, pages 354–367. Masson, Paris, 1992.

[31] M. Bierling. Displacement estimation by hierarchical block matching. In *Proc. SPIE Conf. on Vis. Commun. and Image Proc.*, pages 942–9–51, Boston, MA, November 1988.

[32] R .E. Blahut. *Fast Algorithms for Digital Signal Processing*. Addison-Wesley, Reading, MA, 1984.

[33] T. Blu. Iterated filter banks with rational sampling factors: Links with discrete wavelet transforms. *IEEE Transactions on Signal Processing, Special Issue on Wavelets and Signal Processing*, 41(12):3232–3244, December 1993.

[34] M. Bosi and G. Davidson. High-quality, low-rate audio transform coding for transmission and multimedia applications. In *Convention of the AES*, San Francisco, CA, October 1992.

[35] F. Bosveld, R. L. Langendijk, and J. Biemond. Hierarchical coding of HDTV. *Signal Processing: Image Communication*, 4:195–225, June 1992.

[36] A. C. Bovik, N. Gopal, T. Emmoth, and A. Restrepo (Palacios). Localized measurement of emergent image frequencies by Gabor wavelets. *IEEE Transactions on Information Theory, Special Issue on Wavelet Transforms and Multiresolution Signal Analysis*, 38(2):691–712, March 1992.

[37] R. N. Bracewell. *The Fourier Transform and its Applications*. McGraw-Hill, New York, NY, Second edition, 1986.

[38] K. Brandenburg, G. Stoll, F. Dehery, and J. D. Johnston. The ISO-MPEG-1 audio: A generic standard for coding of high-quality digital audio. *Journal of the Audio Engineering Society*, 42(10):780–792, October 1994.

[39] W. L. Briggs. *A Multigrid Tutorial*. SIAM, Philadelphia, 1987.

[40] C. S. Burrus and T. W. Parks. *DFT/FFT and Convolution Algorithms: Theory and Implementation*. Wiley, New York, 1985.

[41] P. J. Burt and E. H. Adelson. The Laplacian pyramid as a compact image code. *IEEE Trans. Commun.*, 31(4):532–540, April 1983.

[42] J. W. Cassels. *An Introduction to the Geometry of Numbers*. Springer-Verlag, Berlin, 1971.

[43] P. M. Cassereau. A new class of optimal unitary transforms for image processing. Master's thesis, Massachusetts Institute of Technology, May 1985.

[44] P. M. Cassereau, D. H. Staelin, and G. de Jager. Encoding of images based on a lapped orthogonal transform. *IEEE Trans. Commun.*, 37:189–193, February 1989.

[45] A. S. Cavaretta, W. Dahmen, and C. Micchelli. Stationary subdivision. *Mem. Amer. Math. Soc.*, 93:1–186, 1991.

[46] D. C. Champeney. *A Handbook of Fourier Theorems*. Cambridge University Press, Cambridge, UK, 1987.

[47] T. Chen and P. P. Vaidyanathan. Multidimensional multirate filters and filter banks derived from one-dimensional filters. *IEEE Trans. Signal Proc.*, 41(5):1749–1765, May 1993.

[48] T. Chen and P. P. Vaidyanathan. Recent developments in multidimensional multirate systems. *IEEE Trans. on CSVT*, 3(2):116–137, April 1993.

[49] C. K. Chui. *An Introduction to Wavelets*. Academic Press, New York, 1992.

[50] C. K. Chui. On cardinal spline wavelets. In Ruskai et al., editor, *Wavelets and Their Applications*, pages 419–438. Jones and Bartlett, MA, 1992.

[51] C. K. Chui, editor. *Wavelets: A Tutorial in Theory and Applications*. Academic Press, New York, 1992.

[52] C. K. Chui and J. Z. Wang. A cardinal spline approach to wavelets. *Proc. Amer. Math. Soc.*, 113:785–793, 1991.

[53] T. A. C. M. Claasen and W. F. G. Mecklenbräuker. The Wigner distribution - a tool for time-frequency signal analysis, Part I, II, and III. *Philips Journal of Research*, 35(3, 4/5, 6):217–250, 276–300, 372–389, 1980.

[54] R. J. Clarke. *Transform Coding of Images*. Academic Press, London, 1985.

[55] A. Cohen. *Ondelettes, Analyses Multiresolutions et Traitement Numérique du Signal*. PhD thesis, Université Paris IX Dauphine, Paris, France, 1990.

[56] A. Cohen. Biorthogonal wavelets. In C. K. Chui, editor, *Wavelets: A Tutorial in Theory and Applications*. Academic Press, New York, 1992.

[57] A. Cohen and I. Daubechies. Nonseparable bidimensional wavelet bases. *Rev. Mat. Iberoamericana*, 9(1):51–137, 1993.

[58] A. Cohen, I. Daubechies, and J.-C. Feauveau. Biorthogonal bases of compactly supported wavelets. *Commun. on Pure and Appl. Math.*, 45:485–560, 1992.

[59] L. Cohen. Time-frequency distributions: A review. *Proc. IEEE*, 77(7):941–981, July 1989.

[60] L. Cohen. The scale representation. *IEEE Transactions on Signal Processing, Special Issue on Wavelets and Signal Processing*, 41(12):3275–3292, December 1993.

[61] R. R. Coifman and Y. Meyer. Remarques sur l'analyse de Fourier à fenêtre. *C.R. Acad. Sci.*, pages 259–261, 1991.

[62] R. R. Coifman, Y. Meyer, S. Quake, and M. V. Wickerhauser. Signal processing and compression with wavelet packets. Technical report, Dept. of Math., Yale University, 1991.

[63] R. R. Coifman, Y. Meyer, and M. V. Wickerhauser. Wavelet analysis and signal processing. In M. B. Ruskai et al, editor, *Wavelets and their Applications*, pages 153–178. Jones and Barlett, Boston, 1992.

[64] R. R. Coifman and M. V. Wickerhauser. Entropy-based algorithms for best basis selection. *IEEE Transactions on Information Theory, Special Issue on Wavelet Transforms and Multiresolution Signal Analysis*, 38(2):713–718, March 1992.

[65] J. M. Combes, A. Grossmann, and Ph. Tchamitchian, editors. *Wavelets, Time-Frequency Methods and Phase Space*. Springer-Verlag, Berlin, 1989.

[66] T. M. Cover and J. A. Thomas. *Elements of Information Theory*. Wiley Interscience, New York, NY, 1991.

[67] R. E. Crochiere and L. R. Rabiner. *Multirate Digital Signal Processing*. Prentice-Hall, Englewood Cliffs, NJ, 1983.

[68] R. E. Crochiere, S. A. Webber, and J. L. Flanagan. Digital coding of speech in sub-bands. *Bell System Technical Journal*, 55(8):1069–1085, October 1976.

[69] A. Croisier, D. Esteban, and C. Galand. Perfect channel splitting by use of interpolation/-decimation/tree decomposition techniques. In *Int. Conf. on Inform. Sciences and Systems*, pages 443–446, Patras, Greece, August 1976.

[70] Z. Cvetković and M. Vetterli. Consistent reconstruction of signals from wavelet extrema/zero crossings representation. *IEEE Trans. on SP*, 43(3), March 1995. To appear.

[71] I. Daubechies. Orthonormal bases of compactly supported wavelets. *Commun. on Pure and Appl. Math.*, 41:909–996, November 1988.

[72] I. Daubechies. The wavelet transform, time-frequency localization and signal analysis. *IEEE Trans. Inform. Th.*, 36(5):961–1005, September 1990.

[73] I. Daubechies. *Ten Lectures on Wavelets*. SIAM, Philadelphia, PA, 1992.

[74] I. Daubechies and J. Lagarias. Two-scale difference equations I. Existence and global regularity of solutions. *SIAM J. Math. Anal.*, 22:1388–1410, 1991.

[75] I. Daubechies and J. Lagarias. Two-scale difference equations: II. Local regularity, infinite products of matrices and fractals. *SIAM Journ. of Math. Anal.*, 24(24):1031–1079, July 1992.

[76] C. deBoor. *A Practical Guide to Splines*, volume 27 of *Appl. Math. Sciences*. Springer-Verlag, New York, 1978.

[77] Y. F. Dehery, M. Lever, and P. Urcum. A MUSICAM source codec for digital audio broadcasting and storage. In *Proc. IEEE Int. Conf. Acoust., Speech, and Signal Proc.*, pages 3605–3608, Toronto, Canada, May 1991.

[78] N. Delprat, B. Escudié, P. Guillemain, R. Kronland-Martinet, Ph. Tchamitchian, and B. Torrésani. Asymptotic wavelet and Gabor analysis: Extraction of instantaneous frequencies. *IEEE Transactions on Information Theory, Special Issue on Wavelet Transforms and Multiresolution Signal Analysis*, 38(2):644–664, March 1992.

[79] Ph. Delsarte, B. Macq, and D. T. M. Slock. Signal-adapted multiresolution transform for image coding. *IEEE Transactions on Information Theory, Special Issue on Wavelet Transforms and Multiresolution Signal Analysis*, 38(2):897–903, March 1992.

[80] G. Deslauriers and S. Dubuc. Symmetric iterative interpolation. *Constr. Approx.*, 5:49–68, 1989.

[81] R. A. DeVore, B. Jawerth, and B. J. Lucier. Image compression through wavelet transform coding. *IEEE Transactions on Information Theory, Special Issue on Wavelet Transforms and Multiresolution Signal Analysis*, 38(2):719–746, March 1992.

[82] R. A. DeVore and B. J. Lucier. Fast wavelet techniques for near-optimal image processing. In *Proceedings of the 1992 IEEE Military Communications Conference*, pages 1129–1135, New York, October 1992. IEEE Communications Society. San Diego, California.

[83] D. Donoho. Unconditional bases are optimal bases for data compression and statistical estimation. *Applied Computational Harmonic Analysis*, 1(1):100–115, December 1993.

[84] Z. Doğanata and P. P. Vaidyanathan. Minimal structures for the implementation of digital rational lossless systems. *IEEE Trans. Acoust., Speech, and Signal Proc.*, 38(12):2058–2074, December 1990.

[85] Z. Doğanata, P. P. Vaidyanathan, and T. Q. Nguyen. General synthesis procedures for FIR lossless transfer matrices, for perfect reconstruction multirate filter bank applications. *IEEE Trans. Acoust., Speech, and Signal Proc.*, 36(10):1561–1574, October 1988.

[86] E. Dubois. The sampling and reconstruction of time-varying imagery with application in video systems. *Proc. IEEE*, 73(4):502–522, April 1985.

[87] S. Dubuc. Interpolation through an iterative scheme. *J. Math. Anal. Appl.*, 114:185–204,

1986.

[88] D. E. Dudgeon and R. M. Mersereau. *Multidimensional Digital Signal Processing*. Prentice-Hall, Englewood Cliffs, NJ, 1984.

[89] R. J. Duffin and A. C. Schaeffer. A class of nonharmonic Fourier series. *Trans. Amer. Math. Soc.*, 72:341–366, 1952.

[90] P. Duhamel and M. Vetterli. Fast Fourier transforms: a tutorial review and a state of the art. *Signal Proc.*, 19(4):259–299, April 1990.

[91] H. Dym and H. P. McKean. *Fourier Series and Integrals*. Academic Press, New York, 1972.

[92] N. Dyn and D. Levin. Interpolating subdivision schemes for the generation of curves and surfaces. In W. Haussmann and K. Jetter, editors, *Multivariate Approximation and Interpolation*, pages 91–106. Birkauser Verlag, Basel, 1990.

[93] W. H. Equitz and T. M. Cover. Successive refinement of information. *IEEE Trans. Inform. Th.*, 37(2):269–275, March 1991.

[94] D. Esteban and C. Galand. Application of quadrature mirror filters to split band voice coding schemes. In *Proc. IEEE Int. Conf. Acoust. Speech, and Signal Processing*, pages 191–195, May 1977.

[95] G. Evangelista. *Discrete-Time Wavelet Transforms*. PhD thesis, Univ. of California, Irvine, June 1990.

[96] G. Evangelista and C. W. Barnes. Discrete-time wavelet transforms and their generalizations. In *Proc. IEEE Intl. Symp. Circuits Syst.*, pages 2026–2029, New Orleans, LA, May 1990.

[97] A. Fettweiss. Wave digital filters: theory and practice. *Proceedings of the IEEE*, 74(2):270–327, February 1986.

[98] A. Fettweiss, J. Nossek, and K. Meerkröter. Reconstruction of signals after filtering and sampling rate reduction. *IEEE Trans. on ASSP*, 33(4):893–902, August 1985.

[99] P. Flandrin. Some aspects of nonstationary signal processing with emphasis on time-frequency and time-scale methods. In J. M. Combes, A. Grossmann, and Ph. Tchamitchian, editors, *Wavelets, Time-Frequency Methods and Phase Space*. Springer-Verlag, Berlin, 1989.

[100] J. Fourier. *Théorie Analytique de la Chaleur*. Gauthiers-Villars., Paris, 1888.

[101] J. Froment and S. Mallat. Second generation compact image coding with wavelets. In C. K. Chui, editor, *Wavelets: A Tutorial in Theory and Applications*. Academic Press, New York, 1992.

[102] D. Gabor. Theory of communication. *Journ. IEE*, 93:429–457, 1946.

[103] C. Galand and D. Esteban. 16 Kbps real-time QMF subband coding inplementation. In *Proc. Int. Conf. on Acoust. Speech and Signal Processing*, pages 332–335, Denver, CO, April 1980.

[104] C. R. Galand and H. J. Nussbaumer. New quadrature mirror filter structures. *IEEE Trans. Acoust., Speech, and Signal Proc.*, 32(3):522–531, June 1984.

[105] R. G. Gallagher. Variations on a theme by Huffman. *IEEE Trans. Inform. Th.*, 24:668–674, November 1978.

[106] F. R. Gantmacher. *The Theory of Matrices*, volume 1 and 2. Chelsea Publishing Co., New York, 1959.

[107] M. W. Garrett and M. Vetterli. Joint source/channel coding of statistically multiplexed real-time services on packet networks. *IEEE/ACM Trans. on Networking*, 1(1):71–80, February 1993.

[108] C. Gasquet and P. Witomski. *Analyse de Fourier et Applications*. Masson, Paris, 1990.

[109] A. Gersho and R. M. Gray. *Vector Quantization and Signal Compression*. Kluwer Academic Publishers, Boston, MA, 1992.

[110] H. Gharavi. Subband coding of video signals. In J. W. Woods, editor, *Subband Image Coding*. Kluwer Academic Publishers, Boston, MA, 1990.

[111] H. Gharavi and A. Tabatabai. Subband coding of monochrome and color images. *IEEE Trans. Circ. and Syst.*, 35(2):207–214, February 1988.

[112] A. Gilloire and M. Vetterli. Adaptive filtering in subbands with critical sampling: analysis, experiments, and application to acoustic echo cancellation. *IEEE Trans. on SP*, 40(8):1862–1875, August 1992.

[113] I. Gohberg and S. Goldberg. *Basic Operator Theory*. Birkhauser, Boston, MA, 1981.

[114] A. Goldsmith and P. Varaiya. Capacity of time-varying channels with estimation and feedback. To appear, *IEEE Trans. on Info. Theory*.

[115] R. Gopinath. *Wavelet and Filter Banks — New Results and Applications*. PhD thesis, Rice University, 1992.

[116] R. A. Gopinath and C. S. Burrus. Wavelet-based lowpass/bandpass interpolation. In *Proc. IEEE Int. Conf. Acoust., Speech, and Signal Proc.*, pages 385–388, San Francisco, CA, March 1992.

[117] R. A. Gopinath and C. S. Burrus. Wavelet transforms and filter banks. In C. K. Chui, editor, *Wavelets: A Tutorial in Theory and Applications*, pages 603–654. Academic Press, New York, 1992.

[118] A. Goshtasby, F. Cheng, and B. Barsky. B-spline curves and surfaces viewed as digital filters. *Computer Vision, Graphics, and Image Processing*, 52(2):264–275, November 1990.

[119] P. Goupillaud, A. Grossman, and J. Morlet. Cycle-octave and related transforms in seismic signal analysis. *Geoexploration*, 23:85–102, 1984/85. Elsevier Science Pub.

[120] R. M. Gray. Vector quantization. *IEEE ASSP Magazine*, 1:4–29, April 1984.

[121] R. M. Gray. *Source Coding Theory*. Kluwer Academic Publishers, Boston, MA, 1990.

[122] R. M. Gray and L. D. Davisson. *Random Processes: A Mathematical Approach for Engineers*. Prentice-Hall, Englewood Cliffs, NJ, 1986.

[123] K. Gröchenig and W. R. Madych. Multiresolution analysis, Haar bases and self-similar tilings of R^n. *IEEE Transactions on Information Theory, Special Issue on Wavelet Transforms and Multiresolution Signal Analysis*, 38(2):556–568, March 1992.

[124] A. Grossmann, R. Kronland-Martinet, and J. Morlet. Reading and understanding continuous wavelet transforms. In J. M. Combes, A. Grossmann, and Ph. Tchamitchian, editors, *Wavelets, Time-Frequency Methods and Phase Space*. Springer-Verlag, Berlin, 1989.

[125] A. Grossmann and J. Morlet. Decomposition of Hardy functions into square integrable wavelets of constant shape. *SIAM Journ. of Math. Anal.*, 15(4):723–736, July 1984.

[126] A. Haar. Zur Theorie der orthogonalen Funktionensysteme. *Math. Annal.*, 69:331–371, 1910.

[127] P. Haskell and D. Messerschmitt. Open network architecture for continuous-media services: the medley gateway. Technical report, Dept. of EECS, January 1994.

[128] C. Heil and D. Walnut. Continuous and discrete wavelet transforms. *SIAM Rev.*, 31:628–666, 1989.

[129] P. N. Heller and H. W. Resnikoff. Regular M-and wavelets and applications. In *Proc. IEEE Int. Conf. Acoust., Speech, and Signal Proc.*, pages III: 229–232, Minneapolis, MN, April 1993.

[130] C. Herley. *Wavelets and Filter Banks*. PhD thesis, Columbia University, 1993.

[131] C. Herley. Exact interpolation and iterative subdivision schemes. *IEEE Trans. Acoust., Speech, and Signal Proc.*, 1994. To appear.

[132] C. Herley, J. Kovačević, K. Ramchandran, and M. Vetterli. Tilings of the time-frequency plane: Construction of arbitrary orthogonal bases and fast tiling algorithms. *IEEE Transactions on Signal Processing, Special Issue on Wavelets and Signal Processing*, 41(12):3341–3359, December 1993.

[133] C. Herley and M. Vetterli. Wavelets and recursive filter banks. *IEEE Trans. Signal Proc.*, 41(8):2536–2556, August 1993.

[134] O. Herrmann. On the approximation problem in nonrecursive digital filter design. *IEEE Trans. Circuit Theory*, 18:411–413, 1971.

[135] F. Hlawatsch and F. Boudreaux-Bartels. Linear and quadratic time-frequency signal representations. *IEEE SP Mag.*, 9(2):21–67, April 1992.

[136] M. Holschneider, R. Kronland-Martinet, J. Morlet, and Ph. Tchamitchian. A real-time algorithm for signal analysis with the help of the wavelet transform. In *Wavelets, Time-Frequency Methods and Phase Space*, pages 289–297. Springer-Verlag, Berlin, 1989.

[137] M. Holschneider and P. Tchamitchian. Pointwise analysis of Rieman's "non-differentiable" function. *Inventiones Mathematicae*, 105:157–175, 1991.

[138] A. K. Jain. *Fundamentals of Digital Image Processing*. Prentice-Hall, Englewood Cliffs, NJ, 1989.

[139] A. J. E. M. Janssen. Note on a linear system occurring in perfect reconstruction. *Signal Proc.*, 18(1):109–114, 1989.

[140] B. Jawerth and T. Swelden. An overview of wavelet based multiresolution analyses. *SIAM Review*, 36(3):377–412, September 1994.

[141] N. Jayant. Signal compression: technology targets and research directions. *IEEE Journ. on Sel. Areas in Commun.*, 10(5):796–818, June 1992.

[142] N. Jayant, J. Johnston, and B. Safranek. Signal compression based on models of human perception. *Proc. IEEE*, 81(10):1385–1422, October 1993.

[143] N. J. Jayant and P. Noll. *Digital Coding of Waveforms*. Prentice-Hall, Englewood-Cliffs, NJ, 1984.

[144] J. D. Johnston. A filter family designed for use in quadrature mirror filter banks. In *Proc. IEEE Int. Conf. Acoust., Speech, and Signal Proc.*, pages 291–294, Denver, CO, 1980.

[145] J. D. Johnston. Transform coding of audio signals using perceptual noise criteria. *IEEE Journ. on Sel. Areas in Commun.*, 6(2):314–323, 1988.

[146] J. D. Johnston and K. Brandenburg. Wideband coding: Perceptual considerations for speech and music. In S. Furui and M. M. Sondhi, editors, *Advances in Speech Signal Processing*, pages 109–140. Marcel-Dekker Inc, New York, 1992.

[147] J. D. Johnston and A. J. Ferreira. Sum-difference stereo transform coding. In *Proc. IEEE Int. Conf. Acoust., Speech, and Signal Proc.*, pages II: 569–572, San Francisco, CA, March 1992.

[148] JPEG technical specification: Revision (DRAFT), joint photographic experts group, ISO/IEC JTC1/SC2/WG8, CCITT SGVIII, August 1990.

[149] E. I. Jury. *Theory and Application of the z-Transform Method*. John Wiley and Sons, New York, 1964.

[150] T. Kailath. *Linear Systems*. Prentice-Hall, Englewood Cliffs, 1980.

[151] A. Kalker and I. Shah. Ladder structures for multidimensional linear phase perfect reconstruction filter banks and wavelets. In *Proceedings of the SPIE Conference on Visual Communications and Image Processing*, pages 12–20, Boston, November 1992.

[152] A. A. C. M. Kalker. Commutativity of up/down sampling. *Electronic Letters*, 28(6):567–569, March 1992.

[153] G. Karlsson and M. Vetterli. Three-dimensional subband coding of video. In *Proc. IEEE Int. Conf. Acoust., Speech, and Signal Proc.*, pages 1100–1103, New York, NY, April 1988.

[154] G. Karlsson and M. Vetterli. Packet video and its integration into the network architecture. *IEEE Journal on Selected Areas in Communications*, 7(5):739–751, 1989.

[155] G. Karlsson and M. Vetterli. Theory of two - dimensional multirate filter banks. *IEEE Trans. Acoust., Speech, and Signal Proc.*, 38(6):925–937, June 1990.

[156] G. Karlsson, M. Vetterli, and J. Kovačević. Nonseparable two-dimensional perfect reconstruction filter banks. In *Proc. SPIE Conf. on Vis. Commun. and Image Proc.*, pages 187–199, Cambridge, MA, November 1988.

[157] M. Khansari, A. Jalali, E. Dubois, and P. Mermelstein. Robust low bit-rate video transmission over wireless access systems. In *Proceedings of ICC*, volume 1, pages 571–575, May 1994.

[158] M. R. K. Khansari and A. Leon-Garcia. Subband decomposition of signals with generalized sampling. *IEEE Transactions on Signal Processing, Special Issue on Wavelets and Signal Processing*, 41(12):3365–3376, December 1993.

[159] R. D. Koilpillai and P. P. Vaidyanathan. Cosine-modulated FIR filter banks satisfying perfect reconstruction. *IEEE Trans. Signal Proc.*, 40(4):770–783, April 1992.

[160] J. Kovacevic and M. Vetterli. Perfect reconstruction filter banks with arbitrary rational sampling. *IEEE Trans. on SP*, 41(6):2047–2066, June 1993.

[161] J. Kovačević. *Filter Banks and Wavelets: Extensions and Applications*. PhD thesis, Columbia University, Oct. 1991.

[162] J. Kovačević. Subband coding systems incorporating quantizer models. *IEEE Trans. Image Proc.*, May 1995. To appear.

[163] J. Kovačević and M. Vetterli. Design of multidimensional nonseparable regular filter banks and wavelets. In *Proc. IEEE Int. Conf. Acoust., Speech, and Signal Proc.*, pages IV: 389–392, San Francisco, CA, March 1992.

[164] J. Kovačević and M. Vetterli. Nonseparable multidimensional perfect reconstruction filter banks and wavelet bases for $\mathcal{R}^n$. *IEEE Transactions on Information Theory, Special Issue on Wavelet Transforms and Multiresolution Signal Analysis*, 38(2):533–555, March 1992.

[165] J. Kovačević and M. Vetterli. FCO sampling of digital video using perfect reconstruction filter banks. *IEEE Trans. Image Proc.*, 2(1):118–122, January 1993.

[166] J. Kovačević and M. Vetterli. New results on multidimensional filter banks and wavelets. In *Proc. IEEE Int. Symp. Circ. and Syst.*, Chicago, IL, May 1993.

[167] T. Kronander. *Some Aspects of Perception Based Image Coding*. PhD thesis, Linkoeping University, Linkoeping, Sweden, 1989.

[168] M. Kunt, A. Ikonomopoulos, and M. Kocher. Second generation image coding techniques. *Proc. IEEE*, 73(4):549–575, April 1985.

[169] W. Lawton. Tight frames of compactly supported wavelets. *J. Math. Phys.*, 31:1898–1901, 1990.

[170] W. Lawton. Necessary and sufficient conditions for constructing orthonormal wavelet bases. *J. Math. Phys.*, 32:57–61, 1991.

[171] W. Lawton. Applications of complex valued wavelet transforms to subband decomposition. *IEEE Transactions on Signal Processing, Special Issue on Wavelets and Signal Processing*, 41(12):3566–3567, December 1993.

[172] W. M. Lawton and H. L. Resnikoff. Multidimensional wavelet bases. *AWARE preprint*, 1991.

[173] D. LeGall. MPEG: a video compression standard for multimedia applications. *Communications of the ACM*, 34(4):46–58, April 1991.

[174] D. J. LeGall, H. Gaggioni, and C. T. Chen. Transmission of HDTV signals under 140 Mbits/s using a subband decomposition and Discrete Cosine Transform coding. In L. Chiariglione, editor, *Signal Processing of HDTV*, pages 287–293. Elsevier, Amsterdam, 1988.

[175] P. G. Lemarié. Ondelettes à localisation exponentielle. *J. Math. pures et appl.*, 67:227–236, 1988.

[176] A. S. Lewis and G. Knowles. Image compression using the 2-D wavelet transform. *IEEE Trans. Image Proc.*, 1(2):244–250, April 1992.

[177] M. Liou. Overview of the $p \times 64$ kbit/s video coding standard. *Communications of the ACM*, 34(2):59–63, April 1994.

[178] M. R. Luettgen, W. C. Karl, A. S. Willsky, and R. R. Tenney. Multiscale representations of Markov random fields. *IEEE Transactions on Signal Processing, Special Issue on Wavelets and Signal Processing*, 41(12):3377–3396, December 1993.

[179] S. Mallat. Multifrequency channel decompositions of images and wavelet models. *IEEE Trans. Acoust., Speech, and Signal Proc.*, 37(12):2091–2110, December 1989.

[180] S. Mallat. Multiresolution approximations and wavelet orthonormal bases of $L_2(R)$. *Trans. Amer. Math. Soc.*, 315:69–87, September 1989.

[181] S. Mallat. A theory for multiresolution signal decomposition: the wavelet representation. *IEEE Trans. Patt. Recog. and Mach. Intell.*, 11(7):674–693, July 1989.

[182] S. Mallat. Zero-crossings of a wavelet transform. *IEEE Trans. Inform. Th.*, 37(4):1019–1033, July 1991.

[183] S. Mallat and W. L. Hwang. Singularity detection and processing with wavelets. *IEEE Transactions on Information Theory, Special Issue on Wavelet Transforms and Multiresolution Signal Analysis*, 38(2):617–643, March 1992.

[184] S. Mallat and S. Zhong. Wavelet maxima representation. In Y. Meyer, editor, *Wavelets and Applications*, pages 207–284. Masson, Paris, 1991.

[185] S. G. Mallat and Z. Zhang. Matching pursuits with time-frequency dictionaries. *IEEE*

Transactions on Signal Processing, Special Issue on Wavelets and Signal Processing, 41(12):3397–3415, December 1993.

[186] H. S. Malvar. *Optimal pre- and post-filtering in noisy sampled data systems*. PhD thesis, Massachusetts Institute of Technology, August 1986.

[187] H. S. Malvar. Extended lapped transforms: Properties, applications, and fast algorithms. *IEEE Trans. Signal Proc.*, 40(11):2703–2714, November 1992.

[188] H. S. Malvar. *Signal Processing with Lapped Transforms*. Artech House, Norwood, MA, 1992.

[189] H. S. Malvar and D. H. Staelin. The LOT: transform coding without blocking effects. *IEEE Trans. Acoust., Speech, and Signal Proc.*, 37(4):553–559, April 1989.

[190] B. Mandelbrot. *The Fractal Geometry of Nature*. W.H. Freeman and Co., San Francisco, 1982.

[191] J. McClellan. The design of two-dimensional filters by transformations. In *Seventh Ann. Princeton Conf. on ISS*, pages 247–251, Princeton, NJ, 1973.

[192] P. Mermelstein. G.722, A new CCITT coding standard for digital transmission of wideband audio signals. *IEEE Comm. Mag.*, 8(15), 1988.

[193] Y. Meyer. Méthodes temps-fréquence et méthodes temps-échelle en traitement du signal et de l'image. INRIA lectures.

[194] Y. Meyer. *Ondelettes et Opérateurs*. Hermann, Paris, 1990. In two volumes.

[195] Y. Meyer. *Wavelets, Algorithms and Applications*. SIAM, Philadelphia, 1993.

[196] F. Mintzer. Filters for distortion-free two-band multirate filter banks. *IEEE Trans. Acoust., Speech, and Signal Proc.*, 33(3):626–630, June 1985.

[197] P. Morrison and P. Morrison. *Powers of Ten*. Scientific American Books, New York, 1982.

[198] Z. J. Mou and P. Duhamel. Fast FIR filtering: algorithms and implementations. *Signal Proc.*, 13(4):377–384, December 1987.

[199] Z. J. Mou and P. Duhamel. Short-length FIR filters and their use in fast nonrecursive filtering. *IEEE Trans. Signal Proc.*, 39:1322–1332, June 1991.

[200] P. Moulin. A multiscale relaxation algorithm for SNR maximization in 2-D nonorthogonal subband coding. *IEEE Trans. Image Proc.*, 1994. Submitted.

[201] MPEG video simulation model three, ISO, coded representation of picture and audio information, 1990.

[202] F. D. Murnaghan. *The Unitary and Rotations Group*. Spartan, Washington, DC, 1962.

[203] M. J. Narasimha and A. M. Peterson. On the computation of the discrete cosine transform. *IEEE Trans. Commun.*, 26:934–936, June 1978.

[204] S. H. Nawab and T. Quartieri. Short-time Fourier transform. In J. S. Lim and A. V. Oppenheim, editors, *Advanced Topics in Signal Processing*, pages 289–337. Prentice-Hall, Englewood Cliffs, N.J., 1988.

[205] K. Nayebi, T. P. Barnwell III, and M. J. T. Smith. Time-domain filter bank analysis. *IEEE Trans. Signal Proc.*, 40(6):1412–1429, June 1992.

[206] K. Nayebi, T. P. Barnwell III, and M. J. T. Smith. Nonuniform filter banks: A reconstruction and design theory. *IEEE Trans. on Speech Processing*, 41(3):1114–1127, March 1993.

[207] A. Netravali and B. Haskell. *Digital Pictures*. Plenum Press, New York, 1988.

[208] T. Q. Nguyen and P. P. Vaidyanathan. Two-channel perfect reconstruction FIR QMF structures which yield linear phase analysis and synthesis filters. *IEEE Trans. Acoust., Speech, and Signal Proc.*, 37(5):676–690, May 1989.

[209] H. J. Nussbaumer. *Fast Fourier Transform and Convolution Algorithms*. Springer-Verlag, Berlin, 1982.

[210] H. J. Nussbaumer. Polynomial transform implementation of digital filter banks. *IEEE Trans. Acoust., Speech, and Signal Proc.*, 31(3):616–622, June 1983.

[211] A. V. Oppenheim and R. W. Shafer. *Discrete-Time Signal Processing*. Prentice-Hall, Englewood Cliffs, NJ, 1989.

[212] A. V. Oppenheim, A. S. Willsky, and I. T. Young. *Signals and Systems*. Prentice-Hall, Englewood Cliffs, NJ, 1983.

[213] R. Orr. Derivation of Gabor transform relations using Bessel's equality. *Signal Proc.*, 30:257–262, 1993.

[214] A. Ortega, Z. Zhang, and M. Vetterli. Modeling and optimization of a multiresolution image retrieval system. *IEEE/ACM Trans. on Networking*, July 1994. submitted.

[215] A. Papoulis. *The Fourier Integral and its Applications*. McGraw-Hill, New York, 1962.

[216] A. Papoulis. *Signal Analysis*. McGraw-Hill, New York, NY, 1977.

[217] A. Papoulis. *Probability, Random Variables and Stochastic Processes, Second Edition*. McGraw-Hill, New York, NY, 1984.

[218] A. Papoulis. *The Fourier Integral and its Applications, Second Edition*. McGraw-Hill, New York, NY, 1987.

[219] K. K. Parhi and T. Nishitani. VLSI architectures for discrete wavelet transform. *IEEE Trans. on Very Large Scale Integration Systems*, 1(2):191–202, June 1993.

[220] W. A. Pearlman. Performance bounds for subband coding. In J. W. Woods, editor, *Subband Image Coding*. Kluwer Academic Publishers, Inc., Boston, MA, 1991.

[221] W. B. Pennebaker, J. L. Mitchell, G. G. Langdon, and R. B. Arps. An overview of the basic principles of the Q-coder adaptive binary arithmetic coder. *IBM Journal of Res. and Dev.*, 32(6):717–726, November 1988.

[222] A. Pentland and B. Horowitz. A practical approach to fractal-based image compression. In *IEEE Data Compression Conf.*, pages 176–185, March 1991.

[223] C. I. Podilchuk. Low-bit rate subband video coding. In *Proc. IEEE Int. Conf. on Image Proc.*, volume 3, pages 280–284, Austin, TX, November 1994.

[224] C.I. Podilchuk, N.S. Jayant, and N. Farvardin. Three-dimensional subband coding of video. To appear *IEEE Trans. Image Processing*.

[225] B. Porat. *Digital Processing of Random Signals: Theory and Methods*. Prentice-Hall, Englewood Cliffs, NJ, 1994.

[226] M. R. Portnoff. Representation of digital signals and systems based on short-time Fourier analysis. *IEEE Trans. Acoust., Speech, and Signal Proc.*, 28:55–69, February 1980.

[227] J. Princen. The design of nonuniform modulated filter banks. *IEEE Trans. Signal Proc.*, 1994. Submitted.

[228] J. Princen, A. Johnson, and A. Bradley. Subband transform coding using filter bank designs based on time domain aliasing cancellation. In *Proc. IEEE Int. Conf. Acoust., Speech, and Signal Proc.*, pages 2161–2164, Dallas, TX, April 1987.

[229] J. P. Princen and A. B. Bradley. Analysis/synthesis filter bank design based on time domain aliasing cancellation. *IEEE Trans. Acoust., Speech, and Signal Proc.*, 34(5):1153–1161, October 1986.

[230] K. Ramchandran. *Joint Optimization Techniques for Image and Video Coding and Applications to Digital Broadcast.* PhD thesis, Columbia University, June 1993.

[231] K. Ramchandran, A. Ortega, K. M. Uz, and M. Vetterli. Multiresolution broadcast for digital HDTV using joint source-channel coding. *IEEE JSAC*, 11(1):6–23, January 1993.

[232] K. Ramchandran, A. Ortega, and M. Vetterli. Bit allocation for dependent quantization with applications to multiresolution and MPEG video coders. *IEEE Trans. Image Proc.*, 3(5):533–545, September 1994.

[233] K. Ramchandran and M. Vetterli. Best wavelet packet bases in a rate-distortion sense. *IEEE Trans. Image Proc.*, 2(2):160–175, April 1993.

[234] T. A. Ramstad. IIR filter bank for subband coding of images. In *Proc. IEEE Int. Symp. Circ. and Syst.*, pages 827–830, Helsinki, Finland, 1988.

[235] T. A. Ramstad. Cosine modulated analysis-synthesis filter bank with critical sampling and perfect reconstruction. In *Proc. IEEE Int. Conf. Acoust. Speech and Signal Processing*, pages 1789–1792, Toronto, Canada, May 1991.

[236] T. A. Ramstad and T. Saramäki. Efficient multirate realization for narrow transition-band FIR filters. In *Proc. IEEE Int. Symp. Circ. and Syst.*, pages 2019–2022, Helsinki, Finland, 1988.

[237] N. Ricker. The form and laws of propagation of seismic wavelets. *Geophysics*, 18:10–40, 1953.

[238] O. Rioul. Les Ondelettes. Mémoires d'Option, Dept. de Math. de l'Ecole Polytechnique, 1987.

[239] O. Rioul. Simple regularity criteria for subdivision schemes. *SIAM J. Math Anal.*, 23:1544–1576, November 1992.

[240] O. Rioul. A discrete-time multiresolution theory. *IEEE Trans. Signal Proc.*, 41(8):2591–2606, August 1993.

[241] O. Rioul. Note on frequency localization and regularity. *CNET memorandum*, 1993.

[242] O. Rioul. On the choice of wavelet filters for still image compression. In *Proc. IEEE Int. Conf. Acoust., Speech, and Signal Proc.*, pages V: 550–553, Minneapolis, MN, April 1993.

[243] O. Rioul. *Ondelettes Régulières: Application à la Compression d'Images Fixes.* PhD thesis, ENST, Paris, March 1993.

[244] O. Rioul. Regular wavelets: A discrete-time approach. *IEEE Transactions on Signal Processing, Special Issue on Wavelets and Signal Processing*, 41(12):3572–3578, December 1993.

[245] O. Rioul and P. Duhamel. Fast algorithms for discrete and continuous wavelet transforms. *IEEE Transactions on Information Theory, Special Issue on Wavelet Transforms and Multiresolution Signal Analysis*, 38(2):569–586, March 1992.

[246] O. Rioul and P. Duhamel. A Remez exchange algorithm for orthonormal wavelets. *IEEE Transactions on Circuits and Systems II: Analog and Digital Signal Processing*, 41(8):550–560, August 1994.

[247] O. Rioul and M. Vetterli. Wavelets and signal processing. *IEEE SP Mag.*, 8(4):14–38, October 1991.

[248] E. A. Robinson. *Random Wavelets and Cybernetic Systems*. Griffin and Co., London, 1962.

[249] A. Rosenfeld, editor. *Multiresolution Techniques in Computer Vision*. Springer-Verlag, New York, 1984.

[250] H. L. Royden. *Real Analysis*. MacMillan, New York, 1968.

[251] M. B. Ruskai, G. Beylkin, R. Coifman, I. Daubechies, S. Mallat, Y. Meyer, and L. Raphael, editors. *Wavelets and their Applications*. Jones and Bartlett, Boston, 1992.

[252] R. J. Safranek and J. D. Johnston. A perceptually tuned sub-band image coder with image dependent quantization and post-quantization data compression. *Proc. IEEE Int. Conf. Acoust., Speech, and Signal Proc.*, M(11.2):1945–1948, 1989.

[253] N. Saito and G. Beylkin. Multiresolution representation using the auto-correlation functions of compactly supported wavelets. *IEEE Transactions on Signal Processing, Special Issue on Wavelets and Signal Processing*, 41(12):3584–3590, December 1993.

[254] B. Scharf. Critical bands. In *Foundations in Modern Auditory Theory*, pages 150–202. Academic, New York, 1970.

[255] I. J. Schoenberg. Contribution to the problem of approximation of equidistant data by analytic functions. *Quart. Appl. Math.*, 4:112–141, 1946.

[256] T. Senoo and B. Girod. Vector quantization for entropy coding of image subbands. *IEEE Trans. on Image Proc.*, 1(4):526–532, October 1992.

[257] I. Shah and A. Kalker. Theory and Design of Multidimensional QMF Sub-Band Filters From 1-D Filters and Polynomials Using Transforms. *Proceedings of the IEE*, 140(1):67–71, February 1993.

[258] C. E. Shannon. Communications in the presence of noise. *Proc. of the IRE*, 37:10–21, January 1949.

[259] J. M. Shapiro. An embedded wavelet hierarchical image coder. In *Proc. IEEE Int. Conf. Acoust., Speech, and Signal Proc.*, pages 657–660, San Francisco, March 1992.

[260] J. M. Shapiro. Embedded image coding using zerotrees of wavelet coefficients. *IEEE Transactions on Signal Processing, Special Issue on Wavelets and Signal Processing*, 41(12):3445–3462, December 1993.

[261] M. J. Shensa. The discrete wavelet transform: Wedding the à trous and Mallat algorithms. *IEEE Trans. Signal Proc.*, 40(10):2464–2482, October 1992.

[262] Y. Shoham and A. Gersho. Efficient bit allocation for an arbitrary set of quantizers. *IEEE Trans. Acoust., Speech, and Signal Proc.*, 36(9):1445–1453, September 1988.

[263] J. J. Shynk. Frequency-domain and multirate adaptive filtering. *IEEE Signal Processing Magazine*, 9:14–37, January 1992.

[264] E. P. Simoncelli and E. H. Adelson. Nonseparable extensions of quadrature mirror filters to multiple dimensions. *Proc. IEEE*, 78(4):652–664, April 1990.

[265] E. P. Simoncelli and E. H. Adelson. Subband transforms. In J. W. Woods, editor, *Subband Image Coding*, pages 143–192. Kluwer Academic Publishers, Inc., Boston, MA, 1991.

[266] E. P. Simoncelli, W. T. Freeman, E. H. Adelson, and D. J. Heeger. Shiftable multiscale transforms. *IEEE Transactions on Information Theory, Special Issue on Wavelet Transforms and Multiresolution Signal Analysis*, 38(2):587–607, March 1992.

[267] D. Sinha and A. H. Tewfik. Low bit rate transparent audio compression using adapted wavelets. *IEEE Transactions on Signal Processing, Special Issue on Wavelets and Signal*

Processing, 41(12):3463–3479, December 1993.

[268] E. Slepian, H. J. Landau, and H. O. Pollack. Prolate spheroidal wave functions, Fourier analysis and uncertainty principle I and II. *Bell Syst. Tech. J.*, 40(1):43–84, 1961.

[269] M. J. T. Smith. IIR analysis/synthesis systems. In J. W. Woods, editor, *Subband Image Coding*, pages 101–142. Kluwer Academic Publishers, Boston, MA, 1991.

[270] M. J. T. Smith and T. P. Barnwell III. A procedure for designing exact reconstruction filter banks for tree structured sub-band coders. In *Proc. IEEE Int. Conf. Acoust., Speech, and Signal Proc.*, San Diego, CA, March 1984.

[271] M. J. T. Smith and T. P. Barnwell III. Exact reconstruction for tree-structured subband coders. *IEEE Trans. Acoust., Speech, and Signal Proc.*, 34(3):431–441, June 1986.

[272] M. J. T. Smith and T. P. Barnwell III. A new filter bank theory for time-frequency representation. *IEEE Trans. Acoust., Speech, and Signal Proc.*, 35(3):314–327, March 1987.

[273] A. K. Soman and P. P. Vaidyanathan. Coding gain in paraunitary analysis/synthesis systems. *IEEE Trans. Signal Proc.*, 41(5):1824–1835, May 1993.

[274] A. K. Soman and P. P. Vaidyanathan. On orthonormal wavelets and paraunitary filter banks. *IEEE Trans. on Signal Processing*, 41(3):1170–1183, March 1993.

[275] A. K. Soman, P. P. Vaidyanathan, and T. Q. Nguyen. Linear phase paraunitary filter banks: Theory, factorizations and designs. *IEEE Transactions on Signal Processing, Special Issue on Wavelets and Signal Processing*, 41(12):3480–3496, December 1993.

[276] A. Steffen. *Multirate Methods for Radar Signal Processing*. PhD thesis, ETH Zurich, 1991.

[277] P. Steffen, P. N. Heller, R. A. Gopinath, and C. S. Burrus. Theory of m-band wavelet bases. *IEEE Transactions on Signal Processing, Special Issue on Wavelets and Signal Processing*, 41(12):3497–3511, December 1993.

[278] E. Stein and G. Weiss. *Introduction to Fourier Analysis on Euclidean Space*. Princeton University Press, Princeton, 1971.

[279] G. Stoll and F. Dehery. High quality audio bit rate reduction family for different applications. *Proc. IEEE Int. Conf. Commun.*, pages 937–941, April 1990.

[280] G. Strang. *Linear Algebra and Its Applications, Third Edition*. Harcourt Brace Jovanovich, San Diego, CA, 1988.

[281] G. Strang. Wavelets and dilation equations: a brief introduction. *SIAM Journ. of Math. Anal.*, 31:614–627, 1989.

[282] G. Strang and G. J. Fix. *An Analysis of the Finite Element Method*. Prentice-Hall, Englewood-Cliffs, NJ, 1973.

[283] J.-O. Stromberg. A modified Franklin system and higher order spline systems on R^N as unconditional bases for Hardy spaces. In W. Beckner et al, editor, *Proc. of Conf. in honour of A. Zygmund*, pages 475–493. Wadsworth Mathematics series, 1982.

[284] J.-O. Stromberg. A modified Franklin system as the first orthonormal system of wavelets. In Y. Meyer, editor, *Wavelets and Applications*, pages 434–442. Masson, Paris, 1991.

[285] N. Tanabe and N. Farvardin. Subband image coding using entropy-coded quantization over noisy channels. *IEEE Journ. on Sel. Areas in Commun.*, 10(5):926–942, June 1992.

[286] D. Taubman and A. Zakhor. Multi-rate 3-D subband coding of video. *IEEE Trans. Image Processing, Special issue on Image Sequence Compression*, 3(5):572–588, September 1994.

[287] D. Taubman and A. Zakhor. Orientation adaptive subband coding of images. *IEEE Trans. Image Processing*, 3(4):421–437, July 1994.

[288] D. B. H. Tay and N. G. Kingsbury. Flexible design of multidimensional perfect reconstruction FIR 2-band filters using transformations of variables. *IEEE Trans. Image Proc.*, 2(4):466–480, October 1993.

[289] P. Tchamitchian. Biorthogonalité et théorie des opérateurs. *Revista Mathemática Iberoamericana*, 3(2):163–189, 1987.

[290] C. C. Todd, G. A. Davidson, M. F. Davis, L. D. Fielder, B. D. Link, and S. Vernon. AC-3: Flexible perceptual coding for audio transmission and storage. In *Convention of the AES*, Amsterdam, February 1994.

[291] B. Torrésani. Wavelets associated with representations of the affine Weil-Heisenberg group. *J. Math. Physics*, 32:1273, 1991.

[292] M. K. Tsatsanis and G. B. Giannakis. Principal component filter banks for optimal wavelet analysis. In *Proc. 6th Signal Processing Workshop on Statistical Signal and Array Processing*, pages 193–196, Victoria, B.C., Canada, 1992.

[293] F. B. Tuteur. Wavelet transformations in signal detection. In J. M. Combes, A. Grossmann, and Ph. Tchamitchian, editors, *Wavelets, Time-Frequency Methods and Phase Space*. Springer-Verlag, Berlin, 1989.

[294] M. Unser. On the approximation of the discrete Karhunen-Loeve transform for stationary processes. *Signal Proc.*, 5(3):229–240, May 1983.

[295] M. Unser. On the optimality of ideal filters for pyramid and wavelet signal approximation. *IEEE Transactions on Signal Processing, Special Issue on Wavelets and Signal Processing*, 41(12):3591–3595, December 1993.

[296] M. Unser and A. Aldroubi. Polynomial splines and wavelets: a signal processing perspective. In C. K. Chui, editor, *Wavelets: a Tutorial in Theory and Applications*, pages 91–122. Academic Press, San Diego, CA, 1992.

[297] M. Unser, A. Aldroubi, and M. Eden. On the asymptotic convergence of B-spline wavelets to Gabor functions. *IEEE Transactions on Information Theory, Special Issue on Wavelet Transforms and Multiresolution Signal Analysis*, 38(2):864–871, March 1992.

[298] M. Unser, A. Aldroubi, and M. Eden. B-spline signal processing, part I and II. *IEEE Trans. Signal Proc.*, 41(2):821–833 and 834–848, February 1993.

[299] M. Unser, A. Aldroubi, and M. Eden. A family of polynomial spline wavelet transforms. *Signal processing*, 30(2):141–162, January 1993.

[300] M. Unser, A. Aldroubi, and M. Eden. Enlargement or reduction of digital images with minimum loss of information. *IEEE Trans. Image Proc.*, March 1995. To appear.

[301] K. M. Uz. *Multiresolution Systems for Video Coding*. PhD thesis, Columbia University, New York, May 1992.

[302] K. M. Uz, M. Vetterli, and D. LeGall. A multiresolution approach to motion estimation and interpolation with application to coding of digital HDTV. In *Proc. IEEE Int. Symp. Circ. and Syst.*, pages 1298–1301, New Orleans, May 1990.

[303] K. M. Uz, M. Vetterli, and D. LeGall. Interpolative multiresolution coding of advanced television with compatible subchannels. *IEEE Trans. on CAS for Video Technology, Special Issue on Signal Processing for Advanced Television*, 1(1):86–99, March 1991.

[304] P. P. Vaidyanathan. The discrete time bounded-real lemmma in digital filtering. *IEEE*

Trans. Circ. and Syst., 32(9):918–924, September 1985.

[305] P. P. Vaidyanathan. Quadrature mirror filter banks, M-band extensions and perfect reconstruction techniques. *IEEE ASSP Mag.*, 4(3):4–20, July 1987.

[306] P. P. Vaidyanathan. Theory and design of M-channel maximally decimated quadrature mirror filters with arbitrary M, having the perfect reconstruction property. *IEEE Trans. Acoust., Speech, and Signal Proc.*, 35(4):476–492, April 1987.

[307] P. P. Vaidyanathan. Multirate digital filters, filter banks, polyphase networks, and applications: a tutorial. *Proc. IEEE*, 78(1):56–93, January 1990.

[308] P. P. Vaidyanathan. *Multirate Systems and Filter Banks*. Prentice-Hall, Englewood Cliffs, NJ, 1993.

[309] P. P. Vaidyanathan and Z. Doğanata. The role of lossless systems in modern digital signal processing: A tutorial. *IEEE Trans. Educ.*, 32(3):181–197, August 1989.

[310] P. P. Vaidyanathan and P.-Q. Hoang. Lattice structures for optimal design and robust implementation of two-channel perfect reconstruction filter banks. *IEEE Trans. Acoust., Speech, and Signal Proc.*, 36(1):81–94, January 1988.

[311] P. P. Vaidyanathan and S. K. Mitra. Polyphase networks, block digital filtering, LPTV systems, and alias-free QMF banks: a unified approach based on pseudo-circulants. *IEEE Trans. Acoust., Speech, and Signal Proc.*, 36:381–391, March 1988.

[312] P. P. Vaidyanathan, T. Q. Nguyen, Z. Doğanata, and T. Saramäki. Improved technique for design of perfect reconstruction FIR QMF banks with lossless polyphase matrices. *IEEE Trans. Acoust., Speech, and Signal Proc.*, 37(7):1042–1056, July 1989.

[313] P. P. Vaidyanathan, P. Regalia, and S. K. Mitra. Design of doubly complementary IIR digital filters using a single complex allpass filter, with multirate applications. *IEEE Trans. on Circuits and Systems*, 34:378–389, April 1987.

[314] M. Vetterli. Multidimensional subband coding: Some theory and algorithms. *Signal Proc.*, 6(2):97–112, April 1984.

[315] M. Vetterli. Filter banks allowing perfect reconstruction. *Signal Proc.*, 10(3):219–244, April 1986.

[316] M. Vetterli. A theory of multirate filter banks. *IEEE Trans. Acoust., Speech, and Signal Proc.*, 35(3):356–372, March 1987.

[317] M. Vetterli. Running FIR and IIR filtering using multirate filter banks. *IEEE Trans. Acoust., Speech, and Signal Proc.*, 36:730–738, May 1988.

[318] M. Vetterli and C. Herley. Wavelets and filter banks: Relationships and new results. In *Proc. ICASSP'90*, pages 1723–1726, Albuquerque, NM, April 1990.

[319] M. Vetterli and C. Herley. Wavelets and filter banks: Theory and design. *IEEE Trans. Signal Proc.*, 40(9):2207–2232, September 1992.

[320] M. Vetterli, J. Kovačević, and D. J. LeGall. Perfect reconstruction filter banks for HDTV representation and coding. *Image Communication*, 2(3):349–364, October 1990.

[321] M. Vetterli and D. J. LeGall. Perfect reconstruction FIR filter banks: Some properties and factorizations. *IEEE Trans. Acoust., Speech, and Signal Proc.*, 37(7):1057–1071, July 1989.

[322] M. Vetterli and H. J. Nussbaumer. Simple FFT and DCT algorithms with reduced number of operations. *Signal Proc.*, 6(4):267–278, August 1984.

[323] M. Vetterli and K. M. Uz. Multiresolution coding techniques for digital video: a re-

view. *Special Issue on Multidimensional Processing of Video Signals, Multidimensional Systems and Signal Processing*, 3:161–187, 1992.

[324] L. F. Villemoes. *Regularity of Two-Scale Difference Equation and Wavelets*. PhD thesis, Mathematical Institute, Technical University of Denmark, 1992.

[325] E. Viscito and J. P. Allebach. The analysis and design of multidimensional FIR perfect reconstruction filter banks for arbitrary sampling lattices. *IEEE Trans. Circ. and Syst.*, 38(1):29–42, January 1991.

[326] J. S. Walker. *Fourier Analysis*. Oxford University Press, New York, 1988.

[327] G. K. Wallace. The JPEG still picture compression standard. *Communications of the ACM*, 34(4):30–44, April 1991.

[328] G. G. Walter. A sampling theorem for wavelet subspaces. *IEEE Transactions on Information Theory, Special Issue on Wavelet Transforms and Multiresolution Signal Analysis*, 38(2):881–883, March 1992.

[329] P. H. Westerink. *Subband Coding of Images*. PhD thesis, Delft University of Technology, Delft, The Netherlands, 1989.

[330] P. H. Westerink, J. Biemond, and D. E. Boekee. Subband coding of color images. In J. W. Woods, editor, *Subband Image Coding*, pages 193–228. Kluwer Academic Publishers, Inc., Boston, MA, 1991.

[331] P. H. Westerink, J. Biemond, and D. E. Boekee. Scalar quantization error analysis for image subband coding using QMF's. *Signal Proc.*, 40(2):421–428, February 1992.

[332] P. H. Westerink, J. Biemond, D. E. Boekee, and J. W. Woods. Subband coding of images using vector quantization. *IEEE Trans. Commun.*, 36(6):713–719, June 1988.

[333] M. V. Wickerhauser. Acoustic signal compression with wavelet packets. In C. K. Chui, editor, *Wavelets: A Tutorial in Theory and Applications*, pages 679–700. Academic Press, New York, 1992.

[334] S. Winograd. *Arithmetic Complexity of Computations*, volume 33. SIAM, Philadelphia, 1980.

[335] J. W. Woods, editor. *Subband Image Coding*. Kluwer Academic Publishers, Boston, MA, 1991.

[336] J. W. Woods and T. Naveen. A filter based bit allocation scheme for subband compression of HDTV. *IEEE Trans. on IP*, 1:436–440, July 1992.

[337] J. W. Woods and S. D. O'Neil. Sub-band coding of images. *IEEE Trans. Acoust., Speech, and Signal Proc.*, 34(5):1278–1288, May 1986.

[338] G. W. Wornell. A Karhunen-Loeve-like expansion of $1/f$ processes via wavelets. *IEEE Trans. on Info. Theory*, 36:859–861, July 1990.

[339] G. W. Wornell and A. V. Oppenheim. Wavelet-based representations for a class of self-similar signals with application to fractal modulation. *IEEE Transactions on Information Theory, Special Issue on Wavelet Transforms and Multiresolution Signal Analysis*, 38(2):785–800, March 1992.

[340] X. Xia and Z. Zhang. On sampling theorem, wavelets, and wavelet transforms. *IEEE Transactions on Signal Processing, Special Issue on Wavelets and Signal Processing*, 41(12):3524–3535, December 1993.

[341] W. R. Zettler, J. Huffman, and D. Linden. The application of compactly supported wavelets to image compression. In *Proc. SPIE*, volume 1244, pages 150–160, 1990.

[342] J. Ziv and A. Lempel. A universal algorithm for sequential data compression. *IEEE Trans. Inform. Th.*, 23:337–343, 1977.

[343] H. Zou and A. H. Tewfik. Design and parameterization of M-band orthonormal wavelets. In *Proc. IEEE Int. Symp. Circuits and Sys.*, pages 983–986, San Diego, CA, 1992.

[344] H. Zou and A. H. Tewfik. Discrete orthonormal M-band wavelet decompositions. In *Proc. IEEE ICASSP*, volume 4, pages 605–608, San Francisco, CA, 1992.

Index